THE LIFE OF
ARTHUR YOUNG

Memoirs of the

AMERICAN PHILOSOPHICAL SOCIETY

Held at Philadelphia

For Promoting Useful Knowledge

Volume 97

FIG. 1. Arthur Young. Pastel portrait by John Russell. Kindness of the late Col. Sir Edward A. Ruggles-Brise of Spains Hall.

THE LIFE OF
ARTHUR YOUNG
1741–1820

JOHN G. GAZLEY

Hanover, New Hampshire

AMERICAN PHILOSOPHICAL SOCIETY

INDEPENDENCE SQUARE • PHILADELPHIA

1973

92
Y 681 g

International Standard Book Number 0-87169-097-7

To My Wife

Preface

THE primary historical importance of Arthur Young is that he was unquestionably the most significant figure in the English Agricultural Revolution. True, he was neither a great scientist nor a successful practical farmer, but he did more than anyone else to popularize the new ideas and practices. His three English farming tours, which ran to nine volumes, covered England quite thoroughly and described the innovations of the pioneers. During his long tenure as secretary to the Board of Agriculture, from its establishment in 1793 until his death, Young was in an ideal position to propagandize for the new agriculture. He also wrote six of the county surveys which the Board sponsored. On his own Young edited the *Annals of Agriculture,* which began in 1784 and ran for forty-six volumes. Many of the leaders of the Agricultural Revolution contributed to the *Annals,* but Young himself wrote about one-third of all the articles. He also was the author of six more important books in the field, among them one describing two thousand experiments which he had made, another *The Farmer's Kalendar* which ran to ten editions during his lifetime, and the *Political Arithmetic,* a major work in agricultural economics. There are also fifteen pamphlets from his pen primarily in the field of agriculture. In all his agricultural writings he made himself a spokesman for the landed classes, a defender of agrarian interests. And he could write. He was almost always clear and vigorous, and he could turn a neat phrase. He often showed himself a master of irony, and all too frequently of invective.

Arthur Young, however, had other claims to importance. His tours are excellent general accounts of the countries visited, and describe roads and inns, scenery and manufactures, art, country houses, and palaces. His tours to Ireland and France are still used as sources for the study of eighteenth-century conditions, and the French tour is one of the most frequently quoted accounts of pre-revolutionary France and of the early stages of the Revolution.

My article, "Arthur Young, British Patriot," exhibits still another side to Young's writings. As a young man he was a strong mercantilist and supported England's imperial policies during the Seven Years' War. During the 1780's he changed to a *laissez-faire* attitude, attacked colonies, and became quite liberal in general, although agrarian interests always took priority over any general economic theory. In his famous *Travels in France* Young attacked the abuses of the Old Regime and was even somewhat sympathetic to the French Revolution. But in 1792 he again shifted his views, became violently anti-revolutionary and anti-French, and very conservative. His pamphlet, *The Example of France, a Warning to Britain,* is perhaps inferior only to Burke's *Reflections on the Revolution in France* as an influence converting British public opinion to an anti-French and anti-revolutionary attitude. His writings became ever more nationalist and reactionary during the Reign of Terror and the Napoleonic Wars.

Until fairly late in life Arthur Young had not been very religious, and probably not far from Deism. After the death of his favorite daughter in 1797, however, he experienced a complete conversion. Although he published relatively little in the field of religion, the last third of the *Autobiography* and especially his correspondence are typical of the highly emotional and fundamentalist outlook of the Evangelical.

Since Young was primarily a writer, an analysis of his books and more important pamphlets and articles has seemed necessary and since they were very numerous, perhaps more than half of my biography is devoted to them. Of course his *Autobiography* is the major source for his personal life. It includes a great many letters, but there is also in the British Museum Additional Manuscripts 35,126–35,133 a collection of probably no less than 1,500 letters to him. Unfortunately there is no such body of letters by Young. The largest number, more than 100, are in the Burney Papers in the Berg Collection in the New York Public Library, and were written chiefly to Marianne Francis during the last decade of his life. Marianne also wrote numerous letters to Mrs. Piozzi, describing her visits at Young's home at Bradfield. These last letters are in the John Rylands Library at Manchester, England. For his official life, the Letter Books and Minute Books of the Board of Agriculture in the Library of the Royal Agricultural Society are naturally very important. My article, "Arthur Young . . . Some Biographical Sources" lists the less important manuscript sources and it also attempts to evaluate the earlier biographies of Young.

I am certain that the following acknowledgments of thanks will not include many who have helped me during the forty years that I have been working on Young, and I trust that such omissions will be blamed on the lapse of time and my forgetfulness rather than on ingratitude. I have worked in England during sabbatical leaves from Dartmouth College in 1931, 1938, and 1957. At all times I received the complete cooperation of the librarians at the British Museum where I have spent countless hours on the Young manuscripts and other collections. Without the help of Dr. Henry Guppy, Professor Edward Robertson, and Dr. F. Taylor I could never have published my two articles in the *Bulletin of the John Rylands Library*. Similarly without the aid of the late Charles Durant Cassidy and Dr. Frank R. Lewis I could never have completed my article, "Arthur Young and the Society of Arts." The librarians at the Royal Agricultural Society were most kind in making available to me the manuscript records of the Board of Agriculture. I also wish to express gratitude to the Public Record Office for the use of the Chatham Papers and to John R. Gilmour at the Royal Botanic Gardens at Kew which contains some of Sir Joseph Banks's correspondence.

I first visited Bradfield in 1931 and was most hospitably entertained by Miss Ethel Fitzroy, who also gave me pictures of the house and grounds. Through her I met Mrs. Rose Willson, the niece of the last Mrs. Arthur Young, who gave me pictures of the silver dish presented to Young by the Board of Agriculture and of the snuff box given him by Count Rostoptschin of Russia. One of Miss Fitzroy's friends at Bury, Frances J. W. Booth, gave me an old engraving portrait of Young.

In 1938 I received a grant from the Social Science Research Council which made possible a quite systematic search for Young letters in private collections. Willard Connely at the American University Union and Guy Parsloe at the Institute of Historical Research in London provided me with letters of introduction to the descendants of about thirty of Young's closest friends. I also inserted in the London *Times* a letter asking for help in finding letters and memorabilia of Young. Although the positive finds were disappointing, the negative results were important. My most valuable find was unquestionably the eight important letters by Young and a fine pastel portrait of him which were kindly made available to me by the late Col. Sir Edward A. Ruggles-Brise of Spains Hall, Essex. E. N. Hale of Wormley sent me one of Young's bookplates. Mrs. Emily Cooper of Bury allowed me to have a picture of her set of Wedgwood china made for Young and decorated with agricultural implements. Several people permitted me to have copies made of pictures in their possession—H. R. R. Williams of Barnard Quaritch of Young as a young boy with his sister, Bertha Taylor of Lavenham of the old hall at Bradfield, H. I. Jarman of Bury of Young's mother. The grant also made possible a trip to Cambridge and Bury. At Cambridge I was helped by the Rev. C. H. Smyth, John Saltmarch, Rev. C. F. D. Moule, and D. A. Winstanley. Especially pleasant was the hospitality which I experienced in several country houses near Bury where Young was once an intimate—the Duke of Grafton at Euston, F. S. Beauford at Troston, and Orbell Ray Oakes at Nowton.

While at Bury I started to examine the files of the *Bury and Norwich Post* which contained many local news items about Young and his neighbors and also many of his letters to the editor. I did not have time to complete the work, but in 1952 the Dartmouth College Committee on Research kindly made me a grant to get some one to finish it. Professor Charles R. Bagley happened to be at Oxford and secured the services of my former student, Roger H. Sheldon. When other commitments made it impossible for him to complete the job, it was finished by Miss J. Dimock at Bury. The librarians of the Cullum Library at Bury, Mr. H. R. Barker and Mr. Maltby, were most cooperative.

In 1951 I received a grant from the American Philosophical Society to have the Young manuscripts in the British Museum microfilmed.

In my 1957 trip one of the highlights was Mitchelstown in Ireland where Young had served as estate agent for Lord Kingsborough. Professor Vida Henning of Glasgow gave me valuable leads, while Messrs. Thomas O'Keefe, O'Brien, and W. H. Robson treated me most hospitably and helped to identify various buildings connected with Young.

Three American libraries have been of especial help. By far

the most important was the New York Public Library, where Paul North Rice, Dr. John D. Gordan, and John P. Baker have very kindly permitted me to use the Burney manuscripts in the Berg Collection. James T. Babb and Robert F. Metzdorf of the Yale University Library and Richard H. Dillon of the Sutro Branch of the California State Library made available to me further Young letters to Sir Joseph Banks.

During the long years of my research I have been given invaluable help—both practical leads to material and moral encouragement to persevere—by many English and American scholars. Among the former must be mentioned Constantia Maxwell, S. C. Roberts, the late C. S. Orwin, the late Sir Stephen Gaselee, H. Morgan Rees, H. L. Beales, the late George Veitch, the late Robert Mowat, Arthur H. Ridley, F. L. Salter, Sydney Smith, Alfred Cobban, and George E. Fussell. And among the Americans note must be made of Maurice Quinlan, James L. Clifford, Elizabeth R. Foster, Joyce Hemlow, and Frances Childs. All of my Dartmouth colleagues have been very helpful, but especial note must be made of John Adams, Arthur Wilson, and David Roberts. In a special class are several scholars who had aspired to write Young's biography themselves, but generously stood aside for me—John H. Middendorf of Columbia, Richard Wilson of East Anglia, and above all Rodney C. Loehr of Minnesota.

A few miscellaneous examples of help must not be forgotten. Dr. Andrew C. Brown of Edinburgh brought to my attention his valuable work on John Wilson, which I might well have missed. Ernest H. Godfrey of Bedford gave me great help in regard to the Board of Agriculture. Henrietta Taylor gave me an autograph letter by Young. The late Sir Herbert Wilberforce made repeated enquiries to relatives about possible letters from Young to William Wilberforce—unfortunately all in vain. Rev. Herbert Armstrong furnished me with a copy of Young's marriage license and Vera J. Ledger with the Earl of Coventry's will. I was received very hospitably in 1957 at 32 Sackville Street, Young's London home for many years, and again at Bradfield by Mr. Bates.

I wish to acknowledge a further recent and generous grant-in-aid from the Committee on Research at Dartmouth College to help in preparing my manuscript for publication. The maps have been made by Professor Van H. English of the Department of

Geography at Dartmouth College, with the aid of two Dartmouth students, Robert M. Miklas and Mark Potter. Most of my illustrations have been processed by the Dartmouth Photographic Service where Mrs. James D. Gibson has been especially helpful.

JOHN G. GAZLEY

Hanover, N.H.
June 1, 1971.

Contents

Illustrations

THE LIFE OF

ARTHUR YOUNG

I. Boyhood, Youth, and Early Manhood, 1741-1767

ARTHUR YOUNG, the untiring and eloquent apostle for the Agricultural Revolution, was born in London on September 11, 1741.[1] The family estate was at Bradfield Combust, a hamlet in western Suffolk, the name of which originated from a disastrous fire in 1327. Bradfield was about five miles southeast of Bury St. Edmunds, a prosperous market town, a real provincial capital, a center of fashion and culture, a town recommended by Horace Walpole to a young French nobleman because the English language was well spoken there.[2] Bradfield was a tiny village, entirely dominated by the relatively small estate of the same name. Our Arthur's father had been a great planter of trees, chiefly oaks and limes. A fine avenue of the latter led from the old manor house, "a wretched lath and plaster ill-contrived building,"[3] to the small church across the high road from Bury St. Edmunds to London. Arthur continued to plant trees, some cedars and especially larches, so the estate is well wooded. There are three small ponds near the house, in one of which Young took a morning dip for many years, regardless of weather. The largest pond had two small islands in it, and a small rowboat which was the delight of neighboring children and young people. It was also well stocked with fish and Young often held fishing parties there. There is a local myth that a chest lies at the bottom of one of the ponds. There was a "Round Garden" and a "Square Garden" to which Young paid much attention. The barns and stables were still in 1938 much as they were in Young's time. Some of

---

[1] Young, 1898: p. 1, stated that he was born in Whitehall, but Paris, 1820: p. 279, maintained that the birth was in the home of the famous midwife, Mrs. Sidney Kennon, in Clifford Street, and gave the date as September 7. Dr. Paris was Young's physician in his last illness, and his brief article was Young's earliest biography.

[2] de la Rochefoucauld, 1933: p. 21.

[3] Young, 1898: p. 13. For picture of Bradfield in Young's time, see fig. 6.

1

the fields are still known by the names which Young used, one called "Oakey Ley," and another "Little Experiments," where Young conducted hundreds of trials and then published the results. There is a "pub" in the village, known as the "Manger," and some nice old houses. The unpretentious little church is a very attractive building in the Gothic style. Inside and out it is dominated by monuments of five generations of Youngs, including at least three windows in rather bad taste.

The Youngs belonged to the gentry but were certainly not well-to-do. In 1672 an Arthur Young had become owner of Bradfield Combust, but his wife's family, the Canhams, had held the estate since 1620.[4] The Reverend Arthur Young (1693–1759), father of the agriculturist, after being educated at Eton and Pembroke College, Cambridge, settled and probably became curate at Thames Ditton in Surrey.[5] The young clergyman was generously patronized by the two brothers, Richard and Arthur Onslow of Thames Ditton. As Colonel Richard Onslow's chaplain to his regiment, Young went to Flanders during the War of the Austrian Succession. Speaker Arthur Onslow of the House of Commons appointed Young as his private chaplain and later helped him to become a Prebendary at Canterbury. In 1720 Young was instituted as rector at Bradfield Combust and the adjoining Bradfield St. Clare.[6] In 1729 he received the degree of Doctor of Laws and in 1734 published a two-volume work with the forbidding title, *Historical Dissertations on Idolatrous Corruptions in Religion from the beginning of the world, and on the methods taken by Divine Providence in reforming them.* Physically the Reverend Arthur Young was over six feet tall and "remarkably handsome."[7] His writings display his learning but also show that his intellectual interests were chiefly theological. The character and prominence of his friends indicate some social charm, but at home he was taciturn, obstinate, and not always gracious. In his *Autobiography* Arthur Young exhibits no affection for his father although a healthy respect is apparent.

While at Thames Ditton the Reverend Arthur Young fell in love with the wealthy and attractive Anne Lucretia de Cous-

---

[4] Maxwell, 1929: p. xiii, stated that the Youngs had held Bradfield "for over two hundred years," but such a statement is at variance with Young, 1898: p. 2.
[5] Young, 1898: p. 2, stated that he was "elected . . . minister of that parish."
[6] Parish Register at Bradfield.
[7] Young, 1898: p. 3.

maker (1706–1785) from the neighboring village of Weybridge. On their marriage she brought her husband a fortune large enough so that the Bradfield estate was settled upon her for life. Mrs. Young was a woman of much charm of manner, vivacious, active, well read, and fond of conversation. In his *Autobiography* Arthur Young referred to her as "my ever affectionate mother," and wrote that "her kindness and affection for me had never failed during the course of her whole life."[8]

Arthur Young was the youngest of three children. His older brother, John, commonly known as "Jack," was born February 27, 1727/8, and his sister, Eliza Maria, from two to five years later. While Arthur was a young child, Jack Young was at Eton (1739–1747) and then at Cambridge (1747–1750/1), and apparently was a brilliant student. He entered the clergy, became a doctor of divinity, and like his father rapidly acquired church preferments. He was not only fellow at Eton and chaplain to George III, but also prebendary and rector of St. John's Church at Worcester.[9]

As the baby of the family, Arthur was probably somewhat spoiled. Only once was he given corporal punishment by his father and then for an act of cruelty. From at least as early as Christmas, 1750, when he was nine years old, until February, 1758, when he was seventeen, Arthur attended Lavenham School which in his *Autobiography* he called "that wretched place."[10] Although he studied the classics, he complained that he failed to receive a good grounding in them. At one time his father ordered that algebra be substituted for the classics. He would probably have been better educated at the excellent school at Bury St. Edmunds. Lavenham, however, had been his father's first school.

---

[8] *Ibid.*, pp. 77, 126. Defries, 1938: pp. 32–33, and *passim* maintained definitely, but without definite proof, that the de Cousmakers were Jewish. Young, 1898: p. 3, stated of his maternal grandfather, "If ever there existed in human form an Israelite without guile, it was this worthy man. . . ." Such a use of the Biblical phrase, however, does not seem to this author at all conclusive proof that the de Cousmakers were Jewish. It is also true that Young and his favorite daughter, Bobbin, had pronounced Roman noses but their faces were not at all clearly Jewish. The portrait of Young's mother does not look at all Jewish. Nor is there any contemporary evidence that Young was a Jew.

[9] Austen-Leigh, 1921: p. xxiv. Defries, 1938: p. 35, implies that Jack and Arthur played together at Bradfield as children. For picture of Arthur Young as baby, with his sister, see fig. 2.

[10] Young, 1898: p. 7.

The master, the Reverend John Coulter, so favorably impressed Mrs. Young that she threw her influence to Lavenham. It was also not too expensive, for the total costs for a year, including board, amounted only to £17 4s. 4d.[11] There can be little doubt that he was spoiled by Mr. Coulter. He was never punished, was excused from eating pudding which he disliked, had a room to himself, and was permitted to keep a gun and pointer and to go shooting with the master. His mother bought him a little white pony which brought him home every Saturday, bursting with school news, to be petted by an admiring mother and older sister. Unfortunately the weather "or some other circumstance"[12] frequently prevented a prompt return on Monday morning. Inside the beautifully ornamented chantry of the superb Gothic church at Lavenham are several pen-scratched initials, among which one marked A/ looks suspiciously like the work of a little wretch who was to become a famous agriculturist. His Latin training may have been deficient, but he was taught to dance while at Lavenham school and also fell in love with two young ladies of Lavenham.

Although Arthur may not have applied himself too rigorously to his formal studies at Lavenham, he began at that early age to collect books and started to write a history of England. In 1753 Arthur visited London and not only saw the sights which might appeal to any young boy of eleven or twelve, such as the Tower and St. Paul's, but he also saw Garrick, heard Handel's *Messiah,* and had an evening at Ranelagh. At the salon of his aunt, Lady Ingoldsby, he met many celebrities, among them John Wilkes.

In December, 1754, the Rev. Arthur Young received a very considerable bequest from a very old friend, the famous midwife, Mrs. Sidney Kennon, who had brought all the Young children into the world. He was named executor of her will and residuary legatee of her estate which brought him about £5000. With this bequest he made extensive improvements at Bradfield in the summer of 1755. The old part of the house was torn down and then rebuilt on the old foundations, on the old plan and in lath and plaster, with the result that the house "had not a single room free from every fault that could be found, whether as to

[11] *Ibid.,* p. 8.
[12] Betham-Edwards, 1924: p. xxxi.

chimney, doors, windows, or connecting passages."[13] The stables, coach-house, brewery, and barns were rebuilt of brick at the cost of £500. The inconvenience, chilliness, and ramshackle character of Bradfield Hall always embarrassed Arthur, who undoubtedly would have preferred a structure more in accord with the classical tastes of the late eighteenth century.

Several letters in 1755 from his sister, Eliza Maria, who was in London, throw some light upon the teenage Arthur. In one she requested from her brother an exact description of the assembly balls at the Bury Assizes, assuming that he would of course attend. In another she referred to their having often read together and wept over *Gil Blas,* and regretted that he could not have seen Garrick and Mrs. Cibber in *Tancred and Sigismunda.* "You would have been vastly entertained!" She continued: "And now, Mr. Arthur, you being a very good politician, I shall proceed to entertain you with some Parliamentary affairs."[14] It is certainly revealing that Eliza Maria should send a boy of fourteen lace for a pair of ruffles, and that in a subsequent letter she should mention that her aunt had also promised him either a pair of lace ruffles or a guinea. Eliza Maria advised him to take the guinea, "for you have two pair of ruffles which I am sure is as much as you can possibly have occasion for."[15]

In 1756 or 1757 Eliza Maria married John Tomlinson from a family of London merchants. A letter from the young bride to her father in 1757 begged that Arthur be allowed to visit them in London. She reminded her father that her brother must have "his linen washed, stocking mended, &c., and in case he comes on horseback, it may not be amiss to hint to him that he is not to reach London on one gallop, for his impatience may outrun his prudence."[16] The last clause quoted might almost serve as a text for Arthur Young's character. He might have been far more happy, but much less interesting to posterity, had he more often followed his sister's advice.

Eliza Maria's marriage had a major influence on Arthur's career. The Tomlinsons agreed to let him enter their business after he had learned something about trade. Hence he was ap-

[13] Young, 1898: p. 13.
[14] *Ibid.,* pp. 15, 16.
[15] *Ibid.,* p. 19.
[16] *Ibid.,* p. 21.

prenticed for three years to the Messrs. Robertson, wine merchants at Lynn in Norfolk, to whom the Rev. Arthur Young
paid the quite considerable sum of £400 to cover presumably
board and cost of training.[17] Arthur went to Lynn in 1758. It
should be noted that it was not derogatory for a gentleman's son
to become a merchant's apprentice. But Arthur hated the work
from the beginning. "Every circumstance attending this new
situation at Lynn was most detestable to me."[18] His first year
at Lynn was also embittered by the death in childbirth of that
beloved sister whom he described as "a remarkably clever woman,
with much beauty and vivacity of conversation, combined with
much solidity of judgment." The shock had an even more serious
effect upon his mother who never ceased to mourn for her
daughter, and became a victim of religious melancholia. Her
son stated that she never read anything but the Bible after her
daughter's death. Certainly her letters to him are marked by
an almost morbid religiosity.[19]

Furthermore his sister's death destroyed Arthur's prospects of
securing a good place with the Tomlinsons. Every incentive
toward conscientious work thus disappeared. Arthur admitted
quite frankly that his work was perfunctory. Yet the family determined to keep him there, for the apprenticeship premium had
been paid, and there were no other plans for this younger son.
From the point of view of progress towards a livelihood the three
years were wasted and he left Lynn "without education, profession or employment."[20] In retrospect he claimed that the sum
expended at Lynn would have put him through the university
after which he might have been ordained and given the Bradfield living, in which case no biography of Arthur Young would
probably have been necessary.

Although Arthur was unhappy in his work at Lynn there were
compensations. In the first place, he hired separate lodgings and

---

[17] *Ibid.*, p. 22. This account ascribed the decision to put him into business to
his father, but Betham-Edwards, 1924: p. xxxi, maintained that his father wished
him to go to Eton and then the university, but that his mother opposed it.
Betham-Edwards stated the apprenticeship premium at £600.

[18] Young, 1898: p. 22.

[19] *Ibid.* In spite of Young's very clear statement, Defries, 1938: p. 36, wrote:
"No melancholy tendencies showed themselves in the scanty records left of the
lives of Dr. and Mrs. Young."

[20] Young, 1898: p. 24.

thus gained a measure of independence. In the second place, he again fell in love, this time with a Miss Robertson, the daughter of one of his employers. "She was of a pleasing figure, with fine black expressive eyes, danced well, and also sang and performed well on the harpsicord."[21] And in the third place, Lynn was an attractive town to a youth of Arthur's temperament and tastes. It was a busy and prosperous trading center, but more important to Young, it had a theater which was "convenient, very neat, neither profusely ornamented nor disgustingly plain," and a series of assembly rooms, consisting of a "very noble anteroom" which had been the medieval town hall, a ballroom which might have been "elegant" if the musicians' gallery had not been "a mere shelf stuck in between the chimneys," and an adequate card room.[22] It was not customary for merchants' apprentices to attend the monthly assemblies at Lynn, but Arthur Young "spurned" such conventions and danced with the "principal belles." He was certainly well attired, and admits his "great foppery in dress for the balls." He also claimed to be an expert dancer, whose minuets were so graceful that the dancing master at Lynn used them as models for his pupils.[23] It is not difficult to imagine the young beau with his curled wig, his long flaring brocaded or velvet coat with lace ruffles, as he took the floor for the stately minuet with the handsome Miss Robertson in a low-cut dress and wide hoopskirt. Very little is known of his friends at Lynn and unfortunately none of his letters from this period have survived. Almost certainly he was acquainted with the gifted Mr. Charles Burney, organist at Lynn, who had taught Miss Robertson to play the harpsichord so charmingly. Burney's learning and personal charm would surely have attracted Arthur. Very likely he was on friendly terms with Alderman Allen and his family, for it was the alderman's second daughter Martha or "Patty" who a few years later became Mrs. Arthur Young.

The expenses of a young gentleman of fashion left Arthur without means to indulge his mania for book collecting. Hence this boy, only seventeen years old, wrote a pamphlet in 1758, for which he was paid by the publisher in books to the value of ten pounds. Thus Arthur Young's first publication was a political

[21] *Ibid.*, p. 23.
[22] Young, 1768: pp. 44–45.
[23] Young, 1898: p. 24.

tract entitled, *The Theatre of the Present War in North America
with Candid Reflections on the Great Importance of the War in
that Part of the World.* The pamphlet by "A. Y****, Esq." was
fifty-six pages long, was published in London by J. Coote, and
sold for 1*s*. 6*d*.

The first two-thirds of the pamphlet, consisting of four out
of five chapters, described the areas of conflict in America: Cape
Breton and Louisburg; Canada, the St. Lawrence Valley and
Quebec; the "country, forts and settlements" between Quebec
and Fort Frontenac; the Great Lakes, Louisiana, and the Mississippi Valley. Since the budding author had never been to America,
this part of his work merely summarized more detailed topographical works, perhaps those very recently written by Dr. John
Mitchell.[24]

In the last chapter, which consisted of the "candid reflections,"
Young revealed himself a strong patriot, a mercantilist, and an
imperialist. He fully recognized the critical nature of the struggle
and saw the English colonies in real danger from the French
policy of encirclement. He pointed out the necessity of capturing
Niagara, Crown Point, Duquesne, and Quebec, and emphasized
especially the strategic importance of Louisburg. He claimed
that the French could never defeat the English if the colonies
were united under one head. He recognized the vital importance
to England of naval power and hence claimed that the French
possession of colonies, fisheries, and great inland waterways in
America put in jeopardy England's "very being, as a free and independent nation."[25] He also pointed out the economic importance of colonies to the mother country. A nation might almost
dispense with foreign trade if it could develop the potentialities
of its "plantations." The colonies might well produce raw materials sufficient to provide work for all England's unemployed and
also furnish all the necessities and luxuries which could not be
produced in the mother country. "We should consider that the
riches of the plantations, are our riches; their forces, our forces;
and their shipping, our shipping; as they prosper, so will their
mother country prosper of course; hither all their wealth flows
in the end."[26]

[24] Mitchell, 1755 and 1757.
[25] Young, 1758: p. 46.
[26] *Ibid.*, p. 53.

The *Gentleman's Magazine* commented very favorably upon the pamphlet. It described the first part as "a very succinct, yet clear, and even particular account of that part of N. America," and praised the Reflections as being "truely candid, interesting, and judicious."[27] Flattering appraisals, these, of the work of a boy of seventeen by one of the leading periodicals of the day!

It was not surprising, then, that Mr. Coote made a similar bargain in 1759 with the young apprentice at Lynn. The new pamphlet, entitled *Reflections on the Present State of Affairs at Home and Abroad,* was fifty-one pages long and was dedicated to William Pitt in recognition of his "eloquence, public spirit, and heroick [*sic*] resolution, the most shining qualities which can adorn a statesman."[28] The author disclaimed every intention of flattery and there is no trace of obsequiousness. Indeed Young even criticised Pitt who had condemned earlier ministries for sending troops to the continent, and had then continued that very policy when in office. It is hard to believe, however, that the ambitious apprentice, conscious of his ability and total lack of prospects for a congenial career, did not hope that somehow his pamphlet might attract the great man's attention and bring some kind of tangible recognition in the way of employment.

In the first part of the pamphlet Young discussed the continental war and allowed his patriotism to distort his judgment. He blamed the French for starting the war, condemned Maria Theresa for not observing the treaties in which she had ceded Silesia, and justified Frederick the Great in his invasion of Saxony. He attacked the "scandalous state in which the electorate of Hanover was in with respect to its defence,"[29] and condemned the Duke of Cumberland's conduct in the early part of the war. He also demanded a punitive peace settlement against the French.

The more important part of the pamphlet discussed American affairs. It expressed confidence in Pitt's new commanders and emphasized the necessity of seizing Quebec in order to obtain an advantageous peace. The very great value of the colonies was again stressed. Young thought that colonial manufactures in competition with those of England were the chief causes of a possible colonial revolt. England should encourage the colonies

[27] *Gentleman's Magazine* **28** (1758) : p. 542.
[28] Young, 1759: p. iv.
[29] *Ibid.,* p. 11.

to produce goods which she normally imported from other countries such as "hemp, flax, silk, wine, oil, raisins, currants, almonds, indigo, madder, salt-petre, pot-ash, iron, pitch, fur, timber and all other naval stores."[30] He concluded: "This would be the way both to secure the dependence of the colonies, and to reap the benefit of them; and at the same time to promote their growth and prosperity likewise."[31] The last clause sounds rather like an afterthought. Young again urged the necessity of colonial union in the interests of efficient defense. The colonial charters must be revoked, a viceroy appointed to control all armed forces, and a colonial parliament established. True, the colonies might object to surrendering their charters, "but every method ought to be taken to enforce a due obedience."[32] Such arguments, temper, and state of mind differed little from those of George Grenville, Lord North, or George III.

The comment upon Young's second pamphlet by the *Gentleman's Magazine* was anything but favorable. The material was just a newspaper résumé and the reflections those of "a youth of 15." Moreover the serious charge was made that it had been published before under another title, and it was pointed out that, despite the dedication to William Pitt, "prefixed as a new expedient to sell it,"[33] the body of the pamphlet declared that it contained no dedication. In the light of Arthur's youth, the *Gentleman's Magazine's* interpretation of the inconsistency seems unlikely. At the worst of course it suggests downright plagiarism, but Young was more likely guilty only of haste and carelessness, adding the dedication as an afterthought and failing to alter the original draft.

In the *Autobiography*[34] Young claimed that he wrote several other political tracts while at Lynn, but only the two already discussed have survived. Moreover, his earliest biographer, Dr. J. A. Paris, stated that Arthur wrote four novels while at Lynn, and Henry Higgs in his article on Young in the *Dictionary of National Biography* has given their names: *The Fair American;*

---

[30] *Ibid.*, p. 28.
[31] *Ibid.*, p. 28–29.
[32] *Ibid.*, p. 44.
[33] *Gentleman's Magazine* **29** (1759): p. 184.
[34] Young, 1898: p. 24.

Sir Charles Beaufort; Julia Benson, or the Innocent Sufferer; Lucy Watson.[35]

One other important event in Arthur Young's life occurred while he was at Lynn. Shortly after remodeling Bradfield Hall in 1755, the Rev. Arthur Young had been seized with a dropsical complaint which eventually caused his death in 1759. There is no evidence that Arthur felt any very keen grief. He noted, however, that his father died so heavily in debt that it took his mother two years to clear herself.

In their prefaces to the *Travels in France* both Miss M. Betham-Edwards and Miss C. Maxwell state that Arthur left Lynn in 1759, the year of his father's death. But the *Autobiography* declares that he was twenty years old when he left which would place the date in 1761.[36] Moreover, Arthur implied very clearly that he stayed at Lynn for the full term of his apprenticeship which would have ended in 1761.

Young seems to have left Lynn with no definite career plans. He was in London in September and again in December, 1761. The occasion for the first trip was George III's coronation. He probably attended the coronation ceremony in Westminster Abbey and certainly was a spectator at the splendid banquet which followed in Westminster Hall where large crowds were admitted to the galleries. Like many others, during dessert he lowered from the balcony a basket to the table below which was filled by the young Duke of Marlborough. Six days later he was presented at court in a new "full dress suit" especially made for the occasion.[37] It was probably a huge reception and George III probably took no note of the youthful Suffolk gentleman. Their lives, however, were to run in some parallel channels. George

[35] Dr. Paris's statement, 1820: p. 282, was as follows: "encouraged by this composition, he sent him several other manuscripts, among which were four novels. . . ." A recent author, Foster, 1949: pp. 154–158, assigns all four novels to Young without any question, but without any proof. Whether Young actually wrote them is certainly questionable, but it is sure that he did not publish them while at Lynn, for the dates of publication were 1766, 1767, 1768, 1775. In a footnote, p. 155, Foster commented on the first novel: "Perhaps Young hoped that his wife, whose hot temper drove him into the arms of Mrs. Oakes, would grasp the moral of his novel." The novel in question was published in 1766, the year after his marriage, and four years before Mrs. Oakes was born! The doubtful implication that Mrs. Oakes was Young's mistress will be discussed below, pp. 551, 554, 582–583.

[36] Betham-Edwards, 1924: p. xxxii; Maxwell, 1929: p. xiii; Young, 1898: p. 24.

[37] Young, 1898: p. 26. That baskets were actually lowered as Young described is verified by a letter quoted in the *Annual Register* 4: p. 234.

III was but three years older than Arthur and both died in the same year. More important they were to become warm admirers of each other and George III was to go down into history, not only as the muddling tyrant who lost the American colonies, but also more happily as "Farmer George," the patron of the new agriculture.

Young was probably already thinking about establishing a new magazine as early as September. His December trip to London must have been devoted to planning the first number which appeared in January, 1762. Although his success in selling his pamphlets very naturally turned his mind to a literary career, the establishment of a magazine was a fantastic venture for a young man not yet twenty-two and without funds or experience. He boldly solicited Dr. Samuel Johnson to write for the new publication, but Johnson refused and advised that the scheme be abandoned. Young persisted, however, and secured the services of several hack writers. The new publication had the high-sounding title, *The Universal Museum, or Gentlemen's and Ladies' Polite Magazine of History Politicks and Literature for 1762.* It was dedicated to Queen Charlotte, "A Lady of your penetrating genius must certainly be desirous of an acquaintance with the learning of the people over whom heaven has appointed you to reign."[38] Actually there is no evidence that the magazine ever secured any royal patronage nor, frankly, did it deserve any.

It would be interesting to know what articles, if any, were written by Young, but there is no indication who the contributors were. The articles were almost without exception mediocre and some lacked good taste. There were translations of foreign works, some documents relating to current public affairs, and a history of the war, possibly contributed by Young. Several attempts were made at the essay in Addison's or Johnson's manner, under the pseudonyms "The Town" and "The Author." Poetry, book reviews, rebuses, and riddles helped to fill out the monthly quota. Handsome engravings, usually portraits, maps of the West Indies, and once a sheet of music, helped to embellish the publication. Free subscriptions were offered for the best rebuses and riddles and for the best essay on *Tristram Shandy,* and a silver medal for the best poem on love. After five or six numbers Young persuaded

[38] *Universal Museum* 1 (1762): p. ii.

a group of booksellers to take over the scheme: "I fairly slipped my neck out of the yoke."[39]

In September, 1762, Young suffered a lung hemorrhage which forced him to go to Bristol Hotwells, famous for their efficacy in curing consumption. Two of his children died from tuberculosis but, fortunately for British agriculture, Young possessed a tougher constitution or was given better treatment. At Bristol, Young met and played chess with Major-General Sir Charles Howard, who was so attracted by the young man that he offered him a commission in his own cavalry regiment. Young was willing to accept and wrote his mother for her permission, but was not surprised at her refusal, based largely on his weak health.[40]

In 1763 Arthur Young made the greatest decision in his life when he accepted his mother's offer to take a farm at Bradfield. The decision was not made from any innate love of farming, but largely from boredom and desperation. The venture with the *Universal Museum* had failed. His lack of a university education precluded the church. Business he had found distasteful. When his mother vetoed an army career, she probably felt an obligation to suggest a constructive alternative. Her strong religious outlook must have made her anxious over her son's love of London, the theater, and fashionable society, and the resultant increasing debts.[41] His definitely frail health made a London residence and even moderate dissipation undesirable. Moreover, she was undoubtedly lonely and longed to have him near. All these influences combined to make her offer and his acceptance natural.

Young was farming at Bradfield, then, from 1763 to 1767.

---

[39] Young, 1898: pp. 26–27, claimed that he "printed" five numbers of the magazine, but Haslam, 1930; relying on Paris, called it six. Haslam also believed that a poem from vol. 1 of the *Museum*, p. 58, was by Young, but gives no evidence for his attribution. He thought that the articles signed "The Author" were contributed by Young, but they continued long after he abandoned the editorship. Hunt, 1926: p. 12, believed his visit to Johnson occurred at the end of his editorship, but Young seemed to imply that it took place just when the venture was being started. Pell, *Journal of the Royal Agricultural Society* **54:** p. 2, took the same view as Hunt.

[40] Young, 1898: pp. 28–29. These pages imply that the trip took place in 1763, rather than in 1762. Again Young's memory was at fault, for he was farming at Bradfield by September, 1763, and he makes it clear that he did not take up farming until his return from Bristol.

[41] *Ibid.*, p. 26, footnote, stated that his debts in July, 1761, were 5 gs., but that they had risen by the end of the year to £62.

"I had no more idea of farming than of physic or divinity. . . ."[42] So wrote Young many years later, looking back at the beginning of his agricultural career. At first he rented from his mother the home farm of eighty acres and later added another farm. The total amounted to about three hundred acres, divided into thirty-nine fields, twenty-six arable and thirteen grass. The *Autobiography* tells little of his farming experiences at Bradfield but the large two-volume quarto work, *A Course of Experimental Agriculture,* first published in 1770, describes them in great detail.[43]

From the very beginning he was ambitious to be something more than an ordinary gentleman farmer. He began to collect and read books upon agriculture. The preface to *A Course of Experimental Agriculture* reviewed the works of many famous agriculturists, including de Serre, Hartlib, Tull, du Hamel, and his friend Walter Harte, and also showed familiarity with the memoirs of several foreign learned societies interested in agriculture.[44] Such a course of reading naturally opened up endless vistas of improvements. The first half of the eighteenth century had witnessed in England the first steps in the new agriculture which Young did so much to popularize. Jethro Tull's *Horse-hoeing Husbandry* had appeared in 1731 and Turnip Townshend had long since earned his nickname. Drilling, horse-hoeing, turnips, clover, lucerne, marling, and new methods of stock breeding —all were being practiced when Young began his career. Many of these innovations had appeared in the eastern counties near Young's home and presumably were well known among his neighbors.

Young formed "a resolution to try every thing, even those experiments which I was sensible could not answer, but which being recommended by writers of character, I brought to the fair test of experiment alone."[45] Within a period of four years he conducted some thousands of experiments and later took nearly 1800 quarto pages to describe them. Many were small scale, particularly those on the proper time for sowing and the best quantity of seed. Others took whole fields for several years.

[42] *Ibid.,* p. 29.
[43] Young, 1771: 1: pp. xix-xxiii, for a list of the fields on his two farms. My page references are to the four-volume Dublin reprint of 1771.
[44] *Ibid.* 1, pp. xv-xvii.
[45] *Ibid.* 1: p. xvii.

He experimented with the new drill culture for many different crops although he did not own a drill plow until 1766. He experimented with new crops, not only with such well-established ones as clover and turnips, but with potatoes, carrots, cabbages, lucerne, sainfoin, burnet, and Jerusalem artichokes. The influence of his reading upon his experiments is shown by the following: "In autumn 1765, I read Mr. Wynn Baker's experiments on the turnip-cabbage, which seemed so particularly important, that I immediately determined to cultivate so useful a plant."[46] Since the Royal Society of Arts offered premiums of £5 for raising an acre of madder, Young did so in 1765 and again in 1766. In both cases he received the premium, but in the first instance there was a net loss of £16, in the second of only £2.[47] He also experimented with animal husbandry although not on as extensive a scale. In October, 1765, he bought four oxen in Lincolnshire and visited a neighbor in company with a blacksmith to see how they should be yoked, shod, and managed.[48] In the same year he built a paved hog-yard, "with all the adjoining conveniences"— boiling house, copper, pond, pump, cisterns, shed, and troughs— at a total cost of £78 10s. In the following spring he fattened 88 hogs in his new "piggery."[49]

Young was convinced that every experiment must be carefully minuted and the expenses accurately estimated: "The general principle upon which I began and continued this course of experiments, was to keep minutes of everything."[50] The exact date on which the various operations took place is given. The number of plowings and harrowings, and the amount and kind of fertilizer employed are included. Every cost is carefully estimated, including labor and rent, and the whole experiment is always concluded with a final casting up of the profit or loss. Perhaps this emphasis upon careful accounting resulted from his business training at Lynn.

Later in life Young always professed great shame at his temerity in publishing these early experiments. In his *Autobiography* he wrote:

[46] *Ibid.* **4:** p. 271.
[47] *Ibid.* **4:** pp. 253–263.
[48] *Ibid.* **4:** pp. 498, 506. *Cf.* also *Annals* **4:** p. 125.
[49] *Ibid.* **4:** pp. 443–444.
[50] *Ibid.* **1:** p. vi.

And the circumstance which perhaps of all others in my life I most deeply regretted and considered as a sin of the blackest dye, was the publishing the result of my experience during those four years, which, speaking as a farmer, was nothing but ignorance, folly, presumption, and rascality.[51]

In the manuscript *Elements of Agriculture* he was still more severe upon himself:

. . . it is not without the deepest regret, that I reflect on some of those works, in which I hazarded what I was pleased to call experiments, by which I deceived both myself and others. I never think of some of the works of that early period of my life, but I am inclined to set them down as heinous sins against God and man. . . .[52]

As accurate scientific experiments they were undoubtedly failures. His accounting was repeatedly attacked by his critics. He had had no scientific training and what was worse no practical experience as a farmer. Nevertheless one feels that he was too self-critical. He was lacking, true enough, in scientific technique, but he was animated by the scientific spirit. He had clearly mastered the essential principle of modern science that experiments are fundamental to any progress and that they must precede generalizations.

Arthur Young's greatness lay, however, not in his ability to practice agriculture, but rather in his publications on agricultural subjects. In his second year as a farmer he began those agricultural writings which were to continue for more than forty years and were to make him the leading authority of his own time and perhaps the greatest agricultural writer of all time.

He started his agricultural writings modestly by eleven letters to the *Museum Rusticum*, all signed *Y*, and dated from Bradfield, the first on October 2, 1764, the last on April 4, 1765. Most were quite short, some only two or three pages and the longest only fifteen. The first was entitled, "Common Farmers Vindicated from the Charges of being universally ignorant and obstinate; with some Reflections on the present state of Improvements in husbandry."[53] In this short letter Young defended experiments as "the rational foundation of all useful knowledge,"[54] and urged gentlemen to make them, since small farmers could not afford

[51] Young, 1898: p. 30.
[52] Elements of Agriculture, **5:** f. 406.
[53] *Museum Rusticum* **3:** pp. 188–191.
[54] *Ibid.* **3:** p. 191.

to do so. He also declared that none of the principles of the new husbandry were well enough established for the small farmer to adopt. The editors encouraged "this very sensible farmer" to continue to communicate "his sentiments to the public" through the columns of their journal.[55] The resulting stream for the following six months was pretty rapid.

In one letter Young advocated the draining of wet pastures and described his own methods of digging ditches, fencing them with quick-set hedges, and using wooden or stone drains. This emphasis on proper draining became one of his most important maxims for improved agriculture. In the same letter he apologized for his crude style of writing, but maintained that Latin quotations were unsuitable for a publication for practical farmers.[56]

In a contribution entitled "Reasons why Farming so often proves unprofitable,"[57] Young listed the failure of a proper proportion of livestock to land, the lack of sufficient capital, and the common practice of keeping too much land in arable and too little in grass. He also pointed out that gentlemen were particularly apt to fail because they gave too much authority to bailiffs who "are usually dishonest,"[58] were too apt to be carried away by expensive experiments, and failed to keep accurate accounts. Young nevertheless urged gentlemen to keep part of their lands in their own hands and rhapsodized on the pleasures of conducting agriculture along advanced lines:

What can be more amusing than experimental agriculture? trying the cultivation of new-discovered vegetables, and all the modes of raising the old ones; bringing the earth to the finest pitch of fertility, and growing plants infinitely more vigorous and beautiful than any in the common tillage; using the variety of new machines perpetually invented, and observing their effects. . . .[59]

His most interesting essay in the *Museum* urged that a properly equipped traveller should go through Europe to observe agricultural practices in the different countries, from Spain to Russia.[60]

[55] *Ibid.*
[56] *Ibid.* 3: pp. 284–295.
[57] *Ibid.* 4: pp. 264–273.
[58] *Ibid.* 4: p. 268.
[59] *Ibid.* 4: p. 272.
[60] *Ibid.* 4: 58–65.

His ideal traveler must be a practicing farmer with more than
a book knowledge of agriculture, versed in the chief European
languages, able to sketch the various implements and improvements which he saw, and possessed of an ample fortune. He must
be a man of "penetration, quick conception, thoughtful and
attractive," and must possess "a phancy of disposition, patience,
and dexterity."[61] He must observe the methods of cultivation, the
nature of the soil and the plants most suitable for it, the manner
in which lands are rented and the nature of the leases, the condition of the roads, and the treatment of the poor. He must send
back to England samples of seeds and cattle. Above all, he must
keep "an exact and minute journal" of everything he sees and
hears. The result of such a tour by such a man would be "the
most useful book of travels that ever appeared in the world!"[62]
He urged the Royal Society of Arts to back his plan, for they had
the money, could make the proper appointment, and give the
necessary directions. There can be little doubt that Young hoped
that he might be the man chosen. This dream of agricultural
tours eventuated in that magnificent series of travels through
England, Ireland, France, and into Spain and Italy, which were
historically the most valuable of his works.

One article on the relative profit of arable and grass land
aroused criticism by an anonymous contributor who signed himself Ruricola Glocestris. Young replied with some heat, and then
others took up the controversy, another anonymous contributor,
Mago, and the Rev. Thomas Comber.[63] All three critics were
quite courteous and complimented Young on some of his essays,
but they felt that he had been too critical of other authors, and
that his accounts in the above mentioned article were faulty.
Comber wrote bluntly: "I should advise him rather to *throw
away* or *burn* his ledger, than puzzle himself and others in this
unedifying manner."[64] Young's replies were not very convincing
and he lost his temper. For instance he referred to Ruricola
Glocestris: "That he wants to be *set to rights* in his dairy-notions,

[61] *Ibid.* **4:** p. 64.
[62] *Ibid.* **4:** p. 63.
[63] *Ibid.* **4:** pp. 200–201, 231, 274–287; **5:** pp. 47–57, 195–199, 350–359.
[64] *Ibid.* **5:** p. 359. Thomas Comber was rector of Buckworth and Morbane. He
wrote two books dealing with some of Young's later works: *A Free and Candid
Correspondence on the Farmer's Letters,* 1770; *Real Improvements in Agriculture,*
1772. He died in 1778.

or the country of Glocester [*sic*] in its practice of farming, I am sure, is, from his letter, very evident."[65] He was even harsher towards Mago who had criticized Young's friend, Walter Harte: "Mago's objections are scarcely worth notice, and in this case can arise from nothing but his not having seen the book."[66] Although he apologized to the editors for "being, perhaps, too warm,"[67] even his apology had to be edited. This tendency to reply too sharply to criticism remained a failing for many years and certainly lost Young many friends.

Young probably discontinued his communications to the *Museum Rusticum* partly from pique arising from the controversy. He was also urged to stop wasting his talents in "blue-paper Periodical Essays" by the Rev. Walter Harte[68] who became his mentor and friend in 1764 or 1765. Harte had been Pope's friend and had served as tutor to Lord Chesterfield's son on the latter's famous continental tour. He had written some very mediocre poetry and a two-volume life of Gustavus Adolphus. He had been a famous Oxford tutor and vice principal of St. Mary's Hall. He was vain, he was guilty of gross flattery to Pope, and his reports to Lord Chesterfield on his son's progress were not always candid. Nevertheless Harte was good natured, a pleasant companion, and, despite his shortcomings, a man of high character. In 1764 he published *Essays in Husbandry* which is generally regarded as his best work with a less stilted style than usual. Young took up the cudgels for Harte's book in the *Museum Rusticum*[69] while Harte reciprocated by praising Young's articles in the same periodical. In 1765 and in 1767 Young visited Harte at Bath where, as a semi-invalid, he spent most of his time. He wrote: "One hour spent in this gentleman's company, I prized a thousand degrees beyond all the architectural beauties of Bath."[70]

As intimated, Harte was an experienced flatterer. He was "extremely pleased" with Young's proposal for an agricultural tour

[65] *Ibid.* **4:** p. 274.

[66] *Ibid.* **4:** p. 287. Mago's criticism of Young is in **4:** p. 231.

[67] *Ibid.* **4:** p. 287.

[68] Young, 1898: p. 37. There are many references to Harte (1709–1774) in Lord Chesterfield's *Letters to his Son* and some account of him in Samuel Shellabarger's excellent *Lord Chesterfield*, 1935. *Cf.* also *Dictionary of National Biography* and the short life by R. Walsh in the *Works of the British Poets* **29:** pp. 323–330.

[69] *Museum Rusticum* **4:** p. 286.

[70] Young, 1768: pp. 187–188.

throughout Europe, and he "read with delight" Young's remarks on the value of broad-wheeled wagons.[71] Harte urged him not to take seriously the criticisms in the *Museum Rusticum* and to stop sending his writings there. Instead, he suggested, why not collect them and publish them with further additions? Such was undoubtedly the origin of Young's *Farmer's Letters* which really started his reputation as an agricultural writer. Had not Harte encouraged him to persevere in writing on agriculture, the world might never have heard of Arthur Young, for he was sensitive, he had failed in so many things, and the criticisms had wounded him deeply. The volume appeared early in 1767 under the title, *The Farmer's Letters to the People of England . . . to which are added Sylvae: or, Occasional Tracts on Husbandry and Rural Oeconomics.* About three quarters of it, *The Farmer's Letters,* was new; one quarter, the *Sylvae,* consisted of reprints, with no very important changes, of Young's communications to the *Museum Rusticum.*

In March, 1767, the *Gentleman's Magazine* allotted more than two full pages to it. "Too much cannot be said in recommendation of these letters." The review called Young "this able writer," and praised "his masterly chain of reasoning in the true principles of legislation and government."[72] Statesmen, nobility and gentry, and farmers were all urged to read the work. In short the praise meted out to the author, who was still only twenty-five years old, was unstinted.

Since the *Farmer's Letters* was Young's first important book, a fairly complete analysis is desirable. It must be admitted that Young had received more than encouragement from Harte, for many of his leading ideas were in Harte's *Essays on Husbandry.* The work also reveals its author's wide reading with numerous references to many French and English authors—Montesquieu, Rousseau, Vauban, Boulainvilliers, Temple, Child, Petty, and Hanway.

The very first letter begins by stressing the predominant importance of agriculture to the national welfare. "Agriculture is beyond all doubt the foundation of every other art, business or profession; it has therefore been the ideal policy of every wise

---

[71] Young, 1898: pp. 36–37.
[72] *Gentleman's Magazine* 37 (1767): pp. 130–133.

and prudent people to encourage it to the utmost."[73] This glorifi-
cation of the agricultural interest was a permanent theme in
Young's writings, and helps to explain why he was listened to by
the nobility and gentry. It was nothing new. Harte had made
the same claim at the beginning of his book.[74] The French
physiocrats had recently maintained that agriculture alone was
truly productive. Indeed such an emphasis was a part of the
vogue of ruralism so common in eighteenth-century thought as
expressed by Thomson, Cowper, and Rousseau.

It naturally followed that the nation should aid agriculture
as much as possible. Young, again following Harte, was a consistent defender of the bounty upon the export of corn, which
would furnish the incentive to grow enough grain so that famine
need not be feared. Since it is scarcity which makes prices high,
the corn bounty which produces a large supply of grain tends in
the long run to lower domestic prices.

Young consistently advocated that the millions of acres of
waste lands should be brought into cultivation. He urged the
nobility to undertake the task and wrote: ". . . never forget that
there is fifty times more lustre in the waving ears of corn, which
cover a formerly waste acre, than in the most glittering star that
shines at *Almack's*."[75] If the landlords would or could not make
the improvements, however, the state should borrow the money
for such a useful and truly national project. "Happy the monarch
whose reign is adorned by such an event—Yield fame, ye *Edwards*
and ye *Henrys*—acknowledge the infinite difference between *conquest abroad*, and improvement at home."[76] Here was another
of Young's perennial pleas. Marsh and fen lands should be
drained, moors broken up, laid down to grass and eventually
converted into arable.

To Young nothing seemed clearer than that enclosures were
an inevitable preliminary to improvement. In the *Farmer's
Letters* he did not make an elaborate defense of enclosures, for
he considered it unnecessary. "The universal benefit resulting
from enclosures, I consider as fully proved; indeed so clearly, as
to admit no longer of any doubt, amongst sensible and unpreju-

---

[73] Young, 1767: p. 3.
[74] Harte, *Husbandry*, 1770, 2nd ed.: p. 4.
[75] Young, 1767: p. 306.
[76] *Ibid.*, p. 334.

diced people: those who argue now against it are merely contemptible cavillers."[77]

Young also supported large farms, from 100 to 200 acres, because the comparatively large farmer could afford to make improvements and carry on a "spirited" agriculture. He presumably had sufficient capital to fertilize the land properly, to drain and fence it well, to keep sufficient cattle, to grow the new crops, and to employ a large number of workers. Of course enclosures tended to increase the size of farms and thus early in his career Young became a defender of both enclosures and large farms.

In the *Farmer's Letters* Young emphasized the importance to the farmer of keeping at least half of his land in grass. A large amount of grass land was necessary to keep a large stock of cattle; a large stock of cattle was necessary to provide a large quantity of manure; a large amount of manure was necessary to raise good arable crops. He never tired of condemning small farmers who put all their land to the plow in order to raise as much grain as possible for market. They only impoverished the land and decreased the total possible yield.

At this period of his life Young showed very little sympathy with the poor whom he believed to be the chief authors of their own misery. One letter claimed that the wages of the laboring classes were adequate to maintain them decently if they worked steadily. Prices of luxuries were high but the laboring classes should not expect to have luxuries. It comes rather as a shock when he lists as luxuries wheaten bread, beef, mutton, tea, sugar, and butter. He felt that a satisfactory and nutritious diet for the poor might well consist of black bread, cheese, beer, soup, potatoes, fat meat, and rice. In addition of course the laborer should have a warm cottage, tight and decent clothing, soap, candles, and firing. He maintained that the poor could provide themselves with such necessities on current wages and lay by something for sickness and old age. But of course they must work steadily and not waste time and money in tippling at ale houses or drinking tea. He estimated that enough was spent on tea and sugar to provide four million people with bread, and declared that tea

[77] *Ibid.*, p. 91.

"impairs the vigour of the constitution, and debilitates the human mind."[78]

If the laborers could get along so well on current wages, it followed naturally that they should not require poor relief, and yet poor rates were increasing constantly and cutting into the farmer's and landlord's income. The most desirable reform would be to reduce outdoor relief to a minimum. The poor should be forced into Houses of Industry where they would be better cared for than they could care for themselves, and the operation of which would greatly reduce the cost of relief. Of course the poor would object to the Houses of Industry, but he lashed out against such complaints in a bitter passage which well shows his complete lack of sympathy for the poor man's point of view:

As to the *hardship*, there is none in it; for the question ought certainly to be reduced plainly to this; we are to have the burthen of maintaining you, and is it not reasonable that while we maintain you, we should do it where and in what manner we please? We will neither allow you to be vagabonds nor tea-drinkers. . . . People who spent their lives, without laying up enough to support them while old, in a country so full of employment of all kinds as this is, to refuse assistance unless they have it *where* they like, might as well make a bargain for the best green tea and twelvepenny sugar every afternoon, before they accepted of warm cloathing, wholesome food, and a good house over their heads.[79]

This was good vigorous English. It might offend the poor, but they would not read his book. It could hardly fail to appeal to the self-interest of the landlords and farmers to whom the book was addressed. Young also wished to remove another burden upon the landed interest, that of tithes. He proposed that they should be abolished and the rectors compensated with a piece of land of an equivalent annual value to the average receipts from tithes during the past decade.

A few miscellaneous ideas in the *Farmer's Letters* deserve brief mention. In one letter he criticized certain premiums offered for agricultural improvements by the Royal Society for the Encouragement of Arts as impractical, and listed certain desirable experiments which the society might patronize. In a very interesting section on the American colonies he expressed ideas

[78] *Ibid.*, p. 284.
[79] *Ibid.*, pp. 295–296.

quite different from those of his pamphlet eight years earlier. He predicted that they would unite against England, rebel against her, become a powerful nation and eventually conquer all of South America. He thought that the colonies were too expensive to defend and that the same amount of money might much better be expended on reclaiming the waste lands of the British Isles.

On July 1, 1765, Arthur Young married Martha Allen, daughter of a wealthy patrician family of Lynn.[80] As he put it, "the colour of my life was decided."[81] Martha, or "Patty" as she was known to her friends, was slightly older than her husband, having been born on January 31, 1740/1.[82] There are no descriptions of her at the time of her marriage, but Young's earliest biographer wrote: "Mrs. Young possessed all the attractions of person, the accomplishments of mind, and the excellence of heart, to have rendered her a suitable companion for Arthur Young."[83] The marriage took place in the beautiful old church of St. Nicholas. One of the witnesses was Dorothy Robertson, the first girl whom he had courted at Lynn. The terms of the marriage settlement are not fully known, but Arthur Young received outright £1000, while the remainder of his wife's fortune was settled upon her.[84]

The newly married couple boarded with Arthur's mother at Bradfield Hall.[85] Their first child, Mary, was born at Bradfield on December 12, 1766.[86] There can be no doubt, for reasons that later will be examined in detail, that the marriage was not a happy one, and most of Young's biographers have assumed that it was unhappy from the beginning.[87] No proof for such a point of view has been found, while there is evidence that Arthur was very devoted to Martha during the first two years after their

[80] Copy from Parish Records, secured from Rev. Herbert B. J. Armstrong.

[81] Young, 1898: p. 32.

[82] Mrs. Young's tombstone in the Bradfield church gives the year as 1740. Whether this means 1740 or 1741 is uncertain. If it really was 1740 it would have made her nearly two years older than her husband.

[83] Paris, 1820: p. 283.

[84] Young, 1898: pp. 44–45.

[85] Paris, 1820: p. 283. Young, 1898: p. 32, wrote: "We boarded with my mother at Lynn." Young's pen must have slipped, for his mother was at Bradfield and he was farming there.

[86] Young, 1898: p. 43. The date is given on a memorial tablet in Bradfield Church.

[87] Young, 1898: p. 32, footnote 3 by editor Betham-Edwards, and copied by Hunt, 1926: p. 13, note 2.

marriage.[88] Late in 1767[89] Arthur Young, his wife, and baby left Bradfield and took a farm at Sampford Hall in Essex. One reason for the move was friction between Arthur's mother and wife. As he put it, "Finding that a mixture of families was inconsistent with comfortable living, I determined to quit Bradfield. . . ."[90] Probably an equally important reason was lack of financial success. Undoubtedly the experiments were chiefly to blame. He later summed up his experience at Bradfield thus: "Young, eager, and totally ignorant . . . it is not surprising that I squandered much money, under golden dreams of improvements; especially as I connected a thirst for experiment, without the knowledge of what an experiment demands."[91]

[88] Young, 1898; pp. 46–47, and copied below, p. 27.
[89] Many experiments in Young, 1771: 1: pp. 19, 21, 24, were completed as late as September, 1767, definitely at Bradfield. It was probably October, then, before the move took place.
[90] Young, 1898: p. 44.
[91] *Annals* 15: pp. 154–155.

II. Farming and Writing in Essex and Hertfordshire, 1767–1776

Most of Arthur Young's life was spent in his native Bradfield, where he had been brought up and had started his farming. From the autumn of 1767 until late in 1778, however, Bradfield was not his home. He first moved to Sampford Hall in Essex where he remained only six months. A second move took him to Bradmore Farm at North Mimms, Hertfordshire, where he resided until early 1777. In 1776 he made his famous trip to Ireland and in 1777 moved to Mitchelstown in County Cork. Late in 1778 the family estate at Bradfield again became his home and he remained there for the rest of his life. The present chapter will be devoted to the nine years from 1767 when he went to Essex until his first trip to Ireland in 1776.

During these years Young was constantly in financial difficulties and his farming was definitely unprofitable. To make ends meet he resorted to his pen, and an almost steady stream of publications followed—ten books, amounting to eighteen volumes, plus five pamphlets. Naturally much was mediocre or worse, but among the publications of these years were some very valuable works, the three English tours, the *Farmer's Kalendar*, and the *Political Arithmetic*. When he left Bradfield in 1767 he had written one important book, the *Farmer's Letters;* when he returned in 1778 he was the most famous and prolific writer on agriculture in the British Isles.

When Young determined to quit Bradfield he advertised for farms in the newspapers. Apparently the replies were numerous, for he spent six weeks in June and July, 1767, viewing farms all across southern England, from Norfolk to Wales, a trip which became *A Six Weeks' Tour through the Southern Counties of England and Wales* (1768).[1] He finally took Sampford Hall, in

[1] In *Annals* 15: pp. 155–156, Young wrote that the *Tour* resulted from such newspaper advertisements and stated incorrectly that the farm taken after the tour was at North Mims.

26

northwestern Essex, "a noble farm of 300 acres,"[2] which could hardly have rented for less than £200 a year.[3] Young's total capital at this time was probably the £1000 from his wife, and when he was disappointed in getting a loan from a relative, he found himself saddled with a farm much too large for his scanty resources.[4] It was probably late in 1767 when he wrote the following desperate letter to his wife:

Tuesday: 1767

My Dearest,—I am much in hopes I shall have a letter from you tomorrow; if I have not it will be a great disappointment; for when you don't write in huffs your letters are my only comforts. I went to Yeldham's this morning, but he . . . was out, and will not be home of some days. . . . How this terrible affair will end I cannot conjecture, nor what I am to do. . . . I would give my right hand that I had never seen this place, but such reflections only make one the more miserable; and the thoughts at the same time of what you feel with a young child to suckle hurt me more than I can express . . . . I had infinitely rather live in a cottage upon bread and cheese than drag on the anxious existence I do at present. Whichever way I turn my thoughts I see no remedy, nor know who can advise me what step to take. . . . An ill star rose on my nativity; had I never been born it would have been just so much the better for me, for you, and our wretched children; & if anybody was to knock me on the head it would be a trifling favour done to you all three, for most assuredly no good will ever come from my hands.

Adieu! I have scribbled out this paper to but little purpose.

A. Y.[5]

The letter not only reflects Young's acute agony and his deep affection for his wife, but also reveals the birth of a second daughter named Elizabeth but commonly known as Bessy.[6]

The letter, however, was unnecessarily pessimistic, for John Yeldham released him and found another tenant, only stipulating

[2] *Annals* **15**: p. 155.
[3] Young, 1769: p. 79. References to this work are from 2nd ed., 1769. On p. 76 he stated that arable land in that neighborhood rented for 12–16s. per acre, and grass land from 15s. to 1 guinea.
[4] Young, 1898: pp. 44–46. In *Farmer's Guide*, 1771: **1**: pp. 448–449, he estimated that a capital of £1000 would be sufficient only for 100–150 acres.
[5] Young, 1898: pp. 46–47.
[6] The letter quoted above, dated 1767, speaks of "children," but Young, 1898: p. 51, states that Bessy Young was born in 1768.

that Young should meet unpaid bills for seed, tillage, and labor.
Young gratefully wrote: "No person could behave in a more kind
and friendly manner than Mr. Yeldham did on this occasion."[7]
In the *Six Weeks' Tour* he extravagantly praised Yeldham's
culture of barley as "the highest pitch of perfection," and ranked
him "among the first cultivators of his age."[8] In December, 1768,
Yeldham thanked Young for a present of lampreys, some books
and a Westphalian ham, protesting that Young had more than
repaid the favor:

Give me leave to return you my respectful thanks, and to assure
you that in the twenty-six years I have had transactions with
mankind . . . I have scarce ever met with so much gratitude as you
have shown. It will not be in my power ever again to do you any
acceptable service, but for your sake I shall be more ready to do
a kind office than ever; so if I mended your fortune by helping you
off a hurtful contract, you will mend my heart by making me more
in love with mankind, and more ready to seek opportunities of
being useful.[9]

Probably while at Sampford Hall, Young published an anonymous pamphlet, *A Letter to Lord Clive, on the Great Benefits
which may result to the Publick from patriotically expending a
small part of a large private fortune. Particularly in Promoting the
Interests of Agriculture, by forming an Experimental Farm. Containing a practical course of management, with Estimates of the
Expences and Profit. Illustrated with a Plan of the Farm.*[10] In this
little pamphlet of 56 pages Young assumed that Clive would
spend part of his reputedly tremendous fortune made in India
in purchasing land, since "consequence and importance in this
kingdom are more annexed" to land than to money.[11] He urged
Clive to purchase a large estate of waste land which should be
enclosed, brought into cultivation, and turned into a vast experimental farm for crop rotations, various methods of soil treatment,
new crops, and scientific stock breeding. He estimated the initial

[7] *Annals* **15:** p. 155. Young's successor made a fortune on the farm.
[8] Young, 1769: pp. 80–81.
[9] Young, 1898: pp. 47–48.
[10] *Cf. ibid.*, p. 49, where Harte praised the pamphlet "as new, spirited, and
pleasing" and wrote Young that he had shown it to Lord Chesterfield who wished
that he could carry out the scheme himself. *Gentleman's Magazine* **37:** (1767):
p. 596, regarded it as Utopian.
[11] Young, 1767: p. 5.

expense as about £26,000, but predicted that in time the profits might amount to £2000 a year.[12] But Young also had plenty of plans for spending the profits. To be useful the experiments should be published. The remainder of the profits could well be applied for agricultural premiums, and to send a qualified person upon an agricultural tour of Europe, as suggested in that recent publication, the *Farmer's Letters*. To carry out such a scheme would make Clive greater "in retirement" than on the battlefield,[13] and would secure him "more genuine fame than ever Briton yet enjoyed."[14] Unfortunately Clive never attempted to discover the ingenious author to appoint him supervisor of the whole scheme, manager of the experiments, editor of the publications, administrator of the premiums, and traveler extraordinary!

Also while Young was at Sampford Hall, early in 1768, there appeared his much more important work, *A Six Weeks' Tour through the Southern Counties of England and Wales*, published in octavo by W. Nicoll, who had also published the *Farmer's Letters* and the *Letter to Lord Clive,* and whose address Young used at this time when in London.[15] The favorable reception of the *Farmer's Letters* prompted Young to declare that the new work was "By the Author of the Farmer's Letters." The new volume was not reviewed by the *Gentleman's Magazine,* but the *Monthly Review's* lengthy review of eighteen pages, in two successive numbers, was on the whole very favorable.[16] Young again used the letter form and the volume was divided into eight letters. The tour covered about six hundred miles, and was made in a chaise.[17] For the most part he stayed at inns instead of being entertained by the gentry as in his later tours.

The introduction set forth Young's aims: "to display to one part of the kingdom the practice of the other, to remark wherein such practice is hurtful, and wherein it is commendable. To draw forth such spirited examples of good husbandry from obscurity,

---

[12] *Ibid.,* p. 53.
[13] *Ibid.,* p. 2.
[14] *Ibid.,* p. 51.
[15] Add. MSS. 35,126, f. 23. Harte to Young, June 28, 1767, first addressed to Bradfield, then re-addressed to Nicoll.
[16] *Monthly Review* **38:** (1768): pp. 221–233, 276–282.
[17] Young, 1769: p. 120.

and display them the proper objects of imitation. . . ."[18] Later he wrote that in this tour, "for the first time the facts and principles of Norfolk husbandry were laid before the public."[19] He noted the very large fine farms in northern Norfolk and pointed out the use of marl and the cultivation of turnips as two foundations of the Norfolk husbandry. In Suffolk he described at great length the experiments of Mr. Orbel Ray at Tostock and the Rev. Dr. Tanner at Hadleigh in the culture of lucerne.[20] In Essex he praised the fine culture of barley and described the cultivation of hops. In many regions he vigorously condemned the excessive number of draught animals used in plowing: "I cannot give you these vile remnants of barbarity without a great deal of disgust."[21] In Wales he found methods very backward and declared that the farmers had no idea of the advanced methods pursued in the eastern counties. He was of course indignant to see Salisbury Plain "inhabited only by a few shepherds and their flocks," and estimated that if this great uncultivated waste were enclosed it could grow as much grain as all the rest of England exported.[22] He wrote rapturously of potatoes around Ilford, "which afforded me a pleasure superior to that, which any palace could confer; for I found there a husbandry more perfect (that is profitable) than any I ever met with."[23]

In the last letter Young began a practice followed in all his important tours, namely, to draw certain conclusions from the minutes taken en route. On the whole, he found more to praise than to blame, except for the Welsh counties. He strongly condemned the crop rotations where successive crops of grain were taken, and advised as the best possible rotation, where the land would grow turnips, a four-year course: (1) turnips; (2) barley; (3) clover; (4) wheat.[24] Beans or peas might be substituted on lands too heavy for turnips, and carrots or potatoes for turnips when the land tired of that crop. Similarly sainfoin or lucerne might be used instead of clover. Where beans and turnips were well hoed they prepared excellently for grain crops, but when

[18] *Ibid.*, p. x.
[19] Young, 1898: p. 44.
[20] Young, 1769: pp. 54–67.
[21] *Ibid.*, pp. 114–115.
[22] *Ibid.*, pp. 201–202.
[23] *Ibid.*, pp. 246–247.
[24] *Ibid.*, p. 280.

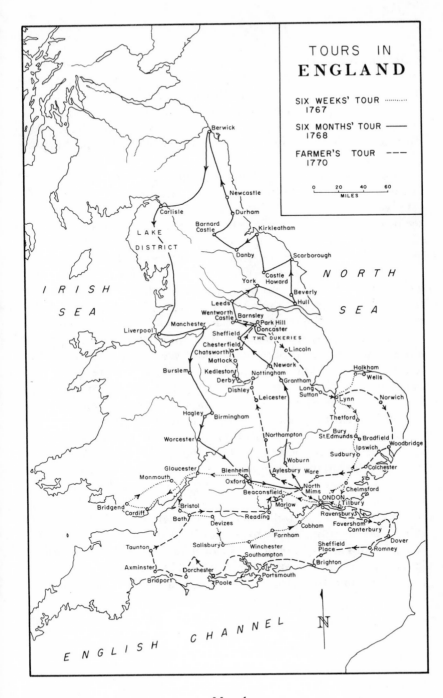

TOURS IN
ENGLAND

SIX WEEKS' TOUR ·········
1767

SIX MONTHS' TOUR ———
1768

FARMER'S TOUR ———
1770

0   20   40   60
MILES

MAP. 1

badly cultivated the land was full of weeds. Much more land could be used for hops and potatoes, both of which yielded immense profits. He was pleased that almost everywhere proper attention was paid to fertilizer, marl in Norfolk, lime in Essex and Wales, animal manure everywhere. He urged the eastern counties to adopt the west country practice of paring and burning old pastures.

Throughout the tour Young paid much attention to the size of farms, rents, prices, and wages, and the latter part of his last letter was really an essay in political arithmetic. Some of his statistics seem amateurish, especially when he tried to show the number of acres per horse, and the number of sheep per acre.[25] But his generalizations upon rents, prices, and wages appear to have much greater validity. He found that bread prices were very uniform,[26] while those of butter and some kinds of meat varied according to the distance from London.[27] In general the wages of "manufacturers" were slightly higher than those of agricultural laborers, except in the eastern counties. He admitted that agricultural laborers who received only 5s. or 6s. a week were underpaid, but felt that 11s. or 12s. was probably too high. He severely condemned some recent riots over the prices of provisions which had been led by highly paid manufacturers of the western counties and agricultural laborers of the eastern counties.[28]

Had Young's *Six Weeks' Tour* been confined to agriculture and political arithmetic, it would have been valuable to the economic historian, but much less interesting to his biographer. No apology was necessary for his comments on the roads. The turnpike from Salisbury to Romsey had "more the appearance of an elegant gravel walk, than of an high-road."[29] There were other good turnpikes, the great north road to Barnet, and some roads leading out of London. But of many, "it is a prostitution of language to call them turnpikes."[30] He declared the muddy turnpike from Bury to Sudbury in Suffolk to be "inferior to

[25] *Ibid.*, p. 302.
[26] *Ibid.*, p. 309.
[27] *Ibid.*, pp. 312, 315.
[28] *Ibid.*, pp. 325–333.
[29] *Ibid.*, p. 305.
[30] *Ibid.*, p. 306.

nothing but an unmended Welsh lane,"[31] and wrote that Norfolk did not possess a mile of really excellent road.[32]

He also commented upon the inns. He called the *Bush* at Wanstead "impertinent and dirty,"[33] and the *White-Lyon* at Cardiff "bad," while the *Angel* in the same town was "worse."[34] Some inns where he stayed in 1767 were still famous in the early twentieth century. The *George* at Winchester he found "dirty and dear; but civil."[35] The *Bell* at Thetford was "good," the *Swan* at Lavenham "civil and reasonable," the *Duke's Head* at Lynn "exceedingly civil and reasonable," and the *Angel* at Bury "very civil and reasonable."[36]

On the other hand, Young felt that he should explain his lengthy descriptions of the homes, gardens, and pictures of the nobility. He pointed out that it would have been stupid not to have seen these places when near them, and stated that his comments might be interesting.[37] Doubtless he realized that they would make the book more palatable to the general reader. He frankly admitted that he was no art connoisseur: "I know nothing of architecture . . . I never speak by rules, but my eyes."[38] And again. "I am no connoisseur in painting."[39] Such limitations, however, did not prevent him from pronouncing his opinion forcibly on the exterior architecture, the arrangement of rooms, the pictures and ornamental statuary, and the gardens.

Among the famous houses visited were Holkham and Houghton in Norfolk, Blenheim and Wilton, and near London, Cobham and Wanstead. He also noticed the architecture at Lynn, Oxford, and Bath. Although Young knew nothing of architecture, he had certain preconceptions. Gothic, in comparison with classic, was "Vile and barbarous."[40] Oxford appeared as only second-rate Gothic. He was no more favorable to baroque and had nothing but contempt for Vanbrugh, "a miserable architect . . . whose

[31] *Ibid.*, p. 258.
[32] *Ibid.*, p. 307.
[33] *Ibid.*, p. 343.
[34] *Ibid.*, p. 342.
[35] *Ibid.*
[36] *Ibid.*, p. 341.
[37] *Ibid.*, p. 2.
[38] *Ibid.*, p. 6, fn.
[39] *Ibid.*, p. 243.
[40] *Ibid.*, p. 123.

buildings are monuments of the vilest taste."[41] Blenheim he called a "quarry,"[42] but his dislike was probably intensified by the high fees charged for seeing it, and the impudence of the servants.[43] The work of Inigo Jones at Wilton he considered "heavy."[44] He decidedly preferred the contemporary neo-classic work of the Adam brothers. The adjectives constantly employed to indicate praise for architecture were "light" and "elegant." He thought Bath much finer architecturally than London: ". . . the most criticising eye must allow that the circus is truly beautiful, and ornamented to that just degree of elegance which . . . lies between *profusion* and *simplicity*."[45] A temple of Bacchus at Cobham consisted "of one handsome room elegantly stucco'd, with a portico of Corinthian pillars, in a light and beautiful taste."[46] He thought Holkham the best arranged house he saw, where people could live grandly and yet comfortably.[47] His passionate love of music was shown when he complained that none of the houses visited possessed an adequate music room, "indispensable" for a fine house.[48]

Young's tastes in painting were similar to his preferences in architecture. His general dislike of Rubens was probably related to his distaste for Vanbrugh. He described a figure of Mary by Rubens at Holkham as "a female mountain."[49] He was fond of portraits and usually praised Van Dyke and Rembrandt. Frequently his highest praise was reserved, however, for second-rate painters—Guido, Parmegiano, Vanderwerff, Maratt. Young's taste was not puritanical: ". . . The colouring of the naked, the soft and delicate expression of the roundness of the breast and hips, . . . are all inimitable."[50]

Young disliked the formal garden and was contemptuous of trees and shrubs trimmed to resemble animals and birds. He admired the work of "Capability" Brown, and wished the garden to imitate nature. He was fond of fine landscapes and striking

[41] *Ibid.*, p. 127.
[42] *Ibid.*, p. 337.
[43] *Ibid.*, pp. 127–128.
[44] *Ibid.*, p. 200.
[45] *Ibid.*, p. 188.
[46] *Ibid.*, p. 227.
[47] *Ibid.*, p. 8.
[48] *Ibid.*, p. 340.
[49] *Ibid.*, p. 17.
[50] *Ibid.*, p. 102.

vistas. He ridiculed a garden at Clifton near Bristol where Gothic and classic ornaments were jumbled together.[51] He found the park at Cobham in a "just taste."[52] His highest praise was given to Persfield near Chepstow, with its fine view of the Wye, "one of the sweetest valleys ever beheld."[53] The paths were arranged merely "as an assistance to view the beauties of nature."[54] But even at Persfield he regretted the absence of cascades: "Nothing has so glorious an effect, as breaking unexpectedly upon a cascade, gushing from the rocks, and over-hung with wood. . . ."[55]

The *Six Weeks' Tour* is thus a curious mixture of agriculture, political arithmetic, and art. In comparison with Young's later English tours it seems thin, amateurish. Nothing is covered very completely and the summary chapter is far from comprehensive. Nevertheless it was successful enough to make the publishers cry for more,[56] and it undoubtedly added to his reputation.

Possibly Young published another book in 1768. As noted above, he is supposed to have written four novels, one entitled "Lucy Watson."[57] There is a novel in the British Museum entitled *The Adventures of Miss Lucy Watson*. It appeared in 1768 anonymously and was published by W. Nicoll, Young's publisher at that time. It is a sentimental, romantic tragedy. The heroine experiences many highly improbable adventures and dies in the end in a poorhouse. There is little internal evidence to show that it was by Young. The hero is a great chess player, and Young was very fond of the game. One character constantly tried to judge whether people's actions were "characteristic." It is interesting that Fanny Burney quotes Young as using this same phrase in the same year, 1768.[58]

While at Sampford Hall early in 1768 Young had the temerity to write twice to James Boswell requesting information on Scotland. The first letter of March 14 started "by apologizing for the impertinence of addressing a Gentleman I have no acquaintance with . . ." He then poured on the flattery: "Your

[51] *Ibid.*, p. 183.
[52] *Ibid.*, p. 225.
[53] *Ibid.*, p. 165.
[54] *Ibid.*, p. 175.
[55] *Ibid.*, p. 177.
[56] Young, 1898: p. 49.
[57] *Cf. supra*, p. 11.
[58] d'Arblay, 1889: 1: p. 7.

Account of Corsica (wch. is become my favorite book) tells
me yt. yo. will not refuse ys. request tho' fro. one so much unknown." After admitting his authorship of the *Six Weeks' Tour*,
Young continued: "I now want a little intelligence of the present
State of a parish or two in Scotland, to complete my inquiries."
Among other things he requested the number of acres, farms, and
livestock, the average rent, the amount of poor relief given,
the acreage sown to various crops, the wages of labor, and the
prices of commodities. He concluded: "I blush at ye extent of
these queries, but, relye on yr. goodness of heart so conspicuous
in yr. works, for my pardon."[59] Surprisingly enough, Boswell
responded that he should "be very happy to contribute, in any
degree towards the information of the authour of the Farmer's
Letters."[60] Young's second letter of April 9, chiefly to thank
Boswell for his courtesy, requested a reply to be sent to "Bradmore Farm North Mims near Hatfield Hertfordshire."[61]

Thus the time of Young's removal to North Mims in the
spring of 1768 is quite definite.[62] The *Six Months' Tour* described one farm around North Mims in such detail that it may
very well have been Bradmore Farm. It contained 100 acres, 60
arable and 40 grass. The rent was £60 per annum. It was stocked
with 8 horses and 4 cows, and employed 8 laborers.[63] Young
never tired of assailing the badness of the soil: "I know not
what epithet to give the soil; sterility falls short of the idea. A
hungry vitriolic gravel. I occupied for nine years the jaws of
a wolf. A nabob's fortune would sink in the attempt to raise good
arable crops . . . in such a country."[64] Young later confessed
that he chose Bradmore Farm largely because the house was
"very neat and small" and had had more than £1,000 expended
upon it.[65] As gentry the Youngs must have a gentleman's resi-

[59] Boswell Papers, Yale University Library.
[60] Add. MSS. 35,126, f. 42.
[61] Boswell Papers.
[62] Young, 1898: p. 49, only states that he moved in 1768 to North Mims,
situated seventeen miles from London, *ibid.*, p. 63. I am indebted to the superintendent of the Hertford Museum, H. C. Andrews, Esq., for the statement that
the map of Hertfordshire, by Dury and Andrews, *ca.* 1766, shows "Bradmoor
Farm" as one and one-eighth mile east-southeast of North Mymms, one-half mile
south of Welham Green, and five-eighths mile west-northwest of Brookmans Park
House.
[63] Young, 1771: 3: p. 156. References to this work are from 2nd ed., 1771.
[64] Annals 15: p. 156.
[65] *Ibid.*, p. 183.

dence suitable to their social status, and apparently they were accepted by the local gentry and nobility. He was invited to join a dining club at Hatfield and there met Samuel Whitbread and Mr. Justice Willes from East Barnet.[66] Unfortunately Young was no more successful financially than at Bradfield or Sampford Hall. Nor was the "vitriolic gravel" entirely to blame. His passion for experimenting proved as expensive as ever. Since the farm was too small to occupy all his time,[67] he dissipated his energies into many other activities. The constant stream of publications certainly consumed much time. He had hardly settled at Bradmore Farm before he started on a six months' tour through northern England,[68] and left the farm to a bailiff. A similar trip was made through the eastern counties in 1770, and he spent nearly half of 1776 in Ireland. Worse still, in 1773 he became a parliamentary reporter for the *Morning Post,* which meant being in London most of the week.[69] Of course his publications brought in large sums of ready money, and he estimated that between 1766 and 1775 he made £3,000 by his writings.[70] Nevertheless he was constantly harassed by financial difficulties while at North Mims.

The proximity of North Mims to London facilitated an intimacy between Arthur Young and the family of Dr. Charles Burney, the musician, whom Young had first known at Lynn. In 1767 the widower Dr. Burney married as his second wife Mrs. Arthur Young's older sister, the widow Mrs. Stephen Allen.[71] In 1768 the Burneys were living in Poland Street in Soho and Young, always fond of music, frequently visited them, especially because Burney's eldest daughter, Hetty, ". . . entertained, or rather, fascinated me, by her performance on the harpsichord and singing of Italian airs. I was never tired of listening . . . and . . . I was thus riveted to her side for six hours together."[72] In

---

[66] Young, 1898: p. 52. Young stated that his friendship was with Samuel Whitbread, the statesman, but he was born in 1758 and Young's friendship was probably with the father, the brewer. Mr. Justice Willes was probably the second son of Sir John Willes, Edward Willes who became a judge of the King's Bench in 1768.

[67] *Ibid.,* pp. 62–63.

[68] *Ibid.,* p. 49.

[69] *Ibid.,* p. 63.

[70] Paris, 1820: p. 290.

[71] d'Arblay, 1889; 1: pp. lxiv, lxv, 2–3. For Dr. Burney's portrait, see fig. 18.

[72] Young, 1898: p. 51.

her biography of her father Fanny Burney wrote that Arthur
Young "was, when in London, all but an inmate of the Poland
Street family. . . ."[73]

Fanny Burney's famous diary contains one of the few contemporary pen portraits of Arthur Young. On May 29, 1768,
while the Burney girls—Hetty, Fanny, Susey, and Charlotte—
were at tea, "that lively, charming, spirited Mr. Young entered
the room. O how glad we were to see him. He was in extreme
good spirits. Hetty sat down to the harpsichord and sang to
him. . . ." When Mrs. Burney returned she easily persuaded him
to join a party to Greenwich the next day, and so he spent the
night there. The girls, including Mrs. Burney's own daughter,
Maria Allen, were up early dressing and primping for the trip:

who should rap at the chamber door but— (my cheeks are crimsoned
with the blush of indignation while I write it) —Mr. Young! I ran
into a closet, and lock'd myself up—however he did not pollute *my*
chamber with his unhallow'd feet, but poor Miss Allen was in a
miserable condition—her Journal, which he wanted to see, in full
sight—on an open bureau. He said he had a right to it as her uncle.
She called Hetty into her room and they were a long time ere they
could turn him out.

Fanny's diary also recounted some of the conversation during
the trip:

Talking of happiness and misery, sensibility and a total want of
feeling, my mama said, turning to me, "Here's a girl will *never*
be happy! *Never* while she lives! for she possesses perhaps as feeling a heart as ever girl had!" Some time after, when we were near
the end of our journey, "and so," said Mr. Young—"my friend
Fanny possesses a very feeling heart?" He harp'd on this some little
time till at last he said he would call me *feeling Fanny*, it was *char-
acteristick,* he said, and a great deal more such nonsense, that put
me out of all patience, which same virtue I have not yet sufficiently
recovered to recount any more of our conversation, charming as
part of it was. . . .[74]

At this time Young was twenty-six years old, had been married
two years, and was the father of two children. Mrs. Young might
have been pardoned a few twinges of jealously at the almost

[73] Burney, 1832: **1:** p. 202.
[74] d'Arblay, 1889: **1:** pp. 5–8. For Young's portrait by George Dance, see fig. 5.

too obvious pleasure which he found in her sister's daughter
and stepdaughters. That his attentions were not really regarded
as impertinent by the young girls, however, is indicated by Fanny's
mature comment, written many years later: ". . . the high, nay,
at that time, the volatile spirits of Arthur Young, though always
kept within certain bounds by natively well-bred manners, and
instinctive powers of pleasing, made him, to the younger group
especially, the most entertaining guest that enlivened the fire
side."[75]

Young spent the latter half of 1768 on his six months tour
through the northern counties. He had advertised the tour in
the newspapers and requested any gentlemen who might allow
him to see their experiments to communicate with the author
of the *Six Weeks' Tour* at Mr. Nicoll's.[76] Consequently he was
entertained by many gentlemen and noblemen, especially in Yorkshire. Mrs. Young accompanied her husband on this trip, which
covered 2,500 miles[77] and was made in a chaise.[78] *A Six Months
Tour through the North of England,* in four octavo volumes, was
not published until very late in 1769.[79] In a very laudatory review in its supplement for 1769, the *Gentleman's Magazine* stated
that the work might well have been conducted at public expense
under the patronage of a patriotic statesman.[80]

The plan resembles that of the *Six Weeks' Tour.* Nearly three
quarters consist of minutes of the tour. Attention is focused
upon agriculture, both on common practices and the improvements of the more "spirited cultivators." The preface defended
Young's practice of describing houses, paintings, and gardens because they might interest some who would be bored with the
agricultural details,[81] but these descriptions were relegated to
the footnotes in small type, often very long and not infrequently
occupying more space on the page than the text. The volumes
were embellished with many plates, some mere diagrams of agri-

[75] Burney, 1832: **1:** pp. 202–203.
[76] Young, 1771: **1:** pp. iii-iv.
[77] *Ibid.* **1:** p. xxi.
[78] *Ibid.* **2:** p. 184.
[79] The title page of the 1st ed. is dated 1770. However, the December, 1769,
issue of the *Gentleman's Magazine* **39** (1769): p. 400, noted it. A letter from
the Earl of Holdernesse, dated December 8, 1769, stated that he had just seen
the 1st ed. *Cf.* 2nd ed., 1771: **2:** *p.* 244 and Add. MSS. 35,126, ff. 66–71.
[80] *Gentleman's Magazine* **39** (1769): pp. 640–642: **40** (1770): pp. 40, 72.
[81] Young, 1771: **1:** p. x.

cultural machinery, others pictures of scenery or parts of the Duke of Bridgewater's Canal. Young acknowledged that he drew some of these pictures[82] which exhibit at least moderate artistic ability. Part of volume three and all volume four are devoted to a topical summary.

It is sometimes assumed that there was little agricultural progress in England before Young wrote, but his own tours show conclusively that the Agricultural Revolution, like all great social and economic changes, was the work of a large number of forgotten men. In this tour he visited at least a score of "spirited cultivators." It is also often assumed that nearly all the improvements took place in southeastern England, but the *Six Months Tour* proves that the northern counties, especially Yorkshire, could boast many innovators.

The northern counties contained many very large estates ruled by their landlords in a highly patriarchal fashion. Outstanding was the Marquis of Rockingham who had cordially invited Young to visit him at Wentworth House. Young responded with fulsome praise. The house was "one of the grandest in England;" the park was "as noble a range of natural and artificial beauty as is any where to be beheld;" the "magnificence of the woods exceed all description;" the temples were "elegant pieces of architecture."[83] Young enthused even more, if possible, over the agricultural achievements of the Marquis who had drained his lands, laid down arable fields to grass, imported hoers to teach his farmers the art of hoeing turnips, introduced many improved agricultural implements, and experimented with various manures.[84] From the *Six Months Tour* the Marquis appears nearly as enlightened as Coke of Norfolk.

In the Cleveland area of Yorkshire Young visited Charles Turner, Esq., of Kirkleatham, who had rearranged his farms into more compact units and taken the worst farms into his own hands when leases expired, improved them and re-let them at higher rents. Turner also experimented with many new crops, especially clover, never before common in that district, and introduced a better breed of cattle, the longhorned Lancashires.

---

[82] For example, plates opposite 2: pp. 183, 310; 3: pp. 118, 219.

[83] *Ibid.* 1: pp. 269–270. Harte had strongly urged Young to praise Rockingham, Add. MSS. 35,126, f. 47.

[84] *Ibid.* 1: pp. 271–316, for Rockingham's improvements.

He built new farmhouses, barns, and cottages, of brick and tile,
two inns, a warehouse, and granaries. He improved the beach
and built bathing machines. He offered employment and proper
housing to all the poor willing to work. "But the idle strolling
part of the poor that can work, but will not, he has little mercy
on; but is sure to punish them in such manner as the law allows
in his acting capacity of a justice of the peace."[85]

Young devoted nearly eighty pages to the improvements of
Simon Scroope, esq., of Danby, who had experimented with new
crops, hollow drains, and many manures. Above all, Scroope was
notable for very extensive experiments with cabbages and his
improvement of nine hundred acres of moorland. In summing
up Scroope's cabbage experiments, Young wrote:

From this day I cannot but suppose the culture of cabbages will
become as common in *England* as turnips; and the same honour, in
all future time, be due to the name of SCROOPE . . . that we now
pay to those of *Weston* and *Tull*, the introducers of turnips and
clover.[86]

During the tour Young and his wife visited many show places,
among them Hatfield House, Woburn Abbey, Burleigh House,
Wentworth Castle, Wentworth House, Castle Howard, Duncombe Park, Raby Castle, and Hagley. He was especially enthusiastic about Wentworth Castle, "surprisingly light and
elegant."[87] He greatly admired Raby Castle, "simply magnificent,
it strikes by its magnitude, and that idea of strength and command one naturally annexes to the view of vast walls, lofty towers,
battlements, and the surrounding outworks of an old baron's
residence."[88] Young also noticed at Raby the fine farm buildings
with a Gothic facade, reproduced in a plate.[89] He did not like
Castle Howard, built by Vanbrugh, nor the ornamental buildings
about the park, "in so heavy and clumsy a stile, as to be perfectly
disgusting."[90] He admired much of the detail at York Minster,
especially the chapter house, but confessed: "I never met with
any thing in the proportion of a Gothic cathedral, that was

[85] *Ibid.* **2:** p. 129. For Turner's improvements, **2:** pp. 98–154.
[86] *Ibid.* **2:** p. 369. For Scroope's improvements, **2:** pp. 344–420.
[87] *Ibid.* **1:** p. 127.
[88] *Ibid.* **2:** p. 428.
[89] *Ibid.* **2:** p. 440.
[90] *Ibid.* **2:** p. 61.

either great or pleasing; the loftiness is ever too great for the breadth, insomuch, that one must bend back the head to be able to view the ceiling."[91] Nor had his taste in pictures changed. At Duncombe Park he described a Rubens, "Nymphs in this master's stile; not tempting ones,"[92] while at Wentworth Castle he commented on Carlo Marratt's mistress, "the lady is beautiful and graceful: *Carlo* had a better taste than *Rubens*."[93] The garden which most appealed to Young was probably Hagley with its urns to Pope and Shenstone, its seat in memory of Thomson, its Grecian portico and its obelisk: " . . . art has added fresh lustre to every feature of nature, and created others which display a pregnant invention, and a pure and correct taste."[94] He deprecated the attempt to restore Fountains Abbey and felt it would be more impressive left "in the wildest and most melancholy state the ravaging hand of time can have thrown it into."[95]

There is considerable evidence in the *Six Months Tour* that Young had fallen into Rousseau's romantic mood. In the Lake District the Youngs explored Derwentwater and Windermere very carefully. He was enchanted by the scenery, made three sketches of Derwentwater, and expressed his fascination with the horrible in nature, in describing the shoreline:

From hence you coast a dreadful shore of fragments, which time has broken from the towering rocks, many of them of a terrible size; some stopped on the land by larger than themeslves, and others rolled into the lake, through a path of desolation, sweeping trees, hillocks, and everything to the water; the very idea of a small shiver against the boat strikes with horror.[96]

His last comment on Keswick was: "What are the effects of a Louis's magnificence to the play of nature in the vale of Keswick! How trifling the labours of art to the mere sport of nature!"[97]

The latter half of the eighteenth century was notable, not only

[91] *Ibid.* 1: p. 180.
[92] *Ibid.* 2: p. 78.
[93] *Ibid.* 1: p. 131.
[94] *Ibid.* 3: p. 296.
[95] *Ibid.* 2: p. 302.
[96] *Ibid.* 3: p. 117.
[97] *Ibid.* 3: p. 127.

for revolutionary changes in agriculture and for a romantic craze for natural beauty, but also for the beginnings of the Industrial Revolution. These too are reflected in the *Six Months Tour*. At Sheffield, Young examined the hardware industries and the tilting mill with its immense hammers run by water power, as well as the large iron foundries and collieries at nearby Rotheram.[98] At Newcastle he was struck with the prepared roadbeds for coal wagons and suggested their possible utility for other industries.[99] He visited the Burslem potteries and paid his respects to Josiah Wedgwood "who not only originally introduced the present cream coloured ware, but had since been the inventor of every improvement."[100] He called the docks "the glory of Liverpool."[101] He described the cloth manufactures at Leeds and noted in passing that a boy of six could earn 1*d*. a day.[102] At Manchester he visited the cotton industries, but found business rather slow because of the friction with the American colonies.[103]

Young was greatly impressed with the efforts of William Danby, Esq., of Swinton, to make the colliers on his own coal mine more self-respecting, and also more dependent, by offering them first a garden and then a field enclosed from the moor. "And by this well-concerted conduct, the whole colliery, from being a scene of idleness, insolence, and riot, is converted into a well-ordered and decently-cultivated colony: It is become a seminary of industry; and a source of population."[104] Young was also instrumental in raising a subscription of about £100 for an especially enterprising collier named James Croft, so that he could leave the mine and concentrate his entire attention upon improved farming on a small scale.[105]

Most interesting were Young's comments on the Duke of Bridgewater's canal. On arrival at Manchester he hired a pleasure boat and spent three days viewing the canal. He was especially impressed by the aqueduct over the Irwell at Barton Bridge: "The

[98] *Ibid.* **1**: pp. 115–116, 122–126.
[99] *Ibid.* **3**: pp. 1–8.
[100] *Ibid.* **3**: p. 255.
[101] *Ibid.* **3**: p. 168.
[102] *Ibid.* **1**: p. 139.
[103] *Ibid.* **3**: pp. 187–194.
[104] *Ibid.* **2**: p. 263.
[105] *Ibid.* **1**: p. xxvii; Young, 1898: p. 55. Danby contributed about £50, Charles Turner about £21, Young and his friends John Arbuthnot and John Wynn Baker a guinea each.

effect of coming at once on to *Barton Bridge,* and looking *down* upon a large river, with barges of great burthen towing along it; and up to another river, hung in the air, with barges sailing upon it, form altogether a scenery somewhat like enchantment."[106] Young enthused over the whole project and praised the "uncommon spirit" of the Duke of Bridgewater who must be "one of the truly great men, who have the soul to execute what they have the genius to plan" and also James Brindley the engineer, "one of the most useful genius's that any age can boast."[107] The account was illustrated by eight plates.

The summary, more than 500 pages long, can be divided into two parts: (1) the technique of agriculture, and (2) its economic and social implications. Part I included chapters on the grain crops, pulse crops, potatoes, cabbages, clover, the relations between the seed sown and the crop harvested, crop rotations, plowing and the proper use of draught animals, livestock, and manures. His comments on many points can easily be guessed. Beans were not hoed as they should be, cabbages were a highly useful crop, too many draught animals were used for plowing, oxen were not utilized sufficiently for this purpose, and most dairies were unprofitable partly because too few hogs were kept.

Nor were his views on the larger implications more novel. He still glorified agriculture as "the foundation . . . whereon the kingdom is built."[108] He continued to support a bounty upon corn exports, and attacked bitterly "the wretched system of eternally stopping the export upon every mob that infests the street."[109] He urged the ministry to appropriate £10,000 annually for the improvement of wastelands and declared that "the greatest disgrace to a kingdom, so truly flourishing, is the existence of so much waste land . . . ."[110] He denied that taxes were too high, but again attacked tithes as essentially "taxes on improvement."[111]

Young strongly emphasized the importance of labor, "Agriculture, arts, manufactures, and commerce, are but so many aggregates of labour."[112] He estimated the average weekly wage for agricul-

[106] *Ibid.* 3: pp. 219–220.
[107] *Ibid.* 3: pp. 240–241.
[108] *Ibid.* 4: p. 367.
[109] *Ibid.* 4: p. 144.
[110] *Ibid.* 4: p. 398.
[111] *Ibid.* 4: p. 336.
[112] *Ibid.* 4: p. 291.

tural labor as 7s. 1d.,[113] which was neither exorbitant nor so low as to oppress the poor. He continued to believe that the average workingman was lazy and would work just enough to procure him a subsistence. He was still bitter against tea drinking and idleness. The following comment on conditions near Oxford was fairly typical: "Poor rates 2s. in the pound; the employment chiefly pilfering and idleness; but all drink tea, and many have their hot rolls with it."[114]

His attitude toward the farmer contained apparent contradictions. In the preface, in expressing gratitude to those who had given him information, he refused to apologize for listing peers and farmers on the same page: "He, who is the BEST FARMER, is with me the GREATEST MAN."[115] And he paid tribute to the common farmer's hospitality. At the same time he felt that the average farmer was so backward that it was useless to expect reform from him. He strongly urged landlords to raise rents to weed out the inefficient farmers and to force better methods. At one point he burst out:

> . . . there is no evil more pernicious to the public, than Great Families . . . letting their estates be rented at low rates, from father to son, by a pack of slovens, rather than not have it to boast, *that their rents have never been raised;* which is nothing more than saying, *my tenants are poor; their husbandry bad; and the state injured in wealth, revenue, and population.* A very patriotic boast! . . . I will venture to assert, that the man who doubles his rental, benefits the state more than himself.[116]

He also believed that large farms were usually more efficient. Since enclosures meant larger farms and higher rents, he favored them, although once he admitted that the small man often suffered by enclosures and also attacked the exorbitant costs of the enclosure process.[117]

On the whole Young painted a fairly bright picture of the country which he had visited, although he was highly critical of many roads. That through the town of Wakefield "ought to be indicted," and that around Askrig was "fit only for a goat

[113] *Ibid.* 4: p. 296.
[114] *Ibid.* 3: p. 339.
[115] *Ibid.* 1: p. xiv.
[116] *Ibid.* 4: pp. 344–345.
[117] *Ibid.* 1: pp. 222–223.

to travel."[118] Nevertheless the general impression was favorable. He maintained stoutly that population was increasing.[119] Toward the end of the work he wrote: "In every part where I have been, I have seen none but the strongest marks of a rich, a happy, and flourishing people."[120]

In 1769 Arthur Young's third child, and only son, was born and was given the name of his father and grandfather.[121] Thus the Youngs had three children in the first four years of their married life, but the fourth and last child only came fourteen years later.

Young's close association with the Society of Arts began in 1769. As noted above, he had received in 1765 and 1766 premiums from that society for raising an acre of madder.[122] In 1768 he had submitted a paper on the rearing and fattening of hogs, which was examined by the Society's Committee on Agriculture February 13, 1769.[123] In submitting his essay he had stipulated that he wished to publish his experiments himself later as part of a large series.[124] At first the Society's Committee on Agriculture insisted that any essay for which they awarded a prize should be published by the Society, but after an exchange of communications with Young they yielded, and on February 17, 1769, awarded him a gold medal which he had chosen instead of a cash prize.[125] Young undoubtedly needed the cash badly but as a gentleman he chose the medal.

On May 17, 1769, Young was elected a member in the Society of Arts.[126] One of his three sponsors was John Arbuthnot who became his closest friend during the North Mims years. Five days after his election he attended the Committee on Agriculture.[127]

[118] Ibid. 4: pp. 424, 426.
[119] Ibid. 4: pp. 414–420.
[120] Ibid. 4: p. 395.
[121] Young, 1898: p. 51.
[122] Cf. supra, p. 15.
[123] Committee Report Books, 1768–1769, ff. 30–31. I was very kindly permitted to use the extensive manuscript records of the Royal Society of Arts by the late C. Durant Cassidy, Esq. The collection consists chiefly of bound volumes of minutes of the Society as a whole, and committee reports. Some miscellaneous material is included in the so-called Transactions and Guard Books. There are also several letter books and some unbound packets of letters. For a study of Young's relations with the Society of Arts, cf. Gazley, 1941: pp. 129–152.
[124] Guard Book B, 1759–1776.
[125] Committee Report Books, 1768–1769, ff. 30, 31, 32, 35, 42; Minute Books 14, f. 53.
[126] Minute Books 14, ff. 74–75.
[127] Committee Report Books, 1768–1769, f. 44.

Early in 1769 Young submitted another paper for a prize on the "Manner of cultivating Coleseed for the feeding of Cattle and Sheep." On May 1 the Agricultural Committee voted that this essay was not deserving of the medal, but on May 24, after his election, the Society asked the Agricultural Committee to reconsider its vote and on June 5 the Committee completely reversed itself. Finally on December 18, it was discovered that the author was no other than Arthur Young, who was awarded a second gold medal![128] Could some pressure have possibly been applied? Although candidates' names were not disclosed before decisions were made, the essays of course were in manuscript. It is difficult to believe that John Arbuthnot did not know very well that the essay on coleseed was written by his friend, Arthur Young.

Young's *Essay on the Management of Hogs* appeared in 1769 and included considerable material not submitted to the Society. Thirty-one experiments from 1765 to 1768 were minuted. Young combatted the prevalent view that a dairy was necessary to raise hogs successfully. Instead of being fed on skim milk, buttermilk, and cheese whey, his hogs were raised and fattened on clover and roots like carrots, parsnips, beets, and potatoes. The second edition of the *Essay on Hogs* (1770) included the prize essay on Coleseed, describing five experiments from 1765 to 1767. Young maintained that coleseed was not a profitable crop for feeding cattle and sheep, by no means as valuable as cabbage or turnips.

A much more pretentious volume by Young, *Letters Concerning the Present State of the French Nation,* was also published by Nicoll early in 1769. This anonymous work stands apart from his major interests, was not influential or popular, and certainly contributed nothing to its author's reputation. The *Monthly Review* pointed out that the author had apparently never been in France, justly called it a compilation, and while admitting that its author had "very enlarged views," nevertheless claimed that geniuses may "form designs too extensive even for themselves to execute."[129] Although the book was hardly more than hack work, it does show Young's views on a wide range of subjects at the age of twenty-seven. The volume contains numerous quotations from French

---

[128] *Ibid.,* ff. 42, 48; 1769–1770, f. 15.
[129] *Monthly Review* **40** (1769): pp. 107–109. Since this review appeared in February, the *Letters* must have been published very early in 1769.

and English authors. The first eight letters deal successively with politics, agriculture, trade and commerce, revenues, military power, social life (really a mélange), the fine arts, and literature. Young thought the unique feature of his work was the exhaustive comparison of France and England in the very long last letter.[130]

Much attention was devoted to the possibility of further wars between the two rivals. Young felt that France's geographical compactness more than compensated for England's insularity. He claimed that the French navy might within five years surpass the British navy. To forestall such a development he even advocated a preventive war if necessary: "She ought on *the political plan*, to take occasion to quarrel with France, whenever the French marine begins to wear a formidable appearance, or threatens in a distant manner to rival her own; by such, and only such means, she can secure to herself the empire of the sea."[131] He considered French negotiators much superior to English diplomats who often lost the fruits of war at the peace conference, partly because the cabinet which concluded a war seldom had begun it and frequently made peace at any price.[132]

In line with mercantilist views that colonies were valuable chiefly for their commerce with the mother country, Young contended that the English should have taken the French sugar islands rather than Canada. Nor should England ever have permitted the French any share in the Newfoundland fisheries.[133] England should continue the next war until her enemy's trade was destroyed and should seize the French sugar islands.[134] On the other hand, he warned the government that its prohibitions upon colonial manufactures would endanger its possession of the continental American colonies. He prophesied that eventually they would become independent anyway and that within a century their population would rise to twenty millions.

Can the English think that they will be able to retain such a number of American subjects in obedience! The monarch, then on the throne of England, will be weak minded, if he does not set sail to

---

[130] Young, 1769: p. iv.
[131] *Ibid.*, p. 434.
[132] *Ibid.*, p. 14 *et seq.*
[133] *Ibid.*, p. 55.
[134] *Ibid.*, pp. 19–22.

America, and take up his residence in the midst of an empire, which seems formed by nature to command the world.[135]

Considerable attention was devoted to internal economic and social conditions. Both countries laid more emphasis upon commerce than agriculture, the "first and original art . . . foundation of all others,"[136] but English agriculture was more progressive than French. Young condemned the French system of taxation because it discouraged agriculture. He declared that there were "not in nature more pernicious methods of raising public money" than by the *gabelle*,[137] and stigmatized the *taille* as "one of the most oppressive and unequal taxes the wit of man could devise."[138] He noticed that class lines were more rigid in France than in England. He lashed out viciously against legal abuses and lawyers in both countries "those savage detestable monsters —those harpies the lawyers. Would to heaven that I could stamp them with an epithet equal to their scoundrel profession."[139]

Much of the volume was devoted to the fine arts and literature. Young declared Versailles "a vast pile of different buildings, without unity, symmetry, elegance, or in short any true taste."[140] In general British architecture was better than French. He declared West to be "the best painter" England had ever produced,[141] and claimed that France had no equal to Hogarth, but he enthused over Boucher, Chardin, Greuze, and especially Vernet. England, however, was superior to France in music, because she had followed Italian modes and had adopted Handel.

The very long chapter on literature included scientists, agricultural writers, and economists. Young highly praised d'Alembert, Bougainvilliers, Buffon, Mirabeau, and Quesnay, but thought Montesquieu overrated. He enthused over du Hamel and Turbilly and considered French agricultural writers better than British. Of Voltaire he wrote, "a more universal genius never lived,"[142] and mentioned especially his *Age of Louis XIV* and

---

[135] *Ibid.*, pp. 439–440.
[136] *Ibid.*, p. 32.
[137] *Ibid.*, p. 86.
[138] *Ibid.*, pp. 90–91.
[139] *Ibid.*, pp. 150–151.
[140] *Ibid.*, p. 155.
[141] *Ibid.*, p. 445.
[142] *Ibid.*, p. 379.

his universal history. His highest praises were reserved for Rousseau, a "man of most irreproachable probity, and great goodness of heart."[143] Of the *Nouvelle Heloise* he wrote:

That celebrated novel—so tender—so agreeable—so natural—so elegant —so characteristical—so philosophical, is one of the finest monuments of the genius of the present age. . . . *the language of love* never before lived in print; the tender pathetic delicacy of the letters on that subject, are not imitations of nature, they are nature herself, they are the breath of inspiration.[144]

Nevertheless France had no historians to compare with Hume or Robertson, no moral philosphers equal to Bolingbroke, Hume, Johnson, or Smith. In the novel England had the greatest of all geniuses—Richardson, ably backed by Fielding, Smollett, and Johnson. English poets were likewise the best. He eulogized the Society of Arts which, if it only published transactions, would excel any academy or learned society in the world.[145]

Young's autobiography entry for 1770 began as follows:

What a year of incessant activity, composition, anxiety and wretchedness was this! No carthorse ever laboured as I did at this period, spending like an idiot, always in debt, in spite of what I earned with the sweat of my brow and almost my heart's blood, such was my anxiety; yet all was clearly vexation of spirit.[146]

The hard work is certain enough, for he published one pamphlet and three books, two of them in two volumes each. Nor were his receipts negligible, for his records show that he received £1,167. Unfortunately he was heavily in debt to a Bury banker. Not only could he borrow no more. There were also insistent demands for immediate repayment.[147] Somehow the bills must be met and he could write at high speed. Hence he rushed into print precipitately and published several works of which he was afterwards ashamed.

Again Fanny Burney's diary contains an excellent account of

---

[143] *Ibid.*, p. 359.
[144] *Ibid.*, p. 355.
[145] *Ibid.*, p. 457.
[146] Young, 1898: pp. 52–53.
[147] *Ibid.*, p. 53. He stated that he had to see his banker at Troston, which was later the seat of Capel Lofft, but there is no indication that he was ever a banker Possibly the banker was James Oakes but there is no proof that he ever had a country estate. John Spink was later Young's regular banker at Bury.

Arthur Young early in 1770. He appeared one afternoon during a shower, "most absurdly dressed for a common visit, being in light blue, embroidered with silver, a bag and sword." Fanny admitted that he "looked extremely well, and looked tolerably *conscious* of it." He had become "all airs and affectation; assumed a coxcombical assurance and indolence joined. . . . He bowed to the ground at entering, then swinging his hat the full extent of his arm. . . ." When they asked after Mrs. Young, he replied that she was somewhere in Soho and would probably arrive soon, as she did, in a sedan chair, full of recriminations towards her husband who had left her after an altercation with a coachman. Mrs. Young thought that the coachman should have been horse-whipped, but instead Young gave him a shilling and dismissed him. Fanny concluded the account with an exclamation: "O rare matrimony."[148]

There also exists a contemporary account of Young's farming in 1770 by a young Scotsman, John Wilson, who stopped in at Bradmore Farm on May 24.

Mr. Young very discreetly show'd me his offices, implements of hus-handry & his fields that are experimentally occupied. The offices are a small Barn with a pig's house by the side of it into which from a chest in the Barn (for holding summer provision) there is a conveyance for pease &c into the swine's trough. Before each cows head, in the cow house, there is a wicket in the wall for giving them their turneps through. Near his scullery he had built two receptables into which are conveyed by a trough or channel his dish washings, kitchen scraps &c &c. They are built with Brick five or six feet underground & as much above & is roof'd. When one reservoir is full it is put into the other & kept till sour as Vergis, when it is mixed with Barley meal and given to his store hogs. In the yard I took nottice [sic] of a large muck hill turned up, and in very good order; but upon nearer examination; found that al-most two thirds of it was clay which had been carried into the yard before winter, & had been turn'd up together with the dung in spring, but so far from incorporating with it, that it was run into lumps ten times tougher & more stubborn than when carried into the yard. His implements of husbandry are so many and various & their several uses & perfections discrib'd [sic] with such Volubility of tongue, that I can say little about them. His soil is a strong gravelly clay, his crops of corn I did not see any & one field of

[148] d'Arblay, 1889: 1: pp. 92–93.

Lucerne he show'd me sown broad cast 20 lb. to the acre, but instead of mowing four times, I doubt much whether it will ever mow once. . . . A field of cinquefoin much in the state of the lucerne, and anoyr. of Burnet very little better. . . .[149]

It is not a very flattering account, and bears out the general impression that Young was a greater writer than practical farmer.

Two letters from Arthur's mother show her worry over him. The first regretted his inability to accept an invitation from Lord Holdernesse who might give him some office. She also referred indirectly to his debts and his lack of religious convictions: "I fear as long as my poor eyes are open I shall never want for something relating to your welfare to vex me extremely. . . ." The second letter passed on from his brother John a compliment from George III on Arthur's pamphlet on the export of corn: "Oh! Arthur, with what capacities are you endowed—with what advantages for being greatly good! But with the talents of an angel a man may be a fool if he judge amiss on the supreme point."[150]

Young's pamphlet, *The Expediency of a free exportation of corn at this time . . .*, by the author of *The Farmer's Letters*, was published by Nicoll probably early in 1770 and went through two editions. Its purpose was to protest against a temporary legislative prohibition upon grain exports, probably that of January, 1770.[151] Young argued that the corn bounty had not raised prices. On the contrary, "the price falls at *home* in proportion to the quantity sent abroad."[152] The value of the bounty to the landed interest consisted, then, not in high but in regular grain prices, and in the exclusion of foreign competition from home markets. He also denied that the bounty had caused a general rise in prices or wages, or the conversion of grasslands to arable. He insisted that the encouragement of export helped ensure an adequate production for domestic consumption. He proposed that a bounty should be given when prices were low, that when they rose considerably

[149] Brown, 1936: p. 157. Wilson's later career shows him a very able man. Arbuthnot, with whom he remained for six weeks, predicted ". . . he will turn out well, very observant and industrious." *Cf.* Add. MSS. 35,126, f. 85.

[150] Young, 1898: pp. 56–57. For the invitation from Lord Holdernesse *cf.* Add. MSS. 35,126, f. 28. Other letters from Holdernesse at this time indicate a very friendly attitude. *Ibid.*, ff. 66–71, 82, 83.

[151] Barnes, 1930: pp. 23–45.

[152] Young, 1770: p. 7.

importation under a duty might be permitted for six months, and that when they became very high importation might be permitted duty free for another six months. But exportation should never be prohibited, for such a policy would discourage production and in the long run raise prices. He also claimed that the corn bounty tended to raise rents for the landlord, for when agriculture is profitable the demand for farms will be steady. The pamphlet closed with a reiteration of the vital importance of the landed interest. If class interests can be harmonized so much the better; but if friction arises it must be made clear that "the landed interest of this country, is of ten times the importance of all other interests, and this in every respect that could come into such an enquiry: wealth; income; population; stability. If interests are ever separately considered, it demands preheminence [sic] and it ought, and must have it."[153]

It is easy to point out flaws in Young's reasoning.[154] If, as he said, the bounty had *not* raised prices of grain or provisions generally, of what point was his elaborate argument that high prices are an accompaniment of prosperity? If the bounty had not raised prices, then it had not helped the workers, for they and their employers are better off when prices are high. He also failed to see that the bounty was no longer necessary to maintain production at a high level, since the growth of cities furnished a market more than large enough to absorb all that English farmers could produce. Young's pamphlet evoked a savage rejoinder by Josiah Wimpey entitled *Thoughts upon Several Interesting Subjects,* in which he accused Young of "misrepresentation, false reasoning, and wilful deceit."[155] In an appendix to the second edition of his pamphlet Young remarked of Wimpey's work: "I know nothing original in the production, but the peculiarity of the abuse. . . ."[156] Wimpey had written that the *Farmer's Letters* could now be easily picked up on the stalls for a shilling, but Young answered that a new edition to sell for five shillings was in the press.[157]

[153] *Ibid.,* p. 41.
[154] Barnes, 1930: p. 27.
[155] *Gentleman's Magazine* **40** (1770): p. 627. This publication said that Young and Wimpey both seemed masters of the subject, but that Wimpey was the first to call names and hence must have the worst of the argument.
[156] Young, 1770: p. 44.
[157] *Ibid.,* p. 66 *et seq.* The 3rd ed. of *The Farmer's Letters,* corrected and enlarged in two volumes, appeared in 1771.

Young's first book in 1770 was *The Farmer's Guide in Hiring and Stocking Farms*, by the author of the *Farmer's Letters*, published by Nicoll in two volumes, more than 950 pages in all.[158] Certainly it was not among his more inspired works. On the whole it is dull and of comparatively little value. The aim was to show how any given amount of capital might best be spent in agriculture. Young always considered a farm as a business venture upon which profits were to be made. Incidentally he regarded nothing less than twenty per cent as a sufficient profit.[159] The prospective farmer must carefully examine his soil, see that his fields were contiguous, and inspect his farm buildings, fences, hedges, ditches, and stone walls.[160] He must see that his lease contained no clauses which would make profits more difficult.[161] He must consider the distance of his farm from market and must enquire into tithes, poor rates, and wages, all as important as the rent. Young felt that most farmers tended to take larger farms than their capital warranted: "Nine out of ten had rather cultivate 500 acres in a slovenly manner, though constantly cramped for money, than 250 acres completely, though they would always have money in their pockets."[162] Successive chapters showed how various amounts of capital from £50 to £20,000 could best be used, with page after page of elaborate computations of dubious value.

Young compared the relative advantages of common and gentlemen farmers. The common farmer certainly had the advantage in labor costs and the costs of teams, and could act as his own bailiff. If the gentleman was to compete, he must be bold in adopting new methods and new crops. Young also appealed to the pleasures of agriculture for the gentleman. It harmonized well with a gentleman's way of living. There was the joy of experimentation, and the satisfaction of encouraging one's tenants and neighbors to make improvements. Such a gentleman farmer

[158] The only evidence for putting this work earlier than the others is that it was reviewed first in the *Monthly Review*.

[159] Young, *Farmer's Guide*, 1770: 1: pp. 342–343.

[160] *Ibid.* 1: p. 40. Young regarded live hedges, ditches, and stone walls good. When he took Bradmore Farm, he states, there was not a single ditch, but he built one for each hedge.

[161] *Ibid.* 1: p. 35. As examples he instanced prohibitions upon breaking up grasslands or upon the cultivation of certain crops. He had himself rejected a clause that turnips were not to be fed on the land.

[162] *Ibid.* 1: p. 98.

"is therein a true patriot, and merits the thanks of his country."[163] He bitterly contrasted the utility of such a man with those who spent their talents in "destroying the human species in physics, tything it in divinity, or ruining it in law."[164]

A very interesting chapter described the farmyard and its buildings. Young never tired of making plans of buildings to conserve manure and urine, or make the feeding of animals easier and thus reduce labor costs. He inserted two elaborate plans, one for a large and a small farmyard, and the other of buildings in which cattle might be fed easily and food boiled and prepared for hogs. These latter details tally closely enough with John Wilson's description of Young's farm buildings at Bradmore Farm to make one suspect that they were actual representations of his own farm.[165]

Young's second book in 1770 was entitled *Rural Oeconomy: or, Essays on the Practical Parts of Husbandry*. Eleven essays comprised about two-thirds of the volume, with the remainder an English translation, presumably by Young, of a French translation of a famous little German book by Hans Casper Hirzel published at Berne in 1761.[166] The "Rural Socrates," the English title of the work, described the agriculture of a Swiss farmer popularly known as Kliogg, many of whose practices and principles resembled Young's. The work was reprinted in Ireland and a second edition was published in 1773. In both style and content the *Rural Oeconomy* resembles *The Farmer's Letters*. He had already expressed most of the ideas in earlier works. The large farm is better than the small one, for "a little farmer is every where a bad husbandman; he cannot afford to do well by his land."[167] Most farmers have too much arable and too little grassland. Large numbers of cattle are necessary for large stocks of manure which in turn are necessary for large crops of grain. Gentlemen farmers should make many experiments: "no amusement in the world equals the forming and conducting experiments in agriculture . . . nor can any business, however important, exceed, in real utility, this amusement."[168]

---

[163] *Ibid.* **2**: p. 474.
[164] *Ibid.* **1**: p. 249.
[165] *Ibid.* **2**: pp. 450–469 for these plans.
[166] Young, *Rural Oeconomy*, 1770: p. 376. *Cf.* article on Hirzel in *Allgemeine Deutsche Biographie*.
[167] *Ibid.*, p. 293.
[168] *Ibid.*, p. 355.

Two essays stand out. Essay 5 begins with a very interesting statement of the current vogue and the rapid progress of agriculture.

Perhaps we might, without any great impropriety, call farming the reigning taste of the present times. . . . The practice gives a turn to conversation, and husbandry usurps something of the territories of the stables and the kennel; an acquisition which I believe, with reasonable people, will be voted legal conquest. . . . no one will dispute there having been more experiments, more discoveries, and more general good sense displayed, within these ten years, in the walk of agriculture, than in an hundred preceding ones. [169]

The farmer is strongly advised to keep accurate accounts. In addition to a cash book and ledger, there must be a minute book of what actually is done each day on the farm, and a sample of the one he used is included.[170] The best methods for hiring and treating labor are discussed in this essay. On the whole Young preferred "labourers" who lived out and were paid by the day or by the piece, rather than "servants," hired and paid by the year who lived in at their employer's expense. The farmer must not treat his laborers too mildly lest they become impudent. He advocated a patriarchal relationship with an elaborate scheme of rewards and punishments. The workers should be summoned to work by a bell, and the bailiff should note every tardiness in his black book. Each laborer should be held responsible for certain machinery, equipment, or livestock which should be inspected quarterly. After the quarterly inspection the workers with good records should be rewarded with gifts of clothes or furniture and with increased wages. On the other hand, those with bad records should be publicly reprimanded, and when very bad, dismissed.

The last essay discussed periodical publications regarding agriculture and foreshadowed the *Annals of Agriculture*, which only appeared fourteen years subsequently. The *Museum Rusticum* was stigmatized as a "book seller's job." The first essential of a

---

[169] *Ibid.*, pp. 173–175.
[170] *Ibid.*, p. 209. There follows an excerpt for one day, June 21:
"Three ploughs in six acres.
A pair of harrows, ditto, covering the turnip seed.
The frosty cow calved.
The wagon to *London* for ashes."
Unfortunately none of Young's manuscript minute books or ledgers have survived.

satisfactory periodical was that each contributor should give his name and place of abode; otherwise his contribution "is not worth a groat."[171] The magazine should appear regularly each month, but should vary in price according to size. The editor should be well known and capable of adding his own material when other contributions were thin. The magazine should be run as a public venture, and all profits should go either to premiums or into a sinking fund. Young was very optimistic about the success of such a journal: "The difficulties at first . . . would be great; but time, perseverence, and perfect disinterestedness, would certainly overcome them all."[172]

Young's last publication for 1770 was *A Course of Experimental Agriculture*, totaling nearly 1,800 quarto pages, in four volumes bound as two. The dedication to Lord Rockingham was signed by Young, so that this was his first work to be openly acknowledged. Its chief importance lies in the light thrown upon Young's farming life at Bradfield in the years 1763–1767.[173] The volumes minute a very large number of experiments, and show that Young was thoroughly imbued with the scientific spirit, faulty as were his methods and accounting. The work is probably the dullest of Young's writings with page after page filled with statistics and accounts. Later he was so much ashamed of this book that he attempted to buy up all the copies he could find. About one-third of the work is devoted to the grain crops, but there are also lengthy sections on the pulse and root crops, cabbages, and the artificial grasses, and comparatively brief sections on plowing, draining, fences, manures, cattle, and agricultural implements. Elaborate experiments were made upon various methods of planting and cultivating, the amount of seed, the best time for sowing, the effects of steeping the seed.

In the latter half of the eighteenth century a fierce controversy raged between the advocates of the old and the new husbandry. The old husbandry meant sowing of arable crops broadcast, while the new, first advocated by Jethro Tull, urged sowing in rows by means of the drill plow and regular cultivation by the horse hoe. Many experiments on many crops in *A Course of Experimental Agriculture* aimed to test the relative merits

[171] *Ibid.*, p. 365.
[172] *Ibid.*, p. 371.
[173] *Cf. supra*, pp. 14–16.

of the two systems. On the whole, Young was very dubious about the drill husbandry. In the first place, he had all kinds of difficulties with the drill plows tried. Once he exclaimed, "but in the name of common sense, what encouragement has any person to engage for profit, in a culture, the implements of which are in such a shape."[174] In the second place, the horse hoe broke the stalks of many crops, especially barley, oats, and pease: "It appears from these experiments, that the new husbandry is so much inferior to the old in the culture of oats, as to be absolutely inexpedient, and but another name for nonsense and absurdity."[175] Beans, however, had tougher stalks and could stand horsehoeing: " . . . the drill culture for beans I have found to merit all the praise that can possibly be given to it. . . ."[176] In the third place, to make horsehoeing possible for certain crops like turnips the interval between rows had to be so great that the final crop was smaller than with the broadcast system. Hence Young drilled the turnip seed in rows much closer together and handhoed the plants. Young especially objected to Tull's advocacy for all plants of a method satisfactory for only a few. "How therefore can any man talk of the drill culture in general? He must particularize his ideas, or he must speak at random."[177]

Somehow Young found time to make his third English tour in 1770. It probably took place in the summer and early autumn, but there is no evidence as to how he traveled and it is uncertain whether Mrs. Young accompanied him as in 1768.[178] The regions covered were basically the eastern and southern counties. He received many invitations in answer to his advertisements and probably stayed chiefly with substantial farmers and country gentlemen. Among the great houses visited were Stow, Newstead Abbey, Kedleston, Chatsworth, Sheffield Place, Milton Abbey,

---

[174] Young, *A Course*, 1770: 4: p. 305.
[175] *Ibid.* 2: p. 186.
[176] *Ibid.* 2: p. 397.
[177] *Ibid.*
[178] Young attended a trial of a horse hoe near London on May 24 (Committee Reports, 1769–1770, f. 62), and addressed a letter to the Secretary of the Society from Bradmore Farm on November 20 (Letters Received, 1770–1773, f. 20). The only evidence that Mrs. Young accompanied him is in John Baker Holroyd's letter in 1771 (Add. MSS. 35,126, ff. 97–98) in which he and Mrs. Holroyd sent remembrances to Mrs. Young. Probably Young and Holroyd met for the first time in 1770 and there is no evidence that the two couples had ever met in London or elsewhere.

and Sion-Hill. Among the places of scenic beauty seen were Dovedale, Matlock, and the Peak. And among the notables visited were some of the greatest farmers of the period, Robert Bakewell of Dishley, the great stock breeder, Mr. Ducket at Petersham, his close friend John Arbuthnot of Morden, and John Reynolds of Addisham. On this trip he also first met John Baker Holroyd, later Lord Sheffield, and visited Edmund Burke at Beaconsfield.

Young was not very active in the Society of Arts in 1770. He attended the Agricultural Committee meeting on January 22, and on May 24 was present at a trial of John Reynolds's Triple Horse Hoe at Kennington.[179] In November Young wrote to Secretary More of the Society, reporting unfavorably on a composition for marking sheep furnished him by the Society.[180]

On October 21, 1770, Edmund Burke wrote his first letter to Young on agricultural matters,[181] requesting information on the use of carrots. He had started carrot culture in 1769, and in 1770 had a fine crop, but had had no success in fattening hogs with them. Was he underboiling or overboiling them? He apologized twice for bothering Young, but excused himself by stating that Young's "knowledge" and "communicative character" must subject him to such requests.

John Arbuthnot's first letter to Young dates almost certainly from 1770.[182] It was carried by Young's servant, to whom Arbuthnot had delivered a double moldboard plow, a horse hoe, some hayseeds, and a peacock. Since Arbuthnot was quite an inventor of agricultural machinery, Young had probably bought the plow and hoe from him, while the peacock was more likely a present. Arbuthnot wrote: "I long to see you, when I hope to hear you have made a pleasant profitable tour, but shun fine houses and beware of flattery." Such a playful sentence could have been written only by a close friend, for Arbuthnot had hit upon two of Young's foibles. One other sentence shows the intimacy between the two families: "Present our best compts. to Mrs. Young whom we hope to see with you. . . ."

---

[179] Committee Report Books, 1769–1770, ff. 23, 62. The *Farmer's Tour* 3: pp. 83–84, has a picture and description of this triple horse hoe.

[180] Letters Received, 1770–1773, f. 20.

[181] Burke, 1844: 1: pp. 245–248.

[182] Add. MSS. 35,126, ff. 84–85. This letter was certainly written after John Wilson left Arbuthnot about August 1, 1770.

"The same unremitting industry, the same anxiety, the same vain hopes, the same perpetual disappointments. No happiness, nor anything like it. . . ." These opening sentences in the *Auto-biography* for 1771 are all too similar to those for 1770. In 1771 he published one pamphlet, the four-volume eastern tour, and his popular *Farmer's Kalendar*. He was thus busy enough. An entry in a memorandum book seems to show a somewhat better financial condition, for his receipts were £697 and his expenses £360.[183] Why then was he so unhappy and disappointed?

In the first place, his feeling toward his wife was becoming one of complete indifference. Again, Fanny Burney's *Diary* is the invaluable source:

Mrs. Young has been on a visit to us for some days. She and her caro sposo . . . are a very strange couple—she is grown so immoderately fat, that I believe she would at least weigh [ ] times more than her husband. I wonder he could every marry her! They have however given over those violent disputes and quarrels with which they used to entertain their friends. Not that Mrs. Young has any reason to congratulate herself upon it, quite the contrary, for the extreme violence of her overbearing temper has at length so entirely wearied Mr. Young that he disdains any controversy with her, scarce ever contradicting her, and lives a life of calm, easy contempt.

Fanny also reported a conversation with Young about relations between husbands and wives. Young speaks first:

"But man and wife can never judge fairly of each other; from the moment they are married, they are too prejudiced to know each other. The last character a man is acquainted with, is his wife's, because he is in extremes: he either loves, or hates her."—"Oh! I don't think that! I believe there are many more who neither love nor hate, than there are who do either."—"It's no such thing!" cried the impetuous creature, "you will find no such thing in life as a medium; all is love or hatred!" I could have said, it is much oftener *indifference,* than either; but I thought it would be too pointed, and dropped the argument.[184]

In the second place, Young's writings were meeting with severe criticisms to which he was always very sensitive. In March, 1771, a letter from Thomas Butterworth Bayley brought to Young's attention a book which attacked him, "I think in a very

---

[183] Young, 1898: p. 57.
[184] d'Arblay, 1889: 1: pp. 114–115.

uncandid disingenuous manner, and it is my opinion, that if you have leisure, *carefully and maturely* to reconsider the subject, you may greatly add to the Reputation you have acquired by your excellent Farmer's Letters, by answering the author of the Considerations."[185] The chief criticisms appeared in the influential *Monthly Review,* many by the Rev. Thomas Comber, his old antagonist in the *Museum Rusticum.* Young had visited him during his six months' tour and had apparently been treated very courteously. In turn Young had written very kindly of Comber's cooperation in procuring him material.[186] Comber lashed out at him viciously, however, in his reviews and did much to tarnish his reputation, especially in connection with *A Course of Experimental Agriculture.*[187]

In the preface to his *Farmer's Tour,* published in 1771, Young referred bitterly to the "miserable cavillers," those reviewers "who read with no other view than to calumniate."[188] He had been accused, he wrote, of being only a pretended farmer without land, of having elaborated his works into many volumes when all he had to say might easily have been compressed into one, of having written only to make money, and of having rushed into print too rapidly. He replied that he had been farming since 1762,[189] that his books had cost him more than they had brought in because they resulted from expensive experiments or journeys, and that although they had appeared rapidly the preparation for many went back ten years. He defended his practice of including many unsuccessful experiments, claiming that "those experiments, which *prove* the notions of some man to be really romantic and absurd, and such as cannot possibly answer, may be as useful to the world as the most brilliant registers of unvarying success."[190]

The appendix to the same work contained an answer to the author of the work which Bayley had brought to his attention, but not very "carefully" or "maturely," as Bayley had urged. On the contrary, Young was pretty "warm" in his defense, and

---

[185] Add. MSS. 35,126, ff. 94–96. The full title was *Consideration on the Policy, Commerce, and Circumstances of the Kingdom.*
[186] Young, 1771: **1:** p. xvii; **2:** pp. 48–49.
[187] *Monthly Review* **44:** pp. 162, 230; **45:** pp. 313, 378, 448.
[188] Young, 1771: **1:** pp. xii-xiii, xviii.
[189] It was really 1763.
[190] *Ibid.* **1:** p. xvii.

declared: "I do not recollect one of an equal number of pages,
that contain a tenth of the falsehoods abounding in that book."
The author had stated that Young really was at York instead of
North Mims, to which Young commented, ". . . one might order
such a fellow to the horsepond, but never altercate the matter."[191]

Young spent nearly ten pages in the appendix trying to refute
the *Monthly Review's* criticism of *A Course of Experimental
Agriculture.* He declared that the reviewer knew nothing of
husbandry and was quite uncandid. Although the points are
pretty technical, Young seems to have the better of the argument
on many. Again, however, he lost his temper, referred to the
"whole range of their dirty annals,"[192] and charged mistakenly
that the reviewer might be his avowed enemy, Joseph Wimpey,
to whom he devoted a long paragraph of abuse.

That Young felt the force of all this criticism there can be no
doubt. In the first place, he later confessed that he should never
have published *A Course of Experimental Agriculture* and was
nearly as severe on it as his critics had been. In the second place,
the entry in Fanny Burney's diary for December 8, 1771, shows
clearly that his reputation was being ruined by the reviewers
indicates that his financial straits were again desperate, and reveals his despondency:

Mr. and Mrs. Young have been in town a few days. They are in a
situation that quite afflicts me, how brought on I know not, but I
fear by extravagance. . . . They seem to have almost ruined them
selves, and to be quite ignorant in what manner to retrieve their
affairs. Mr. Young, whose study and dependence is agriculture, ha
half undone himself by *experiments.* His writings upon this subject
have been amazingly well received by the public, and in his tour
through England he has been caressed and assisted almost universally
Indeed his conversation and appearance must ever secure him wel
come and admiration. But, of late, some of his *facts* have been disputed
and though I believe it to be only by envious and malignant people
yet reports of that kind are fatal to an author, whose sole credi
must subsist on his veracity. In short, by slow but sure degrees, hi
fame has been sported with, and his fortunes destroyed. I grieve
for him inexpressibly; he truly merits a better fate. Too success
ful in his early life, he expected a constancy in fortune, that ha
cruelly disappointed him. His children happily have their mother

[191] *Ibid.* **4:** pp. 467–468.
[192] *Ibid.* **4:** p. 516.

jointure settled upon them. He has some thoughts of going abroad;
but his wife is averse to it. He is an enterprising genius, and I sincerely hope will be able to struggle effectually with his bad fortune;
but how I know not.[193]

In spite of the reviewers, Young received flattering letters from
many prominent men in 1771, which must have soothed his
proud and wounded spirit. On February 28 John Howard, the
philanthropist and prison reformer, replied to Young's letter
about the so-called "Howard potatoes" and then thanked him,
probably for a complimentary copy of Young's census pamphlet
which Howard described as "your very ingenious and usefull
labours."[194] On March 6 John Baker Holroyd thanked Young
for procuring an iron plow, promised to look up Young when
next in London, and added: "I wish you would flatter me with
the prospect of seeing you again in this country."[195] On May 2
Richard Price, the distinguished nonconformist preacher and
publicist, declared that he had read all Young's works "with a
particular Pleasure, approbation, and even admiration," congratulated Young upon his census pamphlet, and urged him to
make tours to Scotland, Ireland, and the Continent.[196] On Sep-
ember 5 Dr. Hunter of York notified Young of his election as
an honorary member of the York Agricultural Society and wrote:
"I have a million of things to say to you upon the subject of
agriculture. It would make me happy to have a person of your
knowledge & application near me."[197] And on December 5 an
unknown Swedish admirer wrote asking questions on fourteen
subjects and promising more questions in his next letter. Fame
thus had its disadvantages! The unknown correspondent also remarked. "The great Linnaeus is thought the first in his science,
but you are to be compared with him in this noble part of
humain [sic] understanding."[198]

Young's friendship with Edmund Burke was marked in 1771
by three very cordial letters from the great statesman to the great
agriculturist. On January 9 Burke thanked Young for some
cabbage seed, announced his intention to try a crop of maize,

[193] d'Arblay, 1889: 1: p. 139.
[194] Add. MSS. 35,126, ff. 92–93.
[195] *Ibid.*, ff. 97–98.
[196] *Ibid.*, ff. 107–108.
[197] *Ibid.*, ff. 109–112.
[198] *Ibid.*, ff. 113–114.

and acknowledged Young's invitation to visit Bradmore Farm.[199] On September 10 he discoursed on carrots, cabbages, the price of hay, pumpkins, buckwheat, the folding of sheep, and the merits of ashes as manure. He highly praised Young's *Farmer's Tour*: "To say there are some errors here and there, is only to say that you have undertaken a work of vast extent and intricacy." Perhaps Young had praised too highly "my feeble infantine attempts in husbandry" and had exaggerated his experiments in the folding of ewes.[200] The third letter, in October, apologized for not having visited Young earlier but expressed hope still to reach Bradmore Farm that autumn. He had now finished the *Farmer's Tour* and praised Young's "synopsis, with which you have so properly and judiciously closed it." [201]

Only two items indicate Young's activity in the Society of Arts for 1771. On March 6 he wrote to the Society about John Brand, "a very ingenious Blacksmith who has long work'd for me," and who had invented a swing plow, "which much exceeds any plough I have yet seen, in cutting a true, regular furrow, well cleared of the loose moulds; or in turning over grass land; at the same time that in strength and duration it is far preferable to all." Young asserted that he had used the plow for several years and offered to "order him to send you one for your inspection and Tryal."[202] On May 27 there took place at John Arbuthnot's farm at Morden a trial of several plows to measure "the force exerted by horses in drawing."[203] Secretary More had invented an ingenious little machine for such tests. Twenty plows were tested, including two of Arbuthnot's, the famous Rotherham plow, the equally famous trench plow of Mr. Ducket, and Mr. Brand's plow. After the tests the Agricultural Committee voted that Mr. Brand's plow was an improvement and deserving of a bounty. Eventually the

[199] Burke, 1844: 1: pp. 248–251.
[200] *Ibid.* 1: pp. 257–261.
[201] *Ibid.* 1: p. 262.
[202] Letters Received, 1770–1773, f. 93. Brand's other inventions were described in *The Farmer's Tour* 2: p. 212.
[203] Committee Report Books, 1770–1773, ff. 54–55. The *Annals* 1: pp. 113–119 also contains an account almost certainly of this meeting, although it is undated. I have drawn part of my material from this account which is at least complacent about the value of such tests, but in *The Farmer's Tour* 4: pp. 476–477, Young criticized severely the value of testing a plow by a trial that is limited to ten minutes or an hour. "I have been present at some of those committe trials, and am clearly of opinion, that not one in ten is worth a groat."

Society approved this vote, Mr. Brand's plow was bought by the Society, and he was awarded a bounty of £20. Young's friendship had indeed been very valuable to the "ingenious blacksmith." It is somewhat surprising to find that Young, who had originally recommended Brand, should have been one of the two judges to test his plow!

Young's first publication in 1771, *Proposals to the Legislature for Numbering the People,* published by Nicoll, was a short brochure of forty-five pages. Young stated that the pamphlet had been stimulated by an argument of the Earl of Chatham based "upon a mere supposition of population" without any definite proof. "I therefore was convinced that an actual enumeration of the people ought to take place."[204] The census should be taken and published every five years. Those best suited to perform the task were the collectors of the land and window taxes. Population should be classified according to occupation. Young rejected the common belief that total population was an accurate index of national strength. He was still expressing strong mercantilist ideas, thinking chiefly of national wealth which was very much more important than population in fighting wars. Indeed a nation might benefit by the loss of men in war, if such losses came from the idle portion of the population. He frankly admitted that he had often been branded as "an enemy to the poor, with all that common-place rubbish of vulgar ideas, which blackens a man as an enemy to humanity, because he sees no use in that sort of population, which increases only to be hanged or starved."[205] Such sentiments were far removed from his later sentimental evangelical solicitude for the poor.

Young's first book in 1771 was the *Farmer's Tour through the East of England,* another four-volume work, published by Strahan and Nicoll. The *Farmer's Tour* is aptly named, for it sticks more closely to agriculture than the earlier tours. There are to be sure the same long footnotes in small type describing houses, pictures, and gardens, but there are fewer accounts of natural scenery, nothing comparable to the description of the Lake District in the *Six Months Tour.* Nor is there much indication in the *Farmer's Tour* that England was experiencing the first stages of the Industrial Revolution, although, of course, the

[204] Young, 1898: p. 58. *Cf.* also Young, 1771: p. 13–14.
[205] Young, 1771: pp. 25–26.

region covered was not the scene of the new industry. The language in the *Farmer's Tour* seems somehow sharper and the use of bitter sarcasm against the adherents of old agricultural methods is more marked. The plates are limited to plans of buildings and diagrams of agricultural implements.

As an art connoisseur Young's tastes had not changed. He did praise Lincoln Cathedral which he thought much finer than York Minster and which he described as "remarkably light."[206] The architecture of his own time, however, appealed to him more, for instance Kedleston, the work of Robert Adam, "one of the finest houses in the kingdom" and the hall "a very noble room, the proportion uncommonly pleasing: the range of pillars is very magnificent."[207] Likewise he praised highly the improvements at Bath. The Circus was "an area no where equalled in the kingdom," and the Crescent when completed would have "a very noble effect."[208]

He still abhorred the formal garden as at Chatsworth, and was especially bitter on the fountains:

As to the water-works, which have given it the title of Versailles in miniature, they might be great exertions in the last age; but in this, the view of Nilus's leaky body, dolphins, sea-nymphs, and dragons vomiting water, trees spirting it from their branches, and temples pouring down showers from their roofs—such fine things as these are now beheld with the utmost indifference—one feels not the pleasure of surprise unmixed with disgust, especially when conducted to four handsome lions, spouting in the full view of the reach of a broad river, whose natural course should eternally silence such *hocus pocus gewgaws*.[209]

However, he found much to admire in the famous grounds at Stow which had recently been modernized. He visited conscientiously the innumerable ornaments in the park—the cave to St. Austin, the Temples of Bacchus and Venus, Queen Caroline's and Lord Cobham's pillars, the Rotunda, the Shepherd's Cave, the Temple of Friendship, the Corinthian Arch, the Gothic Temple, the Temple of Concord and Victory, the Temple of Ancient Virtue, and the Temple of Modern Virtue in ruins—

[206] Young, 1771: 1: p. 447.
[207] *Ibid.* 1: p. 193.
[208] *Ibid.* 4: p. 26.
[209] *Ibid.* 1: p. 212.

and commented upon the views from each.[210] His description of the environs of Roche Abbey illustrated the cult of the romantically "horrible":

. . . and the cliffs are spread with thick woods that throw a solemn gloom over the whole; and *breathe a browner horror* on every part of the scene—all is wild, and romantic; every object is obscure;— every part united to raise melancholy ideas; perhaps the most powerful, of which the human soul is capable.[211]

He criticized the ornamental grounds of the famous banker, Henry Thornton, at Clapham, because the natural was too artificial. There were too many benches and the grotto was too wild for such a gentle stream. He also attacked the taste which would permit "the looking from the lawn on to a bridge, which on crossing, you find, has no connection with the water . . . this is not quite the thing."[212]

Young concluded that agricultural techniques were higher in the region covered in the *Farmer's Tour* than in the area described in the *Six Months Tour*. As he put it, "the culture of this part of the country, if not excellent, is at least many removes from very bad systems."[213] Yet he frequently discovered very bad practices and attacked them bitterly. Shocked when he found manure being burned for fuel in Northamptonshire, Young burst out:

Will ye believe me . . . that this is the constant practice, not only of the cottagers, but of the farmers themselves! No; you will say; *it is impossible; there cannot be such an application of manure any where but among the Hottentots.* I looked attentively at the inhabitants, to see if the guts and garbage of the cows were not very capital ornaments of their persons.[214]

In Yorkshire he could hardly restrain himself when ". . . some very capital slovens assert that twitch is a very good friend óf the farmer; and that they should not be able to get any corn if the land was not full of it. To attempt to reason with such fellows is an absurdity."[215] When he found that the Somersetshire farmers

[210] *Ibid.* 1: pp. 32–42.
[211] *Ibid.* 1: p. 302.
[212] *Ibid.* 2: p. 243.
[213] *Ibid.* 4: p. 383.
[214] *Ibid.* 1: pp. 48–49.
[215] *Ibid.* 1: pp. 250–251.

raised only grains in their crop rotations, he exclaimed: "Bravo! my Somersetshire lads! And what then? Why, then, Sir, we lime on a fallow, and take seven crops more. Incomparable!"[216]

In several places he was almost as severe upon the landlords. In Derbyshire the landlords were so kindhearted about rents that the tenants were not spurred on to better practices. Young declared they were actually worse landlords than the gambler at White's who raised his rents to pay his gambling debts: "Thus is the dice-box in this instance of ten times more value to the nation than the sleeping, dronish state of vegetation in which so many landlords are content to drawl on, and not raise their rents, because their grandfathers did not."[217] He also made a stinging attack upon fox hunting which was rapidly becoming the landlord's favorite sport. The Isle of Wight, he said, was fortunate in having no foxes,

. . . consequently they are without a species of vermin by no means so innocent—the hunters of him; of whom there is too often reason to doubt (at least it is so in my neighborhood) whether the animal that flies, or the brute that pursues, be the greater beast of the two.[218]

In spite of all his criticism, Young's preface listed some seventy enlightened landlords and spirited farmers who were making all sorts of improvements. Only a few outstanding ones can be mentioned. In Yorkshire the agricultural experiments of Colonel St. Leger at Park Hill included trials in sainfoin, burnet, and other artificial grasses, methods of laying down arable to grass, the practice of paring and burning, drilled crops of beans and turnips, the use of covered drains, and various manures.[219] In Somersetshire he visited R. P. Anderton, Esq., of Henlade, who had not only tried the chief artificial grasses, but also had conducted experiments in drilled wheat, barley, oats, pease, beans, and potatoes.[220] As noted above, Young visited Edmund Burke and described his experiments on carrots, deep plowing, draining, drilled beans, potatoes, lucerne, cabbages, manures, the use of oxen, and the folding of sheep. Young praised Burke "for giving

[216] *Ibid.* 4: p. 22.
[217] *Ibid.* 1: p. 162.
[218] *Ibid.* 3: p. 201.
[219] *Ibid.* 1: pp. 257–296.
[220] *Ibid.* 3: pp. 423–483.

so laudable an attention"[221] to the improvement of his country's husbandry. He also admired the house as "regular" and "convenient" with fine views of the neighboring country and gave a detailed account of twenty pictures owned by Burke.

Probably the most lasting friendship which Young formed at the time of the *Farmer's Tour* was with John Baker Holroyd who later, as Lord Sheffield, became a famous patron of agriculture and President of the Board of Agriculture. Young praised the fine park at Sheffield Place, "forming varied lawns well wooded, shelving into winding vales, and commanding very noble sweeps of richly cultivated land."[222] In 1770 Holroyd was attempting to reform the poor laws. Farmers who took poor children as apprentices should not be given regular weekly allowances for them. Old people and the infirm could thus be given more adequate relief, while at the same time rates were cut in half. Young also quoted Holroyd's letter blaming high poor rates upon open commons. Young approved heartily Holroyd's proposed reforms, commended him for his public spirit, and used his arguments to show that enclosures really benefited the poor.[223]

The *Farmer's Tour* is notable for its accounts of the husbandry of some famous practical farmers. Young briefly described the work of Mr. Ducket of Petersham, noted for the invention of the trenching plow and other agricultural implements, one of which Young reproduced in a plate. He also praised highly Ducket's drilled crops and his neat hedges.[224] A longer account was given of the experiments of John Reynolds of Addisham, who had helped to spread the turnip culture through Kent, and was especially famous for introducing the cabbage turnip. The account includes Reynolds's long letter describing his methods and ending with a graceful compliment to Young: ". . . seeing no man on earth is better qualified than Mr. Young to write on the subject of husbandry, it will be an honour done to me, to see my work recorded in his ingenious annals of agriculture." Young reciprocated and declared that Reynolds was "active and spirited, and richly deserves to be had in esteem by all the lovers of good husbandry."[225]

[221] *Ibid.* 4: p. 84.
[222] *Ibid.* 3: p. 145. For a portrait of Lord Sheffield, see fig. 20.
[223] *Ibid.* 3: pp. 143–147, 153.
[224] *Ibid.* 2: pp. 242–247.
[225] *Ibid.* 3: pp. 82, 85.

One of the most celebrated of Young's contemporaries was Robert Bakewell, already famous in 1770 as the breeder of fine cattle and sheep. Bakewell's aim was to raise cattle and sheep for the butcher; his achievement was greatly to increase England's meat supply at a time of rapidly growing population. Stockbreeding, according to Bakewell, should produce small-boned animals which fatten easily in the more valuable joints. The true shape of a beast was "an hogshead, or a firkin; truly circular with small and as short legs as possible; upon the plain principle, that the value lies in the barrel not in the legs."[226] Young saw his famous bull Twopenny which Bakewell let for breeding purposes at five guineas per cow. As a good businessman, Bakewell let his bulls and rams at high rates but seldom sold them. Young was pleased at the gentleness of his animals, at his fine, clean barns, at his methods of feeding. "No where have I seen works, that do their author greater honour."[227]

Robert Bakewell was a common farmer and not a gentleman. John Arbuthnot on the other hand was a gentleman but dependent for his living on farming. His farm at Ravensbury in the parish of Morden, and his various experiments, take up more than half of Volume II, over 300 pages. When Young submitted his account to Arbuthnot for proofreading, the latter questioned whether it was not too long:

. . . but my Good man, surely you spin me out too long, but you best know, don't think me impertinent, but will people have the patience to go through such scenes of damned bad husbandry—many of the courses are so infernal bad. I must beg you to give the reasons which obliged me to pursue them or I fear not much credit will be given to your attentive cultivator as you call him.[228]

The account began: "I proceed to the register of the experiments of John Arbuthnot . . . with the satisfaction of knowing that I shall lay before the public, as useful knowledge as was ever yet received in the walk of husbandry."[229] Young minuted 110 experiments on almost every possible subject, laying down arable to grass, the culture of lucerne and madder, the drill husbandry, crop rotations, manures, and covered drains. Ar-

[226] *Ibid.* 1: p. 112. For a portrait of Robert Bakewell, see fig. 13.
[227] *Ibid.* 1: p. 134.
[228] Add. MSS. 35,126, ff. 105–106.
[229] Young, 1771: 2: p. 251.

buthnot found the drill husbandry very successful for beans, wheat, and turnips, but not for barley or pease. He was certainly outstanding in the culture of madder, which he had studied in Flanders and Holland, and had made profitable on land not exceptionally rich. On a farm of less than 300 acres, 80 were devoted to madder. In another work Young claimed that Arbuthnot was "deeply informed in mathematical mechanics."[230] Surely he was a prolific inventor of agricultural machinery, especially of plows. Many plates gave diagrams for his most famous implements, the great wheel plow, the double moldboard plow, the drill plow, the drain plow, the turnwrest plow, the Barkshire shim, the small and large spiky roller, and the turnip drill.

The *Farmer's Tour* described and evaluated the agricultural systems of several districts where advanced methods were commonly used. None was more famous than Norfolk. In thirteen pages Young summarized the essentials of Norfolk husbandry, all of which were closely connected and absolutely indispensable to each other.[231] The county had been enclosed very completely and cheaply without parliamentary enactments so that more capital was left for improvements. The lands thus enclosed were divided into large farms, for only substantial farmers could afford the new husbandry. Fairly long leases were given to encourage improvements, the first and greatest of which was the application of marl to the light soils. The rotation of crops was (1) turnips, (2) barley, (3) clover, for one or two years, and (4) wheat. Turnips, well hoed, served as the only fallow in the system. If a true fallow be substituted for turnips, fodder would be short. Clover likewise was essential, not only to furnish adequate animal food, but also because the soil was too light to grow wheat until "it is well bound and matted together by the roots of the clover." Much as he found to praise in the Norfolk system, Young criticized some things. The Norfolk farmers failed to hand-hoe their beans and wheat, their sheep were a "contemptible" breed and their horses "indifferent," and they did not grow enough carrots.

Southeastern Suffolk around Woodbridge and Ipswich was another very advanced district. Rents were high and the farmers prosperous, many worth from ten to forty thousand pounds.

[230] Young, 1811: p. 28.
[231] Young, 1771: **2**: pp. 150–163. I have taken the liberty of re-arranging Young's essentials in a different order.

Crop rotations were excellent and two corn crops never grown in successive years. They bought town manures, applied chalk and where possible crag, a composition of powdered shells, better even than marl. They raised carrots on a large scale and produced the fine sorrel breed of horses, short, barrel-shaped, and capable of drawing very heavy loads. Their crops were all hoed so carefully that their fields resembled gardens. "Those who exalt the agriculture of *Flanders* so high on comparison with that of *Britain*, have not, I imagine, viewed with attention the country in question."[232]

The *Farmer's Tour* concluded with some 400 pages of summary, starting with chapters on the crops and grasses so important in the new husbandry—carrots, potatoes, burnet, sainfoin, lucerne, clover, cabbages, and turnips. A second group of chapters discussed the merits of the drill husbandry, rents, the average produce of grain per acre, the importance of hoeing, the proper quantity of seed, the various branches of livestock, and fertilizers.

Much of the summary was devoted to political arithmetic, particularly labor problems. Young believed that wages had increased by about a quarter within the past twenty years, and that they were "high enough for maintaining the labouring poor in that comfortable manner in which they ought certainly to live."[233] Yet poor rates had more than doubled in the past twenty years and were of course unduly high.[234] The best way to reduce them was to force all the able-bodied poor into houses of industry such as that at Nacton in Suffolk which was described at length.[235] Since life in such houses was strictly regimented, only those really in need would apply. Large houses of industry, moreover, could buy supplies in large quantities, and the products made in them could be sold. The total result would be a great drop in poor rates. His whole point of view was consistent. As he put it in one place:

We cannot hesitate to determine, that the poor, generally speaking, have not a shadow of complaint. . . . I would not have the price of labour lowered . . . but I would have industry enforced among the

[232] *Ibid.* 2: pp. 221–222. *Cf.* also 2: pp. 168–178.
[233] *Ibid.* 4: p. 313.
[234] *Ibid.* 4: p. 338.
[235] *Ibid.* 2: pp. 178–190.

poor; and the use of tea restrained. Nothing has such good effects as workhouses."[236]

In another place he wrote: ". . . every one but an ideot [*sic*] knows, that the lower classes must be kept poor, or they will never be industrious."[237]

In 1771 there also appeared the first edition of *The Farmer's Kalendar*, which went through ten editions before Young's death. It was published anonymously by "An Experienced Farmer" and by Robinson and Roberts instead of his usual publishers at this time, Strahan and Nicoll.[238] The volume of about 400 octavo pages is divided into the twelve months of the year, and under each month directions are given of the work to be performed and how it should be done. At the end of each month several blank pages were inserted, upon which the practical farmer could make notes. The introduction claimed that if the book should remind the farmer but once a year of an important occupation forgotten, it would be well worth the price. The author wished to show the farmer that books might be useful, "a truth they will not all acknowledge."[239] The introduction also took issue with Tull's extremist followers who had emphasized tillage almost to the exclusion of manures, which could "never be too much attended to." Careful directions were given for keeping accurate accounts, which Young always thought so essential. An idea of the volume may be gained from listing the subjects covered in April, for instance—barley, white pease, buckwheat, madder, licorice, lucerne, sainfoin, burnet, sheep, cows, horses and oxen, hogs, potatoes, carrots, cabbages, water-furrowing, turnip fallow, woods, hedging, rolling, and hops. Many subjects

---

[236] *Ibid.* 4: pp. 364–365.

[237] *Ibid.* 4: p. 361.

[238] The 10th ed. appeared in 1815 and Amery, 1925: p. 18, notes a 21st ed. in 1862. It is interesting that Young should have published this work anonymously, for everything else written since 1769 had been clearly identified as by the author of the *Farmer's Letters*, if his name had not been openly given. There is also an interesting advertisement appended to *The Farmer's Tour:* "As I have been mentioned in several periodical publications and newspapers, as the author of books which I never wrote: I desire leave to inform the public that I prefix my name to all my works, and therefore am not answerable for any anonymous ones." Yet his next work was published anonymously. Moreover, the *Letters Concerning the French Nation* and *A Course of Experimental Agriculture* were omitted from the list following the above advertisement.

[239] The pages in the introduction are not numbered.

reappeared from month to month, and consequently there was much unnecessary repetition. Young was always a propagandist for his views, and took every opportunity to impress them upon his readers, even to repeating points month after month in almost identical language.[240]

The *Farmer's Kalendar* is pretty dull for the general reader and adds little to Young's views. Only a few miscellaneous points deserve attention. Young attacked vigorously the old maxim that "a penny saved is a penny got." If the farmer saves a penny by failing to drain, manure, or water furrow, he is in reality losing money. The necessity of keeping the horses steadily at work was a cardinal point in his husbandry for they were expensive animals.[241] In one place he described the work of a dairy. The dairy maid must be up every morning by four o'clock and the cows milked by six. "Cleanliness is the great point in a dairy." All the utensils must be boiled daily, and as much cold water as possible should be run through the dairy.[242] In another place he detailed the process of washing and shearing sheep and of winding the wool into fleeces. He fully understood the tricks of the trade: "A man, that understands the business well . . . should take care to turn in all the damaged or ill-looking parts, so as to make as handsome a fleece as he can."[243] He warned the farmer that threshers would steal as much grain as possible and do their work so slovenly that much grain was lost. He recognized that threshing was a "laborious and unhealthy" occupation and urged inventors to produce a cheap threshing mill.[244] He also warned against abuses in gleaning customs, but made a distinction between good business and downright meanness.

. . . make it therefore a law, that no gleaner shall enter a wheat field until it is quite cleared of the crop. . . . But, upon this plan, always desist from turning any cattle into the field, until the poor have gleaned it; for, if a use is made of keeping them out while

---

[240] Fussell, "Farmer's Calendars from Tusser to Arthur Young," *Economic History* 2: pp. 521–535, points out that Young's *Kalendar* is not a very trustworthy source for actual farming conditions because he was so often a propagandist.

[241] Young, 1771: pp. 13, 354–356, 388–389.

[242] *Ibid.*, pp. 164–165.

[243] *Ibid.*, p. 97. Later Young vigorously defended the farmer against wool merchants who attempted to penalize such practices by fines.

[244] *Ibid.*, pp. 382–386.

heaves are there, merely for an opportunity of turning hogs and other cattle in, it is double dealing, and a meanness unpardonable.[245]

To escape his financial troubles Young seriously considered emigration to America in 1772, but his wife was opposed.[246] In March his brother-in-law, an alderman of Lynn, applied to the Earl of Orford for some public position for him. Lord Orford suggested to Lord North a sinecure position as king's waiter, and Lord North replied that he considered Young as one "whose very ingenious and useful writings point out as a very proper object of notice and reward."[247] There is no evidence, however, that he received the post. Early in January Fanny Burney recorded his desperate condition:

He is not well, and appears almost overcome with the horrors of his situation. In fact he is almost destitute. I fancy he is himself undetermined yet what plan to pursue. This is a dreadful trial for him; yet I am persuaded he will still find some means of extricating himself from his distresses; at least, if genius, spirit, and enterprise can avail.

Even at that time his "native vivacity" displayed itself when Fanny asked him whether he was affected. "Affected!" exclaimed he with all his wonted impetuosity, " 'I had rather be a murderer!' "[248] Early in April he apparently had regained his natural buoyancy: "Fortune, I hope, smiles on him again, for he again smiles on the world."[249]

"This year I attended very much the meetings of the Society for the Encouragement of Arts, Manufactures and Commerce, as well as the Committee of Agriculture, of which I was chairman."[250] Thus begins a paragraph of the *Autobiography* for 1772, but Young's memory was at fault, for he was not elected chairman of the Committee on Agriculture until late in 1773.[251] Nor was he present at very many meetings in 1772, although his activity in the Society did increase. On January 12, Young wrote

[245] *Ibid.*, p. 247.
[246] Young, 1898: p. 61; d'Arblay, 1889: 1: p. 139. Her entry is for December 3, 1771.
[247] Young, 1898: pp. 60–61. The term king's waiter applied to certain honorary court functionaries, and also to certain customs officials.
[248] d'Arblay, 1889: 1: pp. 143–144.
[249] *Ibid.*, p. 157.
[250] Young, 1898: p. 59.
[251] Wood, 1913: p. 117; Add. MSS. 35,126, f. 161.

a very long letter to the Society, describing some experiments with Siberian barley seed, which proved much less productive and valuable than the common variety. On January 23 he wrote another very long and somewhat more important and interesting letter, suggesting several experiments for which the Society might offer premiums.[252] The most important were to determine the value of carrots, cabbages, turnips, and potatoes for fattening cattle. Any good farmer could raise large crops of these vegetables, "but gaining these crops is not at present the great object in experimental Husbandry, so much as ascertaining their value, because you will never be able to extend any branch of culture unless you previously show that the product is marketable." The experiments should be made on five animals which must be weighed before and after the experiment, and they must be fed on the crop in question which must also be weighed. Since the experiments would be costly, he suggested a premium of fifty guineas. He also suggested premiums for a drill plow for carrots and the best machines for weighing oxen and washing carrots. He was a little afraid that his proposal for a machine for washing carrots might be ridiculed and wrote that "a worthy friend of mine & a very valuable member [probably Arbuthnot] may smile & think of becoming a candidate by sending a birch broom & a tub," but he insisted the costs of washing carrots in the ordinary way were great. Interestingly enough the Committee on Agriculture accepted his proposals without important changes. Young attended the meetings of the Committee on Mechanics on April 2 and 23. On May 1 he also went to Mr. Fordyce's estate at Roehampton where two plows for cutting water furrows were tested.[253]

Twice during 1772 Young wrote open letters to the *St. James Chronicle*. On March 28 he took issue with certain statements on population by Dr. Richard Price in his recent *Observations on Reversionary Payments*. The letter was admirable in tone, and Young paid full tribute to Price's "ingenious" work, but pointed out that his recognized abilities made his statements all the more dangerous. The second letter, on December 14, attacked

---

[252] Transactions, 1771–1772.
[253] Committee Report Books, 1771–1772, ff. 31, 37. Alexander Fordyce, the banking member of this famous family, had a fine estate at Roehampton. His firm suspended payments in June, 1772, and he absconded.

a proposal to repeal the act for registering corn prices. He swept aside contemptuously the two chief complaints against the act, that the knowledge of varying prices in different regions led to discontent and to higher prices, and that the published prices were not acurate.[254]

Young's only important publication in 1772 was a large quarto volume of more than 500 closely printed pages, entitled *Political Essays concerning the Present State of the British Empire.*[255] Like the *Farmer's Kalendar,* it was published anonymously and Young even quoted from his own works and implied that the name of the author of the *Six Weeks' Tour* or the *Farmer's Tour* was unknown to him.[256] The introduction told how he came to write the book:

In the course of the political part of my reading, as I met with facts that appeared useful, I minuted them under respective heads. This practice I continued until I found my papers of a bulk that surprised me. I then revised and compared my intelligence.[257]

This statement is not the only indication that Young took notes on what he read and systematized them all carefully. His great unpublished agricultural work, *The Elements of Agriculture,* shows clear evidence of the same work methods. The style of the *Political Essays* is almost as heavy as the volume is physically. There are too many long quotations, too many short paragraphs of a single sentence containing one fact. It is a patchwork job. It does, however, show that Young was an indefatigable reader, for he refers to nearly every important eighteenth-century book on agriculture and political economy. It is also valuable because it gives Young's ideas on subjects not covered in his other works.

The volume is divided into six essays, the second of which discussed the constitution. It is thus very important to the biographer, for in no other of his early writings did Young indicate his political attitude. He admitted that the British consti-

---

[254] These letters are re-printed in *Political Arithmetic,* pp. 322–331, 336–342. The analysis of that work, *infra,* pp. 87–90, will cover the ideas expressed.

[255] Young, 1898: pp. 59–60, twice stated that his *Observations on the . . . Waste Lands . . .* was published in 1772, but on p. 63 he correctly attributes it to 1773. He did publish in 1772 a translation from a French work with the title, *Essays on the Spirit of Legislation, in the Encouragement of Agriculture, Population, Manufactures, and Commerce.*

[256] Young, 1772: pp. 91, 143, 162.

[257] *Ibid.,* p. v.

tution provided a great degree of liberty, "the natural birthright of mankind,"[258] but charged that it was seriously defective because a large majority of the people had no vote. He advocated giving the franchise to farmers and even laborers, and pointed out the inconsistency of excluding copyholders. Young greatly feared that the large number of placemen might increase royal power. He pointed out that a virtuous and popular king might, through pensions, sinecures, and honors, much more effectually undermine the constitution than a tyrant whose direct attacks would arouse open opposition.[259] In this essay Young warmly defended the English revolutions of the seventeenth century:

But there is very little reason to paint those civil wars, which are carried on in defence of public liberty, in such horrible colours. Take a nation at large, and its sufferings in them are by no means so terrible as some authors would have us to understand. The great men of prodigious property, may indeed be pretty well stripped; but when we speak of a *nation,* such are but of little consequence.[260]

A man who distrusted that the royal power was becoming too great, who favored an extension of the franchise to farmers and laborers, and who glorified the benefits of revolution was surely no Tory. At a later time he would have been called a Radical!

The third essay devoted to agriculture and about 100 pages long, was largely repetitive of his earlier works. He devoted much space to the scandal of waste lands: "In the whole circle of political economy there is not a more important object than this."[261] He bitterly attacked the wealthy landlord who possessed large pieces of waste land: ". . . possibly as much money is staked by him on a card, or ventured on a horserace, as would fertilize and people every waste acre on his estate." And he suggested a special order of knighthood, to have precedence over baronets and Knights of the Bath, for men who enclosed and put into cultivation a farm of at least 200 acres.[262] Perhaps the most interesting suggestion made in this essay, in the light of the subsequent work of the Board of Agriculture, was that the government should undertake a complete agricultural survey of the

[258] *Ibid.,* p. 19.
[259] Such a stand seems inconsistent from one who had so recently applied for a sinecure.
[260] *Ibid.,* p. 68–69.
[261] *Ibid.,* p. 117.
[262] *Ibid.,* pp. 127–128.

whole kingdom, the nature of the soil and of the substratum under it, the nature of various manures, an exact account of all the breeds of cattle, "in a word, of every circumstance concerning rural economy."[263]

Essays Four, Five, and Six covered respectively the closely related subjects of manufactures, colonies, and commerce. Although he had discarded the older bullionist theories, yet, if mercantilism be contrasted with *laissez faire,* then Young was surely a mercantilist in 1772. He emphasized national wealth and power, he believed in a favorable balance of trade, he considered that colonies had great economic value, and he urged constant state intervention. He repeatedly compared England with her neighbors and assumed that France and England were deadly rivals. He did not fear Spain or Holland, but declared that Dutch power in the east had been founded "upon the most bloody massacres, and the cruellest treachery—the world ever knew."[264]

Young thought that the export of raw wool and its smuggling into France or the Netherlands should be prevented. The export of raw wool would deprive many of the English poor of their employment. Moreover, "as *we* lose this manufacture of so much wool, others must gain it; and, unfortunately the greatest share of it falls to the French . . . our natural enemies."[265] He also urged the government to appoint inspectors of manufactures to assist English producers against severe foreign competition. Young showed himself a great friend to the new machinery. He denied that it created technological unemployment, maintained that machines created work, and concluded "that machines for simplifying work and abridging labour in manufactures are admirable inventions, of prodigious use in rendering commodities cheap, and in employing and maintaining great numbers of people."[266]

The great benefit resulting from colonies is the cultivation of staple commodities different from those of the mother country; that, instead of being obliged to purchase them of foreigners at the ex-

[263] *Ibid.,* p. 161. Page 162 contains an interesting footnote: "Since this was written, the attempt has been partially made by a private gentleman in the *Tours through England,* but this is not the plan I propose."
[264] *Ibid.,* p. 462.
[265] *Ibid.,* p. 182. Later he took a diametrically opposite position when he protested vigorously against the laws restraining the export of wool. Cf. *infra,* pp. 198–199, 215–218.
[266] *Ibid.,* p. 219.

pence possibly of treasure, they may be had from settlements in exchange for manufactures.[267]

With this principle as his starting point, Young found the West Indian sugar colonies most valuable, for they produced such a staple and bought their manufactured goods from England. The middle and northern colonies, on the other hand, were almost worthless. They not only produced no staple, but also bought very little from England because they had manufactures of their own which actually competed with those of the mother country.

Young also made a number of proposals for enhancing the value of the colonies. In the first place, the islands and the southern continental colonies were underdeveloped. Many more staples—vines, spices, tea, coffee, and cocoa—might be introduced and the sugar and cotton cultures might be greatly expanded. He grew lyrical about the possibilities of the uplands of the Carolinas and Georgia:

. . . the planter need but take a walk with his gun, to return loaded with a vast variety of game of the most delicious kinds; he need but row out with his net to return with the utmost plenty of fish, equally pleasing to the palate, and nutricious to the constitution. Every hedge presents him with fruits of a flavour unknown in England. Need we a more pregnant proof than their fatting their hogs with the finest peaches in the universe?[268]

In the second place, he urged that the interior be further opened up, especially the Ohio Valley and the eastern half of the Mississippi basin south of the Ohio. He bitterly condemned the Proclamation of 1763 which had closed the Ohio Valley to settlement, for the Virginians needed this very land to extend their tobacco plantations as the older lands became exhausted.

In the third place, for the middle and northern colonies he suggested drastic changes to destroy their manufactures and to develop raw material staples to be exchanged for English manufactures. No further immigration into these colonies should be permitted. Encouragement should be given to them to produce iron, timber, wool, and ships by bounties and by a regular and large market. On the other hand colonial manufactures should be ruined completely by dumping English manufactured goods

[267] *Ibid.*, p. 274.
[268] *Ibid.*, pp. 310–311.

in the colonies. The fisheries and the carrying trade were also to be closed completely to the colonists.

Behind Young's proposals for the middle and northern colonies lay his fear that they might soon achieve independence and hence the dissolution of the British Empire. They already had most of the essential conditions for independence—a numerous population, extensive manufactures, a flourishing trade, a large number of ships and prosperous fisheries. They could probably force the weaker southern mainland colonies to join them. He thought that a revolt at that time could probably be put down and that Britain was powerful enough "not only to extirpate their trade, their manufactures, their agriculture, but even the very people themselves."[269] Of course England's power to quell a colonial revolt rested absolutely upon her continued naval supremacy. If his proposals were adopted, and if British sea power were maintained, England might control her American possessions for "several ages." Eventually the colonies would almost inevitably become so densely populated that they must turn to manufactures, and at length would become more powerful than the mother country.[270]

For the immediate future, however, Young ardently desired British imperial dominance. The British should gradually take over new colonies, especially in the Pacific, stepping stones from the Falkland Islands to the Philippines. He rejected the argument that Britain already had more than enough colonies. All these islands, he assumed, could produce tropical staples, and hence would increase British trade and British sea power. He laid down minute plans how they might be taken over, and even recognized the value to empire and trade of the missionary:

The chaplain of the fort should be an honest well-meaning clergyman, who should learn their language as soon as possible, a powerful step towards civilizing them, and extending their wants; consequently their demands for manufactures of all kinds would greatly increase. . . .[271]

This whole section glorified the Elizabethan spirit of discovery

[269] *Ibid.*, p. 422.
[270] *Ibid.*, pp. 430–431. *Cf. supra*, pp. 48–49, for similar predictions made three years earlier. However in his earlier work, Young had warned that restrictions upon American manufacturers would lead to a declaration of independence.
[271] *Ibid.*, pp. 461–462.

and adventure, and strongly urged that the southern continent should be carefully explored by an expedition containing botanists, geographers, and mathematicians.

In the last essay on commerce Young attacked the East India Company's monopoly, that "most pernicious charter, which never had ten words of sound reasoning urged in its defence."[272] The East Indian trade should be thrown open to all English merchants and the functions of defense assumed by the government. Free competition among independent merchants would lower the prices of East India goods and increase the export of British manufactures. "Our manufacturers would flourish; our poor set to work; our shipping and seamen vastly increased; the general profit of our commerce enlarged; and our public revenue immensely enriched."[273]

What little is known about Arthur Young during 1773 indicates that marital incompatability and financial worries continued to vex him. Fanny Burney's diary again furnished the only glimpse into his family relations. Sometime about May 1 the Youngs accompanied the Burneys to the theater to see Dr. Goldsmith's new comedy, *She Stoops to Conquer*, which Fanny described as "very laughable and comic."[274] That Young was not as frequent a visitor as formerly is evident from Dr. Burney's letter of October 11, complaining "you have not mumbled tough Beef-steaks, nor eat cold meat, nor taken pot-luck in Qu. Squ. a great while."[275] A little later Young had "a very long, and a very strange conversation" with Fanny which she somewhat indiscreetly but fortunately confided to her *Diary:*

. . . he told me "that he was the most miserable fellow breathing," and almost *directly* said that his *connections* made him so, and most vehemently added, that if he was to begin the world again, no earthly thing should ever prevail with him to marry! That now he was never easy, but when he was litterally [sic] in a plow cart; but that happy he could never be! I am sorry for him—but cannot wonder.[276]

To improve his financial condition, Young applied to John

[272] *Ibid.,* p. 525.
[273] *Ibid.*
[274] d'Arblay, 1889: 1: p. 207.
[275] Add. MSS. 35,126, ff. 157–158.
[276] d'Arblay, 1889: 1: pp. 261–262.

Wilkes for an appointment as professor of trade, political agriculture, etc. The endowment had recently been left by a Mr. Temple of Trowbridge and the appointment lay with the Lord Mayor and Court of Aldermen. Young wrote to Wilkes as a stranger, relying upon his general reputation and his writings to engage Wilkes's support. He claimed that "the design of Mr. Temple is entirely in my line of enquiry" and promised, if elected, "to undertake any public performance that may be thought conducive to the course Mr. Temple meant to patronize."[277] There is no indication that he was the successful candidate.

The year 1773 brought, nevertheless, a considerable and regular augmentation to Young's income. For about three years he served as parliamentary reporter for the *Morning Post* at a salary of five guineas a week. The farm at North Mims "was insufficient to keep me employed,"[278] says the *Autobiography*. His duties as a reporter meant, however, that he left his farm almost entirely to a bailiff and this helps to explain his failure at North Mims. Every Saturday he walked the seventeen miles to North Mims and returned to London, also on foot, on Monday morning. He thus escaped to a considerable extent his wife's nagging but saw little of his young children. Unfortunately there is no record of his social contacts at London except with the Burneys and John Arbuthnot.

As a result of being in London, Young became much more active in the Society of Arts. He attended the Committee on Agriculture quite regularly during the spring of 1773.[279] On March 1 the Committee accepted Young's offer to supply for experimental purposes 100 bushels of clustered potatoes at three shillings a bushel and a week later he was present when some were distributed, among others to Robert Dossie, Jean Hyacinthe de Magellan, the architects Robert and John Adam, Dr. Watson and Dr. Fordyce, Colonel St. Leger, and John Arbuthnot.[280] In

---

.[277] Add. MSS. 30,871, ff. 185–186. This letter is also reprinted in Wilkes, 1804–1805: **5**: pp. 67–68. Young's statement was "not having the pleasure of being personally known to you. . . ." Young, 1898: p. 10, however, stated that in 1753 he met John Wilkes at his aunt's Lady Ingoldsby's, "more than once," but of course Young was only a boy at that time.

[278] Young, 1898: pp. 62–63.

[279] Committee Report Books, 1772–1773, ff. 29, 32, 34, 35, 38, 40, 42, 45. He was present on March 8 and 15, April 26, May 10, 17, and 24, and June 14.

[280] *Ibid.*, ff. 28–29.

October he sent the Society a letter on the marking of sheep[281] and the Agricultural Committee a paper on barley for Dossie's *Memoirs*.[282] As a result of all this activity Young was elected early in December a joint chairman of the Committee on Agriculture.[283]

Young's only publication in 1773 was a pamphlet of 83 pages entitled *Observations on the Present State of the Waste Lands of Great Britain*, by the author of the *Tours through England*, and published by Nicoll for two shillings. A proposed new colony in the Ohio country furnished the occasion for the pamphlet. He repeated the arguments in the *Political Essays* about the value of the Ohio valley for an expansion of the tobacco culture and for the production of hemp. On the whole he pictured the country as an earthly paradise, "a country formed for pleasure, health, and plenty," and pointed out how desirable it might be for small country gentlemen of six or seven hundred pounds a year who could no longer "bring up their families with some decency, keep a tolerable table, dress, and live like gentlemen."[284] It was his own class that Young was describing and it must be remembered that he had himself seriously considered emigration to America.

He then switched his argument abruptly. Why should people complain if emigrants left England or that food prices were high, when nothing was being done to settle and cultivate the great wastes at home? He calculated, as in the *Farmer's Letters* and the *Farmer's Guide*, the tremendous profits from improving wastelands. England's trouble was not that farms were too large, but that estates not used for farming were too extensive. If the government wished to stop emigration why not buy up all wastelands, improve them, re-sell them in small allotments of twenty to thirty acres, and give one free to men with large families. The government should oblige all landlords to sell lands which they would not bring into cultivation. His most novel suggestion was that a group of wealthy men should form a company, subscribe shares of £100 each, and appoint a manager to supervise the opening of wastelands to cultivation. Surely Arthur Young would have been an ideal appointment!

[281] Minute Books 19: f. 2.
[282] Committee Report Books, 1773–1774, f. 2.
[283] Add. MSS. 35,126, f. 161.
[284] Young, 1773: pp. 26, 31.

The *Autobiography* deals with Arthur Young's activity in 1774 in two short paragraphs.[285] He was elected to several learned societies. A letter from Rudolph Valltravers on March 3 requested Young to become an honorary member of the Palatine Society at Lautern for encouraging agriculture, commerce, and industry, and forwarded a letter of thanks for his works. Another letter on May 7 invited Young and his wife to dinner and stated that Valltravers intended to present Young's *Political Arithmetic* to the Elector of the Palatine in person.[286] Sometime during the year he was also elected an honorary member of the Geographic Society of Florence. Much more important, on April 28 he was elected a fellow of the Royal Society.[287] He was always careful in his later years to append the initials F. R. S. to his name, but there is no indication that he was ever active in the Society.

The references in Fanny Burney's diary for 1774 primarily concern Martha Young. She took tea and supper on March 30 with the Burneys, but Arthur was not present. Nevertheless, Fanny remarked, "Her husband is infinitely better; which I must rejoice at." During the conversation Mrs. Young stated that her husband had "some thoughts of going" to Ireland. This is certainly the earliest reference to such a trip. The conversation throws considerable light upon Mrs. Young. "After tea, Mrs. Young, Mrs. Allen, and Mamma talked upon fashions, which is ever an agreeable subject to Mrs. Young, and constantly introduced by her. . . ." Fanny also commented on Mrs. Young's conversation with another member of the company, that she "showed as much too much quickness, as the Doctor did too much dullness. . . ."[288]

Arthur Young attended the meetings of the Society of Arts quite regularly in 1774. He first presided as chairman of the Committee on Agriculture on January 10. He likewise presided on January 17 and February 21, was present on February 28, and on June 20 attended a trial of plows at the Sign of the Tower at Stockwell.[289] On November 9 he wrote that he was willing to serve again as Chairman of the Committee on Agriculture, "if thought worthy of yt honor." Most of the letter concerned the

[285] Young, 1898: pp. 64–65.
[286] Add. MSS. 35,126, f. 163.
[287] Thomson, 1812: appendix, no. 4, p. iv.
[288] d'Arblay, 1889: 1: pp. 282, 285, 289.
[289] Committee Report Books, 1773–1774, ff. 10, 17, 25, 28, 46.

effects of gypsum as a manure which he had tried on three spring
crops, clover, natural grass, and barley:

The result, exactly similar in all three—not the least effect to be
perceived either during ye growth or at the mowing of the crop. I
showed ye trials to several gentlemen, without telling ym what they
were, to know their opinions—no distinctions to be seen by any one.

As ye gyp. is recommended so strongly, & by persons on whom ye
Society depends, there are probably some unrelated circumstances,
either of soil, quantity, or management: Tho' I must remark that with
a manure of any considerable merit, minutiae are un-necessary. Let
a man try night soil even in small quts., he will not want spectacles
to see ye effect let him spread it when, how or in wt manr. he pleases—
Coal ashes tho' not a capital manure, for ye soil sown ye same day
as ye Gyp. were beneficial evidently to ye eye. I do not add yt by any
way of reflection on a manure not sufficiently tried, but merely to
check great expectations.[290]

In the spring appeared Young's *Political Arithmetic,* one of
"my best works, which was immediately translated into many
languages, and highly recommended in many parts of Europe."[291]
Not one of his longest or most pretentious books, it ran to less
than 400 pages and sold for only 5s. 3d.[292] Nevertheless, it still
ranks as one of his best works, and indeed his reputation as an
economist rests primarily upon it.[293] The term "political arithmetic" in the seventeenth and eighteenth centuries was applied
partly to the field of vital statistics, partly to what would be
called today economics. Young's interest in the collection of statistics is evidenced by his pamphlet in favor of a national census
and by his advocacy of the registry of corn prices. Young always
claimed that he was not a theorist but only a practical man.
Certainly he never worked out a system of economics like those
of the Physiocrats or Adam Smith, but his *Political Arithmetic*
is the most complete statement of his general economic philosophy. Instead, however, of discussing the more general problems
of production and distribution, Young's work is concerned primarily, as always, with agriculture. But so broad is his treatment
of agriculture in this work that much space is devoted to such
questions as taxation and population.

[290] Transactions, 1774–1775.
[291] Young, 1898: p. 64.
[292] *Gentleman's Magazine* 44: p. 227.
[293] For a detailed analysis of this work, *cf.* Bonar, 1931: pp. 220–240.

The long first chapter summarized the various factors tending to stimulate English agricultural progress. The second chapter, in which he gave practical advice to foreigners, was much shorter. The third and last chapter attempted to refute certain ideas which Young believed to be erroneous, notably those of the Physiocrats and Dr. Richard Price. There are also several appendices, some mere statistical tables. His two letters to the *St. James Chronicle* were reprinted. There was also a previously unpublished memoir on the corn trade by Governor Thomas Pownall to the Lords Commissioners of the Treasury. The plan of the book leads to much repetition and hence makes a summary rather difficult.

Young's fundamental idea in the *Political Arithmetic* was that national strength depends on national wealth, not on a large population. As he put it, "National Wealth raised by industry, is more advantageous to a nation than an increase of people. . . . But the number of people in a modern state, is by no means the measure of strength: this is wealth alone."[294] And yet Young strenuously denied Richard Price's claim that population had decreased in England since the seventeenth century. National wealth had obviously increased and so population must have increased. For national wealth means plenty of employment and where there is plenty of employment there is sure to be plenty of people. "Is it not evident that demand for hands, that is employment, must regulate the numbers of the people?"[295]

Naturally Young was contemptuous of those who argued that too much money, high prices, the prevalence of large cities with resulting luxury and waste, and the progress of the enclosure movement tended towards depopulation. Easy money and high prices stimulate production. Cities and luxuries likewise tend to create a demand, hence greater production, more employment, and thus indirectly a larger population. Young was not worried by the effects of wars or emigration on population. "You fight off your men by wars—you destroy them by great cities—you lessen them by emigrations—most infallible method of increasing their number—PROVIDED THE DEMAND DOES NOT DECLINE."[296] In general he was very optimistic about England's condition in the eighteenth century:

[294] Young, 1774: p. 269.
[295] *Ibid.*, p. 86.
[296] *Ibid.*, p. 65.

I do not comprehend the amusement that is found in constantly looking at those objects which are supposed to be gloomy—and in regularly lamenting the evils that surround us, though they flow from causes which shower down much superior blessings. When I look around me in this country, I think I every where see so great and animating a prospect that the small specks which may be discerned in the hemisphere, are lost in the brilliancy that surrounds them.[297]

As might be expected, Young dissented vigorously from Price's view that the growth of large farms and enclosures tended towards depopulation. He denied that a country could be really great or prosperous if its land was divided into very small self-sufficient units. It was what the farmer raised for the market which contributed towards national wealth. Of course he emphasized, as always, that improvements could only be made by large farmers, and that no improvements were possible under the open-field system with "every good farmer tied down to the husbandry of his slovenly neighbour."[298] He also denied, but not very positively, that enclosures meant rural depopulation. But what if they did? If a larger agricultural product could be raised by fewer hands, the excess could swell England's industrial laborers, her merchant marine, her army and navy. The net result in any case would be gain. The most important thing is to encourage agriculture, which should never be sacrificed to considerations of population. "The soil ought to be applied to that use in which it will pay most, without any idea of population."[299] The only real hindrances to population growth in England were the Act of Settlement, "the most false, mischievous, and pernicious system that ever barbarism devised,"[300] and the law which stipulated that all cottages must have four acres of land attached to them. Both laws were contrary to the natural liberty of movement and should be repealed.

In discussing taxation, Young again attacked tithes as injurious to improvements and referred to some meetings of "respectable" gentlemen in the previous winter in London to consider how best to alter the tithe laws. He proposed that in lieu of tithes

[297] *Ibid.*, p. 149.
[298] *Ibid.*, p. 199.
[299] *Ibid.*, p. 123.
[300] *Ibid.*, p. 93. *Cf.* also pp. 93–96, 331–335.

the clergy should be given land which they might then either work directly or rent to farmers. He thought the English system of fixed land taxes was the best possible one, and contrasted it with the French *taille* which was harmful because, like English tithes, it was a tax upon improvements. The best taxes were those upon consumption, like customs and excises, from which the English government derived much of its revenue:

. . . the fairest and most equal, and the least burthensome of all others; every class of the people, every individual in the nation bears his share, and that a *voluntary* share, because if he forbears consuming he pays no tax, never advancing a penny unless he buys a taxed commodity, and his very purchase implies an ability to pay . . . .[301]

In the *Political Arithmetic* Young examined the ideas of the French Physiocrats at considerable length.[302] Their leading works were listed and quoted in French. At times he showed a lack of comprehension of their ideas and distorted their conceptions of the net product and of the sterility of commerce and manufactures.[303] His chief objection was undoubtedly to their proposal for a single tax upon land. He quoted Sir James Stewart to deny their contention that all taxes fall eventually upon the land, and declared that customs and excises could not, be shifted from the consumer to the landlord. Their argument that a single tax on land would lower prices did not appeal to Young who regarded high prices as a sign of general prosperity. Young also objected to their advocacy of complete freedom of trade. He not only reiterated his faith in the restrictions upon corn imports and the bounty upon corn exports, both of which were absolutely necessary to insure the farmer "a ready market and a sufficient price."[304] He also declared his adherence to the Navigation Acts

[301] *Ibid.*, p. 214.

[302] *Ibid.*, pp. 209–266.

[303] *Ibid.*, p. 254 *et seq.* On p. 264 he erroneously assigned the *Tableau Économique* to Mirabeau, although on p. 210 in his bibliography he correctly attributed it to Quesnay.

[304] *Ibid.*, p. 28. *Cf.* also pp. 27–46, 193–197. Young favored Governor Pownall's Act of 1773 because it was a permanent measure and because it preserved the bounty whenever corn was cheap at home and whenever export was permitted. Young had discussed the matter with Pownall and found his ideas "perfectly judicious on this point." *Cf. ibid.*, pp. 40–42. For an account of this measure *cf.* Barnes, 1930, pp. 43–45.

and the general theories of the balance of trade. "A general free trade, as there has been no example of it in history, so it is contrary to reason."[305] Young thus clearly showed himself as late as 1774 an adherent to much of mercantilism.

A few minor points must conclude this analysis. In an appendix Young pleaded again for government appropriations to clear wastelands, deplored the money spent for wars, and even declared that the appropriations for the Foundling Hospital might better have gone to a project for reclaiming wastelands which would have given employment to the poor. He closed this section with the exhortation: "WHEN WILL THERE ARISE A MINISTER WITH SPIRIT AND PATRIOTISM SUFFICIENT TO INDUCE HIM TO LET ONE POOR TWENTY THOUSAND POUNDS FOR WASTE LANDS APPEAR? IN THE LONG GRANT OF SO MANY HUNDRED MILLIONS?"[306] In another place he told a pretty story of an experiment in growing wheat in the Royal Gardens at Kew by the young Prince of Wales and one of his brothers under the direction of their tutor, Lord Holdernesse. Such an experiment would impress upon their youthful minds the importance of agriculture for national prosperity and never let them "forget the maxim of a wise sovereign. THE KING'S FAVOR in matters of agriculture, IS AS DEW UPON THE GRASS."[307]

Twice in the volume Young's vanity appeared amusingly. In one place he praised the Society of Arts but suggested that their awards of gold medals should be changed to a model of a plow in gold or silver, or a gold or silver cup.

A man would place such a thing in a glass case, and set it where it might be seen: but a medal, unless a hole is drilled through it, and you wear it pendant from a button hole—is seen by no one—a man must be put to the blush to bring out his medal and show it . . . .[308]

In another place he complained that foreign writers quoting his works had misspelt his name so badly that he hardly recognized it himself. "Even Baron Haller, who composes in English, calls me M. Arthard Joung."[309]

The year 1775 is nearly a blank in Young's life. The *Auto-*

[305] *Ibid.*, p. 262. *Cf.* also p. 234 *et seq.*
[306] *Ibid.*, p. 347.
[307] *Ibid.*, pp. 176–177.
[308] *Ibid.*, p. 173.
[309] *Ibid.*, p. 223.

biography dismisses the year with three lines, with only one positive fact that he spent the winter in London,[310] presumably engaged in reporting parliamentary debates. There are two letters in the British Museum from Arbuthnot, the first of which scolded Young for his "d—d publick spirit" which "will ever make you give others, what you ought to keep for yourself."[311] In the second Arbuthnot promised that he and his wife would soon visit Young and stay as long as possible and he paid high tribute to Young's epistolary style: "am much obliged to you for your letter as every thing that drops from your Pen affords instructive amusement. Your last not least which has induced me to read it several times over in each of which I discovered something new."[312]

Young was very active in the Society of Arts during 1775. He attended thirteen meetings of the Committee on Agriculture, and served as chairman of that committee six times.[313] On January 9 he received a specimen of rhubarb plant for experiment and on March 13 some cabbage seeds.[314] On April 29 he presided over a trial of plows at Stockwell and on May 25 over the Committee on Mechanics.[315]

Ever since 1767 Young's pen had been incessantly busy. But nothing appeared in 1775, unless indeed *American Husbandry*[316]

[310] Young, 1898: p. 65. The most important point in the three lines follows: "From 1766 to 1775, being ten years, I received 3000 L. or 300 L. a year."

[311] Add. MSS. 35,126, ff. 167–168.

[312] *Ibid.,* f. 169. The letter also described Arbuthnot's visit to Young's agricultural rival, William Marshall. "I was at Marshalls where between us two, I saw the worst crops & the worst management I ever beheld his beans just coming into blossom not six inches high & I suppose are now eat up with the bug, but his oxen working incomparably & his ox collars the ne plus ultra—they do him infinite credit."

[313] Committee Report Books, 1774–1775, ff. 19, 28, 32, 35, 42, 45, 46, 52, 54, 1775–1776, 1, 3, 6, 8. The dates when he was present were January 9, February 6, 13, 20, 27, March 6, 13, April 29, May 1, November 13, 20, 27, December 4.

[314] *Ibid.,* 1774–1775, ff. 19, 46.

[315] *Ibid.,* 52.

[316] This work was edited in 1939 by Professor Harry J. Carman, who in his preface discussed the authorship and inclined to feel that Young's claims were stronger than those of other possibilities. In an article in 1940 in *Agricultural History,* Professor Rodney C. Loehr was more skeptical of Young's authorship. The only contemporary review of the work was an unfavorable one in the *Monthly Review* 54: pp. 48–58, to which a note was appended that a correspondent believed that Young was the author. Quite recently a very distinguished American historian Carl R. Woodward (1969: p. 67), concluded that Young was probably the author: "I believe that we are justified in accepting Young as the *presumed* author."

was his work. Surely there is no definite proof that he wrote
this two-volume description of agriculture in British America.
The *Autobiography* and his other writings make no reference
whatever to it.[317] The manuscript letters in the British Museum
throw absolutely no light upon the question. Nor does the text
of *American Husbandry* give away the secret at any point. On
the other hand, there is nothing to disprove absolutely that Young
wrote the volumes. The title page states that they were written
"By an American," but it is obvious throughout the work that
the author was English. Probably the author had never been to
America, although twice he stated that he had seen certain implements at work in New England.[318] The account is singularly
lacking in the specific detail on people and places which a person
familiar with America would probably have mentioned. That
it was published anonymously is no proof that it was not Young's
work. Indeed it seems to have been his habit to cloak with
anonymity any work the reception of which might be doubtful.
His publishers during most of the preceding period had been
Nicoll and Strahan. *American Husbandry* was published by J.
Bow. Nothing can be proved by this point, either, for he gave
The Farmer's Kalendar to a publisher with whom he had no
previous relations. Nor would Young have probably been deterred from writing about American husbandry without a visit
to America. His *Letters Concerning the Present State of the
French Nation* were written without a visit to France. The very
fact that Young published nothing between the spring of 1774
and the time he went to Ireland is in itself suspicious, for he
was just as badly off financially as ever.

The ideas expressed in *American Husbandry*, moreover, are
those which one might expect from Arthur Young at that particular time. In the criticisms of American practices there is
nothing with which Young would not have heartily agreed. The

[317] The extreme brevity of the *Autobiography* for 1775 suggests the possibility
that the account had been cut.

[318] *American Husbandry:* 1: p. 82. Perhaps he was lying consciously; more
likely he was quoting carelessly. The following statement, 1: p. 201. seems disingenuous if the author was English: ". . . but though I cannot give accounts of
which I have such certainty of knowing to be accurate, in the case of Britain,
as in that of Pennsylvania, yet as there are some late writers who are acknowledged
to be of undoubted authority, I shall be able, by means of their works, to draw
up such an account of the profit of husbandry in England, as shall have no
material errors in it." He then used Young's works as his authority.

importance of manures, cattle, ameliorating crops, good fences, proper crop rotations, are all emphasized by the author of *American Husbandry*. Repeatedly reference was made to the need of opening wastelands in England. In one place the author suggested that the farmers of Pennsylvania should form a society to encourage agricultural improvements, similar to the Society of Arts. In another he stated that a large export of grain was necessary for England to avoid scarcity and high prices. This was certainly one of Young's pet theories. In discussing the relations between the mother country and her colonies the author used almost exclusively Young's arguments in his *Political Essays*. The value of colonial staples to the mother country, the desirability of weakening colonial manufactures and fisheries, the possibility of the independence of the colonies are all discussed with the same arguments which Young had employed in the *Political Essays*. Of course it is quite possible that the author had read most of Young's works and shared his opinions. Many acknowledgments were made in footnotes to Young's works. Young had also cited his own works as authorities in the *Political Essays*. Finally the style resembles that of Young very closely. In places the author engaged in the same kind of elaborate theoretical accounts of an imaginary estate over a series of years that Young was so fond of making.[319]

Young's later correspondence with George Washington certainly displayed considerable ignorance of practical agriculture in America. At first glance it seems unlikely that he could have written a work so highly regarded as *American Husbandry*. A more careful examination of that work is inconclusive. In some places the author seems somewhat better informed than Young showed himself nearly twenty years later, but the evidence is not decisive. In summary it seems, therefore, quite possible that Young was the author of *American Husbandry*, but there is no definite proof.

The year 1776, when Young made his first trip to Ireland, marked a new chapter in his life. In the beginning of the *Auto-*

---

[319] *Cf.* Loehr, 1940: pp. 105–106. In one place the author disagreed on a minor point with Young, **1**: pp. 252–253; at another he referred to the author of the *Political Essays* as "a writer who has taken great pains to be well informed." *Cf. ibid.* **2**: p. 4.

biography for that year he summed up graphically the futility of the years at North Mims:

. . . I shall in general remark that the last four or five years of my life had been detestable, my employments degrading, my anxiety endless, every effort unsuccessful, exertion always on the stretch, and always disappointed in the result, uneasy at home, unhappy abroad, existing with difficulty and struggling to live, never out of debt, and never enjoying one shilling that was spent. What would not a sensible, quiet, prudent wife have done for me? But had I so behaved to God as to merit such a gift?[320]

[320] Young, 1898: p. 66.

III. Ireland, 1776-1778

D URING the eight years at North Mimms, Arthur Young had become a nationally known figure through his writings, his tours, and his activity in the Society of Arts. For several years he had spent much time at London engaged in parliamentary reporting. That he was constantly embarrassed financially, that he had failed as a practical farmer at North Mimms, and that his family life was far from happy—all these points should be clear from the previous chapter. Always restless, Young was anxious to try something new. He had come to love traveling and had made a reputation as a traveler interested in more than scenery and country houses. His attractive personality made him a welcome guest. Ireland was near enough to make a trip there not too expensive. When he started for Ireland in June, 1776, Young probably had in mind only the collection of materials for a new *Tour,* but the trip was destined to influence his life far more than he had expected. For when in 1777 he became Lord Kingsborough's estate agent he disposed of his farm at North Mimms. Again when, contrary to expectations, the position in Ireland proved temporary, nothing appeared more eligible than a resumption of farming at Bradfield which remained his home for the rest of his life. *A Tour in Ireland* was published in 1780 and as a single work is only surpassed in permanent value by his *Travels in France.*

The *Autobiography* claims that Young decided to go to Ireland in 1775, but, as mentioned above, Fanny Burney's *Diary* for March 30, 1774, noted that he "had some thoughts of going" to Ireland. Again, according to the *Autobiography,* Young had been urged to undertake the trip by Lord Shelburne, who certainly interested in the new agriculture, desired to improve Irish conditions, and constantly patronized men of talent and learning.[1]

In preparation for his Irish trip Young asked various friends

[1] Young, 1898: p. 67; d'Arblay, 1889: **1:** p. 282.

for letters of introduction, among them Lord Shelburne, Mrs. Vesey, Edmund Burke, Samuel Whitbread, John Arbuthnot, Governor Thomas Pownall, and John Baker Holroyd.[2] The *Autobiography* quotes part of Burke's reply to his request:

He would be very glad to give Mr. Young recommendations to Ireland, but his acquaintance there is almost worn out, Lord Charlemont and one or two more being all that he thinks care a farthing for him. However, if letters to them would be of any service to Mr. Young, Mr. B. would with great pleasure write them.[3]

Burke's letter introducing Young to Lord Charlemont, dated June 4, 1776, paid a graceful compliment to Young's ability and versatility:

To his works, and his reputation, you can be no stranger. I may add, that in conversing with this gentleman, you will find, that he is very far from having exhausted his stock of useful and pleasing ideas in the numerous publications with which he has favoured the world. He goes into our country to learn, if anything valuable can be learned, concerning the state of agriculture, and to communicate his knowledge to such gentlemen as wish to improve their knowledge. . . .[4]

Little is known of Young's activities in the spring of 1776. He was present at the Agricultural Committee of the Society of Arts on February 5 and March 18, and on the latter date presided.[5]

After two weeks visiting relatives and friends at Bradfield and Lynn,[6] Young started for Ireland on June 10 and reached Holyhead on June 19. He cut across England via Coventry, Birmingham, and Shrewsbury, and through Wales via Conway and Bangor.[7] The highlight of the trip was his visit to the industrial district from Birmingham to Wolverhampton, and through Coalbrookdale. On June 12 he reached Birmingham, "that region of Vulcans," and noted a great expansion since 1768. On the

[2] Young, 1780: p. ix.
[3] Young, 1898: p. 67.
[4] Hardy, 1810: p. 185. There is a very favorable sketch of Lord Charlemont as the intellectual leader of Dublin society in Maxwell, 1937: pp. 177–178.
[5] Committee Report Books, 1775–1776, ff. 14, 23.
[6] Both these trips are minuted in the *Annals* 4: pp. 138–190, under the title, "A Tour to Shropshire." At Lynn Young put up at the Duke's Head where the author had the pleasure of spending several nights in 1938. It is surprising that Young stayed at an inn in Lynn, rather than with one of the Allens.
[7] Unfortunately the minutes covering North Wales were lost.

following morning he passed by "Mr. Bolton's great works" and went on to Wolverhampton, where the road for five or six miles, was "one continued village of nailers."[8] Coalbrookdale was famous as the industrial domain of the Darby and Wilkinson families. The valley itself with the surrounding heavily wooded hills was "a very romantic spot" in contrast to "that variety of horrors art has spread at the bottom." He noted especially Darby's cast-iron waggon ways, "the immense wheels 20 feet diameter of cast iron," and Wilkinson's machine "for boring cannon from the solid cast." At the Darby works he "viewed the furnaces, forges, &c. with the vast bellows that give those roaring blasts, which make the whole edifice horridly sublime." He noted that the whole process from mining the ore to the finished product was carried on in the Darby works and pointed out that the substitution of coke for charcoal in the manufacture of iron "must have been of the greatest consequence." Darby employed 1,000 people at this time, and the whole industry seemed highly prosperous.[9]

The "tedious" passage from Holyhead to Dublin took twenty-two hours. Young tarried in the Dublin area just about a week, from June 20 to June 27. He saw the Irish Parliament prorogued, and visited Trinity College with its fine library, the new exchange, the Duke of Leinster's house, the barracks and the Rotunda, an imitation of Ranelagh. On the whole he was impressed "with all those appearances of wealth which the capital of a thriving community may be supposed to exhibit."[10] On June 23 he visited Lord Charlemont's Dublin house and then drove out to his villa at Marino where he especially admired the banqueting room which "has much elegance, lightness, and effect, and commands a fine prospect."[11]

When Young arrived in Dublin he had met Colonel William Burton, aide-de-camp to the Lord Lieutenant, and later Lord Conynham. Colonel Burton was most helpful in providing letters

[8] *Annals* 4: p. 156. The reference is probably to the famous Soho works of Matthew Boulton.

[9] *Ibid.,* pp. 166–168. Perhaps the best account of the industrial achievements of the Darbys and Wilkinsons is in Ashton, 1924.

[10] Young, 1780: p. 2. Page citations to Part II will give that information; otherwise the citation is to Part I.

[11] *Ibid.,* p. 3. Pictures of nearly all the Dublin buildings visited by Young are to be found in Maxwell, 1937.

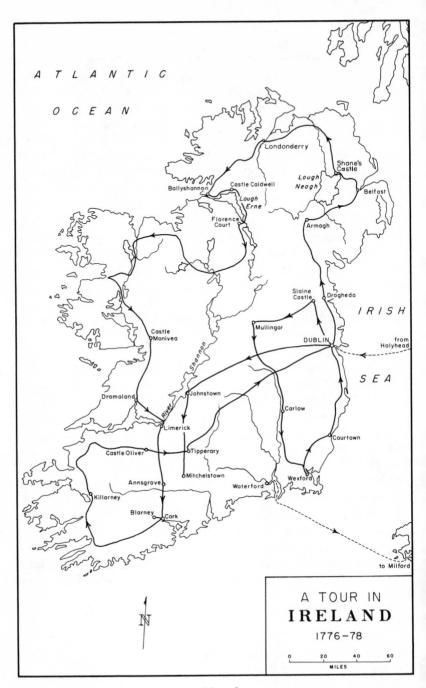

A TOUR IN
IRELAND
1776–78

of introduction and in planning Young's tour. On June 24 he took Young to the seat of the Lord Lieutenant, Lord Harcourt, at St. Wolstans,[12] where he spent two days viewing Dublin's suburbs. He went to Lucan, home of the famous bluestocking, Mrs. Vesey, and to Castle-town, seat of Mr. Thomas Conolly, which Young dismissed rather briefly as "the finest house in Ireland, and not exceeded by many in England. . . ."[13] He also walked through Laughlinstown, the experimental farm subsidized for several years by the Dublin Society and managed by John Wynn Baker, who had died the preceding year. Young had corresponded with Baker for several years and had been much interested in his experiments. The chief causes for his failure, it seems, were a lack of capital and too many experiments. As Young put it:

A man may have all the abilities in the world, write like a genius, talk like an angel, and realy [*sic*] understand the business in all its depths, but unless he has a proper capital, his farm will never be fit for exhibition;—and then, to condemn him for not being a good farmer in practice as well as theory, is just like abusing the inhabitants of the Irish cabbins for not becoming excellent managers.[14]

The similarity of Baker's failure with his own is all too apparent.

Young's trip through Ireland in 1776 was made in a "whiskey," probably with one servant.[15] Fortunately numerous letters of introduction and the well-known hospitality of Irish gentlemen meant that he very seldom had to stay at Irish inns, which proved very bad. On July 18 he wrote: "I was forced to take refuge in a cabbin, called an inn, at Ratoath. Preserve me, fates! from such another."[16]

From June 27 when Young left St. Wolstans until July 17 when he returned to Dublin, Young swung around an inner arc with Dublin as its focus. This trip took him north to the Boyne, west and south by Mullingar and Carlow, reaching the sea at Wexford,

---

[12] Young, 1898: pp. 67–68. *Cf.* also Maxwell, 1925: p. 215, note.

[13] Young, 1780: p. 17.

[14] *Ibid.*, pp. 15–16. Young, 1898: p. 88, spelt middle name "Whyman"; Young, 1771: 1: p. xxvii, spelt it "Whin"; and Young, 1780: p. 15, refers to "John Whyn, baker."

[15] Young, 1898: pp. 68–69. Young, 1780: p. 284 called his conveyance a "chaise" and definitely referred to "my servant."

[16] Young, 1780: p. 94. Young, 1898: p. 69, wrote: "I travelled four hundred miles *de suite* without going to an inn."

and back up along the coast to Dublin. Several days were spent at Slaine Castle,[17] where he again met Colonel Burton. While there he fortunately found John Baker Holroyd in residence on his neighboring Irish estate where he was attempting, without too much success, extensive improvements. On July 6 he was at Rathan visiting Mr. Vancouver, who had been a Norfolk bailiff for seven years before going to Ireland, and who was now improving a piece of bog land for Lord Shelburne and planting turnips on it. Young was particularly pleased "with this improvement from being instrumental in procuring his lordship the person who is executing it."[18] On July 13 he reached the seat of Lord Courtown, near Gowry, where he spent Sunday and attended church and was surprised to see a large congregation, for "this is not often the case in Ireland out of a mass house." Later that day he galloped on the "fine firm beautiful sand for miles."[19]

On July 17 he returned to Dublin passing through a wildly romantic glen, the Dargle, with a beautiful waterfall, his description of which reflects the gloomily romantic tastes of his time: "This horrid precipice, the pointed bleak mountains in view, with the roar of the water, all conspire to raise one great emotion of the sublime. . . . The shade is so thick as to exclude the heavens, all is retired and gloomy, a brown horror breathing over the whole."[20]

On July 18 he set out again on his much longer tour, swinging through an outer arc around the island, lasting almost exactly three months, and ending on October 17 at Waterford. On July 20 Young visited Drogheda and then the site of the battle of the Boyne, where he viewed the obelisk and "indulged the emotions which with a melancholy not unpleasing filled my bosom."[21] The same day he reached Cullen, seat of Lord Chief Baron Forster, a "prince of improvers," who had developed many thousand acres of mountain bog, and had constructed roads, enclosed the

---

[17] Young, 1780: pp. 23–29.
[18] *Ibid.,* p. 59. This probably was Charles Vancouver, who later executed several county surveys. The *Dictionary of National Biography* calls him an American. *Cf.* also Young, 1898: p. 426, note 1, which made the common mistake of confusing him with the discoverer, George Vancouver.
[19] Young, 1780: p. 83.
[20] *Ibid.,* p. 93.
[21] *Ibid.,* p. 98.

land, brought in French and English Protestants, and built houses. A man of large vision, Baron Forster was convinced that the Irish peasants were honest, living among them "without shutters, bolts or bars . . . yet never lost the least trifle. . . ."[22] Naturally Young was enthusiastic: "Such are the men to whom monarchs should decree their honours, and nations erect their statues."[23]

At Armagh, Young was entertained by the great primate, Richard Robinson, Archbishop of Armagh and Baron Rokeby, who showed Young the new palace, school, churches, library, barracks, and market house, built by the archbishop at an expense of £30,000. By giving new leases on condition that houses be rebuilt, the Primate had also indirectly helped to improve the town. As Young said: "He found it a nest of mud cabbins, and he will leave it a well built city of stone and slate."[24] From Armagh to Belfast the country was largely devoted to the linen manufacture which Young described in great detail. From Belfast he went on to Antrim and visited Mr. O'Neill at Shanes Castle on Loch Neagh:

Upon my arrival at the Castle, I was most agreeably saluted with four men hoeing a field of turneps around it, as a preparation for grass. These were the first turnep hoers I have seen in Ireland, and I was more pleased than if I had seen four emperors.[25]

Young reached Londonderry on the evening of August 6 and "waited two hours in the dark before the ferry-boat came over for me."[26]

Five days later he was at Ballyshannon where he was "delighted to see the salmon jump, to me an unusual sight: the water was perfectly alive with them."[27] That night he reached the seat of Sir James Caldwell on beautiful Lough Erne, where he spent a most enjoyable week visiting in turn Caldwell, the Earl of Ross at his "charming" island seat, Belleisle, and Lord Inniskilling at Florence Court. During this week Young the farmer gave way to Young the tourist, and more space was devoted to scenery

[22] *Ibid.*, p. 101.
[23] *Ibid.*, p. 100.
[24] *Ibid.*, p. 104. *Cf.* also Maxwell, 1937: pp. 343–345.
[25] *Ibid.*, p. 128.
[26] *Ibid.*, p. 143.
[27] *Ibid.*, p. 160.

than to agriculture. Sir James Caldwell, one of the most ostentatious men in all Ireland, entertained him "with a politeness and
cordiality that will make me long remember it with pleasure,"[28]
gave him plenty of agricultural information, and showed him
magnificent views of the lake. As was Sir James's custom, when
Young left he was taken on a barge as far as Inniskilling while
his horse and carriage went by land. "Take my leave of Castle
Caldwell, and with colours flying, and his band of music playing,
go on board his six-oared barge for Inniskilling; the heavens
were favourable, and a clear sky and bright sun, gave me the
beauties of the lake in all their splendour."[29] He spent two days
at Belleisle, viewing the deer park three miles away on another
island and Lady Ross's island with its pleasant walks. At Florence
Court more attention was paid to agriculture and the fifty-four
farms into which Lord Inniskilling's estate of 11,000 acres was
divided and on which the rent had quadrupled within thirty
years.[30]

During the latter part of August, Young covered the west
central counties, Cavan, Longford, Leitrim, Roscommon, Sligo,
Mayo, and Galway. In general, it was a pretty backward and
uninteresting country. In county Cavan he first witnessed the
barbarous practice of a plow or harrow being pulled by a horse's
tail. Young was naturally indignant:

> . . . *they very commonly plough and harrow with their horses* DRAW-
> ING BY THE TAIL: . . . they insist, that take a horse tired in traces,
> and put to work by the tail, he will draw better: quite fresh again.
> Indignant reader! this is no jest of mine, but cruel, stubborn, bar-
> barous truth.[31]

Everywhere he saw bogs capable of improvement and noted the
lack of trees in the denuded country. Nevertheless, the picture
was not one of unrelieved gloom, for there were improving landlords, doing what they could to raise standards. Young was
skeptical, however, of the several efforts he witnessed to introduce the linen manufacture: "After all, I see every reason to
assert, that a gentleman, for a shilling he will ever make by

[28] *Ibid.* For further accounts of Caldwell, *cf.* Young, 1898: pp. 69–71; Maxwell, 1940: pp. 36–37.
[29] *Ibid.*, p. 166.
[30] *Ibid.*, p. 171.
[31] *Ibid.*, p. 179.

manufactory, will profit a guinea by the improvement of land; have rascals to deal with in one line, and honest men in the other."[32]

Early in September, Young visited Robert French of Castle Moniva who had converted seemingly hopeless bog lands in Galway into good farming lands. For an account of his improvements French had received a gold medal from the Dublin Society, and Young printed this previously unpublished paper in his tour. French had also built up a flourishing linen industry, partly by maintaining a school where from twenty to forty children were "constantly supported, cloathed, and taught to read and write, and to spin and weave."[33] On September 5 Young entered County Clare and reached Drumoland, the seat of the Irish statesman, Sir Lucius O'Brien, "who had been repeatedly assiduous to procure me every sort of information."[34] When Young left, Sir Lucius accompanied him for a way to show him a famous view of the Shannon River, "a most noble river, deserving . . . fleets of merchantmen . . . instead of a few miserable fishing-boats." The fault was not Ireland's, however, but rather resulted from the English "illiberal spirit of trading jealousy."[35]

Young's first stop in County Cork was at Annsgrove, the seat of Mr. Richard Aldworth, an excellent farmer who grew turnips and applied lime to his land. Young also enjoyed the Aldworths for their "elegant manners and cultivated minds."[36] Mrs. Aldworth read to Young an original manuscript letter from Swift in a manner showing her sympathy with Swift's indictment of everything Irish. Young claimed that the Aldworths stayed in Ireland only from patriotic motives and from their case generalized that people of cultured backgrounds found little congenial company in rural Ireland. In describing Annsgrove, Young wrote: ". . . every thing about the place had a much nearer resemblance to an english than an irish residence, where so many *fine* places want *neatness*, and where, after great expense, so little is found *complete*."[37]

Nearer to Cork, Young visited Mr. St. John Jefferys of Blarney

----

[32] *Ibid.,* p. 194.
[33] *Ibid.,* p. 228.
[34] *Ibid.,* p. 237.
[35] *Ibid.,* p. 243.
[36] Young, 1898: p. 72.
[37] Young, 1780: p. 251.

Castle, who within ten years had transformed Blarney from a village of several mud cabins into a small manufacturing town of ninety houses. Linen, woolen, and stocking industries had been established, a leather and a paper mill built. A church, market house, and four bridges had been constructed. The initial capital and drive had come from Mr. Jefferys with aid from the Linen Board and the Dublin Society. Since Mr. and Mrs. Jefferys were about to leave for France, Young accompanied them to Dunkettle, the home of Dominick Trant, Esq. The next morning the whole party went by boat from Mr. Trant's quay to the ship in the harbor on which the Jefferyses embarked. The trip gave Young the opportunity to admire the water views around Cork harbor. He then returned to Dunkettle with Mr. Trant who was a member of the Irish House of Commons, and over whose beautiful estate with its fine walks and views Young rhapsodized.[38]

Near Cork, Young visited two other improving landlords who were thoroughly conversant with the new husbandry. At Castle Martyr, Richard Boyle, second Earl of Shannon, had grown turnips of prodigious size which had been exhibited by the Dublin Society and had done everything possible to further the spread of their culture. Like Young he was skeptical of the drill husbandry: "I read myself into it, and worked myself out of it."[39] Young especially approved his use of oxen rather than horses and witnessed the practice of having them "draw by the horns" instead of by the yoke, a practice which Lord Shannon had copied from France. Archdeacon John Oliver of Coolmore was also a turnip enthusiast and had brought much wasteland into cultivation. Young praised his "spirited exertions" and concluded: ". . . if a very few improvers in Ireland have gone through more extensive operations, I have not found one more attentive or more practical, and, upon the whole, scarcely any that come near to him."[40] While at Coolmore, Young rode to the mouth of Cork Harbor and concluded that the environs of Cork were preferable as a residence to any place in Ireland. The climate was fine, the views beautiful, the shipping animated, while the size of the

[38] For the Jefferys and Trant, *cf.* Young, 1898: pp. 73–74; Maxwell, 1940: p. 229, notes 123, 133.
[39] Young, 1780: p. 267.
[40] *Ibid.*, p. 281.

town provided a ready market and furnished many articles of convenience.

Much of the trip from Cork to Killarney was through pretty rugged country. At one point his chaise only made the very steep grade with the aid of six laborers, Young's servant, and two passers-by. Young's host at Killarney was the very hospitable Mr. Herbert of Mucruss. Three whole days were spent in sightseeing —the ruins of Mucruss Abbey with its dismal cloisters and heaps of skulls and bones, the hanging woods of Glena, the bold form of Mt. Turk, the island of Innisfallen, "the most beautiful in the king's dominions,"[41] and O'Sullivan's cascade, "to which all strangers are conducted."[42] Only the ascent of Mount Mangerton was prevented by heavy fog. Young devoted nearly twenty pages to the beauties of Killarney which he regarded as the finest lake he had seen, although Lough Erne and Keswick might excel in some details. Young complained, however, of the lack of suitable accommodations for travelers and of exorbitant prices. He recommended a well-built inn with reasonable prices, and especially provided with many indoor amusements to divert the visitors while waiting for good weather which was rare at Killarney.

From Killarney, Young turned north again through counties Kerry and Limerick and along the southern shore of the Shannon estuary almost to the town of Limerick. On the estuary he commented, "Perhaps the noblest mouth of a river in Europe."[43] In general, agricultural practices were pretty poor, although the soil was excellent. At one point he traveled ". . . through a continuation of excellent land, and execrable management."[44] At another point he attacked vigorously the custom of taking ten or twelve crops of oats in succession, and asked, "Were such barbarians ever heard of?"[45]

At Castle Oliver the Rt. Hon. Mr. Oliver invited in neighboring gentlemen interested in agriculture and accompanied Young with a laborer to spade up various soils. At one place Young declared it "the richest soil I ever saw, and such as is applicable

[41] *Ibid.*, p. 295.
[42] *Ibid.*, pp. 297–298.
[43] *Ibid.*, p. 306.
[44] *Ibid.*, p. 305.
[45] *Ibid.*, p. 309.

to every purpose you can wish."[46] The house and park had been largely remodeled by Mr. Oliver who also possessed several fine paintings, one a portrait of Scipio. Young commented:

. . . but Scipio, as in every picture I ever saw of him has no expression. Indeed, chastity is in the countenance so *passive* a virtue as not to be at all suited to the genius of painting; the idea is rather that of insipidity, and accordingly Scipio's expression is generally insipid enough.

Young left Castle Oliver only with reluctance after two nights, "for I found it equally the residence of entertainment and instruction."[47]

The last stage of the tour was from Castle Oliver near Limerick across Tipperary and Waterford counties to the town of Waterford where he took ship to England. Through Tipperary county the soil was very fine, but too much land was used for grazing. At the Earl of Clanwilliam's at Tipperary he met the sons of the two greatest farmers in Ireland who farmed 9,000 and 13,000 acres and paid a rent of £10,000 each, and the first of whom kept 8,000 sheep, 550 bullocks, 180 horses, and employed from 150 to 200 laborers.[48] On October 12 he witnessed Lord de Montalt's changes in his grounds from the formal and artificial tastes of a former age to the natural taste of the eighteenth century. He had opened up "one very noble lawn . . . scattered negligently over with trees. . . ." He also cultivated turnips and was taking his farms into his own hands as the leases expired, and re-letting them for higher rents after making improvements. "This is the true agriculture for profit for a landlord. . . ."[49] Sir William Osborne, at his estate near Clonmell, had gradually turned over a large tract of mountain land to a group of poor peasants. Sir William built the cottages at his own expense and furnished free lime. As a result twenty-two families had settled there and were prospering, although earlier many had been Whiteboys. Young exclaimed:

It shows that the villainy of the greatest miscreants. is all situation and circumstance: EMPLOY, don't *hang* them. . . . by giving property, teach the value of it; by giving them the fruit of their labour,

---

[46] *Ibid.*, p. 313.
[47] *Ibid.*, p. 318.
[48] *Ibid.*, pp. 320–321.
[49] *Ibid.*, p. 322.

teach them to be laborious. . . . Yet in spite of such facts do the lazy, trifling, inattentive, negligent, *slobbering*, profligate owners of irish mountains leave them . . . in the possession of grous and foxes. Shame to such a spiritless conduct![50]

On October 17 Young reached Waterford and loaded his chaise and horses on the *Countess of Tyrone*, expecting to sail immediately. However, the Waterford packets, much to his disgust, only sailed when there were enough passengers to make it profitable. He spent a "beastly night"[51] on board ship, and then, seeing no prospect of leaving immediately, landed again and walked to Ballycanvan, the seat of Cornelius Bolton. Young was immediately attracted by the younger Bolton, a member of the Irish Parliament, "a man . . . of so mild and pleasing a temper that I much regretted that I had him not for a neighbor at Bradfield."[52] Young found the Boltons not only very agreeable hosts, but also enlightened landlords who raised turnips and improved wastelands, settling the poor and making them industrious and prosperous.

After a night at Ballycanvan, Young returned to the ship but had to wait another day before the packet sailed, "and then it was not wind, but a cargo of passengers that spread our sails."[53] After nearly reaching England, they ran into a severe storm which drove them back almost to the Irish coast. The storm raged for thirty-six hours and Young had to take his turn at the pumps. Finally they landed on October 23 at Milford Haven. Later the same day, having reached Haverford West, he described his trip in a letter to his wife:

Haverford West: Oct. 23

My Dearest,—It pleases God that I am once more to embrace you and my children—a passage that is common in eight hours was from Sunday morn eight o'clock till one o'clock this morning Wednesday, thirty-six hours of which, a raging storm; we talk of them at land, but those who have not seen them at sea know not what the very elements are. Pent up in the Irish Channel, the ship ran adrift, wearing to keep free from rocks and sands—the wind did not blow, it was like volleys of artillery; part of the sails were torn into frit-

[50] *Ibid.*, p. 327.
[51] *Ibid.*, p. 334.
[52] Young, 1898: p. 75.
[53] Young, 1780: p. 342.

ters; the waves ran mountain high, while the ship was perfectly tossed on end on them; the cabin window burst open, and deluged everything afloat; the horses kicked and groaned, the dogs howled; six passengers praying, shrieking, and vomiting: every soul sick but myself; the sailors swearing and storming; and the whole—such a scene! The Captain, who has been many voyages, and the pilot thirty-six years, never saw such a storm—to last so long.

It has worried and starved the horses so that I know not what I am to do—shall go with them as far as I can, and if they knock up must leave them and take some fly to be by ye thirty-first; of which send immediate notice to B. . . .

Adieu,
Most truly yours,
A. Y.

Thank God for me. . . . My passage has cost me between 7l. and 8l., which is the very devil, so that I shall come home without a shilling, and the thoughts of coming full swing upon poverty again makes me miserable. Two ships were lost in the storm.[54]

In spite of the sense of urgency in his letter, Young was a professional traveler and he proceeded leisurely enough through North Wales from October 23 to October 30, of course keeping a journal as he went through Carmarthen, Monmouth, and Gloucester to Bath, and then by stage coach to London.[55] Somewhat later Young sent a servant to Bath to bring back his horses and chaise to London. In the chaise there had been a trunk, containing not only a private journal of the Irish trip full of anecdotes, but also his specimens of soils and minerals. When the chaise reached London, however, the trunk was missing. Young believed the servant had stolen it and served a warrant for his arrest. The famous Judge Fielding, brother of the novelist, dismissed the case on the ground that no theft could be proved

---

[54] Young, 1898: pp. 81–82. There is some confusion in the two accounts about the timing of this incident. In the *Tour* it clearly took place in 1776. In the *Autobiography* both the incident and the letter are arranged under 1778, after his final return from Ireland. The letter has no year date, but the account so exactly matches the details in the *Tour* that it seems as though it must have been written in 1776. Moreover, the dates check for 1776, not for 1778.

[55] The account for North Wales, October 23–30, 1776, is very clear in the *Annals* 8: pp. 31–55, although Young admitted that some details were taken from his similar tour in 1778. The *Annals* article continues to p. 88, and includes much material on Shelburne's estate at Bowood.

but only a breach of trust. As Young put it, "and all I got for my pains was abuse from the fellow."[56]

There is little evidence of Young's activities during the last two months of 1776 and the first half of 1777, but he was very active during that period in the Society of Arts. On November 13 Young and Arbuthnot agreed to serve as chairmen of the Committee on Agriculture,[57] and Young frequently presided over that committee in November, 1776, and February, 1777.[58] The last time that he did so was on April 5, 1777, a meeting at the *George* at Morden held to make trials of the Roman yoke for oxen, to the winner of which the gold medal was awarded. At the same meeting there was a trial of a machine for washing potatoes, but the machine "did not answer."[59]

The *Autobiography* for 1777 begins: "This was the first favourable turn that promised anything after ten years' anxiety and misery. . . ."[60] The reference was to the fact that he went again to Ireland, this time as Lord Kingsborough's estate agent. The arrangement was made late in February.[61] His old friend, Mr. Danby of Swinton in Yorkshire, was talking in London to Lord Kingsborough (later the second Earl of Kingston) who mentioned that he needed a good resident agent not only to collect rents but also to supervise improvements at his estate at Mitchelstown in County Cork. When Danby suggested Young as the ideal candidate, Kingsborough was immediately interested. A dinner was arranged at which Young met Lord and Lady Kingsborough and made such a favorable impression that Danby was requested to draw up a sort of contract. Young's duties were stated in very general terms, "to do all the business of an agent and to inspect whatever improvements his lordship is desirous of carrying on." He was to be paid £500 a year and also to receive a retaining fee of £500 before he left England. He was to have a good house rent free and could also take a small farm at the prevailing rents. If Young voluntarily quitted the job before the end of three years, he was to refund "a proportionate part" of the retaining fee, while if Lord Kingsborough should break the agreement

[56] Young, 1898: p. 69.
[57] Minute Books **22:** f. 21.
[58] Committee Report Books, 1776–1777, ff. 1, 4, 6, 10, 12, 15, 23.
[59] *Ibid.*, f. 29.
[60] Young, 1898: p. 75.
[61] *Ibid.,* p. 76.

before a year was up, Young was to receive the year's salary.[62]

The decision having been made, Young had to dispose of his lease at Bradmore Farm and to pack up his belongings. Moving must have been a troublesome business for he had been at North Mims for over eight years, but hope was high for the future. His books "and other effects" were sent by boat to Cork, and he set out for Dublin, probably sometime in July. It was apparently decided that Mrs. Young and the children should not go until he was well settled at Mitchelstown. He spent about two months "in a constant round of Dublin dinners" until the house at Mitchelstown was ready, "whilst a new one was building on a plan and in a situation approved of myself."[63]

Finally on September 24 he left Dublin for Mitchelstown which he reached on October 10. Desirous to enlarge his coverage of Ireland for his book, he took his route through country not visited in 1776. He went south to Kildare, then west through Queens and Kings counties to Lough Derg, formed by the spreading out of the Shannon, then down to Limerick again and back eastwards through counties Limerick and Tipperary. Much of the country was very agreeable, especially in Queens and Kings counties and the region around Lough Derg. As in 1776 Young found hospitable hosts and good farmers willing to furnish him as complete information as possible. Outstanding was Peter Holmes, Esq., of Johnstown on Lough Derg who raised Scottish cabbages and clover, and employed rape cake as a manure. Mr. Holmes invited in others well qualified to give Young information. Always interested in fishing, Young greatly admired Mr. Holmes's well-stocked fish pond, enjoyed eating a twenty-seven pound pike taken from the pond, and watching a fisherman pull in three trout weighing a total of fourteen pounds. Near Johnstown, Young made a four-day stop with Michael Head, Esq., of Derry. Just as he arrived Mr. Head was leaving to dine with a neighbor whose father had improved some mountain land. Young was taken along and the information secured. Mr. Head used as a manure shelly marl dredged from the River Shannon and had improved considerable wasteland. The views of the Shannon and of Lough Derg were magnificent covering a distance of forty

---

[62] Add. *MSS*. 35,126, f. 171. This is a copy of the agreement.
[63] Young, 1898: p. 77. *Cf.* also Young, 1780: p. 3.

miles. From Derry he went to Limerick where he stopped at an inn. "God preserve us this journey from another."[64]

Actually very little is known about the year when Young was at Mitchelstown. Only three pages are devoted to it in the *Auto-biography*.[65] The seven pages in the *Tour* give more detail about Mitchelstown itself than about Young's activities there.[66] Scattered references in the second part of the *Tour* and in the manuscript *Elements of Agriculture* complete the available material. His other biographers rely entirely on the *Autobiography*. No material except that given by Young himself has been found.

Lord Kingsborough's estate was immense, more than sixteen miles long and from five to seven miles broad. The situation of Mitchelstown was "worthy of the proudest capital,"[67] but the town itself before Lord Kingsborough took possession "was a den of vagabonds, thieves, rioters, and whiteboys."[68] A large portion of the estate was wasteland and much of it denuded of trees. The soils were of all kinds, but there was much improvable wetland, and whole tracts of "incomparable" land. Rents only averaged 2s. 6d. an acre. Agriculture was at a low level for the most part. Potatoes and oats were the chief crops. After many successive crops of oats the land was left to "weeds and trumpery; which vile system has spread itself so generally over all the old meadow and pasture of the estate, that it has given it a face of desolation—furze, broom, fern, and rushes owing to this and to neglect, occupy seven-eighths of it."[69] When Lord Kingsborough took over, he found nine-tenths of his estate in the hands of the middleman "whose business and whose industry consists in hiring great tracts of land as cheap as he can, and re-letting them to others as dear as he can." Young was always bitter over this system which broke "that beautiful gradation of the pyramid, which connects the broad base of the poor people with the great noblemen they support," and naturally resulted in "the misery and poverty of the lower classes." The population of the estate was large; "the cabbins are innumerable, and like most

[64] Young, 1780: p. 368.
[65] Young, 1898: pp. 78–80.
[66] Young, 1780: pp. 375–382.
[67] *Ibid.*, p. 375.
[68] *Ibid.*, p. 379.
[69] *Ibid.*, p. 376.

Irish cabbins, swarm with children." Young was also greatly
impressed by the large numbers of hogs:

. . . pigs and children bask and roll about, and often resemble one
another so much, that it is necessary to look twice before the *human
face divine* is confessed. I believe there are more pigs in Mitchelstown
than human beings, and yet propagation is the only trade that
flourished here for ages.[70]

Lord Kingsborough had inherited the Mitchelstown estate two
years earlier, was now only twenty-three years old, and had "just
come from the various gaiety of Italy, Paris, and London."[71]
Young described him thus: "His manner and carriage were remarkably easy, agreeable and polite, having the finish of a perfect
gentleman."[72] He had already built a large mansion with an
accompanying "quadrangle of offices" and a fine garden with
walks and hot houses. He had also constructed three stone farmhouses and several cabins. All this building meant employment
and hence Mitchelstown was "now as orderly and peaceable as
any other irish town."[73] Lord Kingsborough had started to take
back leases from the middlemen and to rent the farms directly
to the occupiers. He had also brought over a skilled nursery man
from England to help start extensive plantations of trees. All in
all, he seems to have been an enlightened and energetic youngster,
anxious to make extensive improvements, as indeed is proved
by the circumstances under which Young was hired.

Probably Young was resident at Mitchelstown for just about
a year from October, 1777, to October, 1778. The clearest statement made by him appears in the *Tour:* "I resided in the county
of Corke, &c. from october to march. . . . I was also a whole summer there (1778). . . ."[74] In another place in the *Tour* he wrote
of spending nine weeks in Dublin, "very busily employed in
examining and transcribing public records and accounts. . . ."[75]
Unfortunately these nine weeks were not definitely dated, but
they probably took place from March to May or June, 1778, and

[70] *Ibid.*, p. 378.
[71] *Ibid.*
[72] Young, 1898: p. 79.
[73] Young, 1780: p. 379.
[74] *Ibid.*, part ii, p. 4.
[75] *Ibid.*, preface, p. xxii. This appears in the Dublin edition, but not in the
London edition where the preface is much shorter.

probably after them he returned to Mitchelstown for the summer.[76]

As land agent for Lord Kingsborough, Young began an ambitious program of improvements. He encouraged his employer to dislodge the middlemen as rapidly as possible. To encourage the cotters and farmers to grow and preserve trees he urged a threefold plan: (1) to give premiums for planting; (2) to insert clauses in the farm leases that trees furnished by the owner must be planted, so many to the acre, and that a fine should be exacted at the end of the lease for any deficiency in this respect; (3) the landlord to provide lumber for repairs at cost.[77] In the spring of 1778, he started to reclaim some wastelands on the southern slopes of the Galtee Mountains. A lime kiln was built which burnt twenty barrels a day. Thirty-four acres were cleared of stones, pared and burnt, limed, and cut into four enclosures surrounded by ditches. Some of the land was sowed with wheat, some with rye, and some with turnips. About one hundred men were employed but they were pretty ignorant and some banks of the ditches collapsed during the autumn rains. Moreover the summer was so dry that the turnip crop failed. The whole business cost £150. The ultimate aim of this improvement had been to raise sheep on the mountains and to feed them during the winter with turnips.

He also encouraged the poor to take over wastelands themselves. He marked out a road and offered portions of the waste to the poor who would fence them in the manner prescribed. The land should be rent free for several years and a guinea would be allowed to each holding to help build a cabin. Apparently there was a ready response and Young was convinced that the poor cotters "wanted nothing but a little encouragement to enter with all their might and spirit into the great work of improvement."[78]

At first the relations between Lord Kingsborough and his agent were friendly and cordial. They explored together the wonderful

[76] There are several perplexing questions about Young's movements in Ireland. In the *Annals* **15:** p. 157, he wrote: "My journies to Ireland occupied the years 1776, 1777, 1778 and 1779." But Young, 1898: pp. 83–84 wrote that he arrived at Bradfield on January 1, 1779 and remained there the entire year. Also in *ibid.*, p. 92 Young referred to a trip from Dublin to London with Hugh Boyd. But there is no mention of Boyd in the accounts of his only certain trips in 1776 and 1778, both of which, moreover, were made directly from Waterford.

[77] Young, 1780: part ii, p. 44.

[78] *Ibid.*, part ii, p. 50.

natural cave on the estate, half a mile long, appearing like "a vaulted cathedral, supported by massy columns," with "very beautiful incrustations of spar, some of which glitters so much, that it seems powdered with diamonds. . . ."[79] That Young was not to lose caste as a gentleman by his new position was evident, for he frequently dined at the castle and afterwards played chess with Lady Kingsborough who regarded him as "one of the most lively, agreeable fellows."[80] That the arrangement was meant to be permanent is evident from the fact that Lord Kingsborough encouraged Young to draw up plans for a new dwelling house. Since this is the only home which Young ever designed, it is interesting to examine his plans. The whole design was to be in the fashionable Palladian style which Young so greatly admired. The house was to be in the middle with yards on either side ending in pavilions. On the main floor there was a hall, breakfast and dining parlors, drawing room and a library, thirty-one by ten feet. On the basement floor were to be wine and beer cellars, pantry, kitchen, servants' hall, and butler's hall. The pavilion on one side was to contain a brewhouse, dairy, and a piazza for cows, while that on the other side should hold stables for ten horses and a coachhouse for two carriages. All in all, not bad for an estate agent![81]

Actually this dream house was probably never built. Local tradition points out his house in almost rural and pleasantly shaded King's Square, a two-story building, connected with and at the end of a block of dwellings, with some nice Georgian exterior decoration. I am indebted to Mr. Thomas F. O'Keefe of Mitchelstown for the information that Young, "offended by the proximity of the Main Street and Market Square to the main entrance to the Castle . . . induced the tenants to depart to Bullock Lane by giving them long leases at attractively low rents and by laying out the New Market Square." Thus Young appears as an early town planner.[82]

[79] *Ibid.*, p. 380. Young also rhapsodized over the views, the cascades, and the waterfalls in the Galtee Mountains on the Mitchelstown estate.

[80] Young, 1898: pp. 78–79.

[81] *Elements of Agriculture* 3: ff. 36–37.

[82] I visited Mitchelstown in Dec. 1957, and was most courteously shown Young's home and that of Major Thornhill by Mr. O'Keefe and Mr. W. H. Robson. The village is a most attractive one and the views of the Galtee Mountains from the New Market Square are magnificent.

Two items from a memorandum book for 1777 also seem to indicate that Young was happy at Mitchelstown: "The year's receipts 1,145 *l.* Wrote *Alcon and Flavia,* a poem."[83] There is also a fragment from a letter to Young from his mother, presumably written in 1777, which is interesting for several reasons:

My memory begins to fail me, but no wonder at 72. That is not the cause of yours doing so, but the multiplicity of business you are engaged in. I attribute it also to being overburthened with your affairs. I can get neither ploughman nor footman to go over to Ireland, so you must see what you can do when you come yourself, which, I am sorry to hear, is not till (next) September. God only knows if I shall live so long as to see you once more. However, to hear you are well and happy is a great comfort to me. . . . I hate now to do anything but sit by the fire and write to you . . . But the happiness of this world, Arthur, is but of a short duration; I therefore wish you would bestow some thoughts on that happiness which will have eternal duration.[84]

At some time while at Mitchelstown Young was joined by Mrs. Young, probably early in 1778.[85] Presumably, but again not certainly, the children accompanied her. A letter from Arbuthnot, unfortunately undated as usual, refers to Mrs. Young's arrival:

I think Madam had a favorable initiation & as first impressions stamp a fixed opinion have no doubt but that the paddy's all appear angels unless she had unfortunately stumbled upon an howl & even that in a fruitful imagination might at least pass for one of her Favourite Handels full chorus's.[86]

As noted above, it seems probable that Young was in Dublin from March until May, 1778, about nine weeks in all, gathering further statistical details for his forthcoming book. Very likely Mrs. Young accompanied him, for Young stated in the *Tour* that, although Dublin lodgings were dear, "we were well accommodated (dirt excepted) for two guineas and an half a week."[87] In going to Dublin he passed through a fine tract of country in Queens County with many fine plantations of trees. Apparently

[83] Young, 1898: p. 77, note.
[84] *Ibid.*
[85] *Ibid.,* p. 78. "1778.—The opening of the year found me at Mitchelstown, where Mrs. Young joined me."
[86] Add. MSS. 35,126, f. 174.
[87] Young, 1780: p. 4.

he spent both nights on the road at inns. The Widow Holland's at Cashel was fine, "clean and very civil." The inn for the second night at Ballyroan was quite different, "kept by three animals who call themselves women: met with more impertinence than at any other in Ireland."[88]

While at Dublin he probably kept pretty busy during the day digging at the records, but the evenings were spent more socially. There were the private entertainments—"a great round of dinners and parties; and balls, and suppers every night in the week," though the rooms were often too small for the company invited. There were two assemblies and two gentlemen's clubs. He also attended the imported Italian comic opera, of which Young was very fond, but the house was usually empty and cold, and the parts were "murdered." As a former parliamentary reporter, he naturally attended the Irish House of Commons frequently and listened to the eloquent speeches of Grattan and other orators. In general Young felt that the level of the Irish house was considerably below the English, a fact which he attributed to the lack of independence. On the whole Young seems to have admired and enjoyed Dublin, all but the dirt. Rooms, persons, cooking, and especially the narrow streets, were all dirty in comparison with England.[89]

Auspicious as had been the beginnings of Young's work at Mitchelstown, friction soon developed with Lord Kingsborough. Unfortunately we have only Young's explanation of the quarrel.[90] His earliest biographer, Dr. Paris, tantalizingly dismissed the subject by saying that Young left Ireland "owing to circumstances which it would be tiresome to detail."[91] Although Young's account shows no bitterness towards Lord or Lady Kingsborough, it naturally lays no share of the blame on himself. One of the middlemen at Mitchelstown was a Major Thornhill, a distant relation of Lady Kingsborough, "a lively, pleasant, handsome man, and an ignorant open-hearted duelist." He and his wife, "an artful designing woman, ever on the watch to injure those who stood in her husband's way, and never forgetting her private interest for a moment," were anxious to take over more farms.

[88] *Ibid.*, p. 383.
[89] *Ibid.*, pp. 4–5.
[90] Young, 1898: pp. 78–80.
[91] Paris, 1820: p. 292.

Since Young was urging Lord Kingsborough to abolish all middle-men on the estate he appeared as an obstacle to the Major's fortunes. Moreover Lady Kingsborough liked and admired her kinsman, the Major. Young felt that Mrs. Thornhill actually concocted the plot which lost him his position. She inspired Lady Kingsborough with jealousy against the children's governess, a Miss Crosby. Young was asked to draw up an engagement by which Miss Crosby could be dismissed with an annual pension. Young was thus much at the castle "in situations which were converted by Mrs. Thornhill into proofs that I was in league with Miss C. for securing the affections of Lord Kingsborough at the expense of his wife." At the same time Mrs. Thornhill suggested to Lord Kingsborough that Young was in love with Lady Kingsborough. "Thus by a train of artful intrigues and deceptions the ladies brought Lord K. to the determination of parting with me. . . ."[92] Naturally Young felt that Lord Kingsborough was lacking in "steadiness and perseverance . . . and was easily wrought upon by persons of inferior abilities."[93] In defense of his Lordship, it must be remembered that he and his wife were both very young and inexperienced.

In the above story Young certainly appears a much injured man. And we have no proof to the contrary. Several purely speculative points must at least be considered, however. It is clear that Young was a lady's man and something of a philanderer. Moreover, he had long since ceased to have any real affection for his wife. We do not know when Mrs. Young returned to England, nor why, but it is quite certain that she left before he did. Futhermore Young had been hired as a *resident* agent. Yet here he was going off to Dublin for nine weeks during his first year of employment, and on a mission with no direct connection with his work at Mitchelstown. Finally, had Lord Kingsborough concluded that Young's plans were going to be very expensive and perhaps not entirely practical?

Whether the above speculations have any validity it is im-

---

[92] Young, 1898: p. 79.

[93] *Ibid.* The article in the *Dictionary of National Biography* on the 2nd Earl of Kingston reveals that he had a very hot temper and that for many years he was separated from his wife. In 1797 one of their daughters was seduced by Col. Henry Fitzgerald, an illegitimate nephew of Lady Kingsborough. As an upshot of this affair, Lord Kingsborough shot Col. Fitzgerald at the Mitchelstown Inn, and was tried for murder by the Irish House of Commons, but was acquitted.

possible to tell without further evidence. At the time of the
decision to dismiss Young, Lord Kingsborough owed him six
or seven hundred pounds. Young offered to return to England
as soon as he was paid, whereupon Major Thornhill came to
Young to say that his lordship was unable to make the payment.
In this dilemma Young suggested a life annuity in lieu of the
cash. Perhaps the case of Miss Crosby brought this expedient to
his mind. At any rate it was agreed that Young should receive
an annuity of £72 for the remainder of his life, which Lord
Kingsborough apparently paid regularly.[94]

After all the business details had been settled, Young had his
books sent to Cork. He himself "stepped into my post-chaise, and,
with a pair of Irish nags" set off for Waterford to take the packet
again to Milford Haven. He had been assured that conditions
at Waterford had changed, but when he arrived there, was told
that the only packet which could take a chaise or horses was
being repaired but would be ready in five days. Actually he
waited twenty-four days, "long enough to have gone round by
Dublin and have reached Rome or Naples."[95] Fortunately the
time was spent pleasantly enough with his old friend, Cornelius
Bolton, Junior, of Ballycanvan. At last he sailed with his chaise,
three horses, and two servants. Apparently he reached England
early in December, followed the same general route as in 1776
through South Wales to Gloucester, and arrived at Bradfield on
January 1, 1779. Another chapter in his life was over. Again he
had failed to establish himself, but he was better off financially
than when he first went to Ireland, he had a welcome addition
to his income from the annuity, and he was free from the wretched
farm at North Mimms.

Although *A Tour in Ireland* was not published until 1780,
it seems wise to analyze and evaluate it in this chapter, and thus
to complete the Irish episode in Young's life. Probably he spent
much of 1779 in preparing it for the press. The first edition was
published in one quarto volume in 1780 in London by Cadell
and Dodsley and was reprinted in two octavo volumes in the same
year in Dublin. A second edition in two volumes also appeared
in London in 1780, published again by Cadell. It is interesting
that a German edition also appeared in 1780, while there was no

⁹⁴ *Ibid.*, p. 80.
⁹⁵ *Ibid.*

French edition until 1800. Back in December, 1776, Young had addressed a letter to Sir Lucius O'Brien asking his advice about the best method of publishing his projected work. He estimated that the work could be brought out in a single quarto volume to sell for 15 to 20 shillings. He believed that the book would not sell well in England and consequently that an Irish publisher would be preferable. But should the copyright be sold to an Irish publisher for a lump sum, or would it be better to publish by selling subscriptions? Young requested Sir Lucius either to suggest reliable Irish publishers to him, or to speak to them himself about the matter.[96] Apparently the subscription plan was the one adopted, for there is a list of several hundred subscribers in *A Tour in Ireland,* following the preface.

Late in August, 1780, the Dublin Society voted to thank Young for his recently published tour and suggested that the second half of the work, the summary, be published separately. Young responded that he was ready to publish the second part separately, or to make an abridgment of the whole, to induce a wider circulation. Nothing came of the proposal, however:

In a few posts I received, under the Dublin postmark, an envelope, inclosing an anonymous essay, cut out of a newspaper; which referred to the transactions of the Society relative to me and condemning pretty heavily my whole publication; and in that unhandsome manner the business ended. I heard no more of them. . . .

Young was disappointed and rather bitter. He believed that a Society with large funds at its disposal might have subsidized an abridgment. He summed up his chagrin thus: "I have but one word to say: *to Ireland I am NOT in debt.*"[97]

The first London edition begins with an eleven-page preface. Part I of the *Tour,* 384 pages long, is the day-to-day diary, which has already been used to follow his journey. Part II, "Observations on the preceding Intelligence," is paged separately, printed in small type, and is 168 pages long. This edition contains only two illustrations, an unnamed waterfall and "An Irish Cabbin," the latter of which may have been drawn by Young.[98]

[96] Historical Manuscript Commission, Report 12, Part x, vol. i, Charlemont Papers, no. 141, p. 335.
[97] *Annals* **15:** pp. 159–161. Most of this material is also in Young, 1898: pp. 87–89.
[98] *Cf.* Maxwell, 1925: prefatory note.

Certainly the *Tour in Ireland* counts among Young's most important works, even though it lacks the charm of his later masterpiece on France. The subject itself, however, was less dramatic, and Young was less mature than a decade later. Moreover, it must be remembered that he lost his private journal of the tour containing many anecdotes which might have added sprightliness. Since the Irish tour is more closely confined to agriculture than the French tour, it has less general interest. Its value to the student of eighteenth-century Ireland is scarcely less, however, than the French tour is to the student of eighteenth-century France. Young himself rated the work highly and declared: "I trust I may say without vanity, that it has stood the test of examination, and received from the best judges the highest commendation."[99] The most recent editor, Professor Constantia Maxwell, declares that the Irish tour "remains our chief authority for Irish economic conditions for the latter part of the eighteenth century."[100]

The preface to *A Tour in Ireland* contains Young's only public statement on the American Revolution, which he blamed on England's colonial policy of commercial monopoly. Had the colonies been given free trade with the mother country, a union would have been formed long ago and the colonies would therefore have contributed naturally to the defense of the empire. He expressed his views thus: "It was not the stamp act, nor the repeal of the stamp act; it was neither Lord Rockingham nor Lord North, but it was that baleful monopolizing spirit of commerce that wished to govern great nations, on the maxims of the counter."[101] He claimed that the war of 1740 was fought to protect English smugglers, the Seven Years' War to exclude the French from the American market, and the American Revolution to preserve the markets gained in 1763. Certainly his views expressed here differ markedly from those voiced in the *Political Essays* in 1772, where he had shown himself a strong adherent of mercantilism, and in the *Political Arithmetic* in 1774, where he

[99] Young, 1898: p. 85.

[100] Maxwell, 1925: p. xv. This edition consists of excerpts and contains nearly all the interesting and valuable passages. Professor Maxwell also used Young's material extensively in her two interesting volumes, *Dublin under the Georges,* 1936, and *Country and Town in Ireland under the Georges,* 1940. There is also an analysis and evaluation of the *Tour* in Haslam, 1930: pp. 129–154.

[101] Young, 1780: p. iv.

had defended the Navigation Acts.[102] There are two possible explanations for the change. Perhaps he had read Adam Smith's *Wealth of Nations*, but no reference to that work appears in the Irish tour. More likely he changed his views after seeing how mercantilism operated in Ireland.

One other comment in the preface on international affairs deserves attention. Young was apparently shocked at the first partition of Poland and argued that at some future time it might be necessary for England to unite with France to preserve western Europe from aggression from the East. As he put it:

It is true we are at present in a war with France, but I must own, the period appears to me fast approaching, when all the western part of Europe will find an absolute necessity of uniting in the closest bands. If the scene which has annihilated Dantzick, was now acting at Hamburgh and Amsterdam, I do not see where the power is to be found, to prevent or revenge it. The consequence of France has been long declining, and the transfer of her exertions from the land to the sea service, may be fatal to the liberties of Europe. If ever the fatal day comes, when that exertion is to be made, all her neighbours would feel it their common interest to second and support her. . . . Then it would appear, that France should have directed all her attention to her army, and Britain to her navy, as the best united means of resisting what Lord Chesterfield very justly terms "new devils" arising in Europe.[103]

Much of Part II of the Irish Tour is devoted to agricultural improvements, with sections on the planting of trees, manures, wasteland, cattle, and wool. At the very end he suggested various crop rotations suitable for Ireland, including crops of turnips, clover, and beans.

Although Young had good reason later to be piqued by the Dublin Society, as noted above, Part II of the *Tour* contains a considerable section on the activities of this model for the Society of Arts.[104] Young was an honorary member of the Dublin Society and probably attended some of its meetings. He praised the Society for many of its activities but felt that it spent too much of its generous parliamentary appropriations upon manufacturing and commerce and too little on agriculture. He especially

[102] *Cf. supra,* pp. 89–90.
[103] Young, 1780: p. iii.
[104] *Ibid.,* part ii, pp. 62–73.

criticized its attempts to foster the silk manufacture. He urged the Society to re-establish a model farm like that formerly operated by John Wynn Baker, but preferably located in a waste-land, not near Dublin. It should not attempt experiments but should rather exhibit practices well established. He also proposed a series of premiums to stimulate Irish agriculture—for the culture of turnips, beans, and flax as parts of rational crop rotations, for the improvement of mountain and bog lands, and for tree plant-ing.

As already noted, Young took a far more liberal stand on com-mercial policy in the Irish tour than in his earlier writings. In Part II several important sections attack illiberal economic poli-cies against Ireland. He condemned the bargain by which the Irish had been induced to discourage their woolen manufactures at the end of the seventeenth century in return for an implicit understanding that the English manufacture of linen would not be encouraged. As a matter of fact, an Irish woolen manu-facture would serve as a healthy competitor to English manufac-tures: ". . . as a fast friend to the interest of my native country, I wish success to those branches of the Irish woollens which would rival our own; . . . it would inspirit our manufactures; it would awaken them from their lethargy, and give rise to the spirit of invention and enterprize."[105] The English, furthermore, had dishonored their part of the bargain, by restricting certain types of Irish linens in order to encourage the English linen industry. Young was also bitter upon the prohibition of Irish meat exports to other countries. He argued that Ireland was a leading customer of England and therefore it was to England's advantage to have Ireland rich, not poor. Furthermore, the richer Ireland became, the larger became the rents of absentee landlords. As he put it, Ireland "is one of the greatest customers we have upon the globe; is it good policy to wish that our best customer may be poor? Do not the maxims of commercial life tell us that the richer he is the better?"[106] Young advocated political union be-tween England and Ireland because he thought that complete free trade could only be attained through union. At the same time, he dismissed the Irish arguments against a union, ap-parently blind to the force of Irish nationalism.

[105] *Ibid.,* part ii, p. 121.
[106] *Ibid.,* part ii, p. 122.

Earlier chapters have shown Young as a champion of the English corn bounty. In Ireland he was very skeptical about bounties in general. The bounties to the linen manufactures were unnecessary and very expensive, even if not particularly mischievous. Bounties might help to establish the Irish fisheries if given in the form of boats and nets to the actual fisherman.

His ire was aroused, however, against the Act of 1759 establishing the bounty on the inland carriage of corn to Dublin. The act aimed to reduce the amount of imported corn into the Dublin market, and hence to encourage tillage in Ireland, by the payment of a bounty in the form of the cost of carriage by land from all parts of the island to the capital. Young admitted readily enough that the objective of the act had been attained. Corn imports had been decreased and corn exports increased. Exports of pork, hogs, and bread had become larger. And certainly tillage had been increased. His objections to the measure started with his belief that it was unnecessary, since corn imports were not excessive. Moreover the cost of the bounty had been very high and was rapidly increasing. Worst of all the act seriously decreased grazing and hence the exports of beef, butter, wool, and yarn. By climate Ireland was better suited to grazing than to tillage. If the land used for tillage had formerly been waste, the measure might have been beneficial, but the increase of corn production had been achieved at the expense of some of the best grazing land in Ireland. Furthermore, the methods employed in corn production were very poor. As Young put the case: "The comparison is not between good grass and good tillage; it is *good* grass against *bad* tillage."[107]

Young drew up elaborate statistics to support his case. The total benefits from the law, consisting of decreased imports of corn and increased exports of corn, bread, hogs, and pork amounted to £62,000. On the other hand the cost of the bounty directly was £47,000 while the decreased exports of beef, butter, wool, and yarn amounted to £159,000. Thus the direct and indirect costs of the bounty were more than three times the gains.[108]

Young also complained that the bounty encouraged land carriage rather than water carriage, a much cheaper mode of transport. Thus horses increased while sailors decreased. "Why

[107] *Ibid.*, part ii, p. 101.
[108] *Ibid.*, part ii, p. 96.

not increase your sailors instead of horses? Are they not as profitable an animal?"[109] The measure benefited only one port, Dublin,
and starved all the others. If a bounty on corn be desired, why
not carry it to the nearest port and then by ship to Dublin?

Young claimed that his attack on the Irish corn bounty was
"the most novel, instructing, and decisively useful part of the
publication." His statistics destroyed the arguments in its favor.
The bounty was cut in half at the next session of the Irish
Parliament, then reduced gradually still more, and finally abandoned. Young thus claimed that he caused great saving to the
Irish nation. "This was much for one individual to effect; and
some reward for such services would not have been much for
the nation to grant."[110] At least the Dublin Society might have
awarded him or have paid for an abridgment of his book.

Young's earlier writings had been generally hostile to the lower
and working classes, always advocating enclosures and arguing
that the poor were the authors of their own misery. He had
always been a liberal, but never a humanitarian. *A Tour of
Ireland* is the first of his writings to recognize that a social problem existed. Of course it is always easier to see abuses abroad
than at home, and it was only much later that Young showed
much sympathy for the English lower classes. But in the Irish
tour he was definitely more friendly to the lower classes than to
the landlords and especially to their agents, the middlemen. The
daily journal gives many examples of benevolently minded landlords, but in his summary the general picture is that of peasants
being persecuted by their landlords.

Young opened his section on the "Labouring Poor" by the
following:

Such is the weight of the lower classes in the great scale of national
importance, that a traveller can never give too much attention to
every circumstance that concerns them; their welfare forms the broad
basis of public prosperity; it is they that feed, cloath, enrich, and
fight the battles of all the other ranks of a community; it is their
being able to support these various burthens without oppression,
which constitutes the general felicity; in proportion to their ease is

[109] *Ibid.*, part ii, p. 97.
[110] Young, 1898: pp. 85–86. Haslam, 1930: p. 148, wrote: "There is no doubt
that, with regard to the Irish tour, Young's greatest success was in connection
with the Bounty on the Inland Carriage of Corn."

the strength and wealth of nations, as public debility will be the certain attendant on their misery.[111]

Although Young criticized absenteeism among the Irish landlords, he was more bitter against the system of sub-letting and middlemen. These men of the lower gentry were those whose habits of "drinking," "fighting," "ravishing" had given Irish society its bad reputation. They were the "tenants, who drink their claret by means of profit rents; jobbers in farms; bucks; your fellows with round hats, edged with gold, who hunt in the day, get drunk in the evening, and fight the next morning."[112] Many were absentees themselves, but the resident middlemen were no better:

Living upon the spot, surrounded by their little under-tenants, they prove the most oppressive species of tyrant that ever lent assistance to the destruction of a country. . . . Not satisfied with screwing up the rent to the uttermost farthing, they are rapacious and relentless in the collection of it.[113]

On the whole, the material condition of the lower classes was not too bad. True, their "cabbins" were wretched, although the mud walls gave more warmth than English lath and plaster. Animals and people lived together in these miserable habitations in a pretty brutish fashion. There was very little furniture as contrasted with English cottages, but Young exclaimed: "I think the comparison much in favour of the irishman; a hog is a much more valuable piece of goods than a set of tea things."[114] The Irish were also much more poorly clothed, and bare feet and legs were general, but Young commented: "as to the want of shoes and stockings I consider it as no evil, but a much more cleanly custom than the bestiality of stockings and feet that are washed no oftener than those of our own poor."[115] Young also believed that the Irish diet of potatoes and milk was healthier than the English diet of bread, cheese, and meat. The point was that the Irish had plenty of potatoes, while the English lacked an abundance of meat: ". . . I have no doubt of a bellyful of the one being better than half a bellyful of the other." He went on:

[111] Young, 1780: part ii, pp. 18–19.
[112] *Ibid.*, part ii, p. 79.
[113] *Ibid.*, part ii, p. 14.
[114] *Ibid.*, part ii, p. 26.
[115] *Ibid.*, part ii, p. 25.

. . . mark the irishman's potatoe bowl placed on the floor, the whole family upon their hams around it, devouring a quantity almost incredible, the beggar seating himself to it with a hearty welcome, the pig taking his share as readily as the wife, the cocks, hens, turkies, geese, the cur, the cat, and perhaps the cow—and all partaking of the same dish. No man can often have been a witness of it without being convinced of the plenty, and I will add the chearfulness, that attends it.[116]

The Irish suffered, however, in comparison with the English because they drank whiskey rather than beer. He protested "their vile potations of whiskey,"[117] so cheap that "a man may get dead drunk for two pence." He urged a heavy excise on whiskey which would not only bring in large revenues, "at the same time that every shilling government got would be half a crown benefit to the publick."[118]

Young apparently liked the common Irish cottar, his cheerfulness, politeness, and love of dancing. True he was a great liar and thief, but Young blamed the landlord for his thievery. The peasant was often accused of stealing wood, but Young pointed out that since the landlord had denuded the country and had done no planting, the peasant could obtain necessary wood in no other way.

They have made wood so scarce, that the wretched cottars cannot procure enough for their necessary consumption, and then they pass penal laws on their stealing, or even possessing, what it is impossible for them to buy. If by another act you would hang up all the landlords who . . . destroy trees without planting, you would lay your axe to the root of the evil, and rid the country of some of the greatest pests in it; but in the name of humanity and common sense, let the poor alone. . . . The honestest poor upon earth, if in the same situation as the irish, would be stealers of wood, for they must either steal, or go without what is an absolute necessity of life.[119]

Young became indignant against the gentlemen who robbed the poor of all human dignity. They could demand almost anything of the poor and expected "unlimited submission," Corporal punishment, even horsewhipping, was not uncommon. On the

[116] *Ibid.*, part ii, p. 24.
[117] *Ibid.*
[118] *Ibid.*, part ii, p. 129.
[119] *Ibid.*, part ii, p. 44.

road the "cars" of the peasant must take to the ditch to make way for the gentleman's carriage. "Knocking down is spoken of in the country in a manner that makes an englishman stare. Landlords of consequence have assured me that many of their cottars would think themselves honoured by having their wives and daughters sent for to the bed of their master."[120]

Young also castigated the penal laws against Catholics. In the first place, the laws had not destroyed Catholicism in Ireland, but had merely strengthened the fanatical religious devotion of the peasants. In the second place, the penal laws had not contributed to the pacification of Ireland, nor had they converted the Catholic Irish into loyal British subjects. Rather religious persecution had only confirmed the Catholics in their hatred and distrust of Protestants. He interpreted the "whiteboy" outrages of the 1760's and early 1770's as activated essentially by religious persecution. Finally the penal laws had led to national poverty and backwardness, not to national prosperity. Since Catholics could not buy land or take a mortgage, they lacked any incentive. Young even charged that the penal laws aimed less at the religion of Catholics than at their property. He wrote: ". . . the scope, purport, and aim of the laws of discovery as executed are not against the catholick religion which increases under them, but against the industry, and property of whoever professes that religion."[121] He pointed out that Irish Catholics were less enlightened and more fanatical than continental Catholics precisely because of the penal laws. He concluded that these laws should gradually be abolished, giving Catholics the right to hold mortgages and buy land, repealing the "abominable premiums on the division of a family against itself,"[122] legalizing public Catholic worship, permitting a Catholic hierarchy to be established, and a Catholic seminary to be opened. Eventually Catholics should be permitted to carry arms and to vote.

Young had always considered tithes an impediment to good farming and the most unfair kind of a tax. In Ireland they added to the oppression of the common people, for they were farmed out to tithe proctors, "a bad set of people,"[123] "very civil to

[120] *Ibid.,* part ii, p. 29.
[121] *Ibid.,* part ii, p. 34.
[122] *Ibid.,* part ii, p. 36.
[123] *Ibid.,* p. 132.

gentlemen, but exceedingly cruel to the poor."[124] Young advocated that Anglican clergymen in Ireland should be given land in lieu of tithes and that residence be enforced. Thus each locality would have one resident gentleman interested in his property, who "could scarcely fail of introducing improvements in agriculture and planting."[125]

[124] *Ibid.,* p. 181.

[125] *Ibid.,* part ii, p. 56. It is ironical that Young's only son should have become an Irish absentee clergyman who fought constantly to increase the tithe return from his living. *Cf.* Gazley, 1956: pp. 362–363.

IV. Farming at Bradfield—
The Annals of Agriculture,
1779-1786

T HE TRANQUIL bosom of my good mother's heritage—my native
Bradfield—once more opened its arms to receive me."[1]

The year 1779 marked another turning point in Arthur Young's
career. He had given up his farm at North Mims. His engagement with Lord Kingsborough had been terminated. On returning from Ireland, he naturally came to Bradfield, but it was
not inevitable that he should remain there. He was thirty-seven
years old with a family of three children, Mary who was twelve,
Bessie ten, and Arthur nine. His financial position was certainly
better than when he first went to Ireland but he still was a poor
man with no assured future. At the same time he was very
famous as the leading agricultural publicist in England. He had
been elected to the Society of Arts and had served as chairman
of its Committee on Agriculture. He had also been elected a
Fellow of the Royal Society and an honorary member of learned
societies in Dublin, York, Manchester, Berne, Zurich, and Mann-
heim.

All these honors, however, brought in no income. Although he
might continue to write, he must do something else. Again as
in 1772,[2] he seriously considered emigrating to America, but
his aged mother entreated him not to go. Martha Young had
opposed emigration in 1772 and it is difficult even to imagine
her an immigrant's wife in America. His mother apparently
suggested that he take over the home farm at Bradfield and
thus he became permanently settled on the ancestral estate.[3]
Travel he always would, for he was a restless soul. And from

[1] *Annals* **15**: p. 161.
[2] *Cf. supra,* p. 75.
[3] Young, 1898: pp. 83–84.

1794 on he spent the winter and spring months regularly in London. But his heart was always at Bradfield.

The early years after his return to Bradfield, 1779–1783, were among the least eventful in Young's life. He made no long extended tours, nor did he publish any important work except the Irish tour. The small farm, "contiguous to the family mansion," which Young took over at Bradfield at Michaelmas, 1779, was gradually enlarged until it reached something between 300 and 400 acres.[4] In 1779 the farm buildings—a barn, a cowhouse, and cart lodge—were "as bad and ill contrived . . . as any in the country."[5]

Young probably spent much of 1779 preparing the Irish tour for the press. Two new honors came to him. The first, his election to the Imperial Œconomical Society of Petersburg,[6] presumably resulted from his earlier aid to Russian students of agriculture in England, and especially to the Rev. Mr. Sambosky, chaplain at the Russian Embassy, whom Young praised warmly for his "knowledge, industry, and indefatigable perseverance."[7] The second was an award of a medal from the Society of Arts for his experiments on the clustered or Howard potatoes.[8]

Young's perennial interest in trees is shown by his measurement of certain trees at Bradfield on four different dates in 1779,[9] and by his planting during this period of seven acres of trees at Bradfield, chiefly larch.[10]

A Tour In Ireland, which appeared early in 1780, certainly enhanced Young's reputation very considerably. The *Gentleman's Magazine* for April and September published two excerpts in each issue—the rise and suppression of the Whiteboys, Young's visit to Castle Caldwell with a plate which is almost certainly Young's work, Archbishop Robinson's improvements at Armagh,

---

[4] *Ibid.*, p. 84.

[5] *Elements of Agriculture* 3: f. 41.

[6] Young, 1898: p. 85.

[7] *Annals* 2: p. 242. During 1779 he went down to the Suffolk coast to purchase horses for Mr. Sambosky.

[8] *Transactions* 3 (1785): pp. 30–99. This is an account of ten experiments on potatoes at Bradmore Farm, 1770–1776. He compared their productivity with other varieties, showed their profit as a fallow crop, and used them to fatten hogs. He also tried various manures for them.

[9] *Annals* 3: pp. 429–430.

[10] *Elements of Agriculture* 5: ff. 107–108.

and the account of drawing oxen by the horns.[11] The review in the *Monthly Review* by Dr. Edmund Cartwright,[12] the inventor five years later of the power loom, appeared in three successive numbers for a total of twenty-six pages.[13] Dr. Cartwright stated: ". . . we shall find no author whose labours can any way come in competition with those of this indefatigable compiler. . . ."[14] Cartwright especially praised Young's strictures on the Irish penal laws and the commercial restrictions. Young's tours, the most valuable of his works, had proved "of very considerable benefit, in introducing many improvements into general use, which were before confined to a particular province, or individual."[15] He concluded: "We have met with little or nothing of that passion for theory and paradox in which this writer sometimes indulged himself. The work before us never could have appeared at a time when it would have been more worthy the public attention than at present."[16]

Young spent at least a month in the winter at London, where he saw much of Arbuthnot and the Burneys, attended the opera, meetings of the Royal Society, and Parliament, and went to many parties.[17] He was also present at five meetings of the Committee on Agriculture of the Society of Arts.[18] At the meeting on February 13 William Marshall (1745–1818), perhaps the ablest of Young's contemporary agriculturists, proposed that the Society should subsidize him £200 a year to enable him to pass a year in each of the six or seven English counties most famous for agricultural progress. After considerable discussion and many questions, the Committee rejected the plan, useful though it might be, because it was against the Society's policy to give direct aid to such proposals. A second motion recommending that individual members encourage the enterprise was also defeated.[19] Only five members of the Committee attended this meeting, of whom Young was certainly the most famous. One would like

---

[11] *Gentleman's Magazine* **50** (1780): pp. 170, 181–182, 430–431.
[12] Nangle, 1934: pp. 8, 226.
[13] *Monthly Review* **63** (1780): pp. 38–45, 97–104, 161–171.
[14] *Ibid.*, p. 38.
[15] *Ibid.*, p. 39.
[16] *Ibid.*, p. 171.
[17] Young, 1898: pp. 91–92. Unfortunately Fanny Burney was not in London.
[18] Committee Report Books, 1779–1780, ff. 13, 20, 23, 27, 31.
[19] *Ibid.*, f. 23.

to know how he voted and what part he played in the rejection of Marshall's scheme. Almost certainly Marshall must have blamed him. They were unfortunately never on very good terms, and this incident could hardly have improved the relationship.

Young also saw much during this stay of the Irish writer, Hugh Boyd (1746–1794), whom he had apparently met in Ireland.[20] Young wrote of frequent breakfasts, morning calls, and dinners at Boyd's residence. Young greatly admired Boyd's ability to "multiply nine figures by eight entirely in his head,"[21] and also his remarkable memory. Once he was at Boyd's for breakfast when Edmund Burke's son called. His father had made a very celebrated speech in Parliament the day before which Boyd and Young had heard. In the heat of the moment Burke had departed from his notes. Now, desirous of publishing the speech, he sent to Boyd for aid. "We set to work to recollect as much as possible his own words, and furnished young Burke much to his satisfaction."[22]

The *Autobiography* for 1780 contains three amusing and very cordial letters from Boyd.[23] That of August 16 is a reply to Young's invitation to visit Bradfield, in which Young urged Boyd to relieve him of the dullness of the country. One sentence of Boyd's reply is interesting: "I should have been happy in being at Bradfield Hall: I long to hear my friend refute himself, to complain with good spirits, and to demonstrate with much wit, that he was extremely dull."[24] The letters show that the families were also intimate, for Boyd referred several times to Mrs. Young with every indication of friendliness.

Young was always a voracious reader and consistently copied out excerpts from everything he read. Thus the *Autobiography* for 1780 includes an excerpt from the letters of the poet Grey. After the excerpt Young commented:

The day is never too long, for I think time spent in reading is always well employed, unless a man reads like an idiot, that is, equally removed from instruction and entertainment. Now the

---

[20] Young, 1898: pp. 92–97.

[21] *Ibid.*, p. 92.

[22] *Ibid.*, p. 93. Young believed that Boyd was the autl   of Junius, a belief shared by several contemporaries, notably by George Chalmers.

[23] *Ibid.*, pp. 93–97. Dates are August 16,   ugust 17, and September 2.

[24] *Ibid.*, p. 96.

Fig. 2. Arthur Young as baby with his sister, later Mrs. John Tomlinson. From Farrer, *Portraits in Suffolk Houses,* opp. p. 380. Kindness of H. R. D. Williams of Bernard Quaritch.

Fig. 3. Martha Young, "Bobbin." Miniature by Plymer. *Autobiography,* opp. p. 265.

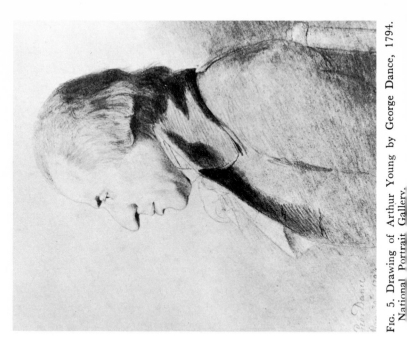

Fig. 5. Drawing of Arthur Young by George Dance, 1794. National Portrait Gallery.

Fig. 4. Engraving portrait of Arthur Young. Gift of Frances I. W. Booth.

FIG. 6. Bradfield Hall in Young's time. *Autobiography,* opp. p. 127.

FIG. 7. Sketch of Bradfield Hall, probably by Mary Young, daughter of agriculturist. Kindness of Bertha Taylor of Lavenham Hall.

135

Fig. 8. Snuff box, originally set in diamonds, presented to Young in 1804 by Count Rostchoptschin, "A Pupil to his Master." Kindness of Mrs. Rose Willson, niece of last Mrs. Arthur Young.

Fig. 9. Silver serving dish, with lamb in high relief on cover, presented to Young by Board of Agriculture in 1811. Kindness of Mrs. Rose Willson.

Fig. 10. Wedgwood breakfast set, decorated with agricultural implements, believed to have been made for Young. Kindness of Mrs. Emily Cooper of Bury St. Edmunds.

FIG. 11. Arthur Young (at extreme left) being presented with medal at Society of Arts. One of wall frescoes by James Barry in Society room. Kindness of late Charles Durant Cassidy.

Fig. 13. Robert Bakewell of Dishley, outstanding breeder of cattle and sheep, "common farmer." From G. E. Fussell, *The English Countryman*, opp. p. 97. Kindness of Hutchinson Publishing Group Ltd.

Fig. 12. Coke of Holkham, Earl of Leicester, eminent landlord, M.P., member of Board of Agriculture, champion of "Norfolk" agriculture. From W. Notestein, *English Folk*, opp. p. 64. Kindness of Harcourt Brace Jovanovich, Inc.

138

Fig. 14. Meeting of Board of Agriculture at 32 Sackville Street. Secretary Arthur Young presumably making a report. Painted by A. C. Pugin and Thomas Rowlandson. From Rudolph Ackermann, *Microcosm of London*, opp. p. 73.

139

Fig. 15. Woburn sheepshearing. Young in group at right. Painting by George Garrard. From the Woburn Abbey Collection, by kind permission of His Grace, the Duke of Bedford.

FIG. 17. Note on inside page to Earl of Fife, illustrating Young's striking handwriting. Gift of Miss Henrietta Taylor.

(ANNALS OF AGRICULTURE.)

SIR,

DURING my absence of eight months on the Continent, some steps having been taken by Government in relation to the Corn Trade and Corn Laws, which will make a new and permanent act necessary—and as such translations are of the first consequence to the Landed Interest, I take the liberty (as Editor of the above work) earnestly to request your answers to the under written Queries, in order that some authority relative to this business may be procured immediately from the Corn Grower, as well as that to which recourse is usually had, the opinion of persons solely concerned in the trade.

I am, SIR,
Your obedient and devoted servant,
ARTHUR YOUNG.

Bradfield-Hall, near Bury, Suffolk,
January 31, 1790.

QUERY I. Is the produce of the wheat crop, on an average of seven years, equal to the mean consumption of your country, district, or vicinity? and if more or less, state it?

II. Supposing an average crop equal to the mean consumption (for twelve months?) of those who usually consume it, what was the proportional produce (for how many months?) of the crop of 1788 which was followed by the high price of the summer 1789?

III. Was the high price of the summer of 1789 owing, in your vicinity, to original scarcity, to export, or to alarm?

IV. How far did the high price of that season influence the prices of other provisions? Meat, butter, cheese, &c. &c.?

V. Were there any circumstances in the distress or relief of the poor worthy of notice?

VI. What were their earnings per diem by the spinning of wool?

VII. Have poor rates risen in consequence of the distress?

VIII. How far was the price of wheat during the preceding winter, in which period the greater part of the produce might be sold, equal to making up for the deficiency of the crop?

IX. What was the state of the crop of 1789, proportionably to a mean one being equal to twelve months common consumption? As far as could be conjectured from its appearance, or the quantities threshed?

X. Did the heavy autumnal rains prevent any land being sown? And if so, has that circumstance tended to raise the price, more than the fate of the seed would tend to lower it?

XI. What was the price of wheat (specify the measure) in your markets for six weeks before the proclamation, fair with a bill of indemnity is now passing, and for six weeks after it?

XII. How far, in your opinion, did the variations in the price, previous and subsequent to that measure, flow from the real quantity that was in the markets, or from any alarm in the minds of the people, or other cause?

XIII. What are the present prices in your markets of

Wheat,	Peas,	Beef,	Butter,	Horses,	Lean Hogs,
Barley,	Beans (horse),	Mutton,	Cheese,	Lean sheep,	Hay,
Rye,	Beans (tick),	Veal,	Lean Beasts,	Lambs,	Straw, &c. &c.
Oats,		Pork,	Cows,		

XIV. What was the price of wool of the last clip? and what has been the operation of the wool act? What effects, and what remedies?

XV. Have any signs of the rot among sheep appeared in consequence of the heavy autumnal rains?

XVI. Lastly, your opinion is requested.—What ought to be the mean price of wheat to pay the farmer the expence of cultivation, and a fair profit on his capital, attention being paid without riding tythes, meeting with another not at present in contemplation, namely, whether, and how far, the price of labour ought to be raised? And attention also being had to the prices of the other products of the soil?

N. B. Wheat has been principally in view, but in the few districts where rye is the common food, the queries are equally applicable to that grain, and to any others proportionably to their being substitutes for the more common object of consumption.

FIG. 16. Questionnaire from Annals of Agriculture.

Fig. 18. Dr. Charles Burney, Young's brother-in-law, eminent musician, founder of Burney clan, including Marianne Francis. Painted by Sir Joshua Reynolds, 1781. National Portrait Gallery.

Fig. 19. Thomas Ruggles, Young's schoolmate. Pastel portrait by John Russell. Kindness of late Col. Sir Edward A. Ruggles-Brise.

FIG. 21. Sir John Sinclair, founder of Board and President, 1793–1798, 1806–1815. Painted by Sir Henry Raeburn. National Gallery of Scotland.

FIG. 20. John Baker Holroyd, Lord Sheffield, President of Board, 1803–1806. Painted by John Downman, 1780. From *The Girlhood of Maria Josepha Holroyd*, opp. p. 112. Kindness of Longman Group Ltd.

143

Fig. 22. Sir Joseph Banks, President of the Royal Society, active member of Board of Agriculture, Australian explorer. Painted by Thomas Phillips, 1810. National Portrait Gallery.

Fig. 23. William Wilberforce, crusader against slave trade, outstanding Evangelical layman. Painted by Sir Thomas Lawrence. National Portrait Gallery.

general occupation of my life—agriculture—has the happy circumstance of giving much employment, and with it exercise, at the same
time that it naturally leads into a course of reading, to which it
gives the air and turn of a study. . . .[25]

Material for 1781 is scanty and the *Autobiography* is not very
clear. Most of the year was again spent at Bradfield. Many projects
seem to have been fermenting in his mind. On Lord Loughborough's suggestion he started to prepare a series of lectures
on agriculture, but nothing came of it.[26] Young had always been
interested in population problems, as seen in his earlier pamphlet,
Proposals to the Legislature for Numbering the People, and
especially in the *Political Arithmetic.* Now in 1781 he wrote a
new article entitled "Of the Population of Different Periods,"[27]
which maintained that population in England had increased,
rather than decreased, during the eighteenth century. Young's
conclusions were based upon his study of baptisms and burials
in a considerable number of parish registers in Suffolk and Norfolk. Since most of these parishes had been enclosed many years
earlier, he concluded that in the long run enclosure did not
lead to depopulation.[28] He also denied that luxury caused depopulation and inserted comparative tables[29] for the thirty years
before 1688 and the thirty years before 1780, which all the
writers had agreed upon as the period of increasing luxury. In
one place he stated epigrammatically that luxury "is nothing but
disseminated wealth, disseminating employment."[30]

In 1781 Young received a rather pathetic letter from Arbuthnot, full of all his difficulties, financial and otherwise. The letter
ended, "if I live I will be with you in the summer," but there
is no evidence that he came.[31] In the *Autobiography* Young stated
that Lord Loughborough secured an appointment for Arbuthnot
under the Irish Linen Board but that Arbuthnot lived only a

---

[25] *Ibid.*, p. 90.
[26] *Ibid.*, p. 97.
[27] *Annals* 7: pp. 429–457. Although dated April 14, 1781, the article was not
published until several years later. Some of the material had been used by Rev.
John Howlett in a pamphlet.
[28] *Ibid.*, pp. 448–449.
[29] *Ibid.*, p. 453.
[30] *Ibid.*, p. 452.
[31] Add. MSS. 35,126, ff. 189–190.

few years longer. "I lost in him by far the most agreeable friend I was ever connected with."[32]

One further event concludes the account of Young's life in 1781 and can best be told in his own words from the *Autobiography*:

In the autumn of this year I spent a month at Lowestoft, where the sea air and bathing agreed so well with me that I do not recollect in my life ever having spent a month with so continued a flow of high spirits, which received no slight addition by the society of a very handsome and most agreeable girl, whose name I have forgotten.[33]

Early in 1782 his aunt, Miss Cousmaker, died at Bradfield and left Young a very considerable legacy—her house, two farms, a carriage and horse, and an annuity in the funds of about £150 a year. She had planned to leave him all her annuities, amounting to £300 a year, "but being much offended with my wife, she gave half of it to another person."[34] She had altered her will without a lawyer's aid, a fact which the lawyers maintained invalidated Young's legacy. Hence the property must pass to the heir at law, a cousin, John Newman Cousmaker, Esq., of Hackney. Poor Young must have been sorely disappointed, but his cousin proved generous, refused to take advantage of a technicality, and drew up a deed giving the property to Young. In the memoirs of his life, published in the *Annals of Agriculture* in 1791, Young publicly acknowledged his gratitude:

An action flowing from true liberality of principle and impossible to be praised too much. Not to take pleasure in acknowledging such proofs of generosity would speak a vice of the human soul, which of all others I hold in the greatest detestation—ingratitude.[35]

On April 22 while in London on the estate business, Young attended a meeting of the Committee of Agriculture of the Society of Arts at the Castle in Stockwall to make some trials of plows and drills.[36] The records indicate this was the last meeting

---

[32] Young, 1898: p. 99. The British Museum contains a bitter pamphlet by R. Stephenson who felt that he had lost his job on the Linen Board to Arbuthnot. The pamphlet pointed out that Arbuthnot had always failed in everything, that he had engaged in smuggling while in France, and accused him of wishing to transform Ireland from a linen manufacturing to a flax growing country.

[33] Young, 1898: p. 100.

[34] *Ibid.*, p. 103.

[35] *Annals* 15: pp. 162–163.

[36] Committee Report Books, 1781–1782, f. 52.

of the Committee that he ever attended. Thus terminated thirteen years of active participation in the Society of Arts.

In 1782 began Young's correspondence with the Rev. Richard Valpy (1754–1836), until recently master at the grammar school at Bury St. Edmunds where Young had sent his only son. The association thus formed had grown into friendship. Now Headmaster of the Reading school, he was to gain great fame as the author of popular Greek and Latin grammars. Young described Valpy in the *Autobiography* thus:

He was a most learned, ingenious and agreeable man, in so much that I greatly regretted his departure, feeling almost sensibly the loss of his society. I have been occasionally connected with him since, and shall always hold him in great estimation for his learning, his talents, and sincerity of friendship.[37]

On February 24, 1782, Valpy wrote that he hoped to be in Bradfield in July and spoke of plans to cross the channel together in the summer of 1783. He encouraged Young to continue with his poetry and promised to serve as a friendly critic:

I hope the unaccountable loss of your manuscript poems will not prevent your muse from further exertion. . . . Poetical images apart, you really should often leave the train of Ceres, & join the choir of the muses. . . . In short, let me insist upon giving you soon my ideas respecting some new Poetical composition of yours. I do not know whether I am a judicious critic, but you know I am a severe one.[38]

Young's social life expanded considerably in 1782, when he became a member of a group of intellectuals who dined every Thursday with Frederick Hervey, fourth Earl of Bristol and Bishop of Derry.[39] The "Earl Bishop," as he was called, possessed all the genius and eccentricities of his famous family.[40] He was an active politician, an outspoken friend of Ireland, an infidel in religion, "a bad father . . . and worse husband"[41] who spent the latter part of his life in open adultery on the continent. He was famous for his two identical palaces, known as Lord Bristol's

---

[37] Young, 1898: pp. 106–107.
[38] Add. MSS. 35,126, f. 199.
[39] Young, 1898: pp. 103–106.
[40] *Cf.* Childe-Pemberton, 1924: **1**: pp. 270–289.
[41] *Ibid.* **1**: p. 2.

"follies,"[42] one at Downhill in Ireland and the other at Ickworth, about three miles from Bury St. Edmunds. Other members of the group were George Ashby, president of St. John's College, Cambridge, Sir John Cullum, author of a famous history of a Suffolk village,[43] and the Rev. John Symonds, professor of history at Cambridge, who became Young's closest friend for the next fifteen or twenty years. Owner of a beautiful new mansion in the Adam manner on St. Edmund's Hill, about one mile out of Bury,[44] Symonds was a typical eighteenth-century clergyman and college professor, easy going, cool, somewhat of a bore, a great traveler, and an amateur farmer. Many of his letters to Young have survived. In age Young was the junior member of the Ickworth dinner parties of which he wrote: "In my life I never passed more agreeable days than those weekly dinners at Ickworth. The conversation was equally instructive and agreeable."[45] The friendship between Young and Lord Bristol seemed easy and cordial. Bristol addressed Young by his Christian name, and gave him a complete set of Rousseau's works, perhaps a somewhat strange gift from an Anglican bishop.[46]

Through his connection with Bristol, Young was also drawn into a lively newspaper controversy. At a meeting at the famous Angel Inn at Bury on July 20, 1782, Lord Bristol proposed a public subscription in Suffolk to build a man-of-war of seventy-four guns.[47] The project developed rapidly and large amounts were pledged, including twenty pounds by Arthur Young, quite a considerable sum for one of his limited means.[48] *The Bury Post and Universal Advertiser* published letters by Young supporting the scheme on September 12 and 26, and on October 17. Young's first letter was actually a reply to Capel Lofft's letter of September

[42] The dinners took place in the old house, not in the new palace with its domed central pavilion in which Bristol never resided.

[43] *History of Hawstead*, 1813, 2nd ed.

[44] This mansion is now known as Moreton Hall. The central core is still as in Young's time and contains some fine Adam decorations on the exterior and two beautiful Adam rooms on the ground floor.

[45] Young, 1898: p. 104.

[46] *Ibid.*, p. 106.

[47] *Ibid.*, p. 101. Young mistakenly dated this plan for 1781. *Cf.* also Childe-Pemberton, 1924: 1: pp. 288–289.

[48] *Bury Post*, August 22, 1782: p. 3. Bristol led the list of contributors with £1000. He also prevailed upon the clergy of his diocese at Derry to give £600. Among Young's friends who made substantial contributions were the Duke of Grafton, £600, Cullum, £50, Ashby and Symonds, £21 each.

12 which was hostile to the plan. Lofft also replied to all Young's letters. In 1783 the letters, together with the list of subscribers, were published by Young at Bury in a pamphlet entitled, *An Inquiry into the Legality and Expediency of Increasing the Royal Navy by Subscriptions for Building County Ships. . . .* Young's adversary in the correspondence, Capel Lofft (1751–1824), lived at Troston Hall, a beautiful old brick mansion about seven miles from Bury. Young's junior by ten years, Lofft was a Foxite Whig, political Radical, and an unbending democrat. Although Young and Lofft were outwardly on friendly terms, their political ideas often clashed and Young frequently suspected Lofft's good faith, apparently without good reason.

Lofft opposed the subscription mainly on constitutional grounds, claiming that only Parliament could grant subsidies and that no monetary aids should be given the crown except by Parliament. He repeatedly declared that parliamentary reform should be the first objective. Such a subscription, "however short of a tax in point of efficiency, exceeds it far in burthen to individuals."[49] In his last letter he declared the war was one "of injustice and of madness in its origin. . . ."[50]

Fundamentally Young's arguments were those of the patriot who burned to defeat the hereditary enemy. He never referred in these letters to the justice of the American cause against Britain, but he was certain of the justice of Britain's cause against France. In his earliest pamphlets written during adolescence in the Seven Years' War, Young had exhibited a flamboyant nationalism.[51] Now again he expressed a florid type of patriotism. His first letter admitted that the building of one ship by Suffolk was not important in itself, but he hoped that many other counties would emulate Suffolk's example and that the result would be a tremendous outburst of national patriotism. "Our enemies have been persuaded, that we are an exhausted, as well as a divided people: . . . Give them some convincing proof to the contrary. Tell them that one noble spirit animates every British bosom." He called for "an electric spark of patriotism, that darts the hidden fire, in a moment, from one extremity of the country to the other."[52]

[49] *Bury Post,* September 19, 1782.
[50] *Ibid.,* October 24, 1782.
[51] *Cf. supra,* pp. 8–10. For a detailed study of Young's nationalism, *cf.* Gazley, 1950, pp. 144–189.
[52] *Bury Post,* September 12, 1782.

Young's second letter denied Lofft's contention that the plan would strengthen the executive power, for Parliament must still man the ship, send it to sea, and pay the sailors. He declared that Lofft was illogical in claiming that parliamentary representation was very imperfect and yet in refusing any popular movement outside such an undemocratic parliament. In his desire to discredit Lofft's arguments, Young completely defended the British constitution: "That bosom which feels not freedom under our constitution, was not born for liberty, but for Bedlam."[53] Young's last letter claimed that Lofft's position almost seemed to desire national defeat in order to prevent the growth of royal power. He declared that was really stating:

. . . that the proposition of building a ship if effectual in this, and imitated in other counties, *will make the war successful and* THEREFORE *I will oppose the measure.* This is admitting all that the friends of the scheme can ask. We have no desire so ardent: the warmest wish of our bosoms goes no further, than as the war is just, so let it be successful. A restless ambition, and not the love of liberty, sent the French to America: is it possible that any friend to his country can languish for her defeat, and sigh over the measures that are to lead her to victory![54]

The whole scheme came to nothing in spite of £21,000 raised in Suffolk. In the first place, no other counties exerted themselves, and Lord Shelburne, the prime minister, who warmly favored the plan, could not raise his own county. In the second place, the need was not really for more ships, but for more seamen. Dr. John Young wrote to his brother Arthur, attacking "your silly Suffolk scheme" on these grounds and quoting Lord Keppel as stating that he already had ten more ships than he needed.[55] Finally, the war was really over and peace negotiations were already in progress.[56]

Sometime in September, 1782, three Russian boys arrived in Suffolk to study agriculture. They had been sent over by Prince

[53] *Ibid.*, September 26, 1782.
[54] *Ibid.*, October 17, 1782.
[55] Young, 1898: pp. 107–109. Very few letters from Dr. John Young to Arthur Young have survived. In this letter, dated October 31 from Eton, Dr. Young accepted Lofft's position that Parliament should supply the government with money, but opposed any radical innovations in the constitution. Suffolk gentlemen should suppress smuggling on their coast rather than raise subscriptions.
[56] Childe-Pemberton, 1924: 1: p. 289.

Potemkin and assigned to the Rev. James Smirnove, the new chaplain at the Russian Embassy. Smirnove himself had earlier studied agriculture under Arbuthnot and Young, and naturally turned again to Young to educate these youngsters. "This I readily did, and took every means to have them well instructed in the English mode of cultivating land."[57] One of the young men may well have been Count Rostoptschin, later famous as governor of Moscow in 1812.

During the years 1781–1786 Young conducted an elaborate series of experiments to test the extent to which phlogiston was the food for plants. The phlogiston theory, which had developed slowly during the seventeenth century, attempted to explain the phenomenon of combustion and asserted that phlogiston, or the fire element, was given off when combustion occurred. The theory was accepted by many eighteenth-century chemists, notably by Joseph Priestley who had published during the 1770's several volumes of his great work, *Experiments and Observations on different Kinds of Air*. After reading this work carefully, Young gained the impression that Priestley believed that phlogiston was the chief food for plants. As early as 1781 he wrote Priestley for advice about his proposed experiments, to which Priestley replied urging him to conduct them.[58] Further letters were exchanged in 1783, in one of which Priestley paid Young a warm compliment:

There is no person I should serve with more pleasure than *you*, because there is no person whose pursuits are more eminently useful to the world. You alone have certainly done more to promote agriculture, and especially to render it respectable in this country, than all that have gone before you.[59]

During his experiments Young consulted Samuel More, secretary to the Society of Arts,[60] and Dr. Isaac Milner, Professor of Chemistry at Cambridge.

In describing his chemical experiments Young was always very

[57] Young, 1898: pp. 102, 125; Add. MSS. 35,126, ff. 203–204. Young again is very indefinite about times. On page 102 it appeared that the boys arrived in 1781; on page 125 it seems to have been 1784. Smirnove's manuscript letter makes it almost certain that it was in September, 1782.

[58] Young, 1898: pp. 99–100.

[59] Add. MSS. 35,126, f. 225.

[60] *Ibid.*, f. 227.

modest if not apologetic. He recognized frankly his want of
scientific training and his lack of resources to construct satisfactory equipment. In 1784 he wrote: "After all, I wish the
reader to consider these minute experiments as nothing more
than weak enquiries; and an insufficient pursuit of hints caught
slowly and with difficulty. My wings are tied from the bolder
flight that my wishes point to."[61] Nevertheless he was indefatigable and in another article wrote: "What is within the power
of my weak and inefficient forces, I shall not neglect."[62] Even
earlier he had written: "It has never been my conduct to
abandon to speculation that which can be brought to the test
of experiment."[63] Back in 1781 he had begun to construct a small
building covered with "rat-work" in which to control his experiments more completely by excluding birds and worms.[64] Later
he was securing "fixed air" from a Bury St. Edmunds brewery,[65]
and ordering some special pots for his experiments from Josiah
Wedgwood.

One of his most ambitious sets of experiments began in 1782
when he planted grain in a number of flower pots, one of which
was left with the earth alone, but to most of which were added
various manures or various gases or "airs,"—"fixed air" (carbon
dioxide), "inflammable air" or phlogiston (hydrogen), and "de-
phlogisticated air" (oxygen). Notes were taken when the sprouts
appeared, several times during the process of growth, and finally
when the grain was cut, the straw weighed, and the number of
kernels counted. Frequently a second or even third crop was
grown to test how long the effects could be noted. Sometimes a
pot might be very disappointing for the first crop, but very effective for the second. Obviously this set of experiments continued in 1783 and even 1784. In these experiments, among
other things, he was attempting to test the effect of charcoal, a
substance acknowledgedly rich in phlogiston. Although Young
knew that the chemists claimed that phlogiston could be freed
from charcoal only through combustion, he surmised that the
processes of fermentation and putrefaction might also produce

[61] *Annals* **3**: p. 127.
[62] *Ibid.* **5**: p. 516.
[63] *Ibid* **1**: p. 169.
[64] *Ibid.*, pp. 140–141.
[65] *Ibid.*, pp. 172–175. Priestley had also used a brewery to procure carbon
dioxide.

phlogiston from charcoal. He was bothered, however, by the fact that although charcoal did benefit the soil, charcoal ashes proved far more beneficial, and by definition charcoal ashes could contain no phlogiston. On the whole he was convinced, however, that he had proved that phlogiston in a volatile state, or "inflammable air," was the chief food of plants.[66]

In 1785 Young conducted a second very ambitious set of experiments to test his general theory. If phlogiston was indeed the food for plants, would it not follow that the better the soil the more phlogiston it would contain? Consequently he made some sixty-odd experiments on soils, plants, and manures. His method was to place a quantity of the substance in a gun barrel, which was heated in a very hot wood fire or in a charcoal furnace, and to collect in jars the gases given off. His title for the experiments was "On the Air expelled from earth, &c."[67] In general he found that the soils which he knew from practice to be the best, yielded the most air, that the vegetables generally regarded as the most exhausting to the soil yielded more air than the less exhausting ones, and that manures yielded large quantities of air, corresponding with their general effectiveness as fertilizers. His summary is worth quoting:

Inflammable air abounds in a most wonderful degree in vegetables . . . . It abounds in those manures which are known to be most fertilizing; and we find it also in soils, in a tolerably exact proportion to their richness: this seems to prove very satisfactorily that this subject is the real FOOD of PLANTS. . . . All the circumstances of the enquiry seem, therefore, to accord, and allow us to suspect . . . . that AIR taken in by the roots, is the true food of PLANTS.[68]

Young was so impressed by his tests that he sent accounts of them to the scientist Richard Kirwan, who wrote on October 24, 1785, that he was "much pleased" with the discovery which ought to be "made publick."[69] Consequently Young submitted his experiments to the Royal Society of which he was a Fellow, in the hope that they would be printed in the *Philosophical Transactions* of that august body. When his paper was rejected, Young was told by a friend that the cause was the dislike of the president, Sir Joseph Banks, for the Portuguese scientist Magellan,

[66] *Ibid.*, pp. 139–189.
[67] *Ibid.* **6:** pp. 265–323.
[68] *Ibid.*, pp. 321–322.
[69] Add. MSS. 35,126, f. 306.

whom Young had unfortunately used as an intermediary.[70] Magellan's own explanation to Young shows something of the factional jealousies within the Royal Society: "I cannot help telling in a whisper, that you may thank the worthy Dr. Blagden, Secret'y to the Soc'y for the refusal of printing your experiments in ye Philal. Transct . . . as he is the director of the blind animal that sits in the chair, and is a great pretender to knowledge on aerial discoveries and investigations."[71] Young was undoubtedly disappointed that his work was not given recognition in the *Philosophical Transactions.* Fortunately he now had his own publication, the newly established *Annals of Agriculture,* and hence his experiments were "made publick" there.[72]

By far the most important event in Young's life in 1783 was the birth of his fourth child, a daughter named Martha Ann but always known as "Bobbin" which "originated in that of 'Robin' which the child gave herself but could not pronounce."[73] A gap of fourteen years had occurred since the birth of his third child. Mary Young was now seventeen, Bessy fifteen, and Arthur fourteen. The arrival of a new baby must have created a stir, and it is little wonder that Bobbin became the special darling of her father.

Sometime during 1783 Young had a severe fever which his friends attributed to his habit of taking an early morning dip in a small pond about two hundred yards from the house. "At this time I was a desperate bather, going into the water every morning at four o'clock each winter, and with or without the obstruction of a thick coat of ice, having often to break it before I could bathe."[74] Naturally his friends remonstrated, but to no avail. As soon as he recovered from his illness, he resumed the habit and continued it for many years.

A letter in 1783 from Robert Bakewell, the famous grazier, paid tribute to Young in terms very similar to those used by Priestley in the letter quoted above: "I shall at all times be happy

[70] Young, 1898: p. 150–151.
[71] Add. MSS. 35,126, ff. 347–348. Date is July 19, 1786.
[72] *Annals* 1: pp. 139–189; 3: pp. 103–127, 476–481; 5: pp. 515–535; 6: pp. 265–323, 442–452.
[73] Young, 1898: p. 110 and note.
[74] *Ibid.* According to Mrs. Rose Willson the pond where he bathed was not the large one in front of the house, but one located on the left of the path leading from the Hall to the church.

in communicating anything in my power that can be in the least serviceable to you or the Public whose best friend you are and to whom I think them more indebted than to all the authors on this subject."[75]

On April 23 Lord Bristol wrote to Young from Londonderry, describing some of his improvements in Ireland, making chaffing remarks about other members of the former dining circle, and inviting Young to visit him in Ireland: "When you are *vapid*, if ever those *petillant* spirits of yours are so, come and imbibe some fixed or unfixed air at the Downhill. . . ."[76]

Two letters also survive from James Barry, the painter of the ambitious murals in the assembly room of the Society of Arts, in one of which Young is being given a gold medal by the Prince of Wales. Young stated that he sat to Barry for the portrait, but exactly when is uncertain. The first letter requested Young to make suggestions and corrections on the pictures, and urged him to speak out boldly, "and the freer and the more extensive your strictures are the more thankful I shall be. . . ."[77] The second letter expressed his pleasure with Young's *Tour in Ireland*.[78] He sympathized with Young's complaint that his services to Ireland had not received proper recognition, but as an Irishman he pointed out that Irish meanness and illiberality could be explained partly by the treatment which Ireland had received from England in times past. Both letters were signed, "Yours affectionately."

If the early years after his return to Bradfield, 1779–1783, were relatively uneventful and unimportant ones in Young's life, such could certainly not be said of the three succeeding years, 1784–1786. As a result of the deaths of his mother in 1785 and his brother in 1786, the whole Bradfield estate became completely his. Foreign students of English agriculture continued to visit Bradfield, more Russians in 1784, a Polish count in 1786, and most important, the two sons of the Duc de Liancourt and their tutor, Maximilien Lazowski, in 1784. Elections to learned societies piled up like honorary degrees in the twentieth century— the Odiham Society[79] and the Royal Society of Agriculture in

[75] Add. MSS. 35,126, ff. 252–253.
[76] Young, 1898: pp. 113–114.
[77] *Ibid.*, p. 115. For Barry's mural see fig. 11.
[78] *Ibid.*, pp. 117–118.
[79] Add. MSS. 35,126, f. 287.

Paris[80] in 1785, the Patriotic Society of Milan and the Geographic Society of Florence in 1786.[81]

Unquestionably, however, the most important development of this period was the publication of the *Annals of Agriculture,* which first appeared in January, 1784. This periodical publication, which ran to forty-six volumes before it finally ended in 1815, certainly constituted one of Young's outstanding contributions to the progress of English agriculture and surely did more than anything in this period to enhance his reputation. It must also have consumed a very large proportion of his time and energy. It is a veritable mine of information for the agricultural historian as well as for Arthur Young's biographer. It was read religiously both by George III and George Washington.[82] At least a quarter of the *Annals* came from Young's own pen, much of it of a potboiling nature, but much also of considerable value.

Young did not begin the *Annals* in 1784 impulsively. As early as 1770 he had proposed in *Rural Oeconomy* a periodical publication devoted to agriculture.[83] In 1771 he corresponded on the subject with John Wynn Baker,[84] and apparently the two men made proposals to several booksellers somewhat later,[85] but nothing came of the plan. Young's residence in Ireland of course postponed any such scheme. The coming of peace in 1783, when men were again thinking of national problems, seemed an opportune time to start such a venture. Young's preface stated that he intended to publish whenever he had sufficient material. If the amount warranted, it would appear every month. The size of each number might vary and the price proportionately.[86] He invited correspondence but from the beginning was very loath to take any unsigned articles.

The new publication had a hard time getting established. At the end of the first volume he published a brief financial statement.[87] The first volume had consisted of five numbers. He

[80] *Ibid.,* f. 311; Young, 1898: p. 135.
[81] Young, 1898: p. 143.
[82] *Ibid.,* pp. 112, 132; *Letters from . . . Washington to . . . Young . . .,* 1802: pp. 9, 10, 11 and *passim.*
[83] Young, 1770: pp. 365–371.
[84] *Annals* 1: pp. 89–90.
[85] *Ibid.,* p. 6.
[86] *Ibid.,* p. 8.
[87] *Ibid.,* pp. 461–467.

had optimistically printed one thousand copies of the first two numbers, but had sold only five hundred of each. Consequently he printed only five hundred copies of the remaining three numbers, but sold only four hundred copies of number 3, and three hundred each of the last two numbers. At the end of volume one, the debit exceeded the credit column, and Young exclaimed bitterly: "If I was to indulge my own feelings only, I would drop the work, and forswear ever taking a pen in hand again, on a subject evidently of no importance in the opinion of the public."[88] In fairness, however, to "several very ingenious gentlemen" who had promised their support, he decided to publish another volume. From the beginning he had disclaimed all idea of personal profit and had promised that increased patronage would mean reduced prices for the subscribers and a larger number of plates.[89] He was willing to continue the project as a public service so long as it paid its way. Volumes two and three went better and it was necessary to reprint some of the early numbers. A second accounting at the end of volume three showed a slight profit which would be completely absorbed, however, by further reprinting expenses. Since his publisher protested against this public accounting as very irregular, there were no more statements of accounts.[90]

The first five volumes were published by H. Goldney of Paternoster Row in London, an inconvenient arrangement which necessitated many trips to London. With volume six Young arranged to have the work printed by J. Rackham of Bury St. Edmunds and sold by one of his old publishers, William Nicoll of London.

From the beginning Young was dependent to a considerable extent upon material from his own pen. He wrote the entire first number himself. Outside contributions came slowly in increasing numbers. An analysis of the first six volumes which cover the years 1784–1786 shows nearly ninety contributors, more than half of whom wrote only one article each.

At first Young relied heavily upon his Suffolk friends and neighbors, and hence the early volumes contain a disproportionate amount of Suffolk material. Very few noblemen contributed dur-

[88] *Ibid.*, p. 464.
[89] *Ibid.*, pp. 7, 462.
[90] *Ibid.*, pp. iv, 249–252.

ing the early years, and Young was rather piqued at their neglect: "I simply state it as a misfortune to the community, that only two of the nobility have condescended to lay before the public any particulars of their agriculture."[91] The contributors in general fell into three main groups, the country gentlemen of Young's class, the Anglican clergy, and the substantial practical farmers, whom Young always distinguished by "Mr." instead of "Esquire." Very few of the contributors in the early volumes were men of national reputation. The contributions of only three men will be covered at this point, one country gentleman, one clergyman, and one practical farmer.

At Young's suggestion, his recent newspaper antagonist, Capel Lofft, contributed three articles, two translations from classical authors and a statistical account of Troston. The articles show a surprising cordiality between the former controversialists. In his first article Lofft wrote, "I confess ambition to occupy a page or two in the Annals of Agriculture,"[92] while Young responded by thanking "my worthy friend."[93] In the Troston article Lofft betrayed a romantic love of nature and described the size of certain trees which he had named for famous people, thus Linnaeus's fir, Shakespeare's mulberry, and Evelyn's elm, and exclaimed, "Oh, that you could find me, or that I could discover a situation and a tree sufficiently picturesque to be Rousseau."[94]

At the end of volume one Young appealed for articles from "the resident clergy who farm their glebes, and are by tythes or compositions, supported by that public, who have a claim for their exertions to serve it in return by something more than preaching."[95] Most prolific of Young's clerical neighbors was Rev. John Symonds who wrote nine articles for the first six volumes. One was a rather heavy-handed satire upon turnip culture in France,[96] and a second a brief account of experiments with different fertilizers upon Symonds's lawn.[97] The others dealt with Italian agriculture, describing at considerable length, and with all the apparatus of eighteenth-century learning, the irriga-

---

[91] *Ibid.*, p. 300.
[92] *Ibid.* 4: p. 75.
[93] *Ibid.*, p. 91.
[94] *Ibid.*, p 320. Lofft's three articles are in 4: pp. 73–91, 205–210, 305–320.
[95] *Ibid.* 1: p. 467.
[96] *Ibid.* 5: pp. 1–17.
[97] *Ibid.* 1: pp. 137–138.

tion systems, the soil, the chief crops, and the climate of that country.[98]

Perhaps the most valuable among the yeomen farmers was Mr. William Pitt (1749-1823) of Pendeford, Staffordshire, who contributed twelve articles to those early volumes. His comments were almost always to the point, and his English clear and forceful. He was the occupier, as he put it, "of a considerable rackrented farm."[99] and an ingenious craftsman who invented his own drill plow which he described with a plate,[100] as well as a machine for weighing swine.[101] He was an enthusiast over turnips and burnet.[102] He opposed Young on a pet subject, the superiority of oxen over horses,[103] and disputed the charge of a gentleman farmer that common farmers were ignorant, obstinate, and unwilling to profit by new ideas.[104]

The most spirited controversy in the *Annals* in the early years concerned fallowing. Was it necessary every few years for the land to remain without a crop and subject only to plowing and cleaning, or could certain "fallow" crops—turnips, beans, carrots, and clover—be substituted for the complete fallow? The storm center of the controversy was Young's neighbor, the Rev. James Fiske of Shimpling who categorically opposed fallowing and wrote with enough warmth to arouse hostility. John Plampin of the same parish was on the other side, as was perhaps the majority of the participants in the discussion. At first the arguments were mild and temperate, but when Fiske was attacked he replied with invective which was answered in kind. One correspondent wrote, "The houses of the Montagues and Capulets seem to be reviving under the respective characters of the FALLOWIST and ANTIFALLOWIST."[105] Young himself, of course, opposed fallowing and often added footnotes or comments, stating his position with more heat than was becoming in an impartial editor.[106] At length one article attacked Fiske in such bitter

[98] *Ibid.* 1: *pp.* 207–219, 405–436; 2: *pp.* 195–216, 254–280; 3: pp. 15–46, 137–166; 5: pp. 317–348.
[99] *Ibid.* 3: p. 192.
[100] *Ibid.* 5: pp. 89–99.
[101] *Ibid.* 6: p. 231.
[102] *Ibid* 1: pp. 393–397; 3: p. 131.
[103] *Ibid.* 4: pp. 423–430; 5: pp. 465–468.
[104] *Ibid.* 4: pp. 472–474.
[105] *Ibid.* 5: p. 440.
[106] *Ibid.* 3: pp. 100–101, 326–329; 4: pp. 276–279; 5: pp. 99–112, 471–476.

terms[107] that Young was forced to call a halt to any further invective, warning "that in future I shall strike out of their letters
whatever tends in the least to ridicule or asperity; controversy may
be useful, but it must be kept within the bounds, not only of
decency, but of politeness."[108]

To the biographer of Arthur Young, his own contributions to
the *Annals* are naturally most valuable, although much of what
he wrote is not important from any point of view. In the early
numbers he inserted accounts of experiments made ten years
before and of tours taken eight years previously. At times he
even resorted to translations of foreign works. Nevertheless many
of his contributions contain considerable merit, and at any rate
an examination of the more important ones is indispensable for
a biography. His major writings in the *Annals* fall into three main
categories. In the first place, there are the journals of his many
trips, which will be used for an account of the year in which they
were made. In the second place, there are the accounts of his
agricultural experiments, the most important of which have
already been analyzed. And in the third place, there are his
miscellaneous articles which cover almost every subject con-
ceivably related to agriculture.

The entire first number of the *Annals* was a general essay by
Young on the state of the nation at the end of the American
Revolution.[109] Perhaps his main thesis was that imperialism was
a bankrupt policy, a costly failure. It had involved England in
two very expensive wars, the latter resulting in the loss of the
American colonies. Furthermore, he predicted that England
might lose her remaining colonies.[110] Canada and Nova Scotia
were not worth keeping, but would probably become independent if they ever became prosperous. The sugar islands might
well fall eventually into American hands. He vigorously condemned Britain's conquests in India, the eventual loss of which
was certain:

It is impossible to hold countries that we treat as we have done our
Indian dominions. A people plundered without shame or restraint,
in order to give us every two or three years the detestable exhibi-

[107] *Ibid.* **5:** pp. 354–359.
[108] *Ibid.*, p. 359.
[109] *Ibid.* **1:** pp. 9–87.
[110] *Ibid.*, pp. 15–19.

tion of men grown great by their flagitious extortions, is but the
repetition of a memento of our future ruin . . . the loss of India!
That day *must* come.—It OUGHT to come—If there is a ruling
Providence that oversees the conduct of nations, . . . we MUST be
driven out of India with abhorrence and contempt, and all the
people of the globe would rejoice at the event.[111]

Even Ireland was already half independent, and in the next
war might either remain neutral or combine with France. An
independent Ireland "would be a monster in politics,"[112] and he
demanded political union as the only answer consonant with
British security.

Young was not despondent about Britain's future, however,
for her trade and manufactures were flourishing, her population
increasing, and her public finances in an essentially sound condition.[113] The great thing was to improve the wastelands. He
analyzed in detail how much more profitably capital could be
employed to bring them into cultivation than to develop sugar
islands, the most lucrative form of colonial enterprise. He definitely proposed[114] that land should be distributed in ten-acre
lots to the poor. The state should build and furnish a cottage
and should provide fencing, seed, and cattle. The occupier should
be given a lease for three generations free from tithes and all
parish charges. In return he should renounce all claims for poor
relief. In making the allotments, preference should be given to
war veterans and holders of pasturage rights on commons suitable
for enclosure. If only he were King of England, exclaimed Young,
so that he might initiate such a beneficial scheme, "making barren
desarts [*sic*] smile with cultivation—and peopling joyless wastes
with the grateful hearts of men . . . rearing the quiet cottage of
private happiness, and the splended turrets of public prosperity."[115]

Such was Young's Utopian vision, but he knew that the chances
for its consummation were slight indeed. Hence he made more
modest proposals for public improvement, which would cost
little if anything. The costs of enclosure should be reduced by a

[111] *Ibid.*, p. 17.
[112] *Ibid.*, p. 22.
[113] *Ibid.*, pp. 26–46.
[114] *Ibid.*, pp. 54–67.
[115] *Ibid.*, p. 61.

general enclosure bill to standardize and simplify the process.[116]
Tithes were impediments to private initiative and he urged a
general commutation.[117] The poor law system encouraged idleness and vice,[118] but the establishment of hundred houses of
industry would be an improvement.

Young also urged an improved educational system for the
gentry and nobility. In the first place, several colleges in each
university should be devoted to agriculture and some crown
lands assigned to them for experimental purposes. In the second
place tutors for the grand tour should be conversant with
agriculture, industry, and commerce. The provinces should be
visited before the capital, so that the young man would have a
general knowledge of a country before he met the important
people in the capital.

Go through Europe on these principles, and on your return take
your seat in the House of Lords or of Commons: you will enter the
lists with others that have made the same tour, and who have drank
deep of a very different knowledge—skilled in the fine arts—criticks
of women—adepts in all the fashionable inanity that graces the page
of a late travelling tutor—if you do not outstrip your competitors
in the race of excellence—of importance—of ambition—of MAN—go
hang or shoot yourself—the sooner you are damned the better.[119]

While discussing wastelands, Young had referred to his hopes
as a young man to superintend a government project for agricultural improvement, but declared: "I am now much older, and
I hope too wise to imagine, that I have any talents that can ever
be of use beyond the limits of my own farm."[120] There is some
bitterness and disillusionment here, and it is hard to believe that
he had completely abandoned such hopes. The insatiable ferment
of his mind was revealed when he had wished that he were king
so that he might execute such a scheme, and later when describing
his ideal grand tour, "If I travelled with a young man . . ."[121]
At the very end, he pointed out that failing government action,
the individual owner and farmer could make his own improve-

[116] *Ibid.*, p. 72.
[117] *Ibid.*, pp. 73–74.
[118] *Ibid.*, pp. 74–76.
[119] *Ibid.*, p. 83.
[120] *Ibid.*, p. 67.
[121] *Ibid.*, p. 82.

ments. The purpose of the *Annals* would be to spread the knowledge, to act as "a universal office of intelligence." In that spirit, "I make the offering of my mite."[122]

In general, Young's ideas in these years seemed to be moving steadily towards *laissez-faire* except where the farmer's interests were concerned. Whenever the merchants or industrialists urged a restrictive policy which would hurt the landed interests, Young employed unanswerable *laissez-faire* arguments against them. Unfortunately he was inconsistent and always discovered reasons why *laissez-faire* principles could not be employed when they might injure the farmer.

Young never tired of attacking the monopolistic aims of the manufacturers and merchants whose narrow-minded views had caused nearly all of England's recent wars. He castigated the East India Company as a noxious monopoly,[123] and opposed the establishment of the Chamber of Manufacturers in 1785 as tending towards monopoly.[124] He favored the relaxation of impediments to free trade, notably in relations with Ireland and France. He pointed out repeatedly that no industry unable to meet foreign competition deserved to survive and emphasized that free trade would mean lower prices to the consumer. The following excerpt from the *Annals* is an almost classic statement of the free trade position:

But till that simple proposition is understoood and admitted, *that there is an advantage in our own fabrics being undersold by foreigners:* till that day comes, all our commercial conduct will be blind. . . . if a frenchman, at Bury, will sell me a shirt better and cheaper than . . . a scotchman, or a Devonshire man, it is certainly my interest to buy it of him. By that cheapness I am enabled to consume so much the more of other commodities; and the capital and industry misemployed in making *dear* shirts, may be turned to making *cheap* hardware and woollens.[125]

He was particularly bitter against the exclusion of French wines in favor of Port wines because he considered French wines better, because the prohibition encouraged smuggling, and because it

---

[122] *Ibid.*, p. 87.
[123] *Ibid.*, p. 119.
[124] *Ibid*. 3: pp. 452–455.
[125] *Ibid*. 4: p. 120. For other *laissez-faire* statements *cf. ibid*. 2: pp. 402–411; 4: pp. 16–28.

reduced the sale of English goods in the French market.[126] He attacked the old mercantilist view that one country prospered only to the extent that its neighbor suffered, and maintained that reciprocal trade enriched both countries.[127]

Young was bitter against the attempts of the manufacturing and commercial interests to prohibit the export of agricultural produce. All such proposals were attempts to monopolize the home supply of raw materials in order to lower the farmer's prices. The export of wool had been forbidden for some time and in 1784 the hat manufacturers attempted to prevent the export of rabbit fur.[128] In 1786 new proposals were made to enforce the prohibitions upon the export of wool, minute regulations designed to prevent smuggling and supposedly to improve the quality of the wool sold by the farmer to the wool merchant.[129] Young condemned "so atrocious an insult on the landed interest."[130] He called upon all the wool raising counties to rouse themselves as Lincolnshire had done under Sir Joseph Banks and Sussex under Lord Sheffield.[131] "Regulation may destroy, but it can never make commerce; and this kingdom has grown great, not *by* her numerous restrictions, but *in spite* of them."[132] The proposals of the wool merchants boiled down to this: "We will buy your wool at a *low* price, and we insist on selling you your cloaths at a *high* one."[133] Young summed up his indignation in one outburst: "If this profligate attempt on the liberty, property, and feelings of the whole landed interest takes effect, the man who keeps a sheep in his possession, must truly want the idea of independence, that *is* the birth right, and *ought to be* the glory of an englishman."[134]

Actually Young was a special pleader for the agricultural interests. When restrictions upon free trade benefited the farmers, Young forgot about *laissez-faire*. At the end of the American Revolution, the West Indian islands wished to buy some food-

[126] *Ibid.* **4:** p. 19.
[127] *Ibid.* **5:** p. 421.
[128] *Ibid.* **2:** pp. 12–17.
[129] *Ibid.* **6:** pp. 506–528.
[130] *Ibid.*, p. 521.
[131] *Ibid.*, p. 527.
[132] *Ibid.*, p. 521.
[133] *Ibid.*, p. 526.
[134] *Ibid.*, p. 520.

stuffs from the American states. In a very favorable review of a pamphlet by Lord Sheffield opposing such a relaxation of trade restrictions, Young insisted that the Navigation Acts be applied against the Americans to insure a monopoly of the West Indian markets to English food producers.[135] Certainly the following is hardly consistent with his tirade against the monopolistic manufacturers:

. . . ardently must every lover of his country hope that those sound principles of navigation, the great foundation of our naval power . . . will get the better of every private and local interest. To supply the West Indies with grain, flour, bread, and all sorts of provisions, ought immediately to be secured exclusively to Great Britain.[136]

In the same review an interesting footnote on Adam Smith again exhibited Young as the plumed champion of the landed interests:

I hardly know an abler work than Dr. Smith's, or one (it is no contradiction) that is fuller of more pernicious errors: he never touches on any branch of rural oeconomy, but to start positions that arise from mis-stated facts, or that lead to false conclusions.[137]

Young was likewise unwilling to accept a completely *laissez-faire* position on the Corn Laws. He admitted that complete free trade in corn was desirable, but declared that every proposal to introduce free trade in corn was brought forward only to deceive the agricultural interests into accepting free imports, for in time of scarcity free exports would never be permitted. His position was well summarized thus:

[Could an absolutely free trade in corn] be positively ensured in all cases whatever, it would be of all others the best regulation. But when prices arose high, petitions would flow in to stop exportation, and such a system would end in this—a free importation, a regulated exportation, and the loss of the bounty to the landed interest.

Such a measure, therefore, should be rejected at the first blush, for it will never be brought forward but as a trick; by men who know, that while they gained a freedom of import, they would soon have an opportunity of destroying that of export.[138]

[135] *Ibid.* 1: pp. 369–388.
[136] *Ibid.*, p. 376.
[137] *Ibid.* 1: p. 380. It is only fair to state that in nearly all his later references to Smith, he was very laudatory.
[138] *Ibid.* 3: pp. 467–468.

Always interested in taxation problems, Young devoted considerable attention to them in the early volumes of the *Annals*. He still believed that excises were the fairest taxes and he still fought every attempt to increase the burden on the agricultural classes. Naturally he bitterly opposed three proposals of this period, the first to abolish the tea tax while increasing the window tax, the second to introduce a tax on tiles and bricks, and the third to substitute a hearth tax for the salt tax.[139]

The lowest classes were exempt from the window tax, while the wealthiest classes hardly felt the burden, which would thus fall most heavily upon "the country gentlemen of small estates,"[140] the men most likely to make agricultural improvements and the very backbone of the country. Already the country gentleman was disproportionately burdened, "his land trebly taxed to church, king, and poor; his house; his windows; his horses; his servants; his sports. . . ."[141] It was especially unjust because this burden on the gentry resulted from wars fought for the mercantile interests, who then largely escaped the necessary taxes to pay for them. He condemned the retention of Gibraltar and continued bitterly:

Cannot our merchants and our mercantile leaders in parliament find some vile rock—some Falklands isle, or Pierre, or Miquelon— some cursed perwannah on the Ganges, or blackened isle that bleeds for sugar—on which to erect their pretentions to future wars? . . . A merchant slips away from taxation; he eludes the statesman's gripe; even your iron fangs Mr. Pitt will not hold him; but the landlord is an easy prey . . .[142]

His prejudice against tea drinking by the masses embittered him the more, for the government proposed to "change . . . the very best mode of taxation for the very worst . . . you demand an enormous one upon his [the gentleman's] windows, . . that his poor neighbors may sip, not their honest ale, but their execrable tea, free from every burden."[143] The same point was made again later: "Cottagers pay nothing to that window tax, yet their inhabitants are great consumers of tea; and there is not

[139] *Ibid.* 2: pp. 301–313; 3: pp. 399–411.
[140] *Ibid.* 2: p. 302.
[141] *Ibid.*, p. 306.
[142] *Ibid.*, pp. 307–308.
[143] *Ibid.*, pp. 311–312. He also proposed in this article to abolish the "rascally game laws, those remnants and rags of feudal barbarism and slavery," and to tax every one that killed or sold game, thus raising a considerable revenue.

in the range of taxation any objects more proper than the *luxurious* consumption of the poor."[144]

The proposed tile and brick tax would raise building costs and hence discourage population. To escape the tax upon tiles, the rural laborers would employ thatch, a practice which "robs the dung hill incredibly" and increases fire hazards. Better to tax thatch or give a bounty for substituting tiles for thatch.[145]

The proposal to substitute a tax on hearths for the salt tax would be "most ruinous to the whole landed interest."[146] He restated his preference for excise taxes with considerable vigor:

Now taxes upon consumption, being blended with the price of the commodity, are paid without being known or felt; he who wishes to consume a bottle of wine, or a pound of salt, knows the price; and if that price, including the tax, is too high for him, he can avoid the whole by desisting from the consumption.[147]

No statesman would dare to place a direct tax on the lower classes and hence the hearth tax would also exempt them. Young suspected that the proposal aimed to relieve the Scottish fisheries which consumed large quantities of salt. Thus it would benefit one commercial interest by greatly increasing the burden upon the small landed proprietor. He calculated what the change would mean to him personally. His family of ten in 1784[148] consumed 125 pounds of salt upon which they paid 11s. 3d. tax. On the other hand, his house had sixteen hearths, and if the proposed hearth tax of 2s. per hearth were adopted he would have to pay £1 12s. "That is to say, an advance of exactly 150 per cent!"[149]

Young also engaged in a spirited controversy in the *Annals* with his friend Lazowski on taxation problems. The whole argument was conducted with great vigor, but with complete good will and friendliness. In a review of Necker's *De l'administration des finances de la France,* Young had contended that the

---

[144] *Ibid.* 3, p. 404.
[145] *Ibid.* 2: pp. 314–315.
[146] *Ibid.* 3: p. 400.
[147] *Ibid.*, p. 402. For similar views expressed a decade earlier, *cf. supra,* p. 89.
[148] In 1784 his immediate family consisted of himself, wife, mother, and four children. Hence his house servants were presumably three. The old, rambling house probably needed the sixteen hearths, for it was always cold and drafty. *Cf.* Young, 1898: p. 13.
[149] *Annals* 3: p. 406.

ability of a nation to pay taxes was proportionate to its freedom. He compared the per capita taxation in France and England, estimating that of the former country at twenty shillings and that of the latter at thirty-seven shillings. Waxing enthusiastic he continued:

But what a great and glorious spectacle of a people does this contrast furnish. Twenty shillings a head excessively burdensome to a nation enjoying the greatest natural advantages of any society in the world, and thirty-seven shillings, comparatively speaking, but as a feather on this unshackled community, where every man follows his own business in the manner he thinks best for his interest![150]

When Lazowski protested against this section of Young's review he had been invited to write a letter to the *Annals*.[151] Lazowski repeatedly praised Young's writings in general and the *Annals* in particular, but in this case he accused Young of prejudice and also took issue with Young's statistics: "I am more astonished that you could be patient enough to calculate at all," and continued a little later, ". . . if you intended to be eloquent, you must be contented: but you are eloquent often enough, you must also be accurate in political disquisitions."[152] Throughout the letter Lazowski implied that Young too often permitted his British prejudices to lead him into unwarranted aspersions upon France. For instance Young had referred to one publication of the French government as "execrable stuff"[153] and to another as "nonsense."[154] He warned Young that Frenchmen reading the *Annals* will never believe that you have much love for them."[155]

In the following number Young made almost no concessions, but took pains to explain that he had meant that English taxes were "a feather" to the lower classes but were far more onerous on the middle and upper classes. He compared in some detail the class structure of the two countries. He admitted that France was wealthy but maintained that the wealth was not sufficiently diffused.

[150] *Ibid.*, p. 510.
[151] *Ibid.* **5**, pp. 70–88.
[152] *Ibid.*, pp. 73–74.
[153] *Ibid.* **4**: p. 436.
[154] *Ibid.*, p. 523.
[155] *Ibid.* **5**: p. 87.

. . . the principles of the government keep the lower classes poor in the midst of all this wealth; and my friend knows very well that it is not the luxury of the few, but the ease and the consumption of the great masses of the people that gives facility to the payment of taxes.[156]

He attacked the "principles, ideas, and prejudices that . . . tend to render industry dishonourable" and "to induce such a general eagerness for employments that confer nobility."[157] The situation was exactly contrary in England: ". . . in England, no man would wish to move from an inferior into a higher class on any motive that relates to taxes; . . . since the lower he decends the more he is in that respect favoured."[158] France was an excellent country for a gentleman or a nobleman, but not for the farmer or peasant. Young's later approbation of the early course of the French Revolution is not surprising when one finds him writing in 1786:

Away with your barbarous ideas that annex honour, and respect, and reward to the classes the most useless of a state! Give ease, and comfort, and consideration to your ragged and oppressed peasantry; let your farmers feel that they have rights on which they can depend, even in the presence of a duke. . . . Let them feel themselves to be men; let them know and depend on their rights; and then leave all to the unshackled efforts of that industry which labours for its own reward. It will cultivate, improve, adorn your kingdom: it will render you great, and your whole people happy. Your revenue, instead of being wrung with difficulty from the hard-earned pittance of starving families, will roll, in copious streams, the overflowings of national prosperity.[159]

As noted above, Young was disturbed by taxation pressures upon English country gentlemen, some so desperate as to consider emigration to America. Emigration might be the answer in some cases, if the individual was willing to abandon "the lustre that elegant society disseminates, the diffusion of knowledge, the graces of literature, the pregnant conversation inspired by taste, and decorated by politeness."[160] Some of the pleasures of a cultivated society might be retained, however, if several families of

[156] *Ibid.*, pp. 173–174.
[157] *Ibid.*, p. 172.
[158] *Ibid.*, p. 177.
[159] *Ibid.*, p. 179.
[160] *Ibid.* **3:** p. 175.

friends emigrated together and took contiguous lands in America. It would be more rational and require less capital for the poor gentleman to abandon his gentility and to become a farmer. True, the "tinsel" aspects of society would be lost, but not any solid tastes in literature, reading, and music. He would only have to renounce his vanity, for fortunately the English farmer suffered no oppression nor even neglect, "not one circumstance in our civil or political liberty that should make a man of sense hesitate ere he engaged in it." Such could be said of no other European country. "Be this distinction the glory and pride of an english-man."[161] The autobiographical article in the *Annals*[162] shows that Young considered such a voluntary change in his own social position, but apparently his own vanity was too great.

Fortunately Young seldom attempted fine writing and composed only one elegant essay in the early volumes of the *Annals,* "On the Pleasures of Agriculture." For the most part his style here was stilted and lacking in spontaneity. In the time-honored tradition of the elegant essay Young studded his with quotations from ancient and modern authors, from Xenophon to Shakespeare to Rousseau. This essay argued not the usefulness of agriculture, but only its pleasures: ". . . let us sue for the favours of a mistress, not the duties of a wife."[163] It was addressed to the gentlemen and rural clergy whose normal amusements were stultifying to the mind, and whose "routine of innanity [*sic*]"[164] was humdrum enough:

Each day brings its breakfast, its dinner, its coffee, and supper; if company comes in, there is a little more eating, a little more drinking, and a rubber or two extraordinary: the servants are orderly; the liveries new; the horses sleek; the equipage clean; and an airing wholesome; so many times a week the news in a coffee-house.[165]

True there were hunting and gaming! And there was no lack of amusements for the young, but when middle age came on the country gentleman might fall victim to the "two vices in the soul not uncommon in age, avarice, and a contracted perversion

[161] *Ibid.*, p. 178.
[162] *Ibid.* **15**: p. 163.
[163] *Ibid.* **2**: p. 457.
[164] *Ibid.*, p. 485.
[165] *Ibid.*, p. 479.

of religion, neither of which could find place in a mind busied in any better pursuit."[166] On the other hand agriculture was one pleasure eminently suitable for middle and even old age. Expense was no obstacle for it might be pursued on twenty as well as a thousand acres. To be sure there was some risk, but surely less than in hunting or gambling. If the gentleman spent three years in learning the rules of the game, and if he applied a capital of £5 per acre, he might expect to find agriculture as "safe to his purse as it is grateful to his imagination." Agriculture was also an all-around-the-year pleasure, especially if a laboratory for conducting experiments were built, the expense of which would be less than that of a "tolerable hunter" or a "brace of pointers."[167] Young also appealed to the ladies to support agriculture as an amusement, and portrayed the gentleman farmer as the ideal husband, ready to plant "a rose or jessamine," or to fix a bench "for viewing the beauties of the landskip, . . . and around it the little pledges, which the same bounteous nature has given to your love . . . mingle their little sports. . . ."[168] Bathos and sensibility!

As a publicist and propagandist Arthur Young approached greatness. He possessed the enthusiasm, the flair for writing, and the necessary contacts. He was especially impressive as a publicist for scientific agriculture. In one article he urged all sovereigns to establish an experimental farm. Every prince built palaces and parks; the establishment of an experimental farm would be an unique achievement, entitling its patron to be called "THE FRIEND OF MANKIND."[169]

A more important article analyzed why agriculture was so much more backward as a science than chemistry or physics.[170] Agricultural experiments often extended over several years and were very expensive. It was very difficult to control all the pertinent factors in an agricultural experiment. Since the ultimate aim of agriculture was profit, nothing was gained by great crops or fine animals if the expenses exceeded the profits. Agriculture also lacked any precise vocabulary. Such terms as sand, marl, clay, were all used loosely. In this article Young also listed some

---

[166] *Ibid.*, p. 468. This quotation certainly exhibits none of the excessive piety and religious melancholia of Young's later years.

[167] *Ibid.*, pp. 485–486.

[168] *Ibid.*, p. 482.

[169] *Ibid.* 4: p. 466. The plan included a small botanical garden and a laboratory.

[170] *Ibid.* 5: pp. 17–46.

of the more important points for experimental study as follows: the reclamation of peat moors and poor blowing sands; the proper crops to banish fallows on heavy stiff soils as turnips had done on sands; the value and proper culture of carrots and potatoes; the best course of crops for various soils; the best breeds of cattle for various purposes; the chemistry of soils and the food of plants.

The *Annals* contain many short articles by Young of a semi-scientific character. His book reviews were too apt to be too sharp. Thus he described one paper in the *Transactions* of the Society of Arts as "the most ridiculous piece of prejudice and absurdity ever published."[171] Thus he classed a work by Thomas Stone, a distinguished agricultural surveyor, "with those which add nothing to the present stock of knowledge"[172] and went on to fill two pages with specific mistakes. Small wonder that Stone became one of Young's chief enemies. More valuable was his description of his own drill plow with a plate.[173] He also published a series of "queries" to his correspondents "On Sowing Wheat," fourteen questions including the time of sowing, how the land is best prepared, the steeps used, the kinds and amount of seed.[174] This particular set of questions did not bring much response, but later Young frequently sent out regular printed questionnaires to a designated list of correspondents.

The *Annals* contained several regular features which increased its value to the farmer. Book reviews, which appeared in the first volume, were not found in every issue, but were quite regular. In the second volume Young began to include tables upon the monthly price of corn by counties and upon the state of foreign exchange.

Although Young did not participate actively in the Society of Arts in 1784, he sent a copy of the *Annals* to the Secretary, Samuel More. He also proposed that the Society should give a premium for an iron oven for baking potatoes for cattle. Young suggested an oven able to bake from twelve to twenty bushels at a time, and specified that "simplicity and cheapness in the construction will be considered as principal parts of its merit."

[171] *Ibid.* **2**: p. 324.
[172] *Ibid.* **4**: p. 353.
[173] *Ibid.* **3**: pp. 55–62.
[174] *Ibid.* **4**: pp. 135–137.

The Committee on Mechanics, to which the proposal had been referred, rejected it, however, because the expense "would be so great as to render it not likely to come into use for the purpose intended."[175] In one of his letters to Secretary More about the *Annals* Young added a humorous note:

In answer to ye Annals yo quote "I would go to ye H. of Commons," whereas I say, "I would *send* to ye H. of C." Have you lived 40 yrs. in Lono & not found out yt a fair lady can make yo a present when ye *go* to her, fro wch yo will be very safe if you only *send*. if you cannot find this out, there are abundance of nymphs that instruct you.[176]

The most interesting event in Young's personal life in 1784 was the visit of François and Alexandre de la Rochefoucauld, the two young sons of the Duc de Liancourt (1747–1827) to Bury St. Edmunds. To complete their education the Duc sent them with their tutor, Maximilien de Lazowski, and two servants to England to learn the language and to gain a knowledge of English life. According to François they were persuaded by Horace Walpole to go to Bury St. Edmunds,[177] but Young stated that Lazowski's interest in agricultural economics led him to Bury as the town nearest to the estate of the leading English authority on agriculture.[178]

The party arrived at Bury early in January and went directly to the Angel Hotel. Although Young was away at the time, they soon met his close friend, Dr. John Symonds, who immediately took them under his wing and brought them to Bradfield shortly after Young's return. Thus began an acquaintance with momentous consequences in Young's life. He left no account of the two young noblemen, but François penned an interesting vignette of him:

He is a man of parts and of wide acquaintance and, what is still more attractive from the social point of view, he is extremely good-humoured, always gay and never out of spirits. He answers your questions in the most charming way and always seems eager to satisfy you.[179]

[175] Committee Report Books, 1783–1784, f. 230.
[176] MSS. Letter at Society of Arts.
[177] de la Rochefoucauld, 1933: p. 21. This little book is based on François' account of the trip for his father, British Museum Add. MSS. 35,108.
[178] Young, 1898: p. 120.
[179] de la Rochefoucauld, 1933: p. 21.

Young's attention was taken especially by Lazowski, the tutor. A gentleman of Polish family, highly cultivated with charming manners, Lazowski had much in common with Young. Essentially from the same class background, they also had the same interests in agriculture and economics. They were both high spirited and both loved argument. They were both fond of society and happy in the companionship of the fair sex. Their mutual encomiums on each other are so much alike that they might almost have been interchanged. Young's comments on Lazowski were as follows:

From that time a friendship between me and Lazowski commenced, and lasted till the death of the latter. He was about forty years of age, and in every respect a most agreeable companion. He soon made rapid progress in the English language, which he spoke not only with fluency, but often with extreme wittiness. There was not in his mind any strong predominant cast, but the grace and facility of his manner, with suavity of temper, made him a great favourite, and being also highly elegant and refined, he often produced impressions which were not easily effaced. From his general conversation in mixed society it was not readily concluded that he could or would attend with great industry and perseverance to objects of importance. But this would have been erroneous, for he exerted the greatest industry in making himself a master of all those circumstances which mark the basis of national prosperity. . . .[180]

And now Lazowski on Young:

The more I live with him the more I am amazed at the quickness of his perception and the profundity of his knowledge in everything concerning agriculture. . . . His interest extends to all branches of trade, to the arts in general and to everything that comes under the head of political arithmetic. What may surprise you even more is that he combines his knowledge with a delightful sensibility. He is enthusiastic about the beauties of J. J. and of Voltaire and can shed tears over *Clarissa*. He is of a cheerful disposition, is always even-tempered and has displayed feelings of real friendship towards me.[181]

Although Young was surely hospitable towards his new friends, there is evidence that they preferred to meet him elsewhere than

[180] Young, 1898: pp. 120–121.
[181] de la Rochefoucauld, 1933: p. xxi.

at Bradfield. One of the few contemporary allusions to Mrs. Arthur Young was made by François:

In spite of all this, I do not enjoy paying him a visit, first because his table is the worst and dirtiest possible, and secondly on account of his wife, who looks exactly like a devil. She is hideously swarthy and looks thoroughly evil: it is rumoured that she beats her husband, and that he good-temperedly puts up with it. Whether this is true I do not know; if it is, it is the husband who is to blame, not the wife. She continually torments her children and her servants and is most frequently ill-tempered towards visitors.[182]

No such contemporary description of what Young had to endure exists except in Fanny Burney's diary of a much earlier period. No wonder he took frequent trips or sought softer female companionship away from home. Of course his flirtations only resulted in more family unhappiness. The *Autobiography* states that he introduced his friend Lazowski to the three beautiful daughters of John Plampin of Chadacre Hall, about three miles south of Bradfield. "I was intimately acquainted with them," wrote Young. Lazowski went there frequently and was much smitten with the youngest of the Plampin girls, very likely the beautiful Betsy who was also Young's favorite and who later, as Mrs. Orbell Ray Oakes, was his most cherished friend. She was only fourteen or fifteen in 1784 while Lazowski was forty and Young forty-two. The girls were flattered by the attention of the older man and the noble foreigners. "They persuaded their father to give a ball, at which the Duke of Liancourt, his two sons, Lazowski and myself were present, and the evening passed with uncommon hilarity till the rising sun sent us home."[183] There is no indication that Mrs. Young was of the party!

On April 21, 1784, Young probably entertained the Frenchmen at Bradfield and wrote the following letter of introduction for the Duke to his brother-in-law, Dr. Charles Burney:

Bradfield Hall
Ap. 21, 1784

Dr. Sir

The Duke de Liancourt at present on his travels in England being

---

[182] *Ibid.*, p. 38.
[183] Young, 1898: p. 154. All that Young said was that Lazowski was attracted by the youngest of the girls, but does not give her name. It seems unlikely that he would have fallen in love with a girl younger than fourteen.

desirous of an introduction to the most celebrated characters in this Kingdom, & having mentioned an inclination of seeing the author of Cecilia I am happy in the opportunity of introducing him to two in one house. In case he should call when Miss Burney is not at home I hope you will be kind enough to oblige me by informing her that I wish him to have some conversation with her; as I think fro. what I have heard him discourse on Works of Imagination in prose, that he has a very correct as well as cultivated taste.

I have not time to read what I have written ten people being talking half french & Half English around me.

I am ever very truly

Yrs.

A. Young[184]

Ever since their arrival in Bury, the closest associate of the French travelers had been Dr. John Symonds. Sometime in the spring he accompanied them to Euston to visit the Duke of Grafton and late in May took them to Cambridge which they enjoyed, but they were surprised that Symonds, "although he is a professor in one of the colleges . . . does not know how the teaching is conducted."[185] What a commentary on eighteenth-century university education! After their return from Cambridge they lived with Symonds at his beautiful home, St. Edmunds Hill. François was grateful for Symonds's never failing kindness. "All that he seemed to want . . . was to give us pleasure and to take all possible trouble about it."[186] There was an excellent library and Symonds corrected their translations and compositions himself. But he was much older than the young Frenchman and was pompous and formal. "Imagine the vanity of an author combined with the ponderousness of a man who falls half asleep after dinner and makes the company yawn by the length of the sessions at table!"[187] He locked up the house at nine o'clock and they could not take a walk after that hour. He held a reception every

---

[184] Burney Papers. These manuscripts are part of the Berg Collection in the New York Public Library. Apparently Fanny did not meet the Duke at this time, but she wrote in her father's Memoirs that Young brought the Duke's sons, Lazowski, and Symonds to call. *Cf.* Burney, 1832: **2**: p. 332.

[185] de la Rochefoucauld, 1933: p. 143.

[186] *Ibid.*, p. 154.

[187] *Ibid.*, p. 149.

Tuesday for tea and cards, but they were so formal that only the older people attended after the second or third time. Young, writing much later, gave a different impression of these occasions, but was twenty years older than François: "Mr. Symonds afterwards gave a weekly ball when the Frenchmen were with him, and these parties were uncommonly agreeable."[188]

The wanderlust was very strong on Arthur Young in the summer of 1784. The first trip, June 28-July 11, was through Essex and Kent with an excursion over to Calais. His companion was his son, Arthur, now fifteen and old enough to start training as an agricultural observer. On the way to Dover they visited several contributors to the *Annals* and made a side trip to Foulness Island off the Essex coast, where he found fine and large crops on a rich salt soil. His reactions to Kentish husbandry were mixed. He had no great opinion of hops: "I wanted no new information to convince me that hop profit is contemptible, or at least a mere lottery . . ."[189] On the other hand their management of beans as a preparation for wheat deserved the highest praise. The "round tilth," a crop rotation of barley, beans, wheat violated one of his chief maxims, "that no two crops of white corn ought ever to come together."[190]

". . . we went over to Calais just to enable us to say that we had been in France."[191] They went from Dover to Calais on July 5 and returned two days later. They put up at "Dessein's celebrated inn,"[192] which "may be reckoned perhaps without injustice the most remarkable thing in Calais."[193] He also succeeded in contacting M. Mouron of Calais, who "received us with the utmost politeness,"[194] and rode over Mouron's famous improvement of a salt marsh which had been conducted with equal spirit and profit. Young's reactions to France were not very favorable. Upon the whole the agriculture was greatly inferior to that of England. The people seemed contented enough, but lacking in "that vivacity and mirth of which I have heard and read so much." In examining men at work, "I thought I per-

---

[188] Young, 1898: p. 154.
[189] *Annals* **2**: p. 77. *Cf.* also Young, 1898: p. 124.
[190] *Annals* **2**: p. 71.
[191] Young, 1898: p. 124.
[192] *Ibid.*
[193] *Annals* **2**: p. 92.
[194] *Ibid.*, p. 81.

ceived a want of vigour and dispatch that could escape no one; perhaps it was prejudice."[195] Although French soldiers were much cleaner and neater than English ones, he was unfavorably impressed by the dominance of the military and characteristically observed:

The worst things that I saw were 4000 troops well dressed, clean, and spruce, and seemed so much superior to the other poor people, that one might collect from it, if other circumstances were wanting, that the government of this kingdom is arbitrary. I do not think there is any danger of the day coming when the character of a farmer shall be in France superior to that of a soldier, and till that day does come, as an Englishman, I have very little apprehension of the utmost efforts which this great, populous, wealthy, and potent country can make, to overturn the balance on which depends the emulation, the merit, and consequently the independence of the european world.[196]

Such was the kind of gratuitous insult to the French which had so naturally irritated Lazowski. Yet two pages later Young was supporting closer trade relations between the two countries:

. . . great, populous, industrious, and wealthy, they are formed to be the reciprocal consumers of each other's products and fabrics; our emulation in every great and liberal sphere that human genius and industry can move in, has at last taken place between them, and is every hour banishing that illiberal jealousy and hatred that once converted rivalry into mutual enmity, to the dishonor of both.[197]

And he apparently saw no contradiction between the two statements!

The return journey took father and son through western Kent, Tunbridge Wells, and Sevenoaks, where they stopped to inspect Knole. It is easy to imagine the father instructing his son in the canons of a "just taste." Young admired the park, the buildings, and the picture collection, especially a Van Dyke, "which is less a representation than life itself; the canvas is alive and speaks a language that will for ever be understood."[198] At the end of the trip Young maintained that Essex ought to rank as high agriculturally as the more famous counties of Kent, Suffolk, and Norfolk. The whole trip had covered 422 miles.

[195] *Ibid.*, pp. 88–89.
[196] *Ibid.*, p. 90.
[197] *Ibid.*, p. 92.
[198] *Ibid.*, p. 98.

Ten days later Young set off upon a five-day trip through Suffolk with Lazowski and his two young French charges. Young and Lazowski traveled in a hired cabriolet while François and his brother rode horseback. There are three independent accounts of this trip, one by Young in the *Annals,* an unpublished manuscript probably by Lazowski, and that by François in the *Mélanges.*[199] The tour aimed to combine general sightseeing with a study of the "high" agriculture of Suffolk. Everywhere Young interrogated farmers and prided himself that his companions would be "convinced of the real importance of attending to the agriculture of the countries through which they may have occasion to travel."[200] François did appreciate that he was making an agricultural tour with a man "who, in all England perhaps, was the most delightful and most useful companion."[201]

Their trip began very early Wednesday morning, July 21. The highlight of the second day out was Mistley, where the whole town was built and owned by Richard Rigby (1722–1788), whose career was a byword in the most corrupt period of English public life. Although the party arrived at the Hall quite early, they were shown over the whole house, and then Rigby drove them through the gardens and estate. François was especially impressed by the exotic trees, the fruit trees "with plenty of peaches and pineapples large enough to make one's mouth water,"[202] the deer and game in the park, and the views of the Stour estuary, with Harwich in the distance. The village with its neat white painted houses was admired, and the fine classical church by Adam led François to exclaim, "I have never seen anything more elegant than this building."[203] Young was not apt to be over friendly to politicians, but he enthused over Mistley:

A new town, of above 40 good brick houses, . . . an elegant church, built by Mr. Adams [*sic*]—an excellent inn—an extensive quay, faced with brick and stone . . . a ship carpenter's yard . . . these are objects that rank in a class abundantly superior to brilliant palaces, and gew-gaw gardens, and when the overflowings of a princely fortune are thus expended, never shall I regret that the service of the public was the source of the wealth thus admirably applied.[204]

[199] *Ibid.,* pp. 105–168; de la Rochefoucauld, 1933: pp. 157–194.
[200] *Annals* **2:** p. 167.
[201] de la Rochefoucauld, 1933: p. 194.
[202] *Ibid.,* p. 168.
[203] *Ibid.,* 164.
[204] *Annals* **2:** p. 111.

And yet Young must have been aware of how this "princely fortune" had been amassed! The same day they visited the cod fisheries and examined a fishing boat and two frigates under construction in the royal shipyard at Harwich. François's comment was interesting: "Some day, perhaps, we shall capture them."[205]

Agriculturally, the height of the tour came when they visited the "Sandlings," southeast of Woodbridge, a district of very light, sandy soil and famous for its culture of carrots and its breed of farm horses. Nowhere in England was the culture of carrots carried on with more vigor. All the farmers declared carrots superior to oats as food for horses. François was skeptical until he offered one to his horse and found that "he ate it greedily."[206] The soil was so loose through part of this country that the horses had great difficulty in pulling the cabriolet even though Young and Lazowski walked. The district also produced a distinctive breed of farm horses, sturdily built, sorrel in color, and with great ugly heads. Young had bought horses from this district for his own use in 1764 and 1776 and for his Russian friend Sambosky in 1779. François expressed his astonishment and admiration at the intelligence of the farmers of this district: "The knowledge which all these farmers possess is incredible—you must see them to realize how simple farmers can talk for an hour on the principles of their calling and on the reasons underlying their various forms of cultivation with a man like Mr. Young."[207] François never quite realized the fundamental differences between an English farmer and a French peasant.

On their last full day out, they visited Heveningham Hall, reputedly the most magnificient house in Suffolk. They arrived at nine o'clock on a Sunday morning and presented their letter of introduction from John Symonds to Sir Gerard Vanneck, the owner. No wonder that "He received us very coldly!" Nevertheless he invited them in, and they breakfasted with him and his sister. During breakfast "the conversation became less chilly every minute, Sir Gerard talked a little, and we realized how friendly he was towards us and how his first greeting had given us an entirely wrong impression of him."[208] After breakfast they were

---

[205] de la Rochefoucauld, 1933: p. 164.
[206] *Ibid.*, p. 181.
[207] *Ibid.*, p. 183. For this part of trip *cf.* also *Annals* **2**: pp. 130, 132.
[208] de la Rochefoucauld, 1933: pp. 185–186.

shown over the whole house, even the bedrooms and kitchen. Then Sir Gerard hospitably provided them with horses for a twelve-mile ride through the park. On returning to the house fine fruits were served to the visitors. Young enthusiastically declared: ". . . it was with great pleasure that I found there had at last arisen a structure in Suffolk deserving the attention of travellers . . . Those who make the Norfolk Tour, will now find it essential to take Heveningham in their way to that county."[209]

On the day they returned to Bury the weather was bad, and "had no mercy on us . . . it rained so much that, to tell the truth, we could hardly think of agriculture." Impatient at the slow progress of the cabriolet, François and Alexandre pushed on by themselves to Stowmarket where "we drank very hot punch, which quickly banished the cold induced by the rain."[210] Young was less disturbed by the weather and even stopped at Stonham Aspal "and took a hasty walk" over Mr. Toosey's "excellently cultivated farm,"[211] with its fine cabbages, Bakewell cattle, and black horses.

The French visitors had been so much pleased with their agricultural tour in Suffolk that they made a similar trip through Norfolk in the last week of September. For most of the trip Young's place was taken by Symonds. François did a good job describing the progressive agriculture of Norfolk, showing himself an apt pupil of his master. When they arrived at Lynn, they found Arthur Young awaiting them. According to François he had "kindly come to introduce us to Mr. Coke, the owner of the finest house in England, which we were anxious to see."[212] Surely a letter of introduction from Young to Coke would have sufficed for such a purpose. It is more likely that the arrival of the party at Lynn on September 29 was planned to coincide with the annual Lord Mayor's Banquet in the handsome old guild hall and that Arthur Young could not resist the temptation to present in person his aristocratic French friends to the notables of the town with which he had such intimate associations. François was duly impressed with the parade, the service presumably in St. Margaret's, the banquet where he was seated next to Lord Orford, and especially the picturesque drinking ceremonial from the

[209] *Annals* **2**: p. 149.
[210] de la Rochefoucauld, 1933: p. 193.
[211] *Annals* **2**: p. 155.
[212] de la Rochefoucauld, 1933: p. 238.

famous King John's Cup. The banquet was followed by a ball which "was certainly a brilliant spectacle," and where he met Mr. and Mrs. Coke, "who were good enough to invite us to Holkham." François broke off his account of the Norfolk tour abruptly "because we enjoyed ourselves too much at Holkham to have time for writing."[213]

It is almost certain that Young partook of the festivities which his friends enjoyed so much, but he also used the opportunity to take notes for a thirty-page article in the *Annals* entitled, "A Minute of the Husbandry at Holkham of Thomas William Coke, Esq."[214] Coke's endeavors on his farm of three thousand acres, four hundred in sainfoin, five hundred in turnips, and three hundred in barley, and which sustained 1,500 sheep and 150 cattle, were "of the most liberal and public-spirited nature."[215] Young pointed out that Coke was not satisfied to carry on the Norfolk tradition of high farming, but was making still greater improvements—the culture of sainfoin to supplement turnips as a food for sheep, and the substitution of various grasses when common red clover failed. Young also praised Coke's use of ducks to eat the black canker worms from turnips, and his practice of working his laborers on Sundays after they had attended church:

His men go to church in the morning, and then immediately to the field, where their useful and honest industry will, I trust, be found as acceptable in the sight of God, as the more common dissipation in an ale house kitchen; to say nothing of the drunkenness, broils, and gaming, which usually take place.[216]

Such praise for Sunday work was a far cry from Young's later rigid Sabbatarianism. In spite of high praise for most of Coke's husbandry, Young criticized the multiple plowing of such light lands, and the summer fallows for land very foul with weeds. Neither practice was necessary, he maintained. He also urged Coke to employ oxen in plowing instead of horses.

Although 1785 was a less important year in Young's life than those imediately preceding or succeeding it, the Young manu-

[213] *Ibid.*, p. 241. It is not known exactly when the French visitors left, but a letter from Bakewell shows they visited him in February, 1785. *Cf.* Add. MSS. 35,126, ff. 276–277.
[214] *Annals* 2: pp. 353–383. For a portrait of Coke of Holkham, see fig. 12.
[215] *Ibid.*, p. 354.
[216] *Ibid.*, p. 379.

scripts contain many interesting letters to him from this year. He was still carrying on a friendly and regular correspondence with the erratic "Earl-Bishop" in Ireland, who was attempting to unite Catholics and Presbyterians against the Anglican oligarchy which ruled Ireland. The letters indicate that Young was less liberal religiously than the bishop and that he distrusted Bristol's flirtations with the Presbyterians. As Bristol put it, "And in this duel of our pens, who would expect a Bishop of the Established Church to be an advocate for the anti-Episcopal Schismatics, called Presbyterians, while a man whose religion lies in his plough and his garden . . . to be so zealous an opponent?"[217] A letter from Dr. Richard Valpy shows that he was still correcting Young's poetry, and that he missed Young acutely: "One of the greatest luxuries that I sigh for in life is that you lived near me."[218]

A rare letter from his brother, Dr. John Young, revealed the latter's new passion for hunting from which he was to break his neck in 1786. It was an expensive taste, but it had improved his health. "I have two very fine horses; the King, who is generally but moderately mounted, will tell you the two best in the hunt." He also passed on the king's praise for Arthur's writings:

Yesterday se'nnight as I returned from the chase the King spoke to me of you in very handsome terms: I find that he reads your publications.

He commended particularly your recent periodical work as being very useful, and was much pleased with your argument to prove that we are not a ruined people, but have great resources.[219]

Such praise was undoubtedly balm to Young's pride, as was a flattering letter from Charles F. Palmer, urging him to write a single volume containing "every thing which a Farmer has need to know. . . . I say you can write that Volume if you will . . ."[220] Unfortunately Young never followed this advice, but permitted his materials for the "essential" work in agriculture to grow into the monumental unpublished ten-volume manuscript, *The Elements of Agriculture.*

[217] Young, 1898: p. 129. Bristol's letters, dated Jan. 15 and March 9, are answers to letters from Young. *Cf.* also Childe-Pemberton, 1924: **2:** pp. 337–375.
[218] *Ibid.,* p. 135.
[219] *Ibid.,* pp. 132–133.
[220] Add. MSS. 35,126, f. 310.

Late in the fall Young received four interesting letters from John Symonds who was in France visiting the Duc de Liancourt. The letters reveal that the initiative for Symonds's heavy-handed satire on turnip culture in France came from Young and that Symonds had rather feared it might offend Lazowski or the Duke. The letters also show that Young was considering a French trip which Symonds advised against, believing that the king's favor might result in a government appointment. Should such an opportunity occur, Young should be at hand and not in France.[221]

Young took a ten-days' trip in March to Robert Bakewell's farm at Dishley in Leicestershire. It would be an interesting tour if for no other reason than that Squire Arthur Young, a gentleman, had as his companion Mr. William Macro, a farmer. True, Macro was more than a common farmer, for he had contributed many articles to the *Annals*. On March 11 Young picked up Macro at Barrow, twelve miles west of Bradfield. They went through Cambridge to Huntington and stopped at Cromwell's birthplace, "the house in which the great wicked man was born."[222] They also stopped at Hazelbeach, the seat of Mr. Ashby, who called in some neighboring graziers and farmers to meet them. Ashby was an outspoken enemy of enclosures, but Young rejected all the arguments against them. However, he condemned the special parliamentary enclosure act as "a composition of publick folly and private knavery, that is a disgrace to common sense,"[223] and urged that the whole procedure should be made cheaper and easier.

Young had last visited Bakewell at Dishley on the *Farmer's Tour,* fifteen years earlier. Bakewell was now sixty years old and at the height of his fame. "I found the enterprizing Mr. Bakewell amidst improvements that will reflect a lasting honour on his name." Young summarized his principles thus:

The leading idea, then, which has governed all his exertions, is to procure that breed which in a given food will give the most profitable meat—that in which the proportion of the useful meat to the quantity of offal is the greatest:—also in which the proportion of the best to the inferior joints is likewise the greatest[224]

The important parts of the animal are thus where the valuable

[221] *Ibid.,* ff. 298–299, 300–301, 304–305.
[222] *Annals* 3: p. 453.
[223] *Ibid.,* p. 459.
[224] *Ibid.,* p. 466. For his earlier visit to Dishley, *cf. supra* p. 70.

joints are located—rump, hip, back, ribs, flank—that is the "backward upward quarters."[225] Young noted one cow "with perfect hillocks of fat on her rump."[226] Bakewell was convinced that no animal could be judged properly by the grazier except by feeling, and declared that, if he were faced by the alternative of "trusting to the eye in the light, or to the hand in the dark,"[227] he would not hestitate a moment to choose the latter. Young was chagrined that Bakewell considered the South Down sheep greatly superior to the Norfolk breed, and had to admit that Bakewell's breed was far superior to either. He still believed that the connoisseur would prefer Norfolk mutton for its flavor and its fine gravy, "but the great mass of mutton eaters, which are in the manufacturing towns, will for ever choose the fattest meat, and give the greatest price for it."[228] Although Young was critical of some of Bakewell's practices, he ended his warmly laudatory account: "The man that has shown experimentally, the talents adequate to this truly national pursuit, is a public man, and ought to be supported by every exertion that can give animation to the efforts of individuals."[229]

Early in 1785 William Pitt introduced into Parliament certain proposals to liberalize Irish trade regulations, proposals which aroused so much opposition both in Ireland and England that they were eventually dropped.[230] In March and April, Young had three letters from George Rose, secretary to the treasury and Pitt's right-hand man, requesting him to come to London to give information to the ministry. Young's own account of this episode in the *Autobiography* follows:

. . . I had a letter from Mr. Rose requesting information relative to the comparative circumstances of the two kingdoms, and Mr. Pitt thought the information so much to his purpose that he desired Mr. Rose to write to me requesting my attendance in town. I accordingly went, and gave Mr. Pitt the information he wished, at the same time answering an abundance of collateral enquiries, for which I received a formal letter of thanks.[231]

[225] *Ibid.*, p. 468.
[226] *Ibid.*, p. 486.
[227] *Ibid.*, p. 469.
[228] *Ibid.*, p. 479.
[229] *Ibid.*, p. 498.
[230] *Cf.* Barnes, 1939: pp. 115–122.
[231] Young, 1898: p. 137. For Rose's letters *cf.* Add. MSS. 35,126, ff. 280, 284, 286.

That Young was not indemnified even for his expenses is clear from John Young's letter: "I see no reason for your being at the expense you allude to for the public and think you ought to be indemnified; you cannot afford these journeys to London, and so I would plainly tell the ministers."[232]

In the *Annals* Young warmly supported Pitt's proposals and vigorously attacked the narrowly monopolistic opposition of the manufacturers. His argument was based on general free trade principles, but his concluding paragraph is perplexing:

It is conviction alone . . . that induces me to take up the pen at present, and by no means a partiality for any minister; a race of men from whom few I believe have less reason to be partial than myself. A race so generally in the habit of doing commercial mischief, that it is seldom they merit support; when, however, they are right, the public good calls upon every friend of his country to promote the measure.[233]

Why did Young write such a paragraph? Was he really angling for public office, or was he bitter for not having been given office? Was it because of his friendship for such opponents of Pitt as Shelburne, Burke, and Bristol? There is no positive answer, but on the face of it, no such disavowal seemed necessary, for his argument was sound and in accord with the most enlightened opinion of the day.

The cordiality of Young's relations with his kinsfolk, the Burneys, is shown by an undated letter from Charlotte Burney to introduce the bearer, M. de Virly, President of the Parlement of Dijon. That Charlotte knew how to approach Young is apparent by her graceful sentence: "You will much oblige me, if you will give him the pleasure of your conversation while he is in your part of the world; this, & *showing him the Lion* of Bradfield is the whole of my petition. . . ."[234] The letter concluded with a note of thanks for an invitation to visit Bradfield that summer, which she feared it would be impossible to accept.

The Young manuscripts contain a letter to Mrs. Arthur Young from Susan Burney Phillips, her stepniece. Mrs. Young was con-

[232] *Ibid.,* p. 132. The date of May 1 shows that Young's trip must have been made in April. Dr. Young informed his brother that he had told the king about the episode.

[233] *Annals* **3:** p. 290.

[234] Add. MSS. 35,126, f. 313.

sidering taking Mary Young to Boulogne to improve both her health and her French, and had requested information about expenses and costs of living. Mrs. Phillips estimated that the two of them might "live comfortably upon a hundred a year at Boulogne provided they can dispense with English cleanliness in lodgings."[235] (François de la Rochefoucauld would hardly have considered the proviso necessary for the Youngs!) Mrs. Phillips advised them, however, to go to some inland town less frequented by the English than Boulogne so that Mary's French might improve more rapidly. There is no other indication of Mary's delicate health at this time, or that the trip was actually made.

There is also an interesting newspaper squib in the *Bury Post* for May 4, 1785. No names were given, but the copy in the Cullum Library at Bury St. Edmunds has a written note at the bottom of the page, "the wife of Arthur Young, Esqr."

A few days since a lady in the neighbourhood, of the Amazonian kind, returning from Brandon in her own carriage, the driver had occasion to alight in order to adjust something that was not right in the harness, which taking up more time than the lady apprehended there was occasion for, she rushed out of the carriage, seized him by the collar, and charged him with having some evil design upon her, and that he should not drive her any more—The man, not in the least intimidated by her threats, said that he would take her at her word, and absolutely left her, refusing to drive her any further, notwithstanding the lady's execrations were now turned to intreaties—In this terrible situation in the middle of a lonely heath, she and her two children were left to themselves. With much difficulty her little boy led the horses to Elden Grayhound, where they procured a person to drive them home.[236]

Certainly the story fits perfectly with François's description of Mrs. Young's treatment of her servants, but the reference to "her little boy" raises questions, for Arthur was at the time fifteen or sixteen years old.

The most important event in the family circle in 1785 was the death of Arthur Young's mother on October 6, at the age of seventy-nine. She had been a widow during the last twenty-six years of her life. She had been devoutly religious, ever since the

---

[235] *Ibid.*, f. 294. The date is June 7 from Upper Litchfield St.
[236] *Bury Post*, May 1, 1785, p. 2.

death of her daughter, and had constantly sought to make her son, Arthur, devout, but to no purpose. He attested her never failing kindness: "She was always extremely fond of me, and ever eager to do what could contribute to my satisfaction, both as to worldly views, but especially as to my eternal interests."[237] John Symonds wrote to his friend: "It must be a great consolation to you to reflect, that you have done all you could to alleviate her cares."[238] Upon her death, Arthur became the owner of Bradfield, an arrangement made amicably with Dr. John Young who had drawn up his mother's will. Arthur was grateful for his brother's generosity, for John Young was entitled to £2000 but refused more than £1200 in the form of a mortgage, "knowing the smallness of the property, and humanely considering that I had a family unprovided for, that he had an ample income and no family at all. . . ." Arthur added. "whether such things happen among relations or strangers, they should be mentioned for the credit of the human heart."[239]

Early in January, 1786, Young spent several very enjoyable days at Rainham, studying "the noble picture of Belisarius by Salvator Rosa," talking farming with the substantial tenant farmers, measuring the circumference and height of trees, and basking in the charms of Lord Townshend's "uncommonly agreeable young wife, equally elegant and beautiful."[240] In the *Annals* Young paid tribute not only to Turnip Townshend, the founder of Rainham's agricultural fame, but also praised his host for the excellent condition of his estate, fine farm buildings, and mutually happy landlord and tenants. After his return home, Young wrote to Townshend on January 30, 1786:

The more I reflect on your Lordship's discovery in feeding cattle and sheep with the trimmings of plantations the more important it appears to me: & I am confirmed in the opinion by that of every person I have mentioned it to. This makes me curious to have

[237] Young, 1898: p. 127. For obituary *cf. Bury Post*, October 19, 1785, p. 2.
[238] Add. MSS. 35,126, f. 301.
[239] Young, 1898: p. 127.
[240] There is some doubt about the date for this trip. Young, 1898: p. 136, assigns it to 1785 as does the account in the *Annals* 5: p. 119. For several reasons I think it probable that this date is a misprint. At the end of the account the date is January 15, 1786 (p. 137), and it seems unlikely that he let a year pass before writing it up. The letter to Lord Townshend referred to below is January 30, 1786 and Townshend's article is dated February 7, 1786.

yr. Lordship's account of it, wch. I hope you will favour me wth. It will convince the world that yr. retirement when in the country is no less dedicated to the service of mankind than yr. more active exertions have been, to promote that of the state. . . .

My most respectable Compts. to Lady Townshend—I hope she will by & by let a cow rival a greyhound.[241]

Young made another trip in January, to the Rev. J. Chevallier and his wife, Young's boyhood sweetheart, Molly Fiske. This trip resulted in three articles for the *Annals*, his own account of the dairy farms in that part of Suffolk,[242] Mr. Chevallier's description of one of his plantations,[243] and Mrs. Chevallier's on the management of her dairy which she described in great detail, with considerable knowledge and competence, and yet with a becoming modesty: "Your request that I would make up the deficiency by sending you such particulars as have occurred to me, was dictated, I am afraid, rather by your partiality than your judgment. . . ."[244]

On January 7, 1786, Young began his famous correspondence with George Washington. He offered to help Washington to procure "men, cattle, tools, seeds, or anything else that may add to yr. rural amusement," and also sent a set of the *Annals*.[245] Washington was undoubtedly familiar with Young's writings and replied on August 6: "You see, Sir, that without ceremony, I avail myself of your kind offer; but if you find in the course of our correspondence, that I am likely to become troublesome, you can easily check me."[246] He also requested Young to obtain two plows, and various quantities of cabbage, turnip, sainfoin, and other seeds. A second letter, dated November 15, further re-

---

[241] Historical Manuscript Commission, Report 11, appendix Part iv, p. 412. Young warmly praised Townshend's article in the *Annals* **5**: pp. 138–144.

[242] *Annals* **5**: pp. 193–224.

[243] *Ibid.*, pp. 145–146.

[244] *Ibid.*, pp. 509–510. I was favored with two very friendly letters, August 5 and August 12, 1938, from J. B. Chevallier, Esq., great-great-grandson of John and Mary Chevallier. Although he possessed no family letters, he enclosed a picture of Aspall Hall, a stately mansion, converted by him into a "School for Farming and Kindred Business for Gentlemens' Sons," which still specializes in dairying. He wrote: "I have always been interested in my ancestress Molly Fiske & like to think that she attracted her quondam admirer to visit and write about Aspall."

[245] Young, 1801: pp. 5–6. Young, 1898: p. 189, mistakenly put the beginning of this correspondence in 1791.

[246] *Ibid.*, p. 7.

quested a complete plan for a farm yard and expressed much pleasure from perusing the *Annals:*

If the testimony of my approbation, Sir, of your disinterested conduct and perseverence, in publishing so useful and beneficial a work (than which nothing, in my opinion, can be more conducive to the welfare of your country) will add aught to the satisfaction you must feel from the conscious discharge of this interesting duty to it, I give it with equal willingness and sincerity.[247]

Young's rather tranquil spring at Bradfield was interrupted very suddenly in May by the shocking news of Dr. John Young's death from an accident while hunting with the king. Dr. Young had recently purchased a new hunter and decided to try him out in the last hunt of the season, although warned that the horse seemed unreliable. About two miles out the horse fell in taking a jump and rolled over with his master, breaking his neck.[248] Dr. Young was fifty-eight years old and was Fellow at Eton, Chaplain to the king, and Prebendary at Worcester. The *Autobiography* reveals him as a man of considerable moral courage, willing even to sacrifice preferment to maintain his standards of conduct. His friendship with the county families—the Graftons, Townshends, and Cornwallises—indicates that Dr. Young was probably a charming companion. His brother mentioned his "eccentric wit." There is no indication that John Young was deeply religious, but rather typical of the easy-going, Latitudinarian clergyman of the eighteenth century, a man of learning and wit, a hunting parson, a man hardly distinguishable from the gentlemen and courtiers who were his friends and associates.

Arthur Young spent some time at Eton settling his brother's affairs, "dining every day with the Provost and Fellows." He was naturally shocked at his brother's sudden death, but there is no evidence he was very deeply affected. There had been thirten years difference in their ages and they were not particularly intimate. Arthur Young was very frankly more concerned at the effect of his brother's death upon his son's career in the church, for Dr. Young had purposely passed by many pieces of preferment in order to advance his nephew. Young's letter to his wife from Eton is revealing:

[247] *Ibid.*, p. 9.
[248] Young, 1898: pp. 138–142. For brief obituary *cf. Gentleman's Magazine* **56**: p. 442.

As I should be sorry to keep from you anything that must give you pleasure in your welfare of your children, I shall report a conversation with Dr. Langford, the under-master, who my brother got the Prebendary of Worcester for by speaking to Lord Sidney.
On his calling on me I lamented the loss—in which he joined warmly—spoke highly of my brother as his friend. I said that my bosom had all the feelings of affection for him, but that the loss to my poor boy was nothing short of ruin. He had no friend left. "No," replied he, "don't say that, for give me leave to say that, feeling as I do the obligations I have been under to Dr. Young, I must be allowed to call myself his friend. If I succeed in life I will be a friend to him, and I hope his progress in his learning will permit me to be so. He said more to the same purpose, and as he is a rising man . . . I hope he will remember it. But the account Mr. Heath gives me is by no means satisfactory, and sorry I am to say that Arthur seems determined to do little for himself. He is now at a crisis, and sinks or swims. I gave Mr. Heath three guineas that he might encourage him with a crown now and then (as from himself) when he did well, but don't write of that to him, and desired him to write me when he was negligent. My brother's affairs turn out very badly. . . .[249]

The above letter is the first indication of Arthur Young's disappointment in his only son, who had gone to Eton as a student in 1785. Unfortunately it was not the last.

From Eton, Young went on to Worcester. Sad as his mision was, Young used the trip for the *Annals*.[250] On the way he stopped at Blenheim which he had viewed nineteen years earlier on the *Six Weeks' Tour*, where he had criticized the architecture very severely and had barely mentioned the park or grounds. Now he concentrated on the grounds which he praised extravagantly, especially the artificial water. Just south of Worcester on the way home he visited the Earl of Coventry's seat at Croome and was pleased with his admirably conducted farm which was also a hobby of Lady Coventry. He enthused about the grounds at Croome and ended with a flamboyant eulogy: ". . . sort of luscious scenery, to which the fancy is apt to add the picturesque idea of nymphs gliding from the lawns and hiding themselves in alcoves, the very bowers of love."[251] Probably the most pleasant

---

[249] *Ibid.*, pp. 142–143.
[250] *Annals* **6:** pp. 116–151.
[251] *Ibid.*, p. 128.

part of the whole trip was his stop at Reading with the Rev. Richard Valpy, who had long been urging him to come. They walked along the Thames and probably discussed poetry. Certainly Valpy introduced him to his father-in-law, Henry Benwell of Caversham who had already contributed two articles to the *Annals* on the bean husbandry and who furnished him with much agricultural information about the whole district. On the way home between Windsor and Bradfield, Young was especially mortified at finding more fallows with poorer crops than elsewhere on his trip. "After having for so many years cried up the husbandry of the eastern counties, will they let the western ones, which were once so inferior to them, excel them?"[252]

In February, 1786, Charlotte Burney, vivacious and witty, beautiful and unaffected, in some ways the most attractive of the Burney girls, was married to Clement Francis, formerly secretary to Warren Hastings in India. On the occasion Arthur Young wrote an amusing letter to the bride:

Dear Madam,—
You know enough of me to be well assured that I can do nothing in a formal or complimentary style—so that if I must either write a letter of congratulation or be guilty of a terrible omission, ye choice is made as soon as thought of, and you must pronounce me guilty.

Besides, congratulate upon what? Upon marrying? To be sure it is a good sort of state for those who know how to make a proper use of it—but how should I know that you are in that number? Many things will be necessary to convince me that you are. Are you disposed to a country life? Or must you be gadding for ever to London?

Will the admiration of a heap of fools weigh in the scale against ye friendship of one worthy man? Are you in the midst of poultry? Do you know your best cow? Are your lambs free from foxes? Are you planting shrubs and making walks? and can you pun as well in Norfolk as in London? Then, on the other hand, there is your husband—What sort of man is he? 'Tis true I hear many excellent things of him; but does he farm hugely? Are his turnips clean? Are his lands forward for beans and oats? Does he plough with oxen? You must confess that these are points much more to the purpose than the common rubbish of character ye common mortals attend to. Hence you see, my friend, that instead of con-

[252] *Ibid.*, p. 150.

gratulating you, I give you a sheaf of reasons for doing no such thing.

Then why write this letter? I'll tell you; it is because Bradfield lies half-way between London and Aylsham; and as your husband has settled in Norfolk (one good point in his character) we must have some farming discourse together. You can tell him I am a plain man not abounding in speeches, and can assure him there is nothing but plain sincerity in the wish that he would make this house his home on every occasion that suits him; he will meet only farmer's fare, but always garnished with a farmer's welcome. Adieu, you see I am as queer a fellow as usual.

Yours, with great truth,

A. Young

Bradfield Hall,
    February 20, 1786[253]

Charlotte began her reply: "I am much obliged to you for your comical & excellent Letter, of non-congratulation, which diverted Mr. Francis & myself very much. But, in the name of conscience, how am I to answer all your questions?"[254]

Dr. Burney and some of his family visited Bradfield in July. His thank-you letter warmly praised the *Annals*, of which Young had sent him a set: "As editor and chief of the *Agricola* family, I think you merit the thanks of every Englishman, not only who loves his country, but who loves his *belly*, for if your discoveries, improvements, and instructions are followed, we may certainly always find upon our island *de quoi manger*." Dr. Burney even declared that, if he were somewhat younger, he would be tempted to abandon London, rent a farm from Young, and "enter myself as your scholar." The letter ended with a note of thanks to the family: "I beg you will present my affectionate compliments to Mrs. Young, and best thanks for the hospitality and kindness with which she treated us at Bradfield; and pray give our hearty love to the gentle, sweet and amiable Miss Bessy."[255] This description of Bessy Young, now eighteen years old, is borne out by every reference of the Burneys.

When Fanny Burney became lady-in-waiting at Windsor to

[253] Hill, 1904: pp. 42–43.
[254] Add. MSS. 35,126, f. 329. Dated March 5, 1786, from Aylsham.
[255] Young, 1898: pp. 144–146.

Queen Charlotte in July, Arthur Young did write a letter of congratulation to her:

My dear Fanny,—

You well know my "lack-lustre" eyes to be dim as an Owl's, yet be assured I have pierced into every News-Paper that came in my way with the glance of an Eagle to discry [sic] a certain name always dear to me, in a point of elevation that gives me very great pleasure. I now venture to join the voice of congratulation with equal sincerity, though less elegance, than you are accustomed to hear or read. . . .

Everybody I see expressed joy at your good fortune.[256]

In August, 1786, Young repeated part of the tour in Suffolk taken in the previous year with Lazowski, probably to show the county to a Polish gentleman, Count Kalaskowski, who spent time at Bradfield.[257] At the end of his account of this trip Young outlined a proposed three-weeks tour of Suffolk and Norfolk to combine the chief sights dear to any tourist with a survey of the agricultural highlights of the two counties. The tour started and ended at Bury St. Edmunds. A whole day should be spent at Cambridge. He defined the center of the true Norfolk husbandry, which of course must be seen, as a triangle with points at Swaffham, Shottisham, and Wells. The three great estates at Rainham, Houghton, and Holkham should be visited. From Holkham the tourist should go to Norwich to see the manufactures, especially the new iron foundry. Stops should be made at the port of Yarmouth and the seaside resort of Lowestoft. After a trip to Sir Gerard Vanneck's "magnificent seat" at Heveningham, the traveler should examine carefully the carrot culture between Saxmundham and Woodbridge. The final stages of the journey would be westward to Long Melford and then north through Bradfield, "where, if you are a good farmer, I shall be glad to see you." "I believe you will not find these three weeks the worst spent in your life."[258]

Twice in the latter half of 1786 Young's advice was sought by government. On July 29 Gab. Leeky, a member of the "Common Council" which was examining the causes for the high prices of provisions, wrote to Young soliciting his opinion.

---

[256] Hill, 1912: p. 15.

[257] *Annals* 6: pp. 217–230; Young, 1898: p. 144. In a letter dated December, 1786, Symonds wished to be remembered "aux Polonais." (*ibid.*, p. 148).

[258] *Annals* 6: p. 230.

In his reply Young apologized for not having answered immediately because of grand jury business, but maintained that government could do nothing to remedy the high prices which were caused primarily by severe frosts which had killed the turnips and thus destroyed large quantities of livestock.

There is no inconsiderable reason for believing that whatever is the price, *ought to be* the price; since price can be formed by nothing but quantity, demand, and competition. If quantity really does not exist, nobody I suppose can expect cheapness. If demand does not exist, the farmers can never expect a good price; and if competition does not take place, trade cannot be free nor the price fair.[259]

His conclusion was perfectly logical from the premises: ". . . doing nothing in this case as in so many others, is the best policy. . . ."[260]

He was also asked to appear before the newly constituted Council for Trade and Plantations and questioned by Lord Hawkesbury on some proposed amendments to the Corn Laws.[261] Apparently Young at first objected to the proposed changes, but eventually agreed that they were fair enough, although he warned that they would be unpopular in the leading corn producing counties. During his testimony he admitted that England's population had grown so rapidly that she could no longer produce all her grain needs. More surprising, he "spoke against the Bounty," which he had always vigorously defended as the very cornerstone of the prosperity of English corn producers.

A letter from Lazowski early in August regretted that Young had not come to France that summer and urged him to make such a trip in the near future. One excerpt from the letter makes a fitting close to this chapter for it graphically portrays Arthur Young in the middle 1780's. After attacking all politicians and wishing them to the devil, Lazowski continued:

I except of my proscription a certain Arthur because he is as good

[259] *Ibid.* 7: p. 43.
[260] *Ibid.*, p. 45.
[261] Add. MSS. 38,347, ff. 121–122 (Liverpool Papers, clviii). There is no date for this memorandum but it is included in a packet dated, August-December, 1786. Charles Jenkinson (1729–1808) became Baron Hawkesbury in 1786 and the first Earl of Liverpool in 1796.

a farmer, than a good companion, because he blows the fire of his laboratory at the same time that he writtes [*sic*] the most agreeable verses, because he instructs the world by his powerful reason, and he is witty enough to have nonsense with the ladies, because he is good for nothing sometimes, and I am very fond of him in those moments and his admirer in the others.[262]

[262] Add. MSS. 35,126, f. 352.

V. France and the French Revolution, 1787-1792

THE ONLY WORK from the ever busy pen of Arthur Young which has been repeatedly re-edited and reprinted is his justly famous *Travels in France*. Every study of the background or early years of the French Revolution quotes from that work which alone among his many writings can be called a "classic." Young's public life from 1787 to 1792 was dominated by his interest in French affairs. Long and extensive trips to France were made in three successive years, 1787, 1788, and 1789. Then followed the preparation of his *Travels* for the press, the first edition of which finally appeared in 1792. Hardly had the *Travels* been published before Young began to react violently against the increasing radicalism of the French Revolution and his first tirade against the Revolution was also written in 1792.

Young must have been badly hurt by the savage review of the first five volumes of the *Annals* in the January, 1787, issue of the influential *Monthly Review*, which had also taken two pot-shots at him in the previous year. In September, 1786, his article for the Bath Society was characterized as "entirely super-fluous."[1] Much more crushing was the review in December of his essay on the clustered potato in the *Transactions of the Society of Arts:*

. . . it appears that Mr. Young is himself but very little conversant in the culture of this valuable plant; that he has been groping his way, to discover facts that were well known to many persons long before he began his experiments; and that he has not yet attained a due degree of knowledge on the subject he treats.[2]

Then in January, 1787, appeared a three-page review of the *Annals*, which blamed Young for being intemperate in language, for toadying to the aristocracy, and for making too many digres-

[1] *Monthly Review* **75:** (1786) : p. 169.
[2] *Ibid.*, p. 421.

sions into the field of politics. No wonder that the practicing farmer refused to subscribe to the *Annals*. The reviewer also ridiculed Young's experiments in agricultural chemistry. No better contemporary critique of Young's weaknesses ever appeared than the following:

In his original essays we perceive the same vivacity of thought, the same quickness of imagination, the same avidity for seizing doubtful facts, the same facility for rearing, upon whatever foundations, structures of stupendous magnificence; the same bias to calculation, the same fondness for political speculations, which distinguish all his other performances, and which render them particularly interesting to those who study agriculture for amusement and recreation.[3]

The embarrassment might have been even greater had Young known that all three reviews were by a noted Scottish agriculturist, James Anderson.[4]

Young was also being attacked early in 1787, and with far less courtesy, by the organized woolen manufacturers who had introduced a bill in 1786 to tighten up the prohibitions upon the export of wool. Although Young had supported the prohibitions upon export in 1772,[5] he now published two bitter articles against the new bill in the *Annals* late in 1786 and two more early in 1787.[6] At times he employed theoretical free-trade arguments. Although denying that there was much smuggling, he declared that the prohibition upon export invited smuggling, especially when the French prices of wool were so much higher. He maintained that the bill aimed further to depress domestic woolen prices, to the detriment of the landed interest and to the profit of the manufacturers. He was particularly caustic upon the minute regulations to prevent smuggling and the extreme penalties against it, and exclaimed, "Do these clauses refer to the policy of the inquisition, or to the conduct of an English farm? . . . And all this to satisfy that vile spirit of monopoly which has so long been the disgrace and the curse of this country."[7]

---

[3] *Ibid.* **76:** (1787): p. 40.

[4] Nangle, 1934: indexes, pp. 1, 2, 144, 211, 226. Anderson (1739–1808) was by no means a contemptible antagonist. Valpy wrote Young February 10 (Add. MSS. 35,126, f. 364): "The Monthly Reviewer uses you ill, in the opinion of all with whom I have conversed. . . . But your character, as a writer, as a farmer, & as a man, is beyond praise or censure."

[5] Young, 1772: p. 182.

[6] *Annals* **6:** pp. 506–528; **7:** pp. 94–96, 150–175, 405–428.

[7] *Ibid.* **6:** p. 515.

Young's articles naturally provoked the woolen manufacturers to attack him personally. A meeting at Devizes, January 10, 1787, stated: "It is however some consolation to reflect that a patient investigation of truth is no distinguished part in Mr. Y's character as a writer." This statement appeared in the *Bury and Norwich Post* on January 24, and the following issue included a vitriolic attack by another spokesman for the woolen manufacturers.[8] Arthur Young always loved a good fight and on February 21 he presided over "a numerous meeting of Land-Owners, Land-holders and Wool-growers" at the Angel Hotel in Bury. The new bill was attacked; a committee was appointed to examine it further; a motion thanked Young "for his liberal and spirited exertions in bringing forward this business;" accounts of the meeting were inserted as advertisements in the newspapers; another meeting was called for March 5 at another famous Suffolk hostelry, the Great White Horse at Ipswich. Among the members of the Bury committee were three of Young's close friends—Thomas LeBlanc, Rev. Roger Kedington, and William Macro.[9] The Ipswich meeting voted to bring the matter before the county Grand Juries at the ensuing Assizes,[10] and requested Young to insert extracts from the bill in the newspapers.[11] The matter was duly presented to the Assizes which asked the sheriff to call a county meeting if the manufacturers attempted to pass the bill at the current session of Parliament. The manufacturers, however, decided not to press their bill at that session, and so the matter rested until after Young's return from France.[12]

While the woolen controversy was boiling, Young wrote a thirteen-page letter to George Washington, enclosing elaborate plans for a barn and other farm buildings.[13] At the very height of the controversy, Young received a letter from the antiquary, George Ashby, written on February 20, asking about the origin of potatoes and requesting him to procure a cow from Mrs. Molly Fiske Chevallier of Aspall. Young found time to answer the

---

[8] *Bury and Norwich Post,* January 24, 31, 1787.

[9] *Ibid.,* February 28, 1787.

[10] *Ibid.,* March 7, 1787.

[11] *Cf. ibid.,* March 7, 14, 1787.

[12] *Annals* **7:** p. 424. Young was congratulated for his exertions in an undated letter from Sir Joseph Banks (Add. MSS. 35,126, ff. 367–368).

[13] *Letters from . . . Washington to . . . Young . . .,* 1802: p. 10. For a diagram of the buildings with an explanation, *cf. Annals* **16:** pp. 149–155. The original letter is in the Library of Congress.

next day, the day of the meeting at the Angel in Bury. He had already written for the cow, made certain suggestions about the origins of potatoes, and requested Ashby to prepare a history of potatoes for the *Annals*.[14]

There is strong evidence that Young's relations with his wife reached a crisis early in 1787. The information came from his future son-in-law, the Reverend Samuel Hoole, who married Bessy Young in 1791. Hoole informed Susan Burney who wrote it to Fanny Burney:

[James Burney] went out of the room for a few minutes, and young Hoole turned to seize the opportunity to tell me he had seen Mr. Young, and been told by him that articles were absolutely drawn up for the separation of himself and his wife, . . . "But," said my Informer—"I do not believe she will quit his house." Tho' all this was not very *new* to me, I could not help shuddering at thoughts of this dreadful connection. . . .[15]

Unfortunately no other information corroborates this letter and certainly the separation never actually took place.

On April 9, 1787, Lazowski wrote to Young, stating that he and the Count François de la Rochefoucauld were planning a trip to the Pyrenees and inviting Young to join their party.

You will learn the French; with us everything will be explained to you; in short I will be with you, and that is enough, I hope. . . . I must not tell you that I shall be another Arthur here for you, not that I presume to say that you will find in me an Encyclopoedia living as I did in you, but your friend, and therefore to your commands in Paris and everywhere.

Young must decide quickly for the party would start shortly after May 15. Each would travel with his own horse, they would cover only twenty or twenty-five miles a day, and his only expenses would be his own. "This was touching a string tremulous to vibrate. I had so long wished for an opportunity to examine

---

[14] Ashby's letter is in Add. MSS. 35,126, f. 366. The author bought Young's reply from an antique dealer in Bury St. Edmunds in 1938. It is covered with annotations by Ashby, who had no great trouble in proving that Young's hasty suggestions about the origin of potatoes were inaccurate.

[15] Johnson, 1926: p. 127. Susan's letter is dated March 25, and the interview with Hoole was on March 21. That Young should have confided such news to Hoole indicates that he was already an accepted suitor for Bessy's hand.

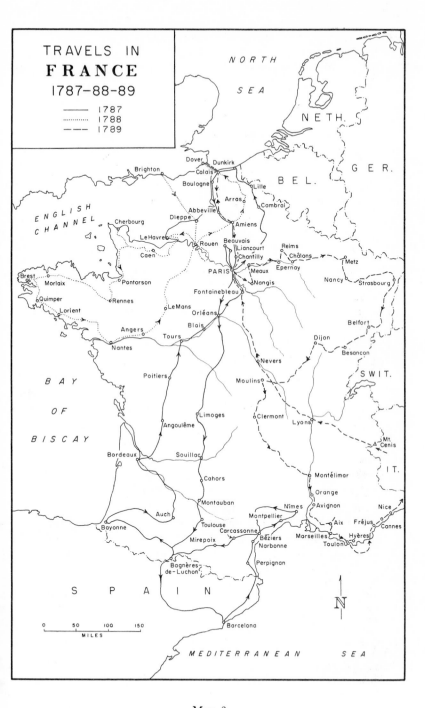

France. . . . This most agreeable plan I instantaneously acceded to, and soon set out for France on horseback."[16]

He left Bradfield on May 9, stopped a day in London, and reached Dover on May 14. Impressed by the magnificent scenery of the coast from Hythe to Dover, and with his mind full of the probably beneficial effects of the recently signed Eden Treaty, he grew lyrical: "Never were a bright sun, and unclouded sky, a clear atmosphere, and a soft delicious day in this month of the poets, more truly in harmony with a scene. I think I shall remember this ride to the last hour of my life."[17] When the packet failed to sail, he embarked at night on a passing boat and arrived at Calais on May 16 after a "villainous passage of fourteen hours."[18]

Nine days later he was in Paris. His mare had been so upset by the boat trip that he gave her a day's rest and spent "an instructive and agreeable evening" with M. Mouron[19] whom he had met at Calais in 1784. His route to Paris lay through Boulogne, Abbéville, Amiens, and Chantilly. He must have been disappointed at the hostility to the Eden Treaty among the woolen manufacturers of Abbéville and Amiens.[20] He noted the heavy farm labor which French women were performing in contrast with English practice.[21] He was impressed with Chantilly's magnificence, the formal gardens, canal, and especially the stables, but found no really fine large rooms in the Chateau and his preference for English gardens was obvious.[22] On May 25, his mare was so ill from dirty French stables and bad treatment by grooms, that he had to leave her to be sent for from Paris, and finish the trip to the capital in a post-chaise.[23]

On arrival at Paris, Young was warmly welcomed by the Duc de Liancourt and his family, and especially by Lazowski. The two days in Paris were hectic. His trunk had gone astray and conse-

---

[16] Young, 1898: pp. 154–157.

[17] *Annals* **7**: p. 574. As usual Young used his trip from Bradfield to Dover to add a few pages for the *Annals*.

[18] Young, 1898: pp. 158–159.

[19] Young, 1793: **1**: p. 7. This is the two-volume Dublin edition. The references will be to **1**: unless specifically noted. For his trip to Calais in 1784, *cf. supra*, pp. 177–178.

[20] *Ibid.*, pp. 11, 13.

[21] *Ibid.*, pp. 11–12.

[22] *Ibid.*, pp. 14–16.

[23] *Ibid.*, p. 16.

quently much of May 26 was spent in buying new things for the trip. Nevertheless he managed a rapid superficial tour of the city with Lazowski who also took him in the evening to his brother's. Claude Lazowski was an inspector of manufactures and at his home Young spent an instructive evening with Broussonet, secretary to the Royal Society of Agriculture, and Desmarets, inspector general of manufactures.[24] When the company broke up, Young went with the Duke's younger son, Count Alexandre, to Versailles where they spent the night. Young saw the *Cordon bleu* conferred on the young Duke of Berry the next morning in the chapel, and also saw the king dine in state. He was disgusted by the excessive ceremony of the court. He noted Louis XVI's awkwardness and inattention, but declared Marie Antoinette "the most beautiful woman I saw today." He criticized the exterior of Versailles for its lack of unity. The Hall of Mirrors was "the finest room I have seen"[25] but the other rooms were mediocre. He was astonished to see poorly dressed people wandering at will through the palace. Late in the afternoon he returned to Paris and was taken by the Duchesse de Liancourt to the opera at St. Cloud.

On May 28 Count François de la Rochefoucauld, now about twenty-two years old, Lazowski, and Young left Paris for Bagnères de Luchon in the Pyrenees. Three servants accompanied them, and they traveled with an English whisky, a French cabriolet, and a saddle horse, presumably Young's mare now fully recovered. The journey took just about three weeks, for they only covered about thirty miles a day.[26] Their route was almost directly south through Orléans, Limoges, and Toulouse. Everywhere except south of Toulouse the roads evoked Young's admiration and astonishment. Not so the inns! He found them superior to English in regard to food and beds, but the rooms were poor, the furniture nondescript, the service wretched, and the maids dirty and unattractive. In Souillac he exclaimed:

It is not in the power of an English imagination to figure the animals that waited upon us here, at the Chapeau Rouge . . . . Some things that called themselves by the courtesy of Souillac women, but in

[24] *Ibid.*, p. 18.
[25] *Ibid.*, pp. 20–21.
[26] *Ibid.*, pp. 20–48.

reality walking dung-hills. But a neatly dressed clean waiting girl at an inn, will be looked for in vain in France.[27]

He found little to praise in French agriculture. Fallows were general and methods backward. In the Sologne he first came upon métayage, "a miserable system, that perpetuates poverty and excludes instruction."[28] At Limoges where they were hospitably entertained by the Bishop there was an agricultural society which had been founded by Turgot, but Young was contemptuous: "This society does like other societies—they meet, converse, offer premiums, and publish nonsense."[29] If only they would establish a model farm! He found many evidences of poverty—women collecting weeds for their cows in their aprons, the lack of shoes and of glass in their windows.

The party arrived at Bagnères de Luchon on June 17 or 18 and very naturally joined forces with the Duc de Rochefoucauld, a cousin of the Duc de Liancourt, and his family. Young went into considerable detail about the daily routine at Bagnères. Instead of drinking the waters, or bathing in them, Young and Lazowski took early morning walks in the mountains, sometimes viewing the beauties of nature, sometimes talking with the peasants. After dinner at 12:30 or 1:00, the fashionable world spent the whole afternoon in playing cards, backgammon, and chess. Early in the evening there was a promenade, followed by supper at 9:00 and an hour's informal conversation. Everyone was in bed by 11:00.[30] Young preferred his native habits. He disliked the continental breakfast and the resultant early dinner for which everyone dressed. Such a routine left no uninterrupted block of seven or eight hours for work or an extended expedition. Young frequently absented himself from the long "mortally insipid" card parties.[31]

During this stay at Bagnères de Luchon, Young had ample opportunity to observe French aristocratic manners in a fashionable watering resort. Everyone came from the privileged classes, several titled families, several gentlemen, about half a dozen young officers, two bishops and three or four abbés. He was very

[27] *Ibid.*, pp. 36–37.
[28] *Ibid.*, p. 25.
[29] *Ibid.*, p. 32.
[30] *Ibid.*, pp. 50–59.
[31] *Ibid.*, p. 52.

favorably impressed by the urbanity of these people, their "invariable sweetness of disposition, mildness of character, and what in English we emphatically call *good temper*." On the other hand they were so polite and mild, that there was seldom an argument, without which conversation "is like a journey on an endless flat."[32]

Young enjoyed the grandeur of the scenery, the high mountains, the wild cascades, and the little villages perched upon almost perpendicular mountain sides. All were described in the rather turgid, romantic style which Young invariably employed in writing of nature. One example will suffice: ". . . the roar of the waters beneath unites in effect with the mountains, whose towering forests, finishing in snow, give an awful grandeur, a gloomy greatness to the scene. . . . All around is great; the sublime of nature, with imposing majesty, impresses awe upon the mind. . . ."[33]

Most of Young's surviving letters to his family during his French tours are to his oldest daughter, Mary. Two were written from Bagnères de Luchon on June 17 and July 7.[34] He constantly urged the family to write more often and at greater length. He was especially irritated that Arthur, now eighteen and a student at Eton, did not write. His devotion to Bobbin appears in nearly every letter. On July 7 he warned that Bobbin not be allowed to "go alone near hedges for fear of snakes." The strained relations with Mrs. Young are reflected in a very typical postscript: "Remember me to yr. mother. I hope she is well."

On July 10 Young and Lazowski started on an eleven days' tour through Catalonia which Young had been told was the garden of Spain. They traveled on mules and had a guide whose French was largely Provençal. Crossing the Pyrenees at Bagnères they struck diagonally southeastward to Barcelona and then north along the coast into France. Altogether they covered 347 miles in Spain.[35] It was certainly an exhausting trip. The first day they traveled along mountain trails and precipices: "The road,

---

[32] *Ibid.*, pp. 54–55.

[33] *Ibid.*, p. 57.

[34] Add. MSS. 35,126, ff. 386–387, 390–391.

[35] The Catalonian tour is an appendix to 1: pp. 591–672. It was also published separately in the *Annals* 8: pp. 195–275. There is much more agricultural detail and much less sprightly travelogue than in the French travels. Most modern editions omit it.

as it is called . . . hangs . . . like a shelf on the mountain side, and is truly dreadful to the inhabitants of plains, from being broken by gullies, and sloping on the edges of precipices; it is, however, passable by mules."[36] For the most part the inns were terrible. The ground floor was usually a stable. Two nights poor Lazowski had to sleep on a table and one night got very little rest, "what with bugs, fleas, rats, and mice."[37] Young bitterly characterized cne inn as "a stinking dirty dreadful hole, without anything to eat or drink, but for muleteers."[38] Both wine and olive oil were rancid, making food and drink difficult. On the plain the heat was excessive. Lazowski was a seasoned traveler, but felt that the day they arrived in Barcelona was "the most fatiguing day he had ever experienced."[39]

Young found Catalonia uninteresting both scenically and agriculturally, a province "whose most striking features are its rocks."[40] The brilliant sunshine on wasteland and rockland soon tired the eye. Where the land was watered, the crops were prodigious, elsewhere paltry. The crying need was government aid for more extensive irrigation. There were few resident landlords and hence no enlightened leadership in the rural areas. Olive trees added nothing to the beauty of the landscape, and the festooned vines, hanging from tree to tree, were disappointing. The two high spots were the monastery at Montserrat and Barcelona. They were hospitably and comfortably entertained at the monastery,[41] but a heavy mist spoiled the famous view. Young found Barcelona admirable.[42] His inn was excellent, the silk and stocking manufactures flourishing, the public buildings fine. He was especially impressed with the fine quay and a new working-class housing development adjoining it. The night they attended the theater, larger than Covent Garden, they witnessed a Spanish comedy followed by an Italian opera.

On July 21 Young and Lazowski passed from Spanish Catalonia into French Roussillon, and reached Perpignan. There they parted. Lazowski returned directly to Bagnères de Luchon, while

[36] Young, 1793: p. 593.
[37] *Ibid.*, p. 607.
[38] *Ibid.*, p. 645.
[39] *Ibid.*, p. 632.
[40] *Ibid.*, p. 670.
[41] *Ibid.*, pp. 625–627.
[42] *Ibid.*, pp. 631–638.

Young took a side trip into Languedoc,[43] only returning to Bagnères on August 6. He continued to travel on his Spanish mule,
north through Narbonne to Beziers, then parallel to the Mediter-
ranean coast through Montpellier to Nimes, then back to Bagnères through Carcassonne and Mirepoix.

On entering Roussillon, Young was struck with the contrast to
Catalonia. The fine roads and bridges and the general atmosphere of "cultivation and improvement" could result only from
government efforts, for the people and the geographic environ-
ment were alike. His *obiter dictum* sounded more like a mercantilist than an adherent of *laissez-faire*. "The more one sees,
the more I believe . . . that there is but one all-powerful cause
that instigates mankind, and that is GOVERNMENT!"[44] On
reaching the Languedoc Canal he exclaimed enthusiastically,
"Here Lewis XIV thou art truly great!"[45] Although the roads
were superb the lack of traffic made them seem extravagant and
even ostentatious. Their military justification apparently never
occurred to him.

He also saw much to please him as an agriculturist. His description of the practice of "treading out the corn" might well be
instanced to prove the general well being of the French peasants
on the eve of the Revolution:

Every soul is employed, and with such an air of cheerfulness, that
the people seem as well pleased with their labours, as the farmer
himself with his great heaps of wheat. The scene is uncommonly
animated and joyous. I stopped and alighted often to see their
method; I was always civilly treated, and my wishes for a good price
for the farmer, and not too good a one for the poor, well received.[46]

At several places he admired the intensive cultivation of very
mediocre land. Assuming with no definite proof that this land
was held in private ownership, he made one of his most frequently quoted statements: "Give a man the secure possession of
a bleak rock, and he will turn it into a garden; give him a nine
years lease of a garden, and he will convert it into a desert."[47]
A sentiment somewhat contradictory to the one quoted above in

[43] *Ibid.*, pp. 59–83.
[44] *Ibid.*, p. 60.
[45] *Ibid.*, p. 65.
[46] *Ibid.*, p. 64.
[47] *Ibid.*, p. 76.

which most improvements had been attributed to government!

Young enthused over Nimes. He grew lyrical over the Roman remains—the "prodigious" Arena,[48] the Temple of Diana, the baths, and the "stupendous" Pont du Gard.[49] His highest encomiums were rightly reserved for the Maison Carré: ". . . it is beyond all comparison the most light, elegant, and pleasing building I ever beheld. . . . There is a magic harmony in the proportions that charms the eye. . . . it is one perfect whole of symmetry and grace."[50] The hotels were also good in these old cities of the south. He usually ate at the *table d'hôte*, partly to save money, partly to converse with strangers, but the French would not talk. Instead of "volubility" he found "taciturnity,"[51] and he was also astonished at their ignorance of England. Once he was even asked whether England had any trees or rivers.

The trip from Nimes back to Bagnères was much less satisfactory. The severe heat nearly prostrated him, and he attempted in vain to hire a carriage of some sort. Many of the inns were even worse than in Spain. St. Geronds was a town of four or five thousand, but its inn was unspeakable:

At St. Geronds go to the Croix Blanche, the most excreable receptacle of filth, vermin, impudence, and imposition that ever exercised the patience, or wounded the feelings of a traveller. A withered hag, the daemon of beastliness, presides there. I laid, not rested, in a chamber over a stable, whose effluviae through the broken floor were the least offensive of the perfumes afforded by this hideous place. . . . Spain brought nothing to my eyes that equalled this sink, from which an English hog would turn with disgust.[52]

Everywhere myriads of flies also tortured Young. Even the medieval beauties of Carcassonne made no impression upon a man whose architectural tastes were neo-classical. In contrasting France with England, Young became unusually peevish and provincial: "What a contrast! This confirms the fact deducible from the little traffic on the roads even around Paris itself. Circulation is stagnant in France."[53]

[48] *Ibid.*, p. 70
[49] *Ibid.*, p. 73.
[50] *Ibid.*, p. 71.
[51] *Ibid.*, p. 72.
[52] *Ibid.*, p. 80.
[53] *Ibid.*, p. 79.

Young reached Bagnères de Luchon on August 6, "not displeased to have a little rest, in the cool mountains, after so burning a ride,"[54] but left there again on August 10, and started on the return trip to Paris which he only reached on September 15. In order to see more he left Luchon before his friends and arrived at Paris later, but they did travel together most of the way from Auch to Orléans.

At Lourdes, where an old castle was used as a state prison, Young commented bitterly on *lettres de cachet.* Declaring that most state prisoners were incarcerated for virtues rather than crimes, he continued:

Oh, liberty! liberty!—and yet this is the mildest government of any considerable country in Europe, our own excepted. The dispensations of providence seem to have permitted the human race to exist only as the prey of tyrants, as it has made pigeons for the prey of hawks.[55]

He was greatly pleased with the peasant proprietors of Béarn living on their enclosed farms. Stone cottages, neat hedges, good farm buildings, clean and pretty country girls—all reflected a high standard of living. Perhaps the peasants owed these blessings to Henry IV. "The benignant genius of that good monarch, seems to reign still over the country; each peasant had *the fowl in the pot.*"[56] He thought that the barren blowing sands beyond Bayonne might have been improved by irrigation. But much of the area was owned by the Duc de Bouillon, and Young expressed his distrust of great noblemen by writing: "A Grand Seigneur will at any time, and in any country, explain the reason of improveable land being left waste."[57]

From August 25 until August 28 Young and his friends were at Bordeaux, the prosperity and beauty of which "greatly surpassed" his expectations. The quays were disappointing, but the streets, public buildings, and private houses were fine. Twice he attended the theater, "by far the most magnificent in France." The environs contained fine country seats. London was certainly a finer city than Paris, "but we must not name Liverpool in competition with Bourdeaux [*sic*]."[58]

[54] *Ibid.,* p. 83.
[55] *Ibid.,* p. 85.
[56] *Ibid.,* p. 87.
[57] *Ibid.,* p. 90.
[58] *Ibid.,* pp. 94–95.

Young again savagely attacked the nobility when he saw some wastes south of Angouleme, belonging to the Prince of Soubise: "Oh! if I was the legislator of France for a day, I would make such great lords skip again."[59] The party spent about a week in the Loire country and visited three famous chateaux, Chanteloup, Blois, and Chambord. At Chanteloup he admired not only the chateau itself and the famed pagoda, but also the "noble cow-house" and the "best built sheep-house I have seen in France."[60] His comments on the rooms at Bloix where the Guises were murdered were interesting: "The character of the period, and of the men that figured in it, were alike disgusting. Bigotry and ambition, equally dark, insidious and bloody, allow no feelings of regret. The parties could hardly be better employed than in cutting each others throats."[61] He liked Chambord much better than Versailles, admired especially its famous staircase, and suggested that it would make an admirable model farm for the turnip culture.[62] The chateau could be used as the director's residence and the barracks turned into stalls for cattle. He visited Denainvilliers, formerly the estate of the noted French agriculturist, du Hamel, whom Young had always revered.[63] Although the present owner was away, Young poked around and was thrilled to find the drill plows which du Hamel had described, and to see a plantation of exotic trees.

Young's letter of September 2 to Mary from Poitiers displayed a deep-seated, almost instinctive, love of his native country:

". . . but that sort of comfort wch. is found so generally in England among the middling & lower classes has hardly any existence in this Kingdom, and it is one of the most capital deficiencies that can be seen in any country; but I believe it is to be found only in England. a country fro. wch. none will travel but to love it the better."[64]

One day after reaching Paris, Young accompanied Count François to the family estate at Liancourt, about forty miles north of the capital. Expecting to stay only three or four days,

[59] *Ibid.*, pp. 99–100. This statement sounded so radical in 1792 that Young apologized for it in a footnote as only expressing the sentiments of the moment.
[60] *Ibid.*, p. 108.
[61] *Ibid.*, p. 110.
[62] *Ibid.*, p. 112.
[63] *Ibid.*, pp. 113–114.
[64] Add. MSS, 35,126, f. 397.

Young remained there more than three weeks. Certainly this visit to an estate which rivaled Versailles in splendor[65] constituted one of the highlights of his trip of 1787, for "the whole family contributed so generally to render the place in every respect agreeable. . . ."[66] The duke was a great builder, an admirer of all things English, and a philanthropist. Hence there was much about which Young could wax enthusiastic—a new plantation "in the English style of gardening," the dairy, "entirely constructed of marble," belonging to the duchess, the cotton manufacture with twenty-five looms, the school where poor girls were taught religion, reading, writing, and spinning, and the fine library of seven or eight thousand volumes.[67] As the guest of the duke at the meeting of the provincial assembly at Clermont, he admired the inclusion at dinner of several prosperous renting farmers and especially of two ladies. He participated in several hunting expeditions, but found French hunting "more tedious than is easily conceived: like angling, incessant expectation, and perpetual disappointment." He enjoyed more the informal dinners after the hunt where they toasted the chase and the plow. "A man is not worth hanging that does not drink a little too much on such occasions. . . ."[68]

While at Liancourt, Young accompanied Lazowski to Ermenonville where Rousseau was buried. Young paid glowing tribute to "that splendid genius," thanks to whom women now nursed their own children and children were free from corsets. Young's description is sentimentally romantic: "The remains of departed genius stamp a melancholy idea. . . . We viewed the scene in a still evening. The declining sun threw a lengthened shade on the lake, and silence seemed to repose on its unruffled bosom. . . ."[69]

A letter to Mary from Liancourt reveals some petty annoyances. There was a question about Arthur's living expenses at Eton. Too bad that the beer had failed, "but yo. seem to have been bad managers about it." A favorite dog had been accused of being a sheep stealer. Mary had requested a French watch but

[65] *Cf.* Dreyfus, 1903: p. 28.
[66] Young, 1793: p. 116.
[67] *Ibid.*, pp. 116–127.
[68] *Ibid.*, pp. 125–126.
[69] *Ibid.*, p. 123.

her father claimed that French watches were worthless, and that he could not afford a gold case with works from some other country, but "I shall bring you some trifle certainly. . . ." There was some dispute with Mrs. Young about Bobbin's education: "You certainly know I coincide entirely with yr. Mother in hoping that no differences may arise any more—they certainly will not from me." Finally, although enjoying Liancourt, he had been "sadly plagued with the tooth ache."[70]

After leaving Liancourt on October 9, for nearly three weeks Young was again a guest of the hospitable family at the Hotel de la Rochefoucauld in Paris.[71] He spent the time very agreeably, visiting the usual tourist sights, meeting with French scientists, and attending the theater nearly every night. He climbed the tower of Notre Dame for the view, saw the Gobelin tapestries, dined with Lazowski at a coffee house in the Palais Royale, admired the Place Louis XV, the tomb of Cardinal Richelieu, the terrace at St. Germain, and especially the Halle aux Bleds. He preferred the French theater over the English in almost every respect, "writers, actors, buildings, scenes, decorations, music, dancing."[72] He again went to Versailles but was disappointed with the Trianon Gardens which had "more effort than nature—and more expense than taste."[73]

Paris seemed much less satisfactory than London for persons of small means. His chief complaint was the dirtiness and narrowness of the streets, plus the fact that young men drove one-horse cabriolets about with complete recklessness. Since the streets lacked sidewalks, one could not walk safely or without fear of ruining one's clothes. "I saw a poor child run over and probably killed, and have been myself many times blackened with the mud of the kennels."[74] Since public conveyances were also unsatisfactory it was desirable to have one's own coach, but this of course added greatly to the expense. Lodgings were both poorer and more expensive than at London. Such high living costs were the more to be regretted because Paris furnished much more stimulation and gave much greater consideration than London to men of letters or science.

[70] Add. MSS. 35,126, f. 398. Dated September 19.
[71] Young, 1793: pp. 127–147.
[72] *Ibid.,* p. 138.
[73] *Ibid.,* p. 143.
[74] *Ibid.,* p. 145.

Probably what Young enjoyed most at Paris, unless it was the theater, was the society of such scientists. He saw quite a bit of M. Broussonet, secretary of the Royal Society of Agriculture, who took him to the "King's cabinet of natural history and the botanical garden."[75] He also went to the École Veterinaire which "does honour to the government of France."[76] Most thrilling were two visits to Lavoisier, to whom he had letters of introduction from Priestley. Mme Lavoisier, "a lively, sensible, scientific lady,"[77] gave him tea and discussed the phlogiston theory with him. Young was greatly impressed with the famous laboratory and its various scientific instruments.

Twice he reported the dinner conversations of a group of men interested in politics. No one admired the ministers, all feared a bankruptcy, all anticipated some revolutionary changes, all felt that a meeting of the Estates General would increase liberty. On the last point Young was skeptical:

. . . but I meet with so few men that have any just ideas of freedom, that I question much the species of this new liberty that is to arise. They know not how to value the privileges of THE PEOPLE: as to the nobility and the clergy, if a revolution added any thing to their scale, I think it would do more mischief than good.[78]

When some of his French friends inveighed against Madame du Barry's extravagance and against royal mistresses in general, pointing out that Frederick the Great had wasted no money on mistresses, Young commented:

No: but he had that which was fifty times worse: a king had better make love to a handsome woman than to one of his neighbour's provinces. The King of Prussia's mistress cost an hundred millions sterling, and the lives of 500,000 men; and before the reign of that mistress is over, may yet cost as much more. The greatest genius and talents are lighter than a feather, weighed philosophically, if rapine, war, and conquest, are the effects of them.[79]

Young left Paris on October 28 and reached Calais on November 8. His route lay north through Picardy and French Flanders,

[75] *Ibid.,* p. 144.
[76] *Ibid.,* p. 139.
[77] *Ibid.,* p. 133.
[78] *Ibid.,* p. 138.
[79] *Ibid.,* p. 142.

then to Dunkirk and south to Calais. He was impressed with the plate glass manufacture at St. Gobain, the Canal of Picardy, the good inns, the Flemish appearing towns, and the all too apparent fortifications. He castigated the narrowly selfish commercial interests of Lille whose proponents so hated the Eden Treaty that they favored war with England. Young's pacifism at this time was expressed in his comment on Dunkirk, "so famous in history for an imperiousness in England, which she must have paid dearly for. Dunkirk, Gibraltar, and the statue of Louis XIV in the *Place de Victoire,* I place in the same political class of national arrogance."[80] Near Dunkirk he visited a little settlement of new houses, each with a garden and several enclosed fields. The soil naturally was "wretched blowing *dune* sand," but it had been improved, and Young concluded, "The magic of PROPERTY turns sand to gold."[81] Since he had to wait three days at Dessein's famous inn at Calais for a wind and a boat, it must have been November 11 before he reached Dover and at least two days more before he saw his darling Bobbin at Bradfield, ". . . and have more pleasure in giving my little girl a French doll, than in viewing Versailles."[82]

While in France in 1787 Young wrote a short essay for the *Annals,* "A Coup d'Œil on the Present Situation of Europe."[83] The occasion was a possible Anglo-French war over Dutch internal affairs. Young maintained that such a war would be frivolous and completely contrary to the best interests of both countries. France could only hope to solve her internal problems by remaining at peace. Victory or defeat might be equally ruinous to both countries, "for the most brilliant victories, and the most ruinous defeats, end nearly, . . . in the same distress and weakness."[84] Such a war would increase the already swollen English national debt, and might very well completely ruin the squires and small gentlemen. The whole article reflects Young's predominant outlook at this time, anti-imperialist, pacifist, and *laissez-faire.*

Probably late in November, Young received a very cordial

[80] *Ibid.,* p. 152.
[81] *Ibid.,* p. 153.
[82] *Ibid.,* p. 155.
[83] *Ibid.,* pp. 673–681. *Cf.* also *Annals* 8: pp. 276–284.
[84] Young, 1793: p. 679.

letter from George Washington, thanking him for sending seeds, plows, and copies of the *Annals* and stating that he was preparing to construct a "barn precisely agreeable to your plan, which I think an excellent one."[85]

During the spring of 1788 Young was very busy leading the fight against the wool bill and spent nine weeks in London on the business. In August and September he was making his second tour in France. For Young's biographer these are the two most important topics for 1788.

As noted earlier, Young had helped to arouse opposition in Suffolk early in 1787 to the very stringent bill to tighten up the prohibition upon the export of raw wool. On returning from France in November, 1787, he found the issue very hot and threw himself again into the conflict. Without doubt Young and Sir Joseph Banks were the leading opponents to the bill. Article after article appeared in the *Annals,* filling a large part of two volumes.[86] He also published two pamphlets in the spring of 1788 on the subject, *The Question of Wool Truly Stated. . .*, and *A Speech on the Wool Bill, that might have been spoken in the House of Commons. . . .*

Young consistently denied that there was any appreciable amount of smuggling, as claimed by the woolen manufacturers. The real purpose of the bill was to depress the price of raw wool; the claim of smuggling was only a pretense. He charged quite openly that the woolen manufacturers had been guilty of a conspiracy to depress the wages of the wool spinners so that many spinners came upon the poor rates, thus adding to the burden upon the landed interests. He depicted the sufferings of the poor spinners in lurid colors, ". . . whole families of honest, industrious children offering their little hands to the wheel, and asking bread of the helpless mother, unable through this *well regulated* manufacture, to give it them."[87] In one article he invented the following imaginary conversation between the spinners and the manufacturers:

[85] *Letters from . . . Washington to . . . Young . . .*, 1802: p. 11. Dated November 1, 1787.

[86] *Annals* **8**: pp. 467–490; **9**: pp. 73–82, 266–376, 458–465; **10**: pp. 9–126, 139–185.

[87] *Ibid.* **9**: p. 269. This material came from a re-print of Young's very long letters in the *Bury and Norwich Post*, February 6, 13.

Mr. Manufacturer, please to buy my wool, it is 10 s. a tod here, and I can get 50 s. abroad for it:
It does not suit me to buy.
Why not?
Because my goods are on hand.
You will not buy?
Certainly not.
Then I beseech you, let me sell to those who will.
By no means: we are the men who give 'encouragement and vigor to your industry, by furnishing you a market.' Have patience, we will buy, when our goods are sold.
But in the mean time, I break and starve.
That is a bad thing, but we also are distressed, and therefore you must be contented.[88]

Young also sent out a questionnaire requesting information about the wages of woolen spinners. The *Annals* summarized the replies of twenty-nine correspondents, from seventeen different counties—clergymen, squires, and farmers—and Sir Joseph Banks and Lord Sheffield. The returns showed that Suffolk woolen spinners received extremely low wages and thus tended to substantiate Young's charges.[89]

He bitterly denied that the interests of the farmer and the manufacturer were identical, and urged the landed interests to combine to protect themselves. He was shocked at the apathy of the landed people when organized opposition to the bill appeared in only two counties. In one place he declared, "Gentlemen. . . you are of no more weight and consideration in the eyes of the government of this country, than the sheep and cattle of your fields,"[90] while in another he exclaimed, "The landed interest, which is a generic term for stupidity, folly, timidity, ignorance, dissipation, and disunion."[91]

Naturally enough, Young was chosen by the Suffolk wool growers to present a parliamentary petition against the bill. Consequently he was in London from April until the final passage of the measure on June 17, "with no other business whatever for my object. . . ."[92] He attended the sessions regularly and gave

[88] *Annals* **9:** p. 363.
[89] *Ibid.*, pp. 353–354.
[90] *Ibid.* **10:** p. 3.
[91] *Ibid.*, p. 252.
[92] *Ibid.*, pp. 8–9.

in the *Annals* detailed reports on the bill.[93] Here his old training as a parliamentary reporter must have stood him in good stead. Lincolnshire's similar petition against the bill was presented by Sir Joseph Banks, president of the Royal Society. As counsel for the petitions Thomas Erskine highly praised Young, "who had been for years employed in the steadiest and deepest researches in these enquiries, and whose perfect knowledge of the subjects he would speak to, was undisputed. . . ." Unfortunately Erskine also stated that Young had inherited "some portion of the talents of the great poet his father,"[94] thus confusing the Rev. Arthur Young with the Rev. Edward Young of the *Night Thoughts*. Young gave witness in the House of Commons on April 22 and 23, and in the House of Lords on June 7, attempting to prove that very little English wool was smuggled into France and introducing letters from Mouron of Calais and from Lazowski to support his evidence.

Earlier Young had been attacked in the newspapers as "wicked pilferer," "malicious pillager," "paltry propagator and retailer of falsehood," and "mercenary writer of a party."[95] And now, Mr. Partridge, counsel for the bill, stated:

. . . for the general prejudice of Mr. Young, little reliance ought to be placed on his evidence. God forbid, however, added Mr. P. that I should attack his evidence as a moral man. I do it not—but upon this business he is unskilled, and wanting knowledge of the subject, his evidence must vanish.[96]

Sometime early in May, Young was burned in effigy by the manufacturing interests, an incident that called forth congratulations from Sir Joseph Banks:

I give you joy sincerely at having arrived at the glory of being burned in effigy; nothing is so conclusive a proof of your possessing the best of the argument. No one was ever burned if he was wrong—the business in that case is to expose his blunders—but when argument is precluded firebrands are ready substitutes.[97]

[93] *Ibid.*, pp. 9–126, 139–185.
[94] *Ibid.*, p. 13.
[95] *Ibid.* **9**: p. 278.
[96] *Ibid.* **10**: p. 36.
[97] Young, 1898: p. 174. Banks mentions no place for the incident. B. H. Latrobe's letter (*ibid.*, p. 173) says it took place in Bury, while Young himself (*ibid.*, pp. 172, 174) says Norwich. I found no account in the *Bury and Norwich Post*.

The measure finally passed both houses by a majority of three
to one, but in very thin houses with less than one hundred voting
in the Commons, and only thirty-three in the Lords.[98] Young was
naturally bitter: "And thus passed a bill the most preposterous,
as a political measure, and the most unjust, and even tyrannical,
as an internal regulation, that had, for many years, been in the
contemplation of the legislature. . . ."[99] One more quote will
summarize Young's whole sense of frustration and bitterness:

I must confess I loose [*sic*] all patience, when I see the possessors of
a rental of twenty millions sterling, a-year, and the occupiers of it
with an hundred millions in their pockets, with all the classes that
depend on them driven, for want of union, into the kennel, and
trampled under feet there, by a combination of tradesmen.[100]

During the long weeks in London, Young probably visited
the Burneys, but the letters and diaries are silent. Hoping to
enlist Charles James Fox in his cause, he obtained a free pass
to the trial of Warren Hastings and apparently attended it fairly
regularly. One day an attempt was made to rob him while walking in Fleet Street, but when his pocket book fell to the street,
Young seized it immediately, and the pickpockets made off.[101]
Young was greatly thrilled when, with his Suffolk friend, William Macro (also a parliamentary witness), he spent a day at
Mr. Ducket's famous farm at Esher. He was again impressed with
Ducket's machinery and especially with his meticulous care in
executing his various agricultural operations. Ducket dined with
them at the *Tun* and they carried on "a very interesting conversation with him to a late hour."[102] "This was a very rich day for
us."[103] On the following morning, April 28, Young attended a
breakfast at the *Ram* in Smithfield, "at what might be called
Mr. Bakewell's levee,"[104] and noted with pleasure that a peer of
the realm was present. The conversation was a good farming one
on the relative merits of Norfolk and South Down sheep.

[98] *Annals* 10: pp. 126, 185.
[99] *Ibid.*, p. 185.
[100] *Ibid.*, p. 408 note.
[101] Young, 1898: p. 164.
[102] *Ibid.*, p. 172.
[103] *Annals* 10: p. 198. From this interview a dispute arose between Young and
Macro on certain aspects of Ducket's work. *Cf. ibid.*, pp. 435–437.
[104] *Ibid.* 15: p. 293.

In April, Young also began his experiments with chicory as an animal food. He had noted its use in France and brought back some seed. The first cutting on July 24,[105] just a few days before he left for France, more than met his expectations. Later he was bitter at "the utter stupidity of the farming world" for their "neglect of this plant."[106]

It was in 1788 that Count Leopold Berchtold visited Young at Bradfield. Young greatly admired his Spartan simplicity, his gift for languages, and his enlightened interest in the Austrian peasants: ". . . of all the multitude of foreigners who frequented my house the most persevering and the most intelligent."[107] In the following year Berchtold dedicated to Young one of his most famous books, *Essay for Patriotic Travellers,* and referred to Young's "unparalleled zeal, ability, and patriotic labours. . . ."[108]

Young's second trip to France was the shortest and least interesting of his three great journeys. He did not go to Paris and confined his journey to the western provinces—Normandy, Brittany, Anjou. He left Bradfield on July 30, reached Calais on August 5, and sailed back from Dieppe to Brighton on October 15. As in 1787, the *Annals* were confided to Thomas Le Blanc, who accepted the job rather unwillingly: ". . . but if you leave the annals unprovided as you did last year it is a very unpleasant task to undertake. . . . it was not conducted last year at all to my satisfaction altho' the deficiency of material was not to be laid to my charge. . . ."[109] He traveled very simply on horseback with his "cloak-bag"[110] or portmanteau, slung on behind him. Before he had been in France a week his mare began to go blind, but she carried him the whole journey without mishap. At Amiens he put up at an inn where Charles James Fox had stayed the night before. He commented in a letter to Le Blanc:

Mr. Fox with a girl in a postchaise slept at this Inn last night, a waiting woman & his valet in a cabriolet & a French courier—the

---

[105] *Ibid.* **10:** pp. 216–217. Later volumes contain frequent references to his cultivation of this plant. *Cf.* **11:** pp. 145–146; **13:** pp. 252–253; **15:** pp. 395–400; **17:** pp. 202–206; **20:** pp. 188–203; **22:** pp. 348–349, 443–444; **24:** pp. 23–29; **26:** pp. 489–495.

[106] Young, 1898: p. 172.

[107] *Ibid.*, p. 170.

[108] Published in English in 2 vols., London, 1789.

[109] Add. MSS. 35,126, ff. 422–423. No date.

[110] Young, 1898: p. 170.

English of the Table d' hote, said he travelled in no stile. It is however a plaguey difft. stile from mine but I have been doing all my life and he has been talking, & therefore my blind mare is but the world's justice, tho' not very poetical.[111]

From Calais he went south through Arras and Amiens to Rouen, where his first impression was unfavorable, "this great ugly, stinking, close, and ill built town, which is full of nothing but dirt and industry."[112] From Rouen he went to Havre, "fuller of motion, life, and activity, than any place I have been in France." He spent several days there, met many "cheerful, pleasing, and well informed" people, and hated to leave a society "which would have made a longer residence agreeable enough."[113] He crossed the Seine estuary by ferry and went to Caen through a very rich country, well enclosed and wooded, and full of fine cattle. Three very pleasant days were spent at Caen with the Marquis de Guerchy, who had been his guest at Bradfield. He met the neighboring nobility and apparently was smitten by the beautiful Norman ladies:

If these French marquesses cannot show us good crops of corn and turnips, here is a noble one of something else—of beautiful and elegant daughters, the charming copies of an agreeable mother; . . . cheerful, pleasing, interesting: I want to know them better, but it is the fate of a traveller to meet opportunities of pleasure, and merely to see to quit them.[114]

One day on the way from Caen to Cherbourg he was quite ill, "not a bone without its aches; and a horrid dead leaden weight all over me. I went early to bed, washed down a dose of antinomial powders, which proved sudorific enough to let me prosecute my journey."[115] He spent one very busy day at Cherbourg visiting the harbor improvements and a neighboring glass factory, but found the town very crowded and expensive: "I was here fleeced more infamously than at any other town in France. . . ."[116] Beyond

[111] Add. MSS. 35,126, f. 427. The "girl" was almost certainly Mrs. Armistead, with whom Fox was living at this time, and whom he took on a continental tour in 1788.
[112] Young, 1793: p. 158.
[113] *Ibid.*, pp. 161–162.
[114] *Ibid.*, p. 165.
[115] *Ibid.*, p. 167.
[116] *Ibid.*, p. 172.

Cherbourg he admired the fine mud houses and barns in the rich Cotentin and remarked that Mont St. Michel was "a most singular and picturesque object."[117]

Young spent just about three weeks in Brittany. He went south to Rennes, then west to Brest, then back east to Nantes. As a whole the province seemed miserably poor. He found Combourg "one of the most brutal filthy places that can be seen,"[118] and noted at Montauban, "A beautiful girl of six or seven years playing with a stick, and smiling under such a bundle of rags as made my heart ache to see her. . . ."[119] He also noticed the complete difference of the Bretons from the French, and their similarity to the Welsh. Near Nantes he traveled over almost unbelievably bad roads to visit the Count de la Bourdonaye, who received him very hospitably. The count, like many others in France, was surprised that Young received no state subvention for his trip, to which Young replied, ". . . whether the minister was whig or tory made no difference, the party of THE PLOUGH never yet had one on its side. . . ."[120]

Nantes impressed Young with its luxury after Brittany's poverty:

Mon Dieu! cried I to myself, do all the wastes, the deserts, the heath, ling, furze, broom, and bog, that I have passed for 300 miles lead to this spectacle? What a miracle, that all this splendour and wealth of the cities of France should be so unconnected with the country![121]

The theater was twice as large as Drury Lane, "and five times as magnificent," while the Hotel de Henri IV was probably "the finest inn in Europe,"[122] and quite cheap. He also noted the *chambre de lecture,* and Wilkinson's factory for boring cannon. He found an ardent spirit of liberty at Nantes: "The American Revolution has laid the foundation of another in France, if government does not take care of itself."[123]

After leaving Nantes Young turned east to Angers and then back north to Rouen, passing through Le Mans. In the Loire

[117] *Ibid.,* p. 173.
[118] *Ibid.,* p. 174.
[119] *Ibid.,* p. 176.
[120] *Ibid.,* p. 185.
[121] *Ibid.,* p. 187.
[122] *Ibid.,* pp. 177–178.
[123] *Ibid.,* p. 190.

valley he witnessed the cheerful activities of the vintage. Further north there was much wasteland, but beyond Le Mans the country and the agriculture greatly improved. He was particularly anxious to see some celebrated agricultural improvements by the Marquis de Tourbilly made about 1760.[124] After considerable difficulty he located the estate, only to find the Marquis dead, but he was hospitably received, shown about the estate by a laborer who had worked under Tourbilly, and even presented with a manuscript by Tourbilly.

A better hotel, some introductions to men who gave him valuable information, and two trips to the theater—all combined to make his second impression of Rouen much better than the first. Although tempted to return directly to England because he had received no letters from home, he finally decided to make a side trip to La Roche Guyon, the home of his friends of 1787, the Duchess d'Anville and the Duc de la Rochefoucauld, where he spent three very agreeable days. He found the way of life at La Roche Guyon very similar to that of an English nobleman of relatively the same income and remarked: "Europe is now so much assimilated, that if one goes to a house where the fortune is 15 or 20,000 *l.* a year, we shall find in the mode of living much more resemblance than a young traveller will ever be prepared to look for."[125] He pointed out, however, that an English nobleman would have invited farmers in to meet him for dinner, and that the ladies would also have been present. Such a thing was almost unheard of in France, where he had only experienced it once, in the previous year at Liancourt.

On returning again to Rouen he had a satisfactory interview with the secretary of the Society of Agriculture and witnessed another performance at the theater. He readily admitted that the French theater was markedly superior to the English in every respect and noted that French audiences exhibited the same generous sentiments he had so often seen in England. After which he remarked: "We are too apt to hate the French, for myself I see many reasons to be pleased with them; attributing faults very much to their government; perhaps in our own, our rough-

---

[124] *Ibid.*, pp. 193–200. Tourbilly's work was *Mémoire sur les défrichements* (1760).

[125] *Ibid.*, pp. 205–206.

ness and want of good temper are to be traced to the same origin."[126]

From Rouen he went to Dieppe and fortunately found the passage boat ready to sail for Brighton, where the Castle Inn there seemed like "fairy land," "But I paid for the enchantment." The next to the last paragraph for his trip in 1788 included a tribute to his blind mare which shows a love of animals not very often expressed by Young:

. . . go on board with my faithful sure-footed blind friend. I shall probably never ride her again, but all my feelings prevent my selling her in France. Without eyes she has carried me in safety above 1500 miles; and for the rest of her life she shall have no other master than myself; could I afford it, this should be her last labour: some ploughing, however, on my farm, she will perform for me, I dare say, cheerfully.[127]

From Brighton he went on to Sheffield Place, "a house I never go to, but to receive equal pleasure and instruction." Some chance remarks thrown out by his hosts convinced him that he had lost a child while in France, so he wrote an almost hysterical letter to Mary:

Dear Mary,

If I have yet a daughter I write to . . . there seems in Lord & Lady Sheffield amidst all their denials, something that fills me with horror.—for I think I see that something has happened. Have my bed aired what I have suffered from the black cloud of anxiety that enwraps me is beyond all expression. God of his infinite mercy! spare me the stroke of the loss of a child!—but this strange silence if you are still living—can be nothing else . . . God bless you.

A.Y.

. . . I dare name no one lest they sh'd have no existence.[128]

Fortunately his fears were entirely groundless.

About two months after returning from France, Young set off again on a month's farming tour, chiefly in Sussex, with his

---

[126] *Ibid.*, p. 203.
[127] *Ibid.*, pp. 209–210.
[128] Add. MSS. 35,126, ff. 448–449. This letter is undated. The Young manuscripts made several guesses, one that it was in 1797 when Bobbin died. The postmark on the back looks like 1788. Young, 1898: p. 210, with references to conversations with Lord Sheffield, make it very likely that it was written from Sheffield Place.

neighbor, the very substantial farmer, William Macro.[129] They traveled in Young's post chaise, left Bradfield on December 12, and were back on January 12, 1789. On the way south they spent much time discussing whether it was more profitable to be a landlord or a farmer, and concluded that a farmer might make as much as eighteen per cent profit, while the landlord who rented his land could only hope for three per cent. Truly the large-scale farmer was a capitalist in Arthur Young's lifetime! The chief purpose of the trip was to examine the famous South Down cattle and sheep of Sussex. On December 17 they arrived at Sheffield Place and stayed there for an entire week. They not only examined Lord Sheffield's own herds and flocks, but through his kindness, those of his tenants and neighbors, including John Ellman of Glynd, one of the most famous sheep men in eighteenth-century England, with whom they stayed three days. On a farm of 580 acres Ellman kept a flock of 1,450 South Down sheep and 140 cattle.

On December 29, while near Battle, they heard that General James Murray, of Beauport Park kept four thousand South Down sheep and fed them largely on potatoes. Young's account follows:

This was sufficient. To come into the country on the search for sheep and potato intelligence, and not see such a man, would not be to make a very wise figure when we returned home. But I had not the honour of being known to the General:—no matter:—4000 sheep fed on potatoes, were an object before which form must give way. I wrote a card, stating our pursuit, and wishes to have it gratified; desiring leave to view his flock. Those who know the General's liberality, and passion for agriculture, will not want to be told what the answer was. We spent five days in his house, and found it the residence of hospitality and good sense.[130]

When Young inquired for an account of the expenses and profits of his sheep, the general replied that he had not separated it from his general accounts, but offered to show Young all his accounts and let him figure it out. Young quickly accepted the invitation and actually spent nearly thirty hours of uninterrupted work on the accounts, aided by the general himself, his bailiff, shepherd, and bookkeeper. The weather was propitious, for the

---

[129] *Annals* 11: pp. 170–304. Much of this very long account is very technical.
[130] *Ibid.*, p. 260. *Cf.* also Young, 1898: p. 171.

snow was so heavy that they left the house only twice during five days, on one of which they "were five feet in drifting snow."[131] Young concluded that the general was making a profit of about seventeen per cent on his potato fed sheep. When they left General Murray on January 3, Young paid him a high tribute, as was befitting to such hospitality to a couple of strangers: "Took our leave of General Murray:—it has been to me a spectacle of singular pleasure, to see the fire and spirit of enterprise. . . . What other pursuit can be found conformable to the feelings of a veteran commander but agriculture? Interesting, philosophical, honourable!"[132] On January 12 they were back in Bradfield, feeling that their tour had "been productive of some of the most important intelligence I had ever an opportunity of laying before the public."[133]

In March, 1788, at a public meeting at Bury St. Edmunds, Young seconded Capel Lofft's motion for the abolition of the slave trade.[134] He also wrote several articles for the *Annals* in 1788 on the subject.[135] He brought out how the trade by its very nature must be inhumane. Only once, when reviewing Rev. John Newton's *Thoughts on the African Slave Trade,* did he wax eloquent and remarked that Newton had painted "a picture of horror, cruelty, misery, and depravity," that must arouse the "unreserved abhorrence which it becomes humanity to feel, and liberality to avow, against this abominable traffic."[136] He denied that the sugar islands were necessary to Britain's prosperity, repeatedly stated that slave labor was much more expensive than free labor, and declared that the abolition of the slave trade would gradually force the planters to employ "cattle and the plough, instead of negroes and the hoe."[137] In the same article Young also maintained that capital invested in the production of slave-made sugar could be much more advantageously employed in agriculture at home, even though such investments were discouraged by land taxes, poor rates, and tithes. When arguing that agriculture was underdeveloped, he listed the tasks still

---

[131] *Ibid.,* p. 261.
[132] *Ibid.,* pp. 293–294.
[133] *Ibid.,* p. 298.
[134] *Bury and Norwich Post,* March 26, 1788.
[135] *Annals* **9:** pp. 88–96, 180–186; **10:** pp. 335–362.
[136] *Ibid.* **9:** p. 182.
[137] *Ibid.* **10:** p. 344.

necessary to bring it to perfection, all of which required larger capital investments. Never did he itemize more completely the program of the agricultural reformers than in the following:

If friability is not given to wet lands by draining, and tenacity to dry ones by claying or marling—If water is not in irrigation conducted over the lands where it might be conducted—If gorze, fern, ling . . . &c. are permitted in fields that would produce, with improvement, better plants—If bog-marsh, or fen, remain undrained—If as much manure is not raised, in the farm-yard as might be—If watercourses and ditches are not kept clean—If bad breeds of cattle, sheep, hogs, and horses are kept where better might be maintained—If bad implements are in use, instead of good—If horses, cows, bullocks, and young cattle, are feeding at large, for want of buildings to stall them, or of labour to attend them—If fallows are continued, for want of force to substitute fallow crops—If a field produces but one crop in the year, which, with additional labour, could produce two—If soils of all sorts, are not applied to the production of most valuable crops of which they are capable—In all these, and in a thousand other cases, the soil is not stocked.[138]

In the last number of the *Annals* for 1788 Young claimed that they were not paying their way. He had sold only 422 copies of volume nine, for £139, while the expenses for that volume had been £210. Only the sale of earlier volumes had helped to balance the accounts. He threatened that if sales did not increase, he might have to cease publication.[139] Probably Young felt particularly pessimistic because his London distributor, William Nicoll, had recently gone bankrupt.[140] He complained that his defense of the landed interest had "provoked a host of enemies," and argued that his unremitting labor on the *Annals* should have furnished him at least with "an exemption from any addition of fresh anxieties, to the decline of a life, chequered with disappointment and infelicity."[141]

France remained the most important factor in Young's life during 1789. Early in June he started his last and most significant trip to the Continent. He was in Paris during the revolutionary events of June, spent the summer in the eastern prov-

[138] *Ibid.*, pp. 357–358.
[139] *Ibid.*, pp. 589–594.
[140] Add. MSS. 35,126, f. 450. This undated letter to Young from Charles N. Cole is listed in the Young manuscripts for 1788.
[141] *Annals* 10: pp. 591–593.

inces and most of the autumn in Italy. Early in January, 1790, he arrived back in Paris and spent most of that month there. At the very end of January, 1790, he was back at Bradfield, after an absence of nearly eight months.

The first five months of 1789 were uneventful with no high-lights. The winter was so severe that he sent out a questionnaire on January 23, requesting information on the temperature, the effects of the frost on various crops and prices, and the methods used to feed livestock.[142] Young's account of how he met the frosts incidentally mentioned that he had 252 sheep, 7 horses, 5 oxen, 28 "neat" cattle, and 20 hogs on a farm of 240 acres exclusive of woods and plantations.[143] He found especially useful his cabbages, "rouen," and turnip tops from a late autumn plant-ing, but he also used hay, turnips, potatoes, and bean meal.

Early in 1789 Young reviewed Joseph Priestley's *Lectures on History and General Policy*. He was pained by a statement from "one of the great ornaments of the age" that the only way to encourage agriculture was to make it subservient to commerce. Young again expressed completely *laissez-faire* ideals: "Would you encourage all? Leave them all to themselves. No prohibitions, no restrictions; and then, no encouragement: Give LIBERTY, and they want it not."[144] With complete inconsistency in the same number Young commented on the importation of American wheat: "My farming readers are desired to note these immense importations. . . . Good wheat selling in Suffolk at 40s. a quarter, and import of foreign corn allowed at the same time!"[145]

In his reply to Young's questionnaire on the frost, one of his Norfolk correspondents, the Rev. William Beevor, complained that Young digressed too often on political subjects like the wool bill. In his reply Young re-stated his conception of the purpose of the *Annals:*

My ideas of such a journal . . . are just the reverse. . . . I con-ceive it ought to treat of Agriculture, under its three great distinc-tions: 1, Practical; 2, Theoretical; 3, Political. I think it would be absurd to suppose that experiment and actual observation do not class in the first—chemical and electrical enquiries, so far as they

---

[142] *Ibid.* **11:** pp. 321–323.
[143] *Ibid.* **12:** p. 226.
[144] *Ibid.* **11:** pp. 378–379.
[145] *Ibid.,* p. 415.

concern the principles of fertility and vegetation, in the second—and the national conduct of all countries, in the encouragement or depression of Agriculture, in the third. [146]

Young probably left Bradfield for his third French journey on June 1. He stayed over two nights in London, secured letters for his Italian trip, attended the opera, and supped with Dr. Burney and Fanny Burney. The opera was so good that he forgot his sheep and pigs at Bradfield, but the conversation at the Burneys' was even better, "how seldom is it that we can meet two characters at once in whom great celebrity deducts nothing from private amiableness."[147] He commented on the Channel crossing: "14 hours for reflection in a vehicle that does not allow one power to reflect."[148] At Calais he took the coach or "diligence" for Paris. His companions were a Swede, a German, a Frenchman and his wife, an affected young Irish girl who taught French, and a young, good-natured Irishman. The French couple fleeced the Irishman of five louis at cards, and Young was completely fatigued by the singing and noise of the lively company. "I lose all patience in such company. Heaven send me a blind mare rather than another diligence!"[149]

When Young arrived in Paris he immediately sought out Lazowski whom he had commissioned to procure lodgings. He was certainly pleased, and probably not very much surprised, to find that his friends insisted that he be their guest. He remained in Paris from June 8 to June 28, a most exciting three weeks to be there. The Estates General had met and was engaged in the long fight by which the Third Estate transformed itself into the National Assembly. Young certainly utilized fully his opportunities to study the opening phases of the French Revolution. As guest of the powerful Rochefoucauld-Liancourt family he met such leaders as the Duc d'Orléans, Abbe Síeyès, and Rabaut St. Étienne. On June 15 he heard speeches to the Third Estate at Versailles by Síeyès, Mirabeau, Mounier, and Barnave, and a week later attended the historic session where some 150 deputies from the clergy joined the Third Estate. On both occasions he was presumably accompanied by Lazowski whose knowledge of

[146] *Ibid.*, p. 643.
[147] Young, 1793: p. 212.
[148] *Ibid.*, p. 213.
[149] *Ibid.*, p. 214.

men and of the situation undoubtedly enhanced Young's appreciation of what was going on. Repeatedly he listened to the incendiary speeches in the gardens at the Palais Royale. He also purchased the magnificent collection of revolutionary pamphlets which now constitutes one of the treasures of the Goldsmiths' Library at the University of London. Somehow he found time to make copious extracts from the Duc de Liancourt's collection of *cahiers*.

Young's first reactions to the course of events were mixed. He certainly had no love for absolute monarchy or the privileges of the nobility and was basically sympathetic with the revolutionary aims to destroy despotism and feudalism:

The spectacle of the representatives of twenty-five millions of people, just emerging from the evils of 200 years of arbitrary power, and rising to the blessings of a freer constitution, assembled with open doors under the eye of the public, was framed to call into animated feelings every latent spark, every emotion of a liberal bosom.[150]

As he left Paris he also expressed in his diary the strong hope that the revolution would lead to the establishment of well-regulated liberty in France:

Such benefits will confer happiness on 25 millions of people; a noble and animating idea, that ought to fill the mind of every citizen of the world. . . . I will not allow myself to believe for a moment, that the representatives of the people can ever so far forget their duty to the French nation, to humanity, and their own fame, as to suffer any inordinate and impracticable views,—any visionary or theoretic systems,—any frivolous ideas of speculative perfection: much less any ambitious private views, to impede their progress, or turn aside their exertions, from that security which is in their hands, to place on the chance and hazard of public commotion, and civil war, the invaluable blessings which are certainly in their power.[151]

Obviously the above statement contains many fears and doubts. Young was seriously disturbed by the violent tone of many of the pamphlets and the Palais Royale harangues, "carried to a degree of licentiousness and fury of liberty, that is scarcely credible."[152]

[150] *Ibid.*, p. 231.
[151] *Ibid.*, pp. 262–263.
[152] *Ibid.*, p. 250.

He discerned no great and responsible leaders and was very critical of the king, queen, and royal family, of Necker, Orléans, and Mirabeau. He compared the actions of the Third Estate to those of the Long Parliament, and predicted civil war and a republic. As a good Englishman, he advised the French to profit from England's long experience. The two upper estates might well unite to form a House of Lords to balance the Third Estate. He ridiculed the idea of a written constitution, "as if a constitution was a pudding to be made by a receipt."[153] He regretted that the liberal party refused the king's concessions at the royal sitting on June 23, for "much was granted to the people in great and essential points."[154] As a former parliamentary reporter he was shocked by the disorder of the Third Estate and even suggested to Rabaut St. Étienne that much time could be saved and much confusion averted by adopting the rules used in Parliament. In summary Young was very sympathetic with the general aims of the liberal party but was greatly worried over the uncompromising temper of the leaders.

Even during the exciting weeks of June, 1789, Young did not forget his primary purpose in visiting France—to study agriculture. As a member of the Royal Society of Agriculture, he attended two sessions of the body. On June 12 he was requested to suggest a subject for a substantial premium offered by the Abbé Raynal, but his proposal for some way to introduce turnip culture into France was rejected. On June 18 he was gratified when M. Brousonnet, the secretary, after consulting Young, proposed Washington as an honorary member, a proposal that was carried unanimously. Nevertheless he expressed skepticism about agricultural societies in general: "I am never present at any societies of agriculture, either in France or England, but I am much in doubt with myself whether, when best conducted, they do most good or mischief. . . ."[155] On June 13 he visited the National Library and was surprised that, among various models of instruments, there was none of a plow or a farm. On the next day he went to the Royal Botanical Garden, admired several promising plants and enthused over M. Thouin, the director, whom he had first met in 1787. On June 16 M. Brousso-

[153] *Ibid.,* p. 260.
[154] *Ibid.,* p. 247.
[155] *Ibid.,* p. 223.

net, with whom he had become very intimate, took him to see a great practical farmer, Cretté de Palieul, ten miles out of Paris. Several other farmers joined them at dinner, and an enjoyable farming conversation ensued on crop rotations and the value of chicory, which Young had started growing after his first trip. He commented:

I never see this plant but I congratulate myself on having travelled for something more than to write in my closet: and that the introduction of it in England would alone, if no other result had flowed from one man's existence, have been enough to show that he did not live in vain.[156]

On June 19 Brousonnet took him to dine with M. Parmentier at the Invalides, after which they visited the experimental grounds of the Royal Society, "of which I shall only say that I wish my brethren to stick to their scientific farming, and leave the practical to those who understand it. What a sad thing for philosophical husbandmen that God Almighty created such a thing as couch (triticum repens)."[157]

While in Paris Young wrote three letters to Mary, on June 11, 20, and 24. In the second he sent a kiss to Bobbin and continued, ". . . tell her to take care of ponds, & snakes & heating herself. . . ." He requested information on conditions on the farm and warned his bailiff to watch the fly in the sheep and to keep the cattle from getting into the corn. A postscript read: "The plainness of the french women is strange. I have not seen 3 handsome ones since I came."[158]

On June 28 Young left Paris and proceeded leisurely eastwards towards Strasbourg. He traveled in a "light French cabriolet for one horse, or gig Anglois,"[159] which he had bought in Paris. Less luxurious than the trip in 1787 in company with Count François de la Rochefoucauld, it was far more comfortable than on the "blind mare" in 1788. Now he could more easily collect specimens of soils and products as well as "regale with clean shirts wch. in a hot climate is the first luxury."[160] He left Paris rather unwillingly. He regretted losing the hospitality of the

[156] *Ibid.*, p. 236.
[157] *Ibid.*, p. 240.
[158] Add. MSS. 35,126, ff. 461–462.
[159] Young, 1793: p. 263.
[160] Add. MSS. 35,126, f. 467.

Rochefoucauld-Liancourt family and especially Lazowski, "whose anxiety for the fate of his country, made me respect his character as much as I had reason to love it for the thousand attentions I was in the daily habit of receiving from him."[161] Life in Paris was comfortable and exciting, while he dreaded French inns and provincial dullness. But his survey must be completed.

His first important stop was at Nangis, southeast of Paris, where he spent three nights at the seat of his friend, the Marquis de Guerchy, whom he had entertained at Bradfield and visited in 1788 at Caen. The Marquise taught him how to make an omelette, "the operation attending which occasioned no little merriment both in the kitchen and parlour."[162] Everyone at Nangis was a politician, even the *perruquier* who dressed him and declared that no more taxes should be paid and that French soldiers would never fire on the people. In vain Young tried to convince his friends to accept the English form of government. Although they willingly agreed that it was the best "the world ever saw," they appealed from "practice to theory."[163] The doctrinaire chaplain of the Marquis was "particularly strenuous for what is called the regeneration of the kingdom, by which it is impossible, from the explanation, to understand any thing more than a theoretic perfection of government, questionable in its origin, hazardous in its progress, and visionary in its end."[164] While at Nangis haymaking was taking place on the English lawn, "and I have had the Marquis, Mons. l'Abbé, and some others on the stack to shew them how to make and tread it: such hot politicians!—it is well they did not set the stack on fire."[165]

At Meaux he visited M. Gibert who had made a fortune from the plow, and who lived as a real farmer, his wife setting the table, and the bailiff and the dairy maid dining with Young and the family. "This is in a true farming style; it has many conveniences, and looks like a plan of living, which does not promise, like the foppish modes of little gentlemen, to run through a fortune, from false shame and silly pretensions."[166] At Chateau

[161] Young, 1793: pp. 263–264.
[162] Young, 1898: p. 187.
[163] Young, 1793: p. 266.
[164] *Ibid.*, pp. 265–266.
[165] *Ibid.*, p. 268.
[166] *Ibid.*, p. 274.

Thiery he was disgusted at not finding a coffee house or newspaper. "What stupidity, poverty, and want of circulation! This
people hardly deserves to be free. . . ."[167] A little further on
he had letters from Brousonnet to M. Le Blanc, upon whom he
counted for further letters for the champagne country. Unfortunately Monsieur and Madame were both away, but fortunately
Mademoiselle was a very pretty young lady who invited him to
await her parents' return. "When persuasion takes so pleasing a
form, it is not easy to resist it." He was hospitably entertained,
gained all the information and letters of introduction which he
could wish, and was delighted to find that Mademoiselle could
sing beautifully while her cousin played an "excellent English
piano forte."[168]

In the champagne country he admired the broad streets and
general air of well-being of Reims, but its cathedral impressed
him less than Amiens. He enjoyed the fine wines: ". . . getting
into Champagne I have excellent wine full of fixed air with the
cork flying up to the ceiling . . . & then ye rheu. left me and
thank God I am very well."[169] At Épernay he drank "prosperity
to *true* liberty in France."[170] At Chalons he met a French officer
who had learned "damme" while in America fighting the British,
who favored force against the French radicals, and who expressed
the common view that the English must be greatly pleased by
the confusion in France as a sort of revenge for French aid to
the Americans. Young remarked: "They feel pretty pointedly
what they deserve."[171]

One day in the country between Chalons and Metz he was walking up a hill to ease the load for his horse and met a poor
peasant woman who complained of poverty, taxes, and feudal
rents. Although only twenty-eight years old, she appeared sixty
or seventy, "her figure was so bent, and her face so furrowed
and hardened by labour." Young continued:

I am inclined to think, that they work harder than the men, and
this, united with the more miserable labour of bringing a new
race of slaves into the world, destroys absolutely all symmetry of

[167] *Ibid.*, p. 275.
[168] *Ibid.*, p. 276.
[169] Add. MSS. 35,126, f. 466.
[170] Young, 1793: p. 277.
[171] *Ibid.*, p. 280.

person and every feminine appearance. To what are we to attribute this difference in the manners of the lower people in the two kingdoms? To GOVERNMENT.[172]

At Metz he was pleased with a cheap and excellent inn, was entertained most hospitably by the Academy of Sciences whose members answered many of his enquiries, and found a public reading room where he consulted a newspaper for the latest news. Indignant at the ignorance and frivolity of the officers at the *table d'hote* at the inn, he exclaimed:

At table d'hotes of officers, you have a voluble garniture of bawdry or nonsense; at those of merchants, a mournful and stupid silence. Take the mass of mankind, and you have more good sense in half an hour in England than in half a year in France—Government! Again:—all—all—is government.[173]

Young found Nancy most attractive. Other towns might be more magnificient, "but there is more equality in Nancy; it is almost all good," while the Place Stanislaus was "superb." When he heard of Necker's dismissal, he inquired what the effects would be at Nancy. To the answer that they must wait on Paris, Young declared, "Without Paris, I question whether the present revolution . . . could possibly have had an origin."[174]

When Young entered Alsace on July 19, he was greatly impressed by its completely non-French character:

Looking at a map of France, and reading histories of Louis XIV, never threw his conquest or seizure of Alsace into the light which travelling into it did: to cross a great range of mountains; to enter a level plain, inhabited by a people totally distinct and different from France, with manners, language, ideas, prejudices, and habits all different, made an impression of the injustice and ambition of such a conduct, much more forcible than ever reading had done. . . .[175]

He arrived at Strasbourg on the day after the news of the fall of the Bastille had reached that city. As he entered he passed a demonstration with trumpets and drums, which so frightened his horse that she nearly trampled "on Messrs. the *tiers etat.*" Young instantly recognized the event's significance:

[172] *Ibid.*
[173] *Ibid.*, p. 283.
[174] *Ibid.*, pp. 285–286.
[175] *Ibid.*, p. 292.

Every thing being now decided, and the kingdom absolutely in the hands of the assembly, they have the power to make a new constitution. . . . It will now be seen, whether they will copy the constitution of England, freed from its faults, or attempt, from theory, to frame something absolutely speculative; in the former case, they will prove a blessing to their country; in the latter they will probably involve it in inextricable confusions and civil wars . . . .[176]

On the next day Young witnessed the sack of the old Hotel de Ville. As he was passing by, he noticed the beginning of the assault and climbed to the low roof of a market stall opposite. "Here I beheld the whole commodiously." Windows were broken, doors forced, and the whole building looted. "From that minute a shower of casements, sashes, shutters, chairs, tables, sophas, books, papers, pictures, &c. rained incessantly from all the windows of the house . . . and which was then succeeded by tiles, skirting boards, bannisters, frame-work, and every part of the building that force could detach." He saw a fine boy crushed to death by some falling object, and was shocked that the troops did not restore order and that some of them even shared in the loot. For two hours he watched the scene, "curious to a foreigner; but dreadful to Frenchmen that are considerate."[177]

While at Strasbourg, Young conversed with scientists, went to the reading room to catch up on the news, and visited the cathedral with its "singularly light and beautiful" tower. The journey from Strasbourg to Toulon took more than seven weeks. Young's route first led in a southwesterly direction through Franche Comté and Burgundy into the Bourbonnais as far as Moulins. The whole country was agitated by peasant uprisings, by the "great fear" of unknown "brigands," by wild tales of outrages and Parisian plots. Young noted repeatedly with astonishment and indignation the scarcity of reliable news. Such a tour at such a time was not exactly pleasant, and Young experienced considerable inconvenience and at times even danger.

At Colmar, Young heard a rumor that the queen was about to blow up the National Assembly and to massacre all Paris. He pointed out the patent absurdity of such a tale, but was not

[176] *Ibid.*, p. 293.

[177] *Ibid.*, pp. 296–297. The building is now the Hotel du Commerce in Gutenberg Place, which dated from the sixteenth century. For more detailed accounts of this event *cf.* Seinguerlet, 1881: pp. 17–26, which uses Young's account and states that a contemporary print substantiates it, and Ford, 1958: pp. 235–253.

believed. "Thus it is in revolutions, one rascal writes, and an hundred thousand fools believe."[178] He arrived at Belfort just one day after serious riots. On July 26 he was forced to buy a revolutionary cockade, but it was pinned on so carelessly that it blew off before he reached the next town, so he was again challenged and questioned, and the people "were very menacing in their manner." With great presence of mind he made an impromptu speech on the steps of the inn, maintaining that in England the nobility rather than the commoners bore the brunt of taxation, after which the mob "gave me a bit of a huzza."[179] He was careful not to lose his cockade again. At Besançon the rumors of outrages were very shocking, "Many chateaus have been burnt, others plundered, the seigneurs hunted down like wild beasts, their wives and daughters ravished, their papers and titles burnt, and all their property destroyed. . . ."[180] Young tried to secure a passport, but no one knew him, and the official was pretty surly in his refusal. Even the theater was disappointing, and Young concluded peevishly: "I do not like the air and manners of the people here—and I would see Besançon swallowed by an earthquake before I would live in it."[181]

Since he had no passport, Young went as rapidly as possible to Dijon where he hoped to find M. de Virly who had visited him at Bradfield. When he reached the gate his passport was demanded, and he was taken to the Hotel de Ville for examination, but, when he claimed friends, was permitted to go to his hotel. To his dismay M. de Virly was out of town. He then sought out M. de Morveau, the celebrated chemist, who should have received letters about him, but none had arrived. Morveau received him courteously, and invited him to call again the next morning, but "I felt very awkwardly."[182] One family at his inn had escaped half naked from their burning chateau, and two gentlemen had been driven from their estates. Naturally Young was pretty nervous that first day at Dijon, but the next morning brought relief, for Morveau had received letters from de Virly and Brousonnet. Now he could obtain his very necessary pass-

[178] *Ibid.*, p. 300.
[179] *Ibid.*, p. 305.
[180] *Ibid.*, pp. 305–306.
[181] *Ibid.*, p. 311.
[182] *Ibid.*, p. 312.

port. There followed two very pleasant days with Morveau and his friend, the learned Mme Picardet, including one dinner, two evening promenades, and a trip to the Dijon Academy of Sciences. He found Morveau "a lively, conversable, eloquent man,"[183] and Mme de Picardet "a very pleasing unaffected woman . . . a treasure to M. de Morveau," for she could discuss chemistry. At the dinner party the conversation centered on the phlogiston theory rather than on the Revolution. Morveau, like Lavoisier, had completely rejected the theory and considered "the controversy as much decided as the question of liberty is in France."[184] During one conversation Morveau denied the existence of the much feared "brigands" and assured Young that the outrages had been performed by the peasants themselves. Surely the days at Dijon mark a high spot in Young's 1789 trip, and must have put him in better humor again.

On leaving Dijon he passed through the rich wine country of Burgundy. He viewed John Wilkinson's very considerable manufactures at La Creusot. The weather was very hot and on August 5 he had a fever and sore throat. He resisted the temptation to rest a day to recover, thinking of the "loss of time, and vain expence." The roads were good, and much land was enclosed, but "villainously cultivated."[185]

He stayed three nights at Moulins, but not because his inn was so attractive, for he "paid extravagantly for the mud walls, cobweb tapestry, and unsavoury scents of the *Lyon d'Or*."[186] On requesting a newspaper at the chief coffee-house, he was told they were too expensive—"I might as well have demanded an elephant." But when the price proved to be high for his coffee, he burst out indignantly, "It is a great pity there is not a camp of *brigands* in your coffee room, Madame Bourgeau."[187] What detained Young at Moulins was the prospect of buying an estate there at an unbelievably cheap price. It was capable of very great improvement and promised considerable profit. House, grounds, stock, and general condition were all excellent, but the fear of the revolution held him back. As he put it, the possibility

[183] *Ibid.*, pp. 317–318.
[184] *Ibid.*, p. 314.
[185] *Ibid.*, pp. 325–326.
[186] *Ibid.*, p. 334.
[187] *Ibid.*, pp. 327–328.

"that in buying an estate I might be purchasing my share in a civil war—deterred me from engaging at present. . . ."[188]

From Moulins, Young went almost directly south to Clermont through Auvergne, and then southeasterly through the Vivarais and across the Rhone to Montélimar. Much of this country was volcanic in origin and Young delighted in the fine lava roads, the fantastic shapes of the mountains, the hexagonal basaltic columns, and the hills cultivated almost to the very top. This portion of his trip, which took about ten days, was also filled with exciting adventures of several kinds.

He disliked Clermont, "one of the worst built, dirtiest, and most stinking places I have met with. There are many streets that can, for blackness, dirt, and ill scents, only be represented by narrow channels cut into a night dunghill."[189] At Clermont he learned of the August 4 decrees abolishing the feudal system. From Clermont he made a side trip to the springs which gushed forth from the volcanic mountains, and hired a peasant village woman as a guide. When they descended to the village the poor woman was arrested by the national guard for guiding a stranger. Determined that she should not suffer on his account, Young accompanied her to the village authorities where the case was heard on the spot. When Young defended her they questioned his intentions. If he was interested only in volcanoes and springs, why was he asking questions about the value and price of land? He only secured her release when he insisted that if they imprisoned her, they must also imprison him and take the consequences. As in many other places, Young was astonished and indignant at the apparent lack of interest of his companions at the *table d'hôte* in the momentous events occurring in Paris. He was certain that everyone in England was discussing them, while he could get no reactions from the French themselves. "The ignorance or the stupidity of these people must be absolutely incredible."[190] He never seemed to realize that people might hesitate in a revolutionary period to answer the questions of an inquisitive foreigner!

On the day after leaving Clermont he had a letter to M. Brés, a physician at Izoire, who invited him to supper, where then

[188] *Ibid.*, p. 333.
[189] *Ibid.*, pp. 335–336.
[190] *Ibid.*, p. 340.

ensued "much animated political conversation." Young found most educated Frenchmen very favorable to the recent happenings at Paris, but he was quite critical of the destructive tendencies of the National Assembly:

. . . they seemed to have a rage for pulling down, but no taste for rebuilding; that if they proceeded much further in such a plan, destroying every thing, but establishing nothing, they would at last bring the kingdom into such confusion, that they would even themselves be without power to restore it to peace and order; and that such a situation would, in its nature, be on the brink of the precipice of bankruptcy and civil war. . . .[191]

He had another encounter with the national guard at Thuytz. He was refused a mule and a guide to view a famous extinct volcano crater near that town. His enquiries seemed so suspicious that his room at the inn was invaded by twenty national guards-men armed with muskets, swords, sabers, and pikes. His passport and papers were demanded and he was accused of being in league with the queen and the Count of Artois to measure their fields so that taxes might be doubled. Only the fact that his papers were in English saved him from arrest. Finally they left him "to the bugs, which swarmed in the bed like flies in a honey-pot."[192] But his troubles had not ended. He missed his way the following morning on the mountain road to Villeneuve-le-Berg, and had to turn around, in which process his mare backed off the road and overturned the chaise. Only a projecting shelf just off the road saved him. He jumped out of the chaise unhurt and some lime-burners extricated his chaise and mare. On reaching Villeneuve-le-Berg he was again sought out at his inn by the national guard who asked him many questions, declared him a suspicious looking person, took his passport, placed a sentinel before his door, and summoned him to the Hôtel de Ville for further questioning. They were astonished at a farmer traveling for agricultural information, but he was treated courteously and eventually his passport was signed and he was offered every "assistance and civility I might want." He was particularly interested in visiting Pradel, the nearby home of the famous early seventeenth-century agricultural writer, Oliver de Serres. Fortunately one of his

[191] *Ibid.*, p. 342.
[192] *Ibid.*, p. 351.

interviewers was a lawyer who had translated Sterne into French, and who took Young to his home, and conducted him in person to Pradel. Although there were few mementoes left, Young was thrilled: "I regarded the residence of the great parent of French agriculture . . . with that sort of veneration, which those only can feel who have addicted themselves strongly to some predominant pursuit, and find it in such moments indulged in its most exquisite feelings."[193]

At Montélimar he had a letter to M. Faujas de St. Fond, a celebrated naturalist and authority upon volcanos. He was received most hospitably, introduced to Abbé Berenger who was working to improve Prostestant-Catholic relations, and to Mme Chainet, wife of a deputy to the National Assembly, an excellent "farmeress" by whom Young was apparently quite smitten: "I was so charmed with the *naïveté* of character, and pleasing conversation of this very agreeable lady, that a longer stay here would have been delicious—but the plough!" Young was delighted with his host: ". . . the liveliness, vivacity, *phlogiston* of his character, do not run into pertness, foppery, or affectation."[194] The whole visit was most pleasurable: "This was one of the richest days I have enjoyed in France; we had a long and truly farming dinner; drank a l'Anglois success to THE PLOUGH! and had so much agricultural conversation, that I wished for my farming friends in Suffolk to partake my satisfaction."[195]

At Orange he admired the Roman ruins and at Avignon seemed most interested in the romantic stories of Petrarch and Laura. Young was more than ordinarily irritated by his French companions at the *table d'hôte* in Avignon, who predicted that the English would attack France during the revolutionary disturbances, but that the national guard would be more than a match for them. Young apparently replied with some heat:

And lastly, I assured them, that should the English attack them at present, they would probably make the weakest figure they had done from the foundation of their monarchy: but gentlemen, the English, in spite of the example you set them in the American war, will disdain such a conduct; they regret the constitution which you are forming, because they think it a bad one—but whatever you

[193] *Ibid.*, pp. 354–355.
[194] *Ibid.*, pp. 358–359.
[195] *Ibid.*, p. 361.

may establish you will have no interruption, but many good wishes, from your neighbour. [196]

Near Avignon he visited the scenes made famous by Petrarch along the banks of the Durance River. He found it "a sort of fairy scene," with well-dressed people promenading along the banks of the stream. Besides, there was an excellent inn. "I walked on the banks of this classic stream for an hour, with the moon gazing on the waters, that will run for ever in mellifluous poetry; retired to sup on the most exquisite trout and craw fish in the world." The next day he visited the famous spring at Vaucluse and again sentimentalized over Petrarch and Laura: "The scene is sublime; but what renders it truly interesting to our feelings, is the celebrity which great talents have given it."[197]

At Aix, Young was entertained two days by the Baron Tour d' Aigues, president of the Parlement of Aix. Young was impressed with the baron's fine agricultural library: "His collection . . . is nearly as numerous as my own."[198] The baron doubted whether the nobility would really be compensated for the loss of their feudal dues and feared they would be left only with "such houses as the mob allows to stand unburnt."[199] Young had noted another effect of the August decrees—the indiscriminate shooting of game all through Provence, to such an extent that five or six times the shot had fallen in or near his chaise. He thus summarized the peasants' reactions to the decrees: "In the declarations, conditions and compensations are talked of; but an unruly ungovernable multitude seize the benefit of the abolition, and laugh at the obligations or recompense."[200]

Young thought the buildings of Marseilles much less magnificent than those of Bordeaux. Nor was the port as fine. He did find, however, plenty of newspapers and was impressed by the teeming population. Since his letter of introduction to Abbé Raynal had not arrived, Young introduced himself. When Young complimented Raynal on his generous gift to the Royal Society of Agriculture for a premium, Raynal declared that he had made similar gifts to other learned societies and also mentioned a project for purchasing models of agricultural implements through the country. Young said that the project "merits great praise; yet it is to be questioned, whether the effect would answer the expense."[201] Young was delighted when Raynal favored some kind of an upper house for France, and also distrusted extreme democracy. Young surprised Raynal when he claimed that the loss of the American colonies greatly increased British prosperity:

. . . for a people to lose an empire—thirteen provinces, and to GAIN by that *loss,* an increase of wealth, felicity, and power! When will the obvious conclusion, to be drawn from that prodigious event, be adopted? That all transmarine, or distant dominions, are sources of weakness and that to renounce them would be wisdom.[202]

Young was not very appreciative of the beauties of the French Riviera. Olives, oranges, and lemons made little appeal to him. The country on the whole was barren, there was too much mountain and wasteland, too little cultivated land. Seldom could he procure milk or butter, edibles which he greatly preferred to olives or oranges. He was greatly impressed with Toulon harbor with its fine men of war, but he was not permitted to see the dockyard in spite of letters of introduction. He had read of the beauties of Hyères, but thought the scenery greatly overrated. Told that it would not be practical to take his mare and chaise into Italy, he decided to sell them. At Toulon he led them into the street and put up a sign that he would sell for 25 louis. The plan worked, and he got 22 louis, more than he had been offered at Aix. Since he had purchased them at Paris for 32 louis, and they had taken him 1,200 miles, he could not complain.[203] For the last part of the way to the Italian border, he hired a peasant woman and her ass while he walked: "Myself, my female, and her ass jogged merrily over the mountains. . . . It is not easy for me to describe, how agreeable a walk of ten or fifteen miles is to a man who walks well, after sitting a thousand in a carriage."[204]

From June 28, when he left Paris, to September 16, when he left France, six of Young's letters to his family have survived.

---

[201] *Ibid.*, p. 377.
[202] *Ibid.*, p. 376.
[203] *Ibid.*, p. 382.
[204] *Ibid.*, p. 388.

None were to his wife, although she wrote him. Four were addressed to Mary—July 9 from Chalons-sur-Marne, July 15 from Nancy, July 27 from Besançon, and August 25 from Montélimar.[205] In the last Young voiced his complete approval of the anticlericalism of the Revolution:

The things the States are doing for France will make it the first country and freest in the world. Tythes are suppressed: and the lower clergy demand openly to have wives, and will probably. Near this place the protestants & catholics joined in the same church in Thanksgiving & Te Deum.

In the letter of July 15 Young suggested that Bobbin was perhaps being spoiled: "I hope Bobbin is not so neglected as last yr. in regard to reading: & that I shall find her to read fluently any easy passage; it is shameful she should not be taught." There is a delightful letter to Bobbin in the *Autobiography* which shows his tenderness towards the little girl who was dearer to him than anyone in the world:

I think it high time to enquire of you how you pass your time—what you do—how Mr. Mag (the pony) does, and the four kittens; I hope you have taken care of them and remembered your Papa wants cats. Do the flowers grow well in your garden? Are you a better gardener than you used to be? . . .

Pray, my little girl, take care, and keep clear from weeds the row of grass I sowed in the round garden. . . . You do not know, my little Bobbin, how much I long to have a walk with you at Bradfield.[206]

By far the most interesting of Young's family letters of 1789, however, was to his only son. From 1785 to 1789 Arthur was at Eton, where he was apparently a good though not a brilliant scholar. During the 1788 French trip the father warned him to return to school promptly.[207] In 1787 he had been disappointed in not receiving more letters from his son,[208] and in 1789 he

---

[205] *Cf.* Add. MSS. 35,126, ff. 466–467, 472–473 for letters from Chalons-sur-Marne and Nancy. The letter from Besançon has been reproduced in manuscript p. 188 in *Autobiography*. In 1931 Miss Ethel Fitzroy, who then lived at Bradfield, permitted me to see the original of the Besançon letter and to copy the Montélimar letter.

[206] Young, 1898: p. 186. Dated from Moulins, August 7. The original is not in Young manuscripts.

[207] Add. MSS. 35,126, ff. 428, 431.

[208] *Ibid.*, f. 390.

expressed his impatience very forcibly to Mary: "I have not had a Lr. from him of 11 months; I suppose because I expressly desired one once a fortnight: but nothing surprises me that comes fro. him; Eton has I hope done so much for his head that it leaves nothing for his heart—God send it may prove so; & yt. I have not impoverished myself for nothing." The postscript added: "Say nothing to Arth: abt. writing; I had much rather have no Lrs. than such as those hints bring: observe this."[209] Obviously relations betwen father and son were far from ideal.

On July 19, 1789, Dr. John Symonds had written that Arthur had been admitted to Trinity College, Cambridge.[210] Of course he would go to Cambridge, for all the Youngs were Cambridge men and it was nearby. Apparently Mrs. Young was anxious that Arthur should go to Trinity and had written to Symonds on the subject. Symonds added: ". . . but the truth is, I was rather at a loss to determine upon any other College." Of course he himself was at Trinity, and he points out, "most certainly there is no college which offers so handsome a premium to young men, who declaim well either in English or Latin; & this is surely very advantageous to an Etonian, who is, or ought to be, particularly formed for it." The fact that the boy could not at first live in the college did not seem to Symonds a severe disadvantage. True it might be somewhat more expensive, but "your wife said, that it would not be difficult to pay the overplus of expence." He went on, "and, as to any danger in young men residing in the town, there is really not more than in College. All depends on the disposition of them."

Symonds's letter almost certainly reached Young before he wrote Arthur from Fréjus. The *Travels* notes that he spent an entire day at Fréjus, "to rest myself—to examine the neighborhood . . . and to arrange my journey to Nice."[211] But this remarkable letter to his son must have taken much of the day, and deserves to be quoted nearly in full:

Frejus, Sep. 13, 89

Dear Arthur,

I received yr. letter wch. I suppose would not have been written if you had not been ordered to do it by yr. Mother, for my repeated

---

[209] Young, 1898: opp. p. 188.
[210] Add. MSS. 35,126, ff. 468–469.
[211] Young, 1793: p. 387.

desire to hear fro. you had little avail. I find you are entered at Trinity & are to go the beg. of October. It is proper for me to give you some advice on yr. entering the world for at Cam. you are so much yr. own master that you may be said to be completely in it.

For four years I shall continue the expence of 100 L a year, after that period you must provide for y'self; my house and table will always be yours, but nothing more; You are therefore to take a resolution & to keep it to make such a use of yr. time as to do much better for yr. self than I cd. do for you, dilligence & perseverance will affect anything if you have done well at Eton—& not one lad in 500 have either been so long at school or at such good schools. I have therefore strong hopes that you will succeed. If you make a brilliant figure in carrying classic prizes you are sure of a fellowship at Trinity; but the society is so numerous that anything less than such a figure does nothing; in such a case all yr. hope is in ye mathematics for acquiring wch. nothing more than dilligence is necessary, every one is sure to succeed that has application sufficient, it is the fashionable study at Cambridge & without it you will do nothing. . . . I shall therefore mention only that algebraic & geometrical knowledge are applicable to all the sciences . . . and that these branches may be of essential use to you in yr. future destination: That destination deserves your most serious attention; a couple of wretched curacies or a lousy living will clothe you but for Gods sake have ambition to do better than that. . . . there is no branch of useful science that has not conducted men to fortune provided there is excellence & superiority; but moderation & a common degree, does nothing. If you attain a high reputation with a good solid character & the french language, a travelling tutorship will be at yr. command. I beg of you consider the importance of yr. time & on no account lose a single day even at Bradfield: and of all things remember (for I confess I fear that most in you) that idle amusive reading is time absolutely lost at yr. age: it gives nothing; but merely kills time that might give every thing. With these views my dear Arthur do not forget that all connection & idle acquaintance of expensive young men who are not celebrated for their talents lead only to expence, loss of time & at last loss of reputation . . . look at the herd of Suffolk boys that are at the university as rocks rising above the water to warn you from the same course: . . . chuse yr. acquaintance as much as possible with people older than yrself. & that will be easy if you gain a character for learning & knowledge . . . . use four yrs. to come well, and with an intrepid preseverance and you will be made for life. . . . There are so few young men to have the courage to do it that those who are steady are sure of the prize: For the first year classics & mathematics your only pursuit after that we will think of other things. I wish to God you would pick up a knowledge of agriculture, by the time I die the world will find out that they might have made a better use of my knowledge & make me offers when too late for me to take them, but it may afford opportunities for a son if he had nothing better, that may be of importance: Such however must not impede your immediate objects—but when at Bradfield you might catch much without loss of time—and the mere habit of gleaning & applying the knowledge that is floating about you is alone of consequence in forming a character. . . . Adieu. I heartily wish you success & am perfectly well assured that it is in your power if you have perseverance and œconomy for without the last the rest is nothing.

A. Y.[212]

Young entered Italy at Nice on September 16 and reentered France on Christmas Day. His route led him through Piedmont and Lombardy to Venice, and then southeastwards through Bologna to Florence. The return trip passed through Modena, Parma, and Turin and thence through the Mont Cenis pass and Savoy to France. Although the Italian trip has less interest than his French tours, it has far more than his short journey into Catalonia. Italian inns were generally better than those of France, and living was much cheaper. But the roads were much poorer and he had little good to say of "that odious Italian race, the *vetturini*,"[213] who provided the horses and carriages for most of his trip:

The world has not such a set of villains as these *vetturini*. . . . Their carriages are wretched, open, crazy, jolting, dirty dung-carts; and as to their horses, I thought till I saw them, that the Irish garrans had no rivals on the globe; but the *cavalli di vetturini* convinced me of the error.[214]

Young approached his Italian trip with high anticipations, for two of his closest friends—Charles Burney and John Symonds— were ardent admirers of Italy. On the day when he arrived in Nice he wrote in his diary: "The first approach to that country so long and justly celebrated that has produced those who have conquered, and those who have decorated the world, fills the

[212] Add. MSS. 35,126, ff. 478–479.
[213] Young, 1793: p. 515.
[214] *Ibid.*, p. 436.

bosom with too many throbbing feelings to permit a bush, a stone, a clod to be uninteresting."[215]

Young greatly enjoyed his four days in Nice. The inn was good, the companions at the *table d'hôte* pleasant, and the promenades fine. Also the English Consul, Nathaniel Green, to whom Young applied for a passport, put himself out to show Young the sights, especially the fine orange and lemon plantations, half-garden, half-farm, and even entertained him with a true English dinner, "with roast beef, plumb pudding, and porter."[216] Young's desire to see the Piedmontese plain led him to go to Milan via Turin rather than via Genoa. The scenery of the maritime Alps was romantic but less so than the Pyrenees. Among his fellow travelers were a Piedmontese colonel, the Chevalier Brun, and his brother, an abbé, both well informed and interesting men. At Cuneo the brothers pushed on a few miles further to visit a third brother, the priest at Centalle. The next morning the three brothers stopped Young's *vetturino* at Centalle and he was invited to have chocolate with them and then prevailed upon to stay for dinner. In order better to answer Young's questions, he was taken to the estates of neighboring landlords. Small wonder that Young was impressed with the kindness of these chance acquaintances of the road. "If I have many such days as this in Italy, I shall be equally well pleased and informed. . . . Take my leave of this agreeable and hospitable family, which I shall long remember with pleasure."[217]

Young felt that Turin's reputed beauty had been greatly exaggerated. The streets were too regular: "Circles, semi-circles, crescents, semi-elipses, squares, semi-squares, and compounds composed of these, mixed with the common oblongs, would give a greater air of grandeur and magnificence."[218] One day he took a horseback ride to the famous Superga where he climbed the dome for the view of the Alps and the Piedmontese plain, "the finest farmer's prospect in Europe."[219] From Turin to Milan he passed through the rice country of Novara and Magenta, "a nasty country, as ill to the eye as to the health."[220] His week at

[215] *Ibid.*, p. 391.
[216] *Ibid.*, p. 397.
[217] *Ibid.*, p. 404.
[218] *Ibid.*, p. 408.
[219] *Ibid.*, p. 411.
[220] *Ibid.*, p. 416.

Milan was very profitable, for he had letters of introduction
to the Abbate Charles Amoretti, Secretary of the Patriotic Society
of Milan of which Young was a member. Young found the Abbate
"agreeable, well-informed, and interesting,"[221] "whose attentions
and assiduity are such as I shall not soon forget."[222] Amoretti took
him to a meeting of the Patriotic Society, but as usual he was
contemptuous of their impractical and frivolous activities: "I
looked about to see a practical farmer enter the room, but
looked in vain. A goodly number of i Marchese, i Conti, i Cava-
lieri, i Abati, but not one close clipped wig, or a dirty pair of
breeches, to give authority to their proceedings."[223] One day
Amoretti took him to a farm of the Marchese di Visconti, chairman of the Patriotic Society, to witness the making of Lodesan
cheese, which Young found very interesting and he commented:
"This day has passed after my own heart, a long morning, active,
and then a dinner, without one word of conversation but on
agriculture."[224] He spent one day in sightseeing, but was disappointed with the Cathedral which "to the eye . . . is a child's
play-thing compared to St. Paul's." He visited several other
churches, the Ambrosian library, the hospital, and da Vinci's
Last Supper, but found the great painting one for artists, "as it
is not a picture for those who, with unlearned eyes, have only
their feelings to direct them."[225] Such candour is refreshing!

All through this part of Lombardy he was astonished at the
wonderful theaters. At Milan he apparently attended the opera
nearly every night. He was delighted with the recently built
Scala Opera House and the accommodations of the pit with its
"broad easy sophas, with a good space to stir one's legs in."[226] He
was greatly surprised at the good theaters at small towns without important manufactures or commerce. At Lodi he attended
a gala night with the Archduke and Archduchess present, and
commented:

. . . . there is not a town in France or England, of double the popu-

[221] *Ibid.,* p. 417.

[222] *Ibid.,* p. 420. For biographical sketch of Amoretti (1740–1816) *cf. Biographie
Universelle.*

[223] *Ibid.,* p. 418. This meeting took place in the famous Brere Palace.

[224] *Ibid.,* p. 420.

[225] *Ibid.,* pp. 424–426.

[226] *Ibid.,* p. 417. Young did not refer to the Scala by name, but almost certainly
he must have meant that building which was built in 1778.

lation, that ever exhibited a theatre so built, decorated, filled, and furnished as this of Lodi. Not all the pride and luxury of commerce and manufactures—not all the iron and steel—the woolen or linen—the silk, glasses, pots, or porcelain . . . ever yet equalled this exhibition of butter and cheese. Water, clover, cows, cheese, money, and music!. . . . The evening would have been delicious to me, if I had had my little girl with me. . . .[227]

Young took exactly two weeks to go from Milan to Venice, with frequent stops on the way. In general he was disappointed, for those whom he wished to see were either away or received him somewhat coldly. In his unsuccessful search at Bergamo for the secretary of the Academy, he had a flirtation with a lady from whom he was asking directions. As he put it succinctly in the *Autobiography:* "At Bergamo I was electrified by the fine eyes of an Italian fair, and just as I was making a nearer approach, impeded in it by the sudden appearance of her husband."[228] For the moment there was danger that the farmer might become a "sentimental traveller."[229] Between Bergamo and Brescia his companions in the *vettura* spent their time in repeating prayers and counting beads, which caused Young to ask: "How the country came to be well irrigated, is a question? Paternosters will neither dig canals, nor make cheese."[230] At Brescia he was again unable to present his letters and when he presented them at Verona his welcome was only perfunctory. On the other hand, at Vicenza the Abbate Pierropan, professor of physics and mathematics, spent an entire day and part of another in entertaining Young, introducing him to the proper people, visiting farms, and interviewing farmers. Naturally Young was much pleased with this "rich day, that pays for the trouble of travelling." He was also much thrilled at Vicenza by the numerous neo-classic masterpieces of his favorite architect Palladio—the rotunda, the Palazzo Raggione, and the Olympian theater, "which pleases all the world; nothing can be more beautiful than the form, or more elegant than the colonnade that surrounds it."[231]

With the limited time at his disposal Young had to choose between seeing Padua and Venice, or going south to Rome. As

---

[227] *Ibid.*, p. 428.
[228] Young, 1898: p. 176.
[229] Young, 1793: p. 435.
[230] *Ibid.*, p. 437.
[231] *Ibid.*, p. 443.

an agriculturist he chose Padua to see the experimental farm of Signore Arduini, reputedly the finest in Europe. On reaching Padua he immediately called on Arduini who was fortunately at home and invited him to view the farm the next day. At the opera that evening the music and acting were not bad, but the house was half empty, the audience shabby, and the musicians dirty. On the next morning he went sight seeing, "viewing buildings, of which few are worth the trouble." Later in the day he visited the famous experimental farm, but his disappointment was indeed deep:

Signore Arduino [*sic*] . . . showed me the experimental farm, as it ought to be called, for he is professor of practical agriculture in this celebrated university. I will enter into no detail of what I saw here. I made my bow to the professor; and only thought, that his experiments were hardly worth giving up the capital of the world. If I keep my resolution, this shall be the last œconomical garden that I will ever go near.[232]

Nor were the interviews on his third day at Padua with two professors at the university too satisfactory. The chemist, Professor Carbury, insisted on talking politics instead of agricultural chemistry, while the noted astronomer, Professor Toaldo, offered to give Young some of his printed works, and then charged for them. That night Young felt pretty ill-natured:

This is the third evening I have spent by myself at Padua, with five letters to it; I do not even hint any reproach in this; they are wise, and I do truly commend their good sense: I condemn nobody but myself, who have, for fifteen or twenty years past, whenever a foreigner brings me a letter, which some hundreds have done—given him an English welcome, for as many days as he would favour me with his company, and sought no other pleasure but to make any house agreeable. Why I make this minute at Padua, I know not; for it has not been peculiar to that place, but to seven-eighths of all I have been at in Italy. I have mistaken the matter through life abundantly,—and find that foreigners understand this point incomparably better than we do. I am, however, afraid that I shall not learn enough of them to adopt their customs, but continue those of our own nation.[233]

[232] *Ibid.*, pp. 445–446. *Cf. Biographie Universelle* for article on Louis Arduini (1739–1833).

[233] *Ibid.*, pp. 448–449. *Cf. Biographie Universelle* for article on Joseph Toaldo (1719–1798).

Granted that he had been grievously disappointed with the lack of hospitality of the professors at Padua. But it was unfortunate that he should have permitted such a sour passage to appear in his printed book, for it could only pain those who had entertained him so hospitably in Piedmont, at Milan and Vicenza.

Young spent the first week of November at Venice, comfortably housed in an excellent inn on the Grand Canal near the Rialto, and with his own gondolier. Much was to be said for Venice, but unfortunately there was no farm land:

. . . thus, for about 7s. 3d. a-day, a man lives at Venice, keeps his servant, his coach, and goes every night to a public entertainment . . . . If cheapness of living, *spectacles,* and pretty women, are a man's objects in fixing his residence, let him live at Venice: for myself, I think I would not be an inhabitant to be Doge. . . . Brick and stone, and sky and water, and not a field nor a bush even for fancy to pick a rose from! My heart cannot expand in such a place. . . .[234]

He soon wearied of sightseeing—palaces, churches, picture galleries. If only there were a guide book which listed the sights in order of importance. His favorite picture was Veronese's "Family of Darius at the Feet of Alexander." He greatly admired the bronze horses on the Cathedral, but they were too far away to be seen clearly. He climbed the Campanile for the magnificent views of the islands. The most impressive churches were St. Georgio Maggiore and St. Maria della Salute. Every evening Young attended the opera or theater. He did not admire Italian ballet and complained that the performers were more agile and energetic than graceful, but he greatly enjoyed most of the operas and plays. He commented especially upon the beauty of the Italian language which seemed just as suited for lofty and terrible sentiments as for those of love and pity.

The five-day trip from Venice to Bologna was made on a covered barge with a dozen people crowded into a dark cabin ten feet square. They slept on mattresses on the floor, "packed almost like herrings in a barrel." The *padrone* cooked in the same cabin and when they changed to a smaller boat not large enough to cook in, he brought his food in a paper, "which he spread on his knees as he sat, opening the greasy treasure, for those to eat out of his lap with their fingers, whose stomachs could bear such a repast." Fortunately much of the route led along the banks of the Po, and Young thus walked much of the way and got his meals at coffee houses. One short stretch was made by a road so bad that seven horses pulled the coach about a mile and a half an hour. Again Young walked. The whole trip must have been pretty ghastly: "The time I passed . . . I rank among the most disagreeable days I ever experienced, and by a thousand degrees the worst since I left England."[235]

The three days at Bologna were pleasant and profitable. Letters to Signore Bignani, a wealthy merchant, resulted in a dinner and a visit to a country estate. Young was even more pleased to find at Bologna an English acquaintance, the Rev. Edward Taylor of Bifrons in Kent. Taylor held a sort of *conversazione,* where Italian and French nobles congregated. Young presented himself the first night and returned the two following evenings.[236] At Mr. Taylor's Young also met the Baron de Rovrure and Madame la Marquise de Bouille who were going to Florence. A party was made up, which also included an Italian merchant and a Scottish tutor named Stewart with his pupil Mr. Kinloch. They jointly hired three *vetturi* and the two-day trip was made over vile roads with stops at miserable inns. Young was disgusted at the accommodations on such an important road: "England and Italy have a gulph between them on the *comforts* of life, much wider than the channel that parts Dover and Calais."[237]

Unquestionably his two-week stay at Florence was the most pleasurable part of Young's Italian trip. Young, the French baron and marquise, the Scottish gentleman and his protégé, all took lodgings together and were later joined by Young's Milanese friend, the Abbate Amoretti. The group was singularly congenial:

There was not one in the party which any of us wished out of it; and we were too much pleased with one another to want any addition . . . Half a dozen people have rarely been brought together, by such mere accident, that have better turned the little nothings of

---

[235] *Ibid.,* pp. 467–468.
[236] *Ibid.,* p. 481. For Taylor's agricultural experiments and picture collection, *cf.* Young, 1771: **1:** pp. 52–60.
[237] *Ibid.,* p. 485.

life to account (if I may venture to use the expression) by their best cement—good humour.[238]

Fortunately for Young the English Ambassador at Florence was Lord John Augustus Hervey, who "in the most friendly manner, desired I would make his table my own, while I was at Florence, —that I should always find a cover, at three o'clock. . . ."[239] Lord Hervey's rooms were always warm "from carpets and good fires,"[240] while most Italian homes were cold. At Lord Hervey's Young met many Englishmen, most of them much wealthier than he. "I was, according to a custom that rarely fails, the worst dressed man in the company; but I was clean, and as quietly in repose on that head, as if I had been either fine or elegant."[241] At an earlier age Young would have been much bothered, but now he was nearly fifty and able to rest on his solid reputation to compensate for lack of finery.

Young also found at Florence more men interested in scientific agriculture than at any other Italian city, and was hospitably received by several. Signore Tartini, secretary of the Georgofili Society of which Young was an honorary member, presented him with books and accompanied him to the royal "oeconomical garden," operated by Signore Zucchino in a manner superior to any similar institution in Italy. Zucchino, a close friend to Paolo Balsamo, professor of agriculture in Sicily, who was now studying agriculture near Bradfield, showed Young Balsamo's account of the Bradfield farm. Young was especially impressed with Signore Fabbroni who took him to the natural history and scientific collections and who twice entertained Young at his conversazione, where Young found good talk, interesting people, and Signora Fabbroni who was not only intelligent, but also "young, handsome, and well made," a worthy subject for a Titian Venus.[242]

Young was overwhelmed with the Florentine art collections and felt that "to view them with an attention adequate to their merit, one ought to walk here two hours a day for six months."[243]

[238] *Ibid.*, pp. 514–515.
[239] *Ibid.*, p. 499. Hervey was a son of Young's old friend, the Earl Bishop.
[240] *Ibid.*, p. 505.
[241] *Ibid.*, pp. 498–499.
[242] *Ibid.*, p. 497. *Cf. Biographie Universelle* for article on Jean V. M. Fabroni (1752–1822).
[243] *Ibid.*, p. 493.

He grew lyrical over the Venus de Medicis: "You may view it till the unsteady eye doubts the truth of its own sensation: the cold marble seems to acquire the warmth of nature, and promises to yield to the impression of one's hand. Nothing in painting so miraculous as this."[244] He declared one of Titian's nudes "probably the finest picture of one figure, that is to be seen in the world,"[245] his eyes were "rivetted" on Raphael's portrait of Julius II, and he stated that he had never understood "ideal grace in painting"[246] until he saw Raphael's Madonna della Sedia. He probably reflected the art tastes of the eighteenth century when he never mentioned the Cathedral. More surprisingly he never noted the works of Leonardo, Michelangelo, or Donatello.

The only thing which Young disliked about Florence was its climate. Early in his stay it rained incessantly, and later turned very cold. Of one day in the country he wrote that he "never felt such a cold piercing wind in England," while a few days later he complained that the wind "drives ice and snow to your vitals. And this is Italy, celebrated by so many hasty writers for its delicious climate!"[247]

Young left Florence and his friends with considerable regret on December 2. Two days later he was back at Bologna, having crossed the Apennines with some difficulty for the roads were icy and the landscape covered with snow. At Bologna he visited the Institute which he found overrated. Young never appreciated scientific museums where machines were neatly arranged and cataloged. Why were they not being used to carry out more experiments? Moreover, museums were always deficient in agricultural implements. He wondered why Renaissance Italy produced so many more fine public and private buildings than Holland or England in the eighteenth century, and called upon historians to furnish an answer rather than studying wars and battles.

For two days of the week's return trip from Bologna to Turin along the south bank of the Po, Young fortunately had as his traveling companion his friend, the Abbate Amoretti, whom he had met again at Bologna. While he was again at Turin, the

---

[244] *Ibid.*, pp. 489–490.
[245] *Ibid.*, p. 492. He identified this picture only as "the woman, lying on a bed."
[246] *Ibid.*, p. 503.
[247] *Ibid.*, pp. 502, 505.

English ambassador, Mr. Trevor, not only entertained him hospitably, but introduced him to the Portuguese Minister, Don Roderigo de Souza Continho, "one of the best informed men I have any where met with."[248] There was a dinner of intellectuals with the conversation confined strictly to agriculture and its politics. Young only wished that the Portuguese minister was his neighbor in Suffolk, which was the highest praise he could give.

On December 21 he left Turin and passed through Savoy to France over the Mont Cenis pass. He found the process of "rammissing" down the mountain much less thrilling than he expected, but there were amusing moments as when a girl on a mule was thrown into a snow bank head first, "the girl's head pitched in the snow, and sunk deep enough to fix her beauties in the position of a forked post. . . ."[249] He spent the day before Christmas at Chambery and of course visited Charmettes, the famous home of Mme de Warens and Rousseau, walking along the paths which Rousseau had described and carefully identifying everything—the house, road, vineyard, and gardens—which Rousseau had immortalized. He even took the trouble to extract a death certificate of Mme de Warens from the Rector and inserted the text in his *Travels*. Just before crossing into France, he expressed his regrets at leaving Italy, probably never to return:

. . . what country can be compared with Italy? to please the eye, to charm the ear, to gratify the enquiries of a laudable curiosity, whither would you travel? In every bosom whatever, Italy is the second country in the world—of all others, the surest proof that it is the first.[250]

From his three months in Italy only five of Young's letters home have survived.[251] From Florence he wrote cryptically about bad tidings of his old friend John Arbuthnot: "I am greatly concerned for Mr. Arbuthnot, tho his silence made him dead to us from ye time he went to Ireland; I never knew a family which was the center of every mild & agreeable virtue so shattered into nothing by a man's failure."[252] Even more disquieting was

---

[248] *Ibid.*, p. 533. John Hampden-Trevor was ambassador at Turin, 1783–1798.
[249] *Ibid.*, p. 537.
[250] *Ibid.*, pp. 542, 545.
[251] Add. MSS. 35,126, ff. 482–483, 486–487, 490–491, 492–493, 494–495. Brief excerpts from the last two are in Young, 1798: p. 185.
[252] Add. MSS. 35,126, f. 492.

the news of the death of William Macro, to whom he paid a touching tribute in a letter from Turin: "But the death of poor Macro grives [*sic*] me to the heart; nobody knew his value better if so well as I did. He was worth forty squires & a dozen Lords: . . . I regret him more than I can express; & shall have many a heavy hour for his loss."[253] He was running short of money and had to make special arrangements with a banker at Turin, probably through the good offices of Mr. Trevor. The costs of his trip until he reached Florence were mentioned in a letter from that city: "I took £100 with me & it lasts exactly 6 months buying books included."[254] From Bergamo he referred whimsically to his language difficulties: "At this place there is not a word of french to save one's soul; so I am forced to talk Italian, as a Spanish cow does french, but I have tried hard to get a little, for I must have it to read their agriculture."[255]

Young received several letters from his wife, but complained bitterly to Mary that he could not read or understand parts of them, and that they contained much inconsequential detail. At Florence a letter from his wife had "no date & not a word of Bobbin in it; what a way of writing is this to a man 1400 miles fro. home! There is much in her Lr. I cannot read what can she mean by this *one quarter of an hairs breadth doubled with yr. bailiff wrote them nor care a pin.*"[256] Likewise he found fault with her letters received at Turin: ". . . they are so crosswritten & so cram'd and topsy turvy that like the oracles of old they may be made to speak whatever is in the readers head, alley croaker, or Paradise lost are all one. . . . How can a person of common sense and not insane think of retailing on paper the minutiae of a serv'ts conduct down to his saying *I don't care;* &c &c and this to a man in Italy. . . ."[257]

On the return trip to Paris Young spent two nights at Lyons, which he found neither impressive nor beautiful. Attempts to gain agricultural information from the famous agricultural writer, Abbé Rozier, were unsuccessful, for he was too theoretical. Fortunately he then visited a Protestant minister, B. S. Frossard,

---

[253] *Ibid.*, ff. 494–495. *Cf. Annals* 13: pp. 162–163, for a brief obituary on Macro.

[254] Add. MSS. 35,126, f. 493.

[255] *Ibid.*, f. 487.

[256] *Ibid.*, f. 492.

[257] *Ibid.*, ff. 494–495.

who not only gave Young much information himself, but also arranged a dinner party with Young and M. Roland de la Platière, inspector of manufactures at Lyons and later famous as the Girondist leader. Young enjoyed that day: "We discussed these, and similar subjects, with that sort of attention and candour that render them interesting to persons who love a liberal conversation upon important points." Young's reference to the beautiful and even more famous Madame Roland is tantalizingly vague: "This gentleman, somewhat advanced in life, has a young and beautiful wife. . . ."[258] It is not even absolutely certain that Young met her.

In deference to an Englishman, with whom he had been traveling since he left Turin and who wished to return home as rapidly as possible, Young agreed to go from Lyons to Paris by post chaise. In general he detested that mode of travel, but now it would give him more time in Paris to study the Revolution. During Young's final stay in Paris, Jan. 3–20, 1790, he was again the guest of the Duc de Liancourt. A few days after his arrival, he accompanied the Duke and Lazowski to Liancourt to advise on the conduct of a farm which the Duke proposed to operate on improved English principles. While at Liancourt Young noted with approval that the poor had appropriated some waste land on the estate for their own use: "I heartily wish there was a law in England for making this action of the French peasants a legal one with us."[259] Young always was an advocate of bringing wasteland into cultivation!

Young spent most of his time in Paris studying the progress of the Revolution. One morning he walked in the Tuileries Garden and saw the royal family, carefully watched by the National Guard. The king seemed in excellent health and spirits, but the queen's countenance showed the effects of what she had endured. Young also noted the little dauphin digging in his garden with hoe and rake. He was glad that the populace raised their hats tó the royal family, but was shocked that the gardens were open to everyone. He disagreed with his friends that such treatment of the royal family was really essential to preserve liberty, and questioned, "on the contrary, whether it were not a very dangerous step, that exposes to hazard whatever had been gained."[260]

[258] Young, 1793: pp. 547–548.
[259] *Ibid.*, p. 559.
[260] *Ibid.*, p. 555.

As Grandmaster of the Wardrobe, Liancourt had apartments in the Tuileries and twice a week gave dinners to a distinguished company of deputies. Young was often the guest at such dinners and also at the evening salon of the Duchess d'Anville. Among the celebrities he thus met were the Deputy Le Chapelier, Volney the celebrated traveler, Bougainville the circumnavigator, and Condorcet the last of the Encyclopoedists. He attended the National Assembly several times and heard speeches by Mirabeau, "truly eloquent—ardent, lively, energetic, and impetuous."[261] He criticized the practice of permitting the audience in the galleries to applaud or to hiss, "an indecorum which is utterly destructive of freedom of debate."[262] One evening some friends took Young to the Jacobin Club, where he was introduced and formally elected as a kind of honorary member.[263]

Young spent much time examining books, pamphlets, and papers, deciding which to purchase, and taking notes on the less important. He also copied some public records, presumably the *cahiers,* of which his finished work showed such a careful study. Fortunately the Duc de Liancourt had a very fine collection of material, going back to 1787. Such work took "many hours a day, with what I borrow from the night."

Just before leaving Paris, Young inserted in his diary a short comparison of life in France and England. Like nearly all travelers, he found French food very much better and confessed that even roast beef at Paris was as good as at London. He noted the universal custom of serving some kind of a dessert in France. On the whole, he found the French cleaner in their persons. They never drank from the same glass after others had used it. Table linen was changed more frequently, and napkins were universally used even by the very poor. All bedrooms in France were fitted with a kind of bath known as a *bidet,* but "their necessary houses are temples of abomination."[264] He praised the French custom of having married sons and daughters living in the same house with their parents. That such an arrangement worked was evidence of the good temper of the French people. "Nothing but good humour can render such a jumble of families agreeable, or

[261] *Ibid.,* pp. 559–560.
[262] *Ibid.,* p. 566.
[263] *Ibid.,* p. 579.
[264] *Ibid.,* p. 583.

even tolerable."[265] The French paid less for their clothes and fashions changed much less frequently. Fewer people lived beyond their means. His conclusion is interesting:

On comparison with the English, I looked for great talkativeness, volatile spirits, and universal politeness. I think, on the contrary, that they are not so talkative as the English; have not equally good spirits, and are not a jot more polite. . . . I think them, however, incomparably better tempered. . . .[266]

The last day in Paris he spent with his friends. He praised Liancourt warmly for his part in the Revolution, consistently espousing the popular cause and just as consistently opposing all unnecessary violence and bloodshed. His last evening was spent with Lazowski, "he endeavouring to persuade me to reside upon a farm in France, and I enticing him to quit French bustle for English tranquility."[267]

His last letter to Mary was written from Paris on January 10. His family was apparently bothering him with all sorts of commissions, some of which Young could not understand: "You should have told me what sort of ribbon wd. have suited you best, but as you say nothing I suppose it is fancy ones; Bobbin shall have what she wants tell her but what is a sash? I suppose a broad ribbon; you should always explain things to a man whose head is occupied." At the end he requested to "have my bed laid in directly & the parlour aired for 3 or four days."[268]

Young left Paris on January 20 and was back at Bradfield ten days later after an absence of eight months. The journal for his trip ends: ". . . the greatest satisfaction I feel at present, is the prospect of remaining, for the future, on a farm, in that calm and undisturbed retirement, which is suitable to my fortune, and which, I trust, will be agreeable to my disposition."[269] Such sentiments were natural and probably sincere when written, but Young could never have been happy to remain permanently at Bradfield without frequent sojourns away from the home which he nevertheless loved so deeply.

[265] *Ibid.*, p. 584.
[266] *Ibid.*, p. 586.
[267] *Ibid.*, p. 588. It is not certain that Young ever saw Lazowski again, although he may have come to England in the early 1790's.
[268] Add. MSS. 35,127, ff. 1–2.
[269] Young, 1793: pp. 588–589.

Young did spend most of 1790 at home in relative quiet. It took much of the spring and summer to prepare the French travels for publication. He made a three-hundred-mile trip, probably in the early spring, into Sussex to purchase cattle, some for Liancourt, and a bull for himself. Lord Sheffield's hospitable home at Sheffield Place was almost certainly his base during the search. The most important purchases were made from a Mr. Kenward of Fletching, from the finest herd of cattle that he saw.[270] He probably stopped at London for a day to attend a committee sitting on the proposed new corn law.[271] He was also at Houghton in the spring to view Lord Orford's fine herd of South Down sheep, lately purchased from the Macro estate.[272] In the autumn he went into Norfolk to witness "the greatest drilling experiment that perhaps was ever made; 300 acres of turnips drilled by Mr. Cook, with his own well-known drill, and managed under his personal direction, against 300 other acres sown broadcast."[273]

As always Young was busy at Bradfield in 1790 with various agricultural experiments. He continued his experiments with chicory and still found the plant excellent as animal food. His trials with French gypsum as a manure were less fortunate. In 1788 he had printed in the *Annals* a paper by Samuel Powel, president of the Philadelphia Society for Promoting Agriculture, on the use of French gypsum as a fertilizer in America.[274] Young had consequently bought a ton at Rouen, had it ground at Bury into a fine powder, and had used it in two grass fields, totally without result. "At present I only suspect, that *on my lands,* gypsum, as a manure, is good for nothing. . . ."[275] Early in 1790 he purchased in London a machine for weighing cattle and had a shed built for it by a wheelwright "that I keep in constant employment." Young considered the total cost of about £20 a good investment: "I have no doubt of profiting much more than

[270] *Annals* **15:** pp. 427–434. Opposite page 427 is a picture of Young's Sussex bull. The idea that he was purchasing cattle for Liancourt is gained from a letter to Sheffield, quoted below, which I have been kindly permitted to copy and use by the University of London, Collection of Autograph Letters, f. 121.

[271] *Annals* **13:** p. 459.

[272] *Ibid.* **15:** p. 306.

[273] *Ibid.* **16:** p. 181.

[274] *Ibid.* **10:** pp. 315–324.

[275] *Ibid.* **14:** pp. 318–319.

that sum per annum by it. . . . no grazier ought to be without it."[276] It was immediately useful to weigh cattle that he was fatting with various types of food.

Young's chief public interest in 1790 lay in the agitation over a new Corn Law. The harvests of 1789 and 1790 were both bad. Late in 1789 an Order in Council prohibited grain exports and permitted importation with low duties. The basis for the Act of 1791 was the report, or "Representation," published in March, 1790, by the Committee of the Privy Council on Trade and Plantations.[277] Before Young left Paris he probably received a letter from John Symonds, stating that Lord Hawkesbury and George Chalmers had requested him to confer with them on the corn trade.[278] As a result he apparently saw Hawkesbury in London *en route* to Bradfield.[279] The day after his return he composed a questionnaire with sixteen questions relative to the crop of 1789, how poor it had been, the causes for the high prices, the effects of the Order in Council, the current prices, and finally what would be a reasonable price to give the farmer a fair return on his capital.[280] Twenty-seven replies to his questionnaire and a rather perfunctory summary by Young were printed in the *Annals*.[281] Young was quite bitter over the Order in Council, which he considered unnecessary since there had been very little export anyway. Since bread was not made of barley or beans, the Order in Council should not have included similar provisions for them as for wheat. The true aim of the Order in Council was to depress prices, to the loss of the farmer and to the profit of the brewers and distillers.

[276] *Ibid.*, p. 163.
[277] *Cf.* Barnes, 1930: pp. 53–60.
[278] Add. MSS. 35,127, ff. 3–4.
[279] University of London, Collection of Autograph Letters, f. 121.
[280] *Annals* 13: pp. 185–187. The author has in his possession, through the kind generosity of Miss Henrietta Taylor, the copy of this questionnaire sent to Lord Fife, on which Young wrote the following brief note:
"My Lord,
I hope you have not forget [*sic*] the Queries I took the liberty of troubling you with So great a planter cannot serve the publick more effectually than by instructing it
Yr. Lordships
Obliged & Devoted
A Y"
For photographs of this questionnaire and the note in Young's handwriting, see figs. 16 and 17.
[281] *Ibid.* 14: pp. 75–79.

Lord Hawkesbury sent the "Representation" to Young with a request for his comments.[282] He printed it in the *Annals*,[283] with scattered footnotes, some favorable but more critical. He felt that the producers' interests were being sacrificed to those of the consumers. Since the government was thinking primarily of riots and commotions, its chief objective was to keep prices low. He was very anxious for a permanent law, upon which the producer could count. "I am clear that a bad *permanent* plan is better than a *varying* good one."[284] Thus he objected to the proposal that the Privy Council could issue temporary Orders in Council. Since prices in the eastern producing counties were usually and naturally lower than in the western consuming counties, he disliked all proposals to regulate exports and imports on the basis of national average prices. Rather such decisions should be reached for each locality on the basis of local prices. He feared that the new law would be pushed through hastily without giving the agricultural interests time to make their weight felt. The account of his visit to the Committee is interesting:

I dropped one day into the committee room, the only day I had in town, and I thought myself in the midst of the Highlands,— every face was Scotch, and they were not present for nothing, for they added five pages to the bill.—Let the Norfolk and Suffolk members give an account of what their attendance was!!![285]

Throughout his discussion of the proposed Corn Law, Young exhibited hostility towards Parliament of an almost revolutionary intensity. At one point he wrote:

God send the time may not soon arrive, when the nation shall feel the mischiefs of our shallow policy, of sacrificing the landed interest on every occasion, to favour the trading and manufacturing ones: a policy well worthy of assemblies, filled with gamblers, fox-hunters, and nabobs.[286]

He expressed similar bitterness on several less vital matters. When a measure to expedite enclosures was defeated, he exclaimed:

282 Add. MSS. 35,127, f. 21.
283 *Annals* 13: pp. 352–410.
284 *Ibid.*, p. 392.
285 *Ibid.*, p. 459.
286 *Ibid.*, pp. 181–182.

If the angel Gabriel, from heaven, appeared, with a proposal to promote agriculture, he would be scouted.

If an imp, from somewhere else, appeared, with one to fetter it, in favour of the loom or the counter, he would be heard with applause.[287]

When it was proposed to lower fees on private bills, the measure was opposed because any such reduction would overburden the Commons with more business than it could handle. Young himself favored three important classes of private bills—enclosures, turnpikes, and canals—and felt that such an attitude by the Commons prevented progress. Again he lashed out:

It is not at all proper to meet at nine, ten, or eleven o'clock, for the benefit of the publick, and to enable their constituents to have their business done, but to load the passing of bills with such fees as shall deter from demanding them . . . And we are to suppose, that this wise and awful senate is right in meeting to do the national business at four o'clock in the afternoon, as it would be prejudicial to weak nerves to be out of their bed before two. . . .[288]

Young reserved his real fury, however, for tithes, "the grand obstacle to agricultural improvement," and declared: "I almost dare hazard a wish, that we were put on a footing with our neighbours in that respect."[289] Again he wrote:

If something is not done to ease this pinching point, a contest will be raised by and by in this kingdom, hot enough to have dreadful effects. It is the fashion of the period in England to GRANT *nothing;* the consequence of which may be, *all* will be TAKEN.[290]

The following letter from Young to Lord Sheffield contains references to several of his interests early in 1790.[291]

My Lord,
On my return I fd. ye favour of the 4th June: I shall publish the ox. and I do beseech you to reflect on sheep and timber according to yr. promise.

I want to buy some capital Sussex oxen for the Duke de Liancourt[292]

[287] *Ibid.* **14:** p. 313.
[288] *Ibid.* **13:** pp. 350–352.
[289] *Ibid.*, p. 202.
[290] *Ibid.* **14:** p. 9.
[291] University of London, Collection of Autograph Letters, f. 121.
[292] Young's advertisement in the *Bury and Norwich Post*, February 17, 1790, for a bailiff to go to France was also probably for Liancourt.

have you any of the very first rate? I was at his house at Paris for 3 weeks & dined with 30 deputies twice a week. Sir J.[293] read me a part of yr Letter: Your Lordship has good ideas of them. Mr. Trevor who treated me with great and kind attention at Turin desired particularly to be remembered to you & Mrs. Trevor (who is nervous) equally to Lady Sheffield: he introduced me to the Portuguese minister, one of the best informed men I have met with—I went no further than Venice & Florence. In point of trade & manuf France is absol. ruined. Are you a convert yet to ye Corn T[torn]

I am my Lord

Yr. much obliged
& Devoted

A Young

I hope Lady S. & Miss H. are perf. well

I had a conf. with Lord Hawkesbury on ye new corn bill wch. appary. will [pasted over] governt. one—& so, not too good for ye L. Int. Two bulls also & some cows.

During the summer and autumn of 1790 John Symonds took a long tour to the east coast of England, through southern Scotland, and down the west coast. He wrote Young long letters in the form of a journal. One in early September, written from Rydal Water, near Ambleside, when he was visiting Bishop Richard Watson of Llandaff, has an amusing reference to Young:

As we were returning to Bowness, the boatman asked us, whether I would not pay a visit to Mr. Young's station? "Mr. Young's station, said I! one hath heard of Roman stations in history and in Autonius's Itinerary, but who ever heard of Mr. Young's station?" "Sir, replied the boatman, it is on the mountain behind Bowness, and it is called both Brant fell, & Mr. Young's station, but is now better known under the latter name. Every traveller, Sir, goes thither and carries with him the Northern Tour." Upon this I answered that it was improper for me to visit it, for what atonement could I make to Skiddon . . . & Rydal hag, whom I had neglected? The Bishop smiled, and said, "You must submit, or never think of seeing again Mr. Young." This Episcopal mandate was too peremptory to be disobeyed . . . and I followed my guide in a quest of this station. Whilst I was climbing the heat of the weather & a foolish desire of seeing the lake induced me to look back upon it; at which my

---

[293] Probably Sir Joseph Banks.

guide said "that I had been guilty of a great impropriety; for Mr. Young had given positive orders that no one should look back." "And do all travellers comply with these orders" I asked? "Yes, Sir, all the gentlemen, but not so easily the Ladies; for yesterday a party was here, & began to be refractory, till one more sober than the rest said, that Mr. Young would turn every one of them into a pillar of salt.[294]

In the fall Young had a very serious illness, "a violent fever, which brought me to the brink of the grave." His physician afterwards informed Young that during one of the attacks of delirium he "broke forth into one of the most eloquent and sublime prayers he ever heard, to his utter amazement."[295] Young attributed his recovery largely to the skill of his physician, Dr. William Norford of Bury:

To that skill, united with assiduous attention, I owe the power of being able to write these lines; and I should esteem myself guilty of an injustice, if I named him without seizing, with pleasure, the opportunity of testifying, to that keen sagacity which discriminates a distemper, and the deep experience that dictates the prescription which attacks the malady.[296]

Very typically Young prevailed upon Norford to contribute two articles for the *Annals*.[297]

While recuperating from his sickness Young penned a little autobiographical essay, "Memoirs of the last Thirty Years of the *Editor's* Farming Life."[298] Much use has already been made in preceding chapters of this major primary source for Young's biography. Perhaps his physical debilitation at the time helps to explain the complaining and self-pitying tone of the article. True, he admitted imprudence, especially in writing too much inferior work. But Young gives the impression constantly that he has not received his just deserts, that he had spent his life in semi-public pursuits with no adequate compensation. He complained that the Dublin Society gave him no reward for his work on Ireland. He

---

[294] Add. MSS. 35,127, ff. 52–53. These letters contain much interesting material on agriculture, architecture, and manufacturing improvements, for Symonds had covered much of this ground in 1762 and makes interesting comparisons.

[295] Young, 1898: pp. 187–188. *Bury and Norwich Post* mentioned Young's illness, November 17, and his recovery, December 22.

[296] *Annals* **15:** p. 369. *Cf.* article on Norford in *Dictionary of National Biography*.

[297] *Ibid.* **14:** pp. 441–444; **15:** pp. 60–66.

[298] *Ibid.* **15:** pp. 152–197.

complained of the lack of support of the *Annals,* that his total subscription was still only 350.[299] Especially he complained of the burden of taxes and tried to show that the government took £219 out of a total rental from his estate of £229.[300] If another war came, he would probably lose his estate entirely, and he was very pessimistic about the prospect: "While I am writing the fleets are in motion, that mark a British interference in the perplexed labyrinth of northern politics; just escaped from a war at three million expence, for the vile otters of Nootka Sound."[301] England was a fine country for the poor and the very rich, but not for the middle class of small gentlemen. He mentioned the temptation to emigrate to France or America: "However, to one or the other, or some asylum where I shall not be flayed alive by tythes, taxes, and rates, I must look, since ruin alone waits such properties as mine in England."[302] The following quotation shows the very unhealthy state of his mind:

But this is the language of complaint, disappointment, and regret; what do you expect to get by that? The question of a friend.—To get!—to get what I have hitherto gotten—the reward which a bosom conscious of feeling for the general good, enjoys in the rectitude of its own emotions.

Am I not old enough to know that . . . the great and powerful . . . never listened to tales of complaint. I am not so young; the buffetings of five-and-twenty years . . . have hardened me to both hopes and fears.—and let me add, that he who has grown grey under neglect, is a fool if he hath expectation, and a poltroon if he hath hope.[303]

The year 1791 was not significant in Young's life, but it is heavily documented. On January 2 Young wrote to Sir Joseph Banks that he considered the pending Corn Bill "very passable."[304] Much more rabid than Young, Banks replied on January 15: "How you can consider the Present Corn Bill passable I confess amazes me one would think that during your illness your apothe-

[299] *Ibid.*, p. 171.
[300] *Ibid.*, p. 190. As so often was the case, Young's figures were questionable.
[301] *Ibid.*, p. 180.
[302] *Ibid.*, p. 176.
[303] *Ibid.*, pp. 178–179. Young, 1898: p. 188 realized that he had gone too far when he referred to "that melancholy review of my past life, which is printed in the 'Annals.' "
[304] Royal Botanic Gardens at Kew, Banks Correspondence, 2: f. 26.

cary had dosed you with some of Ld. Hawkesbury's Pills."[305] So
in another letter Young tried again to explain his position, but
it is still not very clear:

You must mistake me about the Corn bill, or my intellect must be
in want of pills; tho not Lord H's.

The question lies in a nutshell—The part of the bill that is already
old Law is decided; & probably will excite no debate. This bill
was brought in on the rejection of the Norfolk & Essex district
bill and the question shall the prices be ascertained by districts;
or—will you give governt. the power of substituting the general
price of the Kingdom, at their pleasure? That is giving uncertainty
as the leading feature where permancy should alone be found. This
is ye grand objection.

As governt. will have wheat as low as they can the next objection
is their not giving the farmer a recompense in oats and beans: but
there never was such a recompense thought of; & therefore it would
be a new proposition.

But absolute permanence is the only pillar of Corn police and
for this reason I think the bill bad—but on comparison with the
system that has governed us for 20 yrs. past (yet more uncertain)
I think it not only passable, but good.[306]

On January 25 Young wrote Washington requesting information about agricultural conditions in the United States, and
sending him two copies of the *Annals,* some seeds and some
manufactured wool from a fleece which Washington had sent
him. It seems unpardonable for Young to request agricultural
information from the President of the United States, even in
return for various favors rendered. But there is no indication
that Washington resented the request. On the contrary he wrote
to correspondents and furnished Young with a very valuable
account of agricultural conditions in Pennsylvania, Maryland,
and Virginia.[307]

Young's political opinions in 1791 were just as radical as in
1790, if not more so. He attacked Parliament for neglecting the
landed interests and claimed that the benches were too often
empty when such subjects as wool, corn, enclosures, tithes, and
poor laws were being considered. He continued:

---

[305] Add. MSS. 35,127, f. 80.
[306] Banks Correspondence, **2:** f. 28. Article in *Annals* **15:** pp. 79–85, adds little.
[307] *Letters from . . . Washington to . . . Young. . . ,* 1802: pp. 17–50.

A little French discipline would, on many occasions, be very useful in this country. The men who thus turn their backs on the most important business of the kingdom, assure us that the constitution wants no reform, and that the best representation of one hundred millions a year is by the agents of manufacturers, and by the returns of pot-wallop boroughs.[308]

When a correspondent implied that British oak was necessary to maintain the navy and national security Young sneered in a footnote: "For the security of the Eastern Indies—Mr. Harries is too good a patriot to wish success to such abominable speculations."[309] A footnote to another article attacked all colonization: "The very principles of colonization are rotten, no wonder they are to be bolstered up with prohibitions, the whole family of which are rotten also."[310] One article claimed that Englishmen were actually purchasing lands in the Crimea, where there were no restrictions upon enclosures and no corn laws, prohibitions upon woolen exports, tithes, or poor rates. The same article pointed out that if the revolution went well in France, that country would "offer advantages to cultivators to be found no where else."[311]

Young's sharpest barbs were again leveled against tithes as a focus to express all his grievances. An analysis of the tithe arrangements in a Leicestershire enclosure bill was followed by several pages of vitriolic attack, not only upon tithes, but upon parliamentary abuses in general:

Because our constitution was good, on comparison with those of our neighbours, therefore the grossest and most horrible oppressions are voluntarily to be submitted to, because to remedy them would be to innovate! Our properties are to be so devoured by taxes . . . as, in effect, to reduce whole classes of the people to poverty;— and we are to be left groaning under the slavery of tythes, after they have ceased through half Europe, because any change in the constitution of parliament, or the church, would be dangerous. . . .

It behoves the government of this country to take warning in time. The progress of liberty is rapid and epidemical; it has effected revolutions in Ireland, America, and France:—The oppression of tythes

[308] *Annals* 15: p. 394.
[309] *Ibid.*, p. 556.
[310] *Ibid.* 16: p. 302.
[311] *Ibid.* 15: p. 552.

has been almost swept out of Europe: the period advances with celerity when the people of England will be brought to say, with one voice, WE WILL NOT PAY THEM. It will then be seen whether the bayonet will be brought to the levy.[312]

An anonymous reader, "J. S." indignantly protested that the value of the *Annals* was greatly lessened in the minds of moderate men "by the violence with which you express yourself on some subjects, and indecency with which you often arraign the acts of the legislature."[313] Young's violence even induced a very frank rebuke from Lazowski:

. . . I have been much hurted in reading some of your annals. Why, Arthur begins to play about properties, about church, about every thing which may call forth the riots in his country. this is the manner of our great men to get fame and glory, and you wish to be famous also. such way is easy and must be yielded to our low rattlers, to the lovers of the rights of man; I know nothing so pregnant with mischief but to take every thing, the man, for instance, in an abstract way. . . . if you are missatisfied with your glorious constitution, with your truly practical liberty, go to the very hell, sense has forsaken your head.[314]

The language may not have been very correct, but it was vigorous and clear enough.

The evidence for Young the agriculturist is even more abundant for 1791 than for Young the politician, but much of it is unimportant and uninteresting. In July he was busy transplanting cabbages on quite a large scale:

As soon as heavy showers set in, I got all the men & women that could possibly be procured, to go to work in the rain. A number drew the plants and loaded one horse carts, which laid the plants in heaps through the fields; children carried them in baskets and dropt them on the ridges, and women dibbled them in: and I gave beer and gin to keep them at work.[315]

Altogether fifty-three hands were engaged in the operation, and in three days they planted twenty acres.

[312] *Ibid.,* pp. 577–578.
[313] *Ibid.* **16**: p. 277. Could this possibly have been his friend, John Symonds?
[314] Add. MSS. 35,127, f. 128. In an earlier letter of June 27, ff. 105–106, Lazowski was apparently hurt at some complaint of Young's about money matters, but had added, "What happy man I would be, was I with you at Bradfield. . . ."
[315] *Elements of Agriculture* **7**: f. 112. *Cf.* also *Annals* **18**: p. 106.

Young's most important trip in 1791 was in the summer into the central counties. For ten days he was the guest of the Duke of Grafton at Wakefield Lodge in Whittlewood Forest. The Duke saw to it that Young was furnished with all available information. He described the common husbandry at great length, but there was little notable either in the husbandry or in his description. Whittlewood Forest was very profitable to the Duke, but Young claimed that from a national point of view forests were bound to disappear in a rapidly developing country like England. Regrets for the decline of forests were absurd, for "Rough, waste, and barbarous countries, are the proper nurseries of timber: it is a produce inconsistent with a high degree of cultivation and improvement. To complain of its scarcity, is to reprobate national prosperity. . . ."[316] The crown ought to sell all the remaining royal forests. Young praised the Duke highly for the excellence of his fences and hedges, his kindness to the poor, the fine farm buildings and the abundant crop of hay. He also greatly admired the lawns, landscapes, and the artificial water, "of which it is sufficient to say, that it was formed by Brown. . . ."[317]

On the way from Wakefield Lodge to Birmingham he spent a day at Stratford sightseeing in connection with "our divine poet," visiting the town hall with its pictures, the church with its epitaph and Shakespeare's rather dilapidated birthplace.

He also went out of his way to visit a certain Mr. Boote, who had gained the gold medal of the Society of Arts for some experiments in drilling, hoping to get valuable material for himself and the *Annals* on that vexed subject.

Mr. Boote politely showed me his farm, but expressing some dissatisfaction at any thing concerning it being made public, I left my notes with him . . . on his promising to explain himself by letter. After I returned home I received one from him, in which he used this expression, *relying on your honour not to publish my mode of husbandry.*

My readers will rest satisfied, after this, that I shall not publish one syllable more about a farm which has excited so much attention to drill ploughs, and drilling. The public will draw their own conclusions. Many of the crops are very fine.[318]

[316] *Annals* **16:** p. 512.
[317] *Ibid.,* p. 522.
[318] *Ibid.,* p. 529.

Seldom had Young met such treatment in England and he did not trouble to hide his resentment.

At Birmingham Young was greatly impressed by the tremendous changes since his last visit in 1768, but was surprised that more power machinery was not used. The great improvements were the new canals and he found the canal port "a noble spectacle, with that prodigious animation, which the immense trade of this place could alone give." Altogether he judged that Birmingham was probably "the first manufacturing town in the world."[319] He also noted that Birmingham flourished mightily, even though the wages were probably higher than anywhere in England.

While at Birmingham, Young passed Joseph Priestley's home, recently wrecked in the anti-French riots, and walked over the ruins of Priestley's laboratory,

the labours of which have not only illuminated mankind, but enlarged the sphere of the science itself; which has carried its master's fame to the remotest corners of the civilized world; and will now, with equal celerity convey the infamy of its destruction to the disgrace of the age, and the scandal of the British name. . . . These are the principles that instigated a mob of miscreants—I beg pardon;—of '*FRIENDS and Fellow Churchmen,* attached to CHURCH AND KING'—to act so well for the reputation of this country.[320]

At Birmingham, Young had met Robert Bakewell by appointment and together they traveled leisurely for nearly a week to Dishley through some of the great sheep and cattle country, also notable for some fine irrigation and draining. They visited several farms which had been drained by Joseph Elkington who was doing notable work in the region. Young was moved to lyrical eloquence by the great improvements in many fields in this whole district: "What trains of thought, what a spirit of exertion, what a mass and power of effort have sprung in every path of life, from the works of such men as Brindley, Watt, Priestley . . . Arkwright, and let me add my fellow traveller Bakewell!"[321] They spent several days with Sir Robert Peel's partner, Mr. Wilkes, who had purchased an estate at Measham for £50,000, on which he conducted cotton manufactures, engaged in ambitious irrigation schemes, pared and burned land on a large scale, had

[319] *Ibid.*, p. 532.
[320] *Ibid.*, pp. 530–531.
[321] *Ibid.*, p. 546.

his own brick making works, raised fine crops planted with Mr. Cooke's drill, and was a spirited breeder.[322]

At the *Queen's Head* in Ashby de la Zouche Young dined with some of the great breeders of cattle and horses, but especially of the sheep known as the New Leicesters. Quite recently about sixteen of the greatest breeders had formed the Tup Society. Its members were hard-headed businessmen, who had made breeding a big business operation. They seldom sold their sheep, and never their rams or tups. Rather they let out their tups at fantastic prices to men who wanted to improve their breeds. Young was told that Bakewell was getting one thousand guineas a season for renting one of his tups which would cover one hundred ewes. He was shown a bull which was let out at not less than twenty-five guineas a cow. The members of the society were very secretive. Prize cattle and sheep were not shown, even to Young, on the ground that the prospective hirer might not then be willing to take an average tup. Apparently they only set minimum prices for letting their rams, for as Young said: "The tup master does not *ask* a price, the hirer *bids*." Young admitted that the Society was regarded with hostility in some quarters "as a knot of monopolists, associated to humbug the public,"[323] but he defended the practices of the society. He summarized his position thus:

All these questions turn but on one point; do they tend to raise prices? If they do, they are right and laudable; for it is already sufficiently proved that price and improvements go hand in hand. And can any one be surprised that more care and attention should be paid to breeding animals that *let* at 500 and 1000 guineas, than to such as are *sold* for five?[324]

Young also devoted considerable attention to Bakewell's cabbage culture, his use of Cook's drill, and his irrigation schemes. He remarked, however, "But the great object at Dishley is LIVE STOCK."[325] As much as Young admired the New Leicester breed,

[322] *Ibid.*, pp. 547–565.

[323] *Ibid.*, pp. 589–590.

[324] *Ibid.*, p. 597. Young was bitterly assailed in an anonymous letter for political liberalism and for being taken in by the Tup Society "With respect to yourself, Sir, it very plainly appears that the Tup Society have gained some ground . . . and that you now believe a ram let for a thousand guineas, is a cheap bargain." *Cf. ibid.* 17: p. 197.

[325] *Ibid.* 16: p. 587.

he still found them deficient in several points—their wool brought only a low price; their mutton was inferior in flavor to several other breeds, and they were too delicate to stand folding. The ideal sheep, he went on, should have four qualities: good mutton; fine wool; hardiness for the fold; and a well-formed carcass.

Early in 1791 Young published in the *Annals* a quite detailed account of his own experience in regard to sheep. He had originally had Norfolk sheep, but in 1784 had begun to raise South Downs, partly through Lord Sheffield's influence. By 1791 his flock of 350 was predominantly South Down, with a slight mixture of Bakewell and Spanish. The South Downs stood folding very well, their fleeces were somewhat heavier than Norfolk, the mutton was just as good and much better in summer, and a larger number could be kept on a given amount of ground. In this article Young gave a fairly detailed account of his farm, which consisted of 240 acres, 70 of which were put to corn, and which supported 45 horned cattle and 10 horses in addition to the 350 sheep.[326] To prevent the rot, he depended largely on his sheep-yard, "a large enclosure . . . defended by buildings from prevalent winds, with a barn always open for them to lie in or out of at pleasure; the barn and all the yard surrounded with racks for hay, and both well littered with leaves, straw, or stubble."[327]

Early in October, 1791, Young was the guest of Mr. John Ellman of Glynd, for the Lewes Fair. Again the chief object of interest was sheep and South Downs in particular. Among other distinguished farmers present were Mr. Boys of Betshanger and the famous Mr. Ducket of Surrey. Young took copious notes on his trip, which fill nearly fifty pages of the *Annals*.[328] Full of his Leicestershire experience, he was rather critical of the Sussex sheep-men for emphasizing wool rather than mutton. On the return trip he visited Mr. Ducket at Easher. After praising some rather technical improvements, Young remarked:

That thought had as much merit, and a success as much proportioned to the merit, as a new evolution or movement giving a victory to the banners of Frederick the Great—for THE GREAT doubtless he will

---

[326] *Ibid.* **15:** p. 298. *Cf. supra* p. 227 for account of his livestock in 1789.
[327] *Ibid.,* p. 304.
[328] *Ibid.* **17:** pp. 129–178.

continue, as his excellence was in cutting throats; but I would give my vote much readier for Ducket the Great.[329]

When Young returned from his summer trip, he probably found awaiting him a letter from Sir Joseph Banks that the king had offered him a Spanish Merino ram.[330] Don, as he was named, arrived at Bradfield early in November.[331] Eventually his picture appeared in the *Annals,* and an accompanying description, together with a public expression of gratitude to George III. Such a gift showed clearly that the king was not hostile to the agricultural interests, as so much recent legislation had been. Young also stated that the king's act was a much truer criterion of greatness than military glory:

I believe the period is advancing, with accelerated pace, that shall exhibit characters in a light totally new; . . . that shall pay more homage to the memory of a prince that gave a ram to a farmer, than for wielding the sceptre—obeyed alike on the Ganges and the Thames.[332]

In the summer Young had spent ten days with the Duke of Grafton at Wakefield Lodge. In November he was again the duke's guest for five days at Euston. Although a considerable company was present, Young went for a specific purpose, to take "the level of the Duke's river for four miles, in order to see how much land he might water . . ." There is little evidence that Young had been particularly friendly with Grafton before this time, although his brother, John Young, had been, and of course John Symonds was an intimate. But the friendship was to grow. Young summed up the duke's character and personality:

The character of this Duke is original; he is uncommonly sensible, there is no stuff in him; he is cold, silent, reserved, and even at times sullen, and he is removed from all that ease and suavity that render people agreeable; yet there is such a solid understanding, and so much learning and knowledge on certain topics, that one must value him in spite of our feelings.[333]

A major family event in 1791 was the marriage of Young's

[329] *Ibid.,* p. 164.
[330] Add. MSS. 35,127, f. 115.
[331] *Bury and Norwich Post,* November 16, 1792.
[332] *Annals* 17: p. 530.
[333] Young, 1898: p. 193.

second daughter, Bessy, to the Rev. Samuel Hoole. Bessy Young
was twenty-three years old. She had always been a favorite with
the Burneys and in 1786 Dr. Burney had written, "and pray give
our hearty love to the gentle, sweet, and amiable Miss Bessy."[334]
The bridegroom was the son of John Hoole, a very learned trans-
lator of Tasso and Ariosto, and an intimate friend of Dr. Samuel
Johnson. He had probably been courting Bessy for some time,
for he had informed Fanny Burney of an impending separation
between Arthur Young and his wife as early as 1787.[335] Young
wrote of his son-in-law: "He is a very sensible, moral man of
strict integrity, and always behaved to my daughter with much
tenderness."[336] Dr. Burney's letter of congratulation showed his
approval:

I must now hasten to congratulate you and Mrs. Young on the marriage of our dear and worthy girl Bessy. The match, indeed, is not
splendid for either in point of circumstances; but they are quite as
likely to scramble happily through life, with good hearts and wishes
limited to their means, as the richest peers and peeresses in the
land.[337]

The young couple settled at Abinger in Surrey, where Arthur
Young visited them in October during his trip to the Lewes
Fair.[338]

The year 1792 was one of the most important in Young's life.
It witnessed the publication of his *Travels in France,* his most
important and famous book by far. It also witnessed a sharp and
decisive change in his opinions. Until the middle of 1792 Young
was definitely a liberal and a reformer, if not a radical. Until then
he was on the whole favorable to the events in France. Suddenly
in the summer he began to react violently against everything
connected with the French Revolution, and not long thereafter
started to oppose every reform or change in Britain. Before the
summer of 1792 Young was almost a pacifist; after that time he
became primarily a nationalist consecrated to ultimate British

---

[334] *Ibid.,* p. 146. Bessy does not seem to have been a very interesting girl. Her
letter to her father, October 29, 1789, is quite insipid, very different from the
sparkling letters of the Burney girls. *Cf.* Add. MSS. 35,126, f. 448.

[335] *Cf. supra,* p. 200.

[336] Young, 1898: p. 189.

[337] *Ibid.,* p. 198.

[338] *Annals* 17: p. 159.

victory, at whatever cost, over all the ideas and works of the Revolution. From the summer of 1792 until his death twenty-eight years later Arthur Young was definitely a conservative.

On January 18 Young wrote a long and very technical reply to Washington's letter of December 5, 1791. Young found American agricultural conditions very difficult to understand:

American products, it is true, are shocking, and mark a management which, thank God, we know nothing of. Such crops would not be found in any part of this kingdom. The observation, that in America farmers look to labour much more than to land, is new to me; but it is a calculation which I cannot understand, for, exactly in proportion to the dearness of labour, is the necessity of having good crops. . . .[339]

Young was busy fatting cattle in the spring of 1792, but wasn't very satisfied that the business was profitable. Nor was he sure what would be more so, unless it was dairying, which a gentlemen could not carry on very satisfactorily.

When a man cannot fix on the only thing his farm is well adapted to (a dairy, which I cannot do, for reasons not necessary to mention, the case of thousands as well as myself), he is sure to grope for sometime in the dark before he is well satisfied with his path.[340]

He was more pleased with his management of sheep in a spring after very severe frosts had completely spoiled many turnips. Fortunately his large supply of cabbages and twenty-five acres of rouen saved the day. Unfortunately none of his neighbors copied this practice which he had been carrying on for ten years.

Young had vigorously condemned the slave trade in 1788. In the spring of 1792 the subject was again being agitated. Young apparently did not attend the meeting at Bury.[341] His daughter Bessy wrote that the Hooles so strongly opposed slavery that they had abandoned sugar and were using honey instead. She doubted whether enough people would follow their example to make it worth while, but her arguments had been overruled, "till at last I became a convert, and though I am very well

[339]*Letters from . . . Washington to . . . Young . . .*, 1802, p. 120.

[340]*Annals* **18:** p. 104. Young is referring of course to the fact that Mrs. Young as a lady could not supervise a dairy as a farmer's wife might.

[341]*Bury and Norwich Post*, February 15, 1792. Meeting held on February 10. Capel Lofft was a speaker. For Young's stand in 1788, *cf. supra*, p. 225.

reconciled, I shall not be sorry to hear of some other means being found of cultivating it."[342] In the *Annals* Young urged the government to experiment on some small island with sugar production by free labor. It was unfair to the capital invested in the slave trade to abolish it without some such experiment. If England alone abolished the trade, it would only ruin Liverpool and Bristol, for just as many slaves would be imported from some other country. He obviously feared that the agitation might lead to a slave rebellion and hence he thought troops should be stationed in the islands as a precaution. All in all, his article was not very forthright against slavery, in spite of his statement, "I am not pleading for the slave trade, which I truly abhor."[343]

Perhaps one reason why Young was lukewarm in the antislavery crusade was his preoccupation with his own crusade against mad dogs. Young became thoroughly aroused by an epidemic of hydrophobia in the winter of 1791–1792. On February 15 a mad dog attacked some cows of William Green at Bradfield, and on the following day Young wrote to the *Bury Post* calling for a meeting at the Angel Hotel in Bury on February 29.[344] The meeting "was both numerously and respectably attended," and Young submitted a series of resolutions, noting the seriousness of the situation, claiming that many useless dogs were kept, urging the Justices of the Peace to refuse poor relief to any one with a dog, requesting landlords and farmers not to employ any laborer possessing a dog, and asking the high sheriff to call a further meeting to adopt a petition praying Parliament to institute a tax on dogs. The meeting voted thanks to Arthur Young "for his very humane attention, and liberal exertions to remove an evil of such immediate consequences."[345]

In the speech introducing his resolutions, Young vigorously attacked the "scandal of paupers" keeping dogs:

. . . thus the farmer pays for the keeping of dogs, whose principal occupation is, at the best, to do mischief to his live stock, and perhaps, by madness, to an amount that may be his ruin; if this is not an abuse that calls for an immediate reply, none such exists.

[342] Add. MSS. 35,127, f. 147.
[343] *Annals* 17: p. 527.
[344] *Bury and Norwich Post,* February 22, 29, March 7, 14, 21, 28, 1792, and *Annals* 17: pp. 533–564.
[345] *Bury and Norwich Post,* March 7, 1792.

At the same time he pointed out that it was unfair to prevent the poor from having dogs unless the wealthy would also keep their own dogs carefully chained during the emergency: "A man that will not determine to do this steadily, has no right to expect that from his poor neighbour which he will not do for himself."[346]

At the second meeting on March 24,[347] held at the Bury Assembly House to petition Parliament for the dog tax, Young seconded the motion. Strenuous opposition was offered, however, by Capel Lofft, Young's old antagonist of a decade earlier on the county ship question,[348] who proposed that the whole matter be postponed by moving adjournment. Young in turn attacked Lofft's arguments, after which the resolution was passed with Lofft voting alone in the negative.

Somewhat later Lofft addressed twelve questions on the subject to Young, which were inserted in the *Annals,* along with Young's replies. Lofft showed himself a lover of dogs, a friend of the poor, and a warm humanitarian. Young's replies emphasized the seriousness of the crisis and the very real danger to livestock and persons. The temper of the two men may be seen in a brief quotation from each:

Lofft: "Is an animal of the sense, courage, affection, and faithfulness of the dog, fitter to be destroyed than protected?"

Young: "It is very easy to find an animal of more general, and more important use, than a dog; it is very easy to find a sheep, a cow, a hog, a horse, and an ass."[349]

Young wrote another article in the *Annals* on the subject. When it was pointed out that the prime minister would be embarrassed to add another tax after recently removing some taxes, Young was aroused to one of his most radical outbursts:

---

[346] *Annals* 17: pp. 537, 539.

[347] *Bury and Norwich Post,* March 28, 1792. After one of these meetings, Young received a mysterious, undated, and unsigned letter (Add. MSS. 35,127, ff. 178–179). In Young's writing at the bottom is "Miss Fergus." "I expected to have seen you on Wednesday, after you had finished with the *mad dogs,* & was a little angry & very much mortified at the disappointment. . . . You cannot possibly be more sick & disgusted than I am with the unmeaning round of frivolities in which I move; the great question is how I am to extricate myself from it . . ." Young probably first met Miss Fergus on October 24, 1791 at Sir Thomas Gage's. *Cf.* Young, 1898: p. 191.

[348] *Cf. supra,* pp. 148–150.

[349] *Annals* 17: pp. 555, 560.

Are we to view our children expiring in the tortures of the hydrophobia, and then listen with patience to those who insult us with a
minister's convenience, his delicacy, or his objections!. . . . That the
minister's private motives are more likely to weigh with the House
of Commons, than the interest and direct demands of the people of
England: this throws the business on its right issue, and we come
at once to the source the nation has to complain of, the representation of the people being such, that that house cannot *feel* for
the people. It is a senate that feels for itself. To this cause we may
look for tythes, tests, poor rates, game laws, charters, corporations,
immense and ruinous taxes, wars, national debts, monopolies, prohibitions, restrictions, and all the mischiefs of the commercial
system.[350]

Nor is the above the only instance of Young's radicalism in
the spring of 1792. Pitt's widely praised speech on the state of
the nation only aroused him to fury because Pitt had paid no
attention to the landed interest:

This the speech of a great minister at the close of the 18th century! No: it is a tissue of the common places of a counting-house,
spun for a spouting-club, by the clerk of a banker: . . . This the
reach of mind and depth of research, to mark the talents framed to
govern kingdoms! These big words to paint little views—and splendid periods that cloath narrow ideas! These sweepings of Colbert's
shop—These gleanings from the poverty of Necker![351]

When a correspondent remonstrated against Young's attacks
upon Pitt, he reiterated his views in a footnote:

Of what account, except to boys and girls, are *glorious speeches,
and splendid investigations* from a minister, as proofs of his regard
to the agricultural interest, unless backed by leading and powerful
ACTIONS? Where are the facts? Will you go to wool? or rabbit
wool? or corn? or malt-taxes? or forests? produce the great leading
features of his administration, so far as agriculture has any concern;
and if these are all blanks, or worse than blanks, of what avail all that
Mr. Eliot has said in his favour? *so far as the plough is concerned—*
beyond which I presume not to open my lips.[352]

In another brief article Young attacked the hat manufacturers
for complaining of lax enforcement of the prohibition upon the

[350] *Ibid.*, pp. 482–483.
[351] *Ibid.*, p. 373.
[352] *Ibid.* **18:** p. 66. The correspondent was Francis P. Eliot of Elmhurst Hall.

export of rabbit fur. Young lashed out against them and "our trading legislature who never conducts itself, but on the principles of the counter."[353] He also protested against the greatly increased poor rates, which he blamed on the idleness of the poor, the graft and inefficiency of the poor law officers, and Parliament:

Doubtless, there are little farmers in this parish, who are heavily and cruelly burthened, either to support sturdy beggars, who can, but will not work, and who are richer in fact than themselves, or to contribute to the illicit profits of men, who thrive by the abuses thus tolerated by the legislature of a country which calls. itself free![354]

He bitterly attacked a bill to keep the New Forest intact under royal control and government inspection with the object of providing oak for the royal navy. Young proposed that the forest should be destroyed, the land enclosed and divided into private holdings. Again he practically accused the ministry of conscious dishonesty:

The ministers, who bring in this bill, are not fools—they know mankind better—they know themselves better—they do not bring it forward, therefore, for oak, and would laugh at the infantile credulity that gave credit to such a profession; the crop they look for is of a different nature and growth—inspection—control—commission—view—examinations—and, by consequence, officers, and appointments in plenty.[355]

On May 30, 1792 the *Bury and Norwich Post* noted that "this day is published" Young's *Travels in France*. Printed in Bury by J. Rackham and sold in London by W. Richardson, it was a very substantial quarto volume and sold for £1. 5s.[356] Apparently some subscribers to the *Annals* had hoped that he would publish it in that periodical. His answer in the *Annals* seems a bit disingenuous:

---

[353] *Ibid.* 17: p. 503.

[354] *Ibid.*, p. 499.

[355] *Ibid.*, p. 580.

[356] The first edition, reprinted in two fat octavo volumes in Dublin in 1793, is the one I have used. The second edition appeared in 1794 in two quarto volumes, the first at London, the second at Bury St. Edmunds. A first French edition appeared in 1793 and a second in 1794, and a German edition in 1793–1795. Betham-Edwards' edition first appeared in the Bohn Library in 1889, that of Okey in Everyman's Library in 1915, and the excellent one by Maxwell in 1929. The most complete modern edition is in French in three volumes by Sée, 1931.

I have but one possible objection; which is that of crouding too much with foreign intelligence a journal so well supported by practical English correspondents. . . . my present intention therefore is, to make it a separate work, in case I should be tempted to publish it; for while these Annals, nobly supported as they have been by the correspondence of the first farmers of the age, have failed of ensuring more than a precarious existence, it should seem that no work of mine, on a subject truly important, has any great chance of success. The bookseller will read, perhaps commend the book—shake his head—and wish me a good night. I shall, however, write the book, and certainly either burn or publish it.[357]

Such a statement could hardly have deceived the most naive reader. He had also been pessimistic, however, at the end of 1790 in his "Memoirs" in the *Annals:*

I . . . shall finish the register of these travels, which I have worked at all the preceding summer (1790); and this undertaking will terminate perhaps with the same good fortune that pursues me through life; my MS. when finished, no bookseller will purchase, and it will rest on the shelf no bad *memento* to remind me, that here ought to finish forever such errors of conduct.[358]

His actual bargain with the bookseller appears both in the *Autobiography* and in a letter to Sir Joseph Banks whose advice he requested:

Richardson ye bookseller has been here, and offers me 6*s*. a book for all sold of my Travels printing1000—1 Vol. 4to price 1.1.0 he to run all hazards & ye copyright after to be mine for a new bargain. The offer I think seems fair; but I have been so long out of the world that I know no more of these matters than if I had never been concerned in them. Would you have the goodness to give me yr. opinion?[359]

Like many authors Young found his original manuscript much too long for one quarto volume and was forced to prune very vigorously:

. . . when Rackham's compositor came to cast off the MS. he found enough for two large quarto volumes, since which discovery I had to strike out just half of what I had written; and the advantage will be very great to my work. I read the books as they are wanted

---

[357] *Annals* **13**: pp. 156–157.
[358] *Ibid.* **15**: p. 177.
[359] Banks Correspondence, **2**: f. 44. Date is May 7, 1791.

for the press again and again, reducing the quantity every time till I got it tolerably to my mind, but yet not to the amount of half. The work is certainly improved by this means.[360]

The basic plan of Young's *Travels in France* is the same as his English and Irish tours, that is, a diary of his itinerary, followed by an analytical survey. According to his own account, posterity must thank an anonymous friend, very possibly Dr. John Symonds, for having saved those parts of the diary which distinguished it from all his other work and make it one of the greatest travel books ever written.

When I had traced my plan, and begun to work upon it, I rejected, without mercy a variety of little circumstances relating to myself only, and of conversations with various persons which I had thrown upon paper for the amusement of my family and intimate friends. For this I was remonstrated with by a person, of whose judgment I think highly, as having absolutely spoiled my diary, by expunging the very passages that would best please the mass of common readers; in a word, that I must give up the journal plan entirely or let it go as it was written. . . . The high opinion I have of the judgment of my friend, induced me to follow his advice; in consequence of which I venture to offer my itinerary to the public, just as it was written on the spot . . .[361]

Thus instead of cutting the diary, he cut the supporting details from many chapters of the survey. Only the specialist and scholar ever consult the survey, but the diary remains a masterpiece to be enjoyed generation after generation by the general reader.

Since the diary was used as the basis of my account of his trips to France in 1787, 1788, and 1789, no further attention seems necessary. In the Dublin edition of 1793, which has been used for this work, the second volume of 571 pages is devoted to the analytical survey. The twenty-two chapter headings follow in order: "Of the Extent of France; Of the Soil, and Face of the Country; Of the Climate of France; On the Produce of Corn, the Rent, and the Price of Land in France; Of the French Courses of Crops; Irrigation; Meadows; Lucerne; Sainfoin; Vines; Of In-

[360] Young, 1898: pp. 192–193.
[361] Young, 1793: pp. 4–5. I have no proof that the adviser was Symonds, but they were very close, and it seems reasonable to believe that Symonds was well acquainted with the manuscript.

closures in France; Of the Tenantry, and Size of Farms in France; Of the Sheep of France; Of the Capital Employed in Husbandry; Of the Price of Provisions, Labour, &c.; Of the Produce of France; Of the Population of France; Of the Police of Corn in France; Of the Commerce of France; Of the Manufactures of France; Of the Taxation of France; On the Revolution of France."[362]

With the exception of the last chapter, which will be treated separately in some detail, much of the survey is very technical and not very interesting. On the whole, except in the culture of vines and lucerne, he found France backward agriculturally. He was shocked at the amount of wasteland, the lack of irrigation, the wretched crop rotations, and the ignorance of the importance of grassland. He was astonished that enclosure did not usually imply advanced agricultural methods as in England. He found the peasants' holdings much too small for real agricultural improvement. He claimed that the small peasant proprietor in France was much worse off than the English agricultural laborer. He believed that France was overpopulated, especially in the rural districts. He found nothing good in the all too common system of métayage: "This subject may be easily dispatched; for there is not one word to be said in favour of the practice, and a thousand arguments that might be used against it."[363] Basically French agriculture was understocked, both in livestock and in capital. Overpopulation, the smallness of peasant holdings, and the lack of capital were all closely related factors which helped to explain French agricultural backwardness.

Although Young was on the whole friendly to the Revolution when he wrote, and although he said almost nothing good about the Old Regime, nevertheless he feared that certain policies of the National Assembly would aggravate rather than remedy the evils in French agriculture. He feared that the new taxation system would be based on the physiocratic notions of a land tax as the chief reliance for revenue. He had long ago expressed his opposition to the physiocratic doctrines in the *Political Arithmetic*.[364] He still felt that excise taxes upon luxuries constituted

---

[362] In the 1794 edition the survey was much longer and there were also chapters on silk, cattle, plants, wastelands, coals, woods, tillage and implements of husbandry, manures, and the English farm of the Duc de Liancourt.

[363] Young, 1793: **2**: p. 241.

[364] *Cf. supra* pp. 87, 89–90.

the best kind of taxes: ". . . the best taxes are those on consumption; and the worst those on property."[365] He feared that the National Assembly would prohibit the export of corn and take very strong measures against speculation. Young had always favored a free export of corn. More surprisingly he actually defended corn speculation as an automatic regulator of supply and demand, and expressed himself as so often in superlatives: "The multitude NEVER have to complain of speculators; they are ALWAYS greatly indebted to them. THERE IS NO SUCH THING AS MONOPOLIZING CORN BUT TO THE BENEFIT OF THE PEOPLE."[366] He disliked the democratic tendencies to make peasant holdings still smaller and to discourage the breaking up of commons. He felt that the National Assembly had erred in accepting the obligation of the state to provide for the poor: "I cannot but be persuaded, that the poor ought to be left to private charity, as they are in Scotland and Ireland, to an infinitely better effect than results from the rates in England."[367] In general Young displayed the same individualistic, *laissez-faire* attitudes that he had followed since 1780. Private initiative should be unhampered by government interference. He was still a free trader and an anti-imperialist. "It would be right for every country to open her colonies to all the world on principles of liberality and freedom; and still it would be better to go one step further, and have no colonies at all." He even opposed navies:

. . . the possession of many sailors, as instruments of future wars, ought to be esteemed in the same light, as great Russian or Prussian armies; that is to say, as the pests of human societies; as the tools of ambition; and as the instruments of wide extended misery.[368]

Several statements in the survey, inserted almost as asides, throw interesting sidelights on Young's life. In one place he reveals that he was growing maize at Bradfield.[369] In another he referred to his private library: "my library abounds more with French georgical authors, as well as those branches of political oeconomy which tend to elucidate such questions, than any other

---

[365] Young, 1793: **2:** p. 489.
[366] *Ibid.,* p. 403.
[367] *Ibid.,* p. 320.
[368] *Ibid.,* pp. 427–428.
[369] *Ibid.,* p. 29.

I have had the opportunity to examine. . . ."[370] The following shows something about Young's tastes and standard of living:

Should the good genius of THE PLOUGH ever permit me to be an importer of Champagne, I would desire Mons. Quatrefoux Paret-claine, merchant of Epernay, to send me some of what I drank in his fine cellars. But what a pretty supposition, that a farmer in England should presume to drink Champagne even in idea!—the world must be turned topsy-turvy before a bottle of it can ever be on my table. Go to the monopolizers and exporters—go to—and to—and every where—except to a friend of the plough.[371]

The survey concluded with an admirable essay on the French Revolution, not only moderate in tone, but based on a very careful study of the *cahiers*. In analyzing the causes of the Revolution, Young maintained that the evils of *lettres de cachet* had been exaggerated, although he attacked severely the general administration of justice. He brought out clearly the injustices of the tax exemptions of the privileged classes and especially of the notorious *gabelle* or salt tax. He was equally severe upon the feudal exactions upon the peasantry, but felt that tithes were less onerous than in England. The first part of the essay concludes that the evils of the Old Regime fully justified the Revolution:

The true judgment to be formed of the French revolution, must surely be gained, from an attentive consideration of the evils of the old government: when these are well understood—and when the extent and universality of the oppression under which the people groaned—oppression which bore upon them from every quarter, it will scarcely be attempted to be urged, that a revolution was not absolutely necessary to the welfare of the kingdom.[372]

He then discussed the results of the Revolution upon various classes. True the nobles had suffered severely, but their *cahiers* indicated an attachment to the abuses of the Old Regime which made him rather unsympathetic. He was worried that tenants were refusing to pay rents. He sympathized with the artisans who had gained very little from the Revolution. Although the Revolution had benefited the lower clergy it had ruined the higher

---

[370] *Ibid.*, p. 47.
[371] *Ibid.*, p. 210.
[372] *Ibid.*, p. 517.

clergy. In general the French clergy were probably less deserving of their fate than many of their English brethren, and he penned a very severe indictment of eighteenth-century English parsons:

One did not find among them [French clergy] poachers or fox-hunters, who, having spent the morning in scampering after hounds, dedicate the evening to the bottle, and reel from inebriety to the pulpit. . . . A sportsman parson may be . . . a good sort of man and *an honest fellow;* but certainly this pursuit, and the resorting to obscene comedies, and kicking their heels in the jig of the assembly, are not the occupations for which we can suppose tythes were given.[373]

The chief beneficiaries of the Revolution seemed to have been the peasants, the improvement of whose lot largely compensated for the excesses of the revolution.

Go to the aristocratical politicians at Paris, or at London, and you hear only of the ruin of France—go to the cottage of the *metayer,* or the home of the farmer, and demand of him what the result has been—there will be but one voice from Calais to Bayonne.[374]

Young willingly admitted the atrocities of the Revolution as well as the mistakes of the National Assembly. He complained that the new constitution gave the vote to those without property, and failed to note that the tax qualification for voting really disfranchised the poor. The new French constitution was still an experiment which might or might not be successful, but the English constitution was no longer in the experimental stage, and in certain respects it must be judged a failure.

What can we know, experimentally, of a government which has not stood the brunt of unsuccessful and of successful wars? The English constitution has stood this test, and has been found deficient; or rather, as far as this test can decide any thing, has been proved worthless; since in a single century, it has involved the nation in a debt of so vast a magnitude, that every blessing which might otherwise have been perpetuated is put to the stake; so that if the nation do not make some change in its constitution, it is much to be dreaded that the constitution will ruin the nation.[375]

Young also analyzed the effects of the Revolution on the

[373] *Ibid.,* pp. 523–524.
[374] *Ibid.,* p. 530.
[375] *Ibid.,* p. 535.

neighbors of France. In the despotic countries similar violent revolutions might occur. If drastic reforms were not made in England, the government might not last twenty years. His own program of reforms would definitely class him as a radical:

The means of making a government respected and beloved are, in England, obvious: taxes must be immensely reduced; assessments on malt, leather, candles, soap, salt, and windows, must be abolished or lightened; the funding system, the parent of taxation, annihilated for ever, by taxing the interest of the public debt . . . tythes and tests abolished; the representation of parliament reformed, and its duration shortened; . . . the utter destruction of all monopolies . . . and, lastly, the laws, both criminal and civil to be throughly [*sic*] reformed.[376]

Naturally Young rejected the conservative idea that it would be dangerous to make reforms lest they lead to revolution. "Can such ignorance of the human heart, and such blindness to the natural course of events be found, as the plan of rejecting *all* innovations lest they should lead to greater?"[377] Young was also very critical of Burke at this time. Twice he attacked him by name,[378] while at another point he remarked, concerning the constructive results of the Revolution: "The men who deny the benefit of such events, must have something sinister in their views, or *muddy in their understandings.*[379]

Nevertheless Young did fear that reform in England might get out of hand, necessary though it was. He was afraid of the mob, whether a radical one or a conservative one such as had wrecked Priestley's home at Birmingham. The remedy was the creation of a militia of property, a proposal which appeared first in this essay and to which he reverted repeatedly in following years:

. . . a national militia, formed of every man that possessed a certain degree of property, rank and file as well as officers. Such a force, in this island, would probably amount to above 100,000 men; and would be amply sufficient for repressing all those riots, whose objects might be, immediately or ultimately, the democratic mischief of transferring property.[380]

[376] *Ibid.,* pp. 538–539.
[377] *Ibid.,* p. 537.
[378] *Ibid.,* pp. 534, 547.
[379] *Ibid.,* p. 549.
[380] *Ibid.,* pp. 540–541.

There were two addenda to the chapter on the French Revolution. The first, entitled "1792," attempted to bring up to date the information about the effects of the Revolution upon agriculture, commerce, manufactures, and finance. This first addendum expressed a little more skepticism about some specific policies, but it was in no way hostile to the Revolution. The second addendum, dated April 26, 1792, had no heading and was only three pages long. It was added just before the final printing, and after the declarations of war between France on the one hand and Austria and Prussia on the other. While many predicted that France would be easily defeated, Young was doubtful. France had many excellent border fortresses behind which the raw troops of the new government might fight very well. He discounted the danger of civil war, for opposition in time of war became treason and would be ruthlessly suppressed. He even maintained that an Allied victory would be against British national interests. He had not forgotten the partition of Poland and doubted whether Englishmen would really like Prussians and Austrians in military control of the Low Countries. In case of a decisive Allied victory which would upset the balance of power, England might join with France. The *Travels in France* ended with the following:

Should real danger arise to France . . . it is the business, and direct interest of her neighbours, to support her. The revolution, and anti-revolution parties of England, have exhausted themselves on the French question; but there can be none, if that people should be in danger:—WE hold at present the balance of the world; and have but to speak, and it is secure.[381]

In less than six months Young had completely shifted his position.

Young's *Travels in France* were quite widely reviewed. Strangely enough, the *Gentleman's Magazine* seems to have neglected it entirely. The half page in *The New Annual Register* for 1792 was quite favorable and declared that the diary was "written with great freedom and spirit, although sometimes negligently and incorrectly." The survey contained some attitudes from which the reviewer differed, but he felt it was "of great value and importance, and replete with information, deserving

---

[381] *Ibid.*, p. 571. This second addendum of April 26 was omitted from the 1794 edition.

the serious attention" of both France and England.[382] The *Annual Register* for 1792, which did not appear until 1798, printed a review of fourteen pages, with copious excerpts. The first paragraph showed the general attitude towards Young:

Mr. Young, who has been long a distinguished writer on the subject of agriculture, appears now before the public in a new light, as a traveller of no mean abilities, possessed of much judicious reflection and accurate discrimination. Every publication of this gentlemen is entitled to a considerable degree of respect. Although he may not uniformly challenge the approbation of his readers, he seldom fails to entertain them, and never grows tedious or insipid. His very foibles afford amusement; and even the little traits of personal vanity, which occasionally betray themselves, and which he seems by no means anxious to conceal, excite rather the smile of complacency than the sneer of contempt.[383]

By far the most exhaustive review was in *The Monthly Review,* which ran installments in January, February, and March, 1793, to a total of forty pages.[384] *The Monthly Review* had frequently treated Young pretty roughly and he must have awaited its judgment with considerable impatience and perhaps trepidation. As in most eighteenth-century reviews, there were long excerpts from the book, but there was also much comment. The first installment in January was given the place of honor as the first book noticed in the new year. The first page of the review was nicely calculated to please and at the same time enrage the author.

How often have we . . . lamented the choice of this indefatigable writer, in his literary walk! Had he persevered in the cultivation of polite literature, with which, if we recollect, he set out, he might have acquired honours equal to his natural talents.

Providentially, however, for the public, Mr. Young in agriculture, as Mr. Burke in politics, has been eminently serviceable; not altogether through the intrinsic value of his own writings, but by provoking and exciting men of more judgment though of less splendid imaginations, and thus drawing them forth, perhaps reluctantly, into the public service. What the alchymists were to chemistry and true philosophy, Mr. Young and Mr. Burke have been to agriculture and politics.

[382] *New Annual Register* (1792), pp. 279–280.
[383] *Annual Register* (1792), Account of Books, p. 200.
[384] *The Monthly Review* **10** (1793): pp. 1–13, 152–168, 279–290.

The work of Mr. Young, now before us, possesses, comparatively with his tours in England, considerable merit. . . .

It is, however, our duty to apprize those whom it may concern, that the leading object of our author's pursuit, . . . was not *agriculture* but *political oeconomy;* and how he could hold out the former in preference to the latter, and thus do great injustice to his work, is to us a matter of surprise. All the world must know (if, as Mr. Y. intimates, all the world read his books,) that he cannot write successfully on *agriculture;* while in *political arithmetic* he has deservedly gained considerable credit. . . .[385]

The first installment covered the trips of 1787 and 1788. Two other short quotations must be made. "The author's sensations and feelings, in situations that are interesting, are such as many men experience, but which few can report on so well as Mr. Young."[386] "We honor our author exceedingly, for admiring so warmly (in his riper years), the amiable part of the sex."[387] The second installment dealt with the trip of the revolutionary year 1789:

. . . his information is good, his adventures are singular, and his observations are lively, acute, and frequently just . . . the political sentiments of Arthur Young, like those of Thomas Paine, though often theoretical, are sometimes practical, and at this time deserve the attention of all ranks of society.[388]

The last paragraph of this second installment is more genuinely favorable to Young than was the custom of *The Monthly Review:*

We cannot, however, take our leave of this part of Mr. Young's performance, without thanking him, in the most unequivocal manner, for the entertainment which he has afforded us; . . . and . . . we promise, henceforward, never more to be out of temper with Mr. Young and his writings. He has, in these journals, evinced such goodness of heart, and such honesty of disposition, that we can pass over, with patience, the foibles and errors of judgment which he may evince, and to which, indeed, all men are more or less liable.[389]

Unfortunately the reviewer did not keep his promise even to

[385] *Ibid.*, pp. 1–2.
[386] *Ibid.*, p. 3.
[387] *Ibid.*, p. 10.
[388] *Ibid.*, p. 153.
[389] *Ibid.*, p. 168.

the next month, for the concluding installment is full of acerbity. Not that it was entirely unfavorable, particularly to those chapters treating of political arithmetic rather than agriculture. For instance, the reviewer commented on the chapter on the courses of crops, "What a falling off is here!—but no wonder. In the last chapter, we heard Mr. Young speaking in the character of a *political arithmetician;* in this, he figures in the character of a farmer!"[390] At the very end the reviewer described Young's style as "the frank, open, undressed, (or, shall we say? the *stark naked*) manner in which the author's remarks and reflections are introduced. . . ." The final paragraph urged that the diary be republished separately and predicted that it "would be perused with avidity, by all ranks of readers; as containing perhaps more entertainment and information than any thing heretofore published of the kind." The reviewer continued:

The *second part* we will not hesitate to recommend, strongly, and sincerely, to the most serious attention of the rising Republic; to whom also, we . . . think, in the same sincerity, its author might be of essential service, in assisting to establish a suitable system of *Political Economy;* and, shall we add? in giving them at least, a *Relish for Agriculture.* [391]

Perhaps the most interesting agricultural article by Young in the *Annals* in early 1792 was on the farmer's cart, embellished with a plate showing the cart used at Bradfield.[392] Ireland had made Young an enthusiast for the small one-horse cart, since a single horse could draw a much greater proportionate weight than when harnessed in a team. Young performed all the work at Bradfield with five of these carts.

There was considerable traveling back and forth among the gentlemen agricultural improvers on the east coast in the spring and summer of 1792. Late in May, Young went into Bedfordshire with Thomas Ruggles, author of *On the Police and Situation of the Poor,* which appeared in serial form through ten volumes of the *Annals* from 1789 to 1793. As a former schoolmate at Lavenham, Ruggles was a very old friend and beginning about 1789 or 1790 they became very intimate. Young was a member

[390] *Ibid.,* p. 283.
[391] *Ibid.,* p. 290.
[392] *Annals* 18: pp. 178–192.

of a new farming club at Long Melford formed by Ruggles in 1791.[393] Late in June Young visited Ruggles for a week at one of his estates, Spains Hall in northern Essex. Unfortunately Young's account of his stay is confined very rigorously to agriculture, for one would like to know Young's reaction to the charming Elizabethan mansion.[394] Probably Young did not appreciate it, for his taste was definitely classic. He was more interested in the old fish ponds, urged his friend to renovate them, and tried to prove to Ruggles that an acre of fish pond might be as profitable as an acre of wheat. One day Young and Ruggles visited Mr. Allen Taylor, who raised great crops of beans and cabbages and who had an ancient oak tree which he named "Young's Oak." They also made a little side trip down to the coast, stopping at the monument to the celebrated botanist, John Ray, recently repaired by the combined exertions of Ruggles, Sir Joseph Banks, and Sir John Cullum. They made still another trip to Castle Hedingham where Lewis Majendie had greatly improved his estate since Young's last visit. He was an enthusiast about drilling for wheat, cultivated white clover seed of a very fine quality, fatted Welsh heifers with great profit, and was landlord of some of the richest hop lands in England. Young described the whole trip as "an excursion equally agreeable and instructive."[395]

A week after his return Young had as guests at Bradfield two distinguished farmers whom he had often visited, John Boys of Betshanger and John Ellman of Glynd, who were making a month's tour, visiting many famous agriculturists—Mr. Ducket, Robert Bakewell, Coke of Norfolk—and Arthur Young. Their account of Young's farm is the most circumstantial contemporary description in existence. Of course they were not completely impartial, for their account was to appear in the *Annals*. They arrived on July 9 in time for tea and met "with the most polite and friendly reception." After tea they visited his cattle tied in

---

[393] *Ibid.*, pp. 220–228. For the farming club *cf.* Young, 1898: pp. 193–194.

[394] *Annals* 18: pp. 391–444. In 1938 I was hospitably entertained at Spains Hall by Colonel Sir Edward A. Ruggles-Brise, Bart., who kindly permitted me to copy twelve manuscript letters from Young to Ruggles, and who presented we with a copy of the very fine pastel portrait of Young which hung in the dining room. In some ways the best of all Young's portraits, it has been reproduced as fig. 1, frontispiece, in this book. Russell's portrait of Ruggles is also reproduced as fig. 19. At least one of Spains Hall's fishponds was in existence in 1938.

[395] *Ibid.*, p. 444.

stalls, including some fine oxen and a bull of the Sussex breed. They also saw "ploughing implements of various descriptions out of number, which Mr. Young seems to have procured at a great expence, with a view of ascertaining which are the most useful."[396] On the following day they saw their host's South Down ewes and "a very strong crop" of chickory. It is hardly surprising that Young's best ram was a South Down "bred by Mr. Ellman." The King's Spanish Merino had "a very pretty little carcass, and fine wool," but also "immense horns, and a throat like a southern hound." The Iceland ram was "by far the worst we ever saw." They commented on Young's cabbage planting which was at its height: "The whole of this operation is both contrived and executed well." They passed through his experiment field with its many varieties of grasses, and were shown a sample of his plowing with oxen, "much superior to any ploughing we ever saw at home." Their summary was very favorable: ". . . his crops in general very good; land clean and in good order; which is a little extraordinary considering the wetness of the soil, and how much Mr. Young is from home."[397]

In August, 1792, Young took the initiative in attempting to establish a wool fair for Suffolk and Norfolk, comparable to the Lewes Fair in Sussex. On August 6, he wrote an open letter to the *Bury and Norwich Post,* pointing out that wool prices in Suffolk and Norfolk lagged behind those of Sussex and other counties which had a fair, and called for a meeting at the Angel Hotel in Bury. At this meeting Young opened the business with a speech elaborating on the advantages for the flock masters from such a fair. The profit of the middleman or wool broker might be eliminated, for the manufacturers could buy directly from the flock masters at the fair. A fair would also stimulate the production of better wool, for all wool would be sold in the open, and comparisions could more easily be made. Twenty-eight flock masters attended the meeting, many of them men with large flocks, and it was decided that a fair should be held annually at Thetford at the end of July. The meeting also voted thanks to Young "for his attention to the interests of the country."[398]

---

[396] *Ibid.* **19:** pp. 89–90.
[397] *Ibid.,* pp. 93–94.
[398] *Ibid.* **18:** p. 621. For Young's letter *cf. Bury and Norwich Post,* August 8, 1792.

The summer of 1792 was important in Arthur Young's life, however, not because he visited Thomas Ruggles or because John Boys visited him, but rather because it marked a decisive turning point in his outlook on public affairs. On April 26 he penned that addendum to the *Travels in France,* predicting that England might support France in the war which had just begun. In May appeared the *Travels in France,* an important book which stamped its author as a liberal and as pro-French, but on August 20 he wrote a short article for the *Annals* which started as follows:

The fearful events which are at present passing in France, with a celerity of mischief that surpasses equally all that history has to offer, or fancy to conceive, afford a spectacle interesting to every man who possesses PROPERTY; and to none more than to farmers. The quarrel now raging in that once flourishing kingdom, is not between liberty and tyranny, or between protecting and oppressive systems of government; it is, on the contrary, collected to a single point,—it is alone a question of property; it is a trial at arms, whether those who have *nothing* shall not seize and possess the property of those who have *something.* A dreadful question—a horrid struggle which can never end but in the equal and universal ruin of all. . . .[399]

From the date of that article until his death in 1820 Arthur Young was a conservative in politics and an anti-French patriot. There is one piece of evidence to show that the change may have begun a couple of months earlier. The *Annals* contain a long excerpt from a French committee report dated April 26, 1792, and he had commented pretty bitterly over the proposal to divide uncultivated commons and heaths among the poor: "Such schemes . . . are the right and genuine fruits of a democracy;— that is, of a government by representatives chosen by people of no property—of a government founded on the PERSONAL *rights of man.*"[400]

The above quotations make it very clear that what aroused Young to oppose the revolution was the danger to private property. He now attacked the whole concept of "equality," for after the abolition of the feudal system, what else could it mean than equality in property? "The word is absurd if it attaches not to property, for there can be no equality while one man is rich and

[399] *Ibid.,* p. 486. This article and the two following ones were entitled, "French Events Applicable to British Agriculture."
[400] *Ibid.,* p. 148.

another poor."[401] He now blamed the French Constitution of
1791 for having established representation upon a personal
rather than a property basis: "IF PERSONS ARE REPRE-
SENTED, PROPERTY IS DESTROYED." He pointed out that
opinions must change as conditions change. The Revolution
had been an experiment, which had failed. He expressed his
idea in good farming idiom: *"the thing is tried; that method
of drilling has been experimented and found good for nothing;
the crop did not answer."*[402] In this first article he denied that
he opposed all parliamentary reform and stated that he favored
a reform to abolish some rotten boroughs and give the franchise
to men who "possess" property of at least £100 a year. Again
he urged his militia of property to protect property rights against
revolution at home. One regiment of one thousand cavalry in each
county "would give certain and permanent security against the
mischievous example of France."[403] Although this article contains
no reference to the overthrow of the French monarchy on August
10, it seems likely that Young was influenced by that event, since
he wrote his article on August 20.

To his surprise several people wrote privately to him protesting his article and accusing him of changing his outlook. As
a result he wrote a second article for the *Annals,* denying that
he had changed his principles.

My principles I certainly have not changed, because if there is one
principle more predominant than another in my politics, it is the
principle of change. I have been too long a farmer to be governed by
any thing but events; I have a constitutional abhorrence of theory,
and of all trust in abstract reasoning; consequently I can rely on
nothing, but experience; in other words, on events.

He went on to state very positively his attitude at that moment, *"that the little finger of a French democracy, established on
personal* representation, is a more odious tyranny than the heavy
arm of Turkish despotism."[404] He showed how some clauses of

[401] *Ibid.,* p. 487.
[402] *Ibid.,* p. 492.
[403] *Ibid.,* p. 494.
[404] *Ibid.,* pp. 582–583. *Cf. ibid.* 19: pp. 154–169 for Gamaliel Lloyd's protest,
dated October 28, 1792, against Young's first article. Very mild in tone, it did
point out that under the Constitution of 1791 only the active citizens voted, and
hence that Young erred when he stated that voting was by person. To which
Young replied in a footnote, p. 157, "True;—but they vote with the pikes of the
mob at their throats."

the Declaration of the Rights of Man had been violated or perverted: ". . . the right of resistance against oppression, became the power to oppress; the right to liberty crammed every prison on suspicion; the right to security fixed it at the point of the pike; the right to life became the power to cut throats." This second article referred more specifically to the events of August 10 and the weeks following. He also attacked by name many radical English reformers, Mackintosh, Cartwright, and especially Paine, "that prince of incendiaries."[405] He protested against the mass meetings favoring the Revolution and the "licentiousness of publication."

On October 12 Young wrote a third article for the *Annals*,[406] specifically mentioning the September Massacres and the aboliton of the monarchy, and attacking the requisitioning of supplies and food. He declared that the French system had actually become anarchy and predicted the ultimate rise of a dictator. Again he attacked the English sympathizers with the French Revolution—Paine, Cooper, Sheridan, and even his old friend, Priestley. He maintained that the Revolution had "brought more misery, poverty, devastation, imprisonment, bloodshed, and ruin of France, in four years, than the old government did in a century." Young very strongly opposed the proposal of the English pamphleteer, Cooper, that a national system of education be established in England:

I always held the indiscriminate instruction of the poor to be a mischievous business. . . . I do not find on my farm, in the village, or its vicinity, that those are the best ploughmen and carters who are the deepest adepts of the Right of Man. If there must be hewers of wood and drawers of water, why preach equality.[407]

Note has been taken above of John Boys's account of Arthur Young's agriculture in 1792. There also exists for that year a more important and vastly more interesting account of Arthur Young by Fanny Burney. In June, Young had invited Dr. Charles Burney and his daughter Fanny to visit Bradfield. In his reply

[405] *Ibid.* **18:** p. 588.
[406] *Ibid.* **19:** pp. 36–51.
[407] *Ibid.*, pp. 48–49.

of July 17,[408] Charles Burney declined the invitation, partly because of poor health and partly because of prior plans. He promised, however, to send his youngest daughter, Sarah, to Bradfield for a visit and she probably spent several months there.
Young's letter of June 18 to Fanny, her reply of July 17, and
his answer to her, have all survived. Together they reveal the
wit, gaiety, and charm of Arthur Young in middle age.

Bradfield Farm, June 18th, 1792
What a plaguy business 'tis to take up one's pen to write to a
person who is constantly moving in a vortex of pleasure, brilliancy,
and wit,—whose movements and connections, are, as it were, in
another world! One knows not how to manage the matter with
such folks, till you find that they are mortal, and no more than
good sort of people in the main, only garnished with something
we do not possess ourselves. Now, then, the consequence—

Only three pages to write, and one lost in introduction! To the
matter at last.

It seemeth that you make a journey to Norfolk. Now do ye see,
if you do not give a call on the farmer, and examine his ram (an
old acquaintance), his bull, his lambs, calves, and crops, he will
say but one thing of you—that you are fit for a court, but not for
a farm; and there is more happiness to be found among my rooks
than in the midst of all the princes and princesses of Golconda.
I would give an hundred pound to see you married to a farmer
that never saw London, with plenty of poultry ranging in a few
green fields, and flowers and shrubs disposed where they should
be, around a cottage, and not around a breakfast-room in Portman
Square, fading in eyes that know not to admire them. In honest
truth now, let me request your company here. It will give us all
infinite pleasures. You are habituated to admiration, but you shall
have here what is much better—the friendship of those who you loved
long before the world admired you. Come, and make old friends
happy.

A. Young[409]

---

[408] Young, 1898: pp. 214–215. Burney commented very favorably on the *Travels
in France*, which they were reading in the evenings. "No one can accuse you of
drowsiness, like old Homer and such folks; you are always awake, and keep your
readers so. . . . Though an enemy to the old tyranny, you neither reason about
the rights of man like Wat Tyler or even Tom Payne."

[409] d'Arblay, 1842: **2:** p. 355.

Fanny's witty reply was written on the same day as that of her father:

Chelsea College, July 17, 1792

Nay, if you talk of your difficulties in fabricating an epistle to me, please to consider how much greater are mine in attempting to answer it. You! a country farmer, the acknowledged head of "the *only art worth cultivating*," as you tell us,—the contemner of every other pursuit, the scorner of all old customs, the defier of all musty authorities, the derider of all fogrum superiors,—in one word a Jacobin. You afraid? and of whom? a Chelsea pensioner? One who, maimed in the royal service, ignobly forbears, spurning royal reparation? one who, though flying a court, degenerately refrains from hating or even reviling kings, queens, and princesses? One who presumes to wish as well to manufactures for her outside, as to agriculture for her inside? One who has the ignorance to reverence commerce, and who cannot think of a single objection to the Wool Bill? One, in short, and to say all that is abominable at once, one who in theory is an aristocrat, and in practice a *ci-devant* courtier?

And shall a creature of this description, the willing advocate of every opinion, every feeling you excommunicate from "your business and bosom" *dare* to write to *you?* Impossible!

Whether I shall come and see you all or not is another matter. If I can I will.

P. S. Will Honeycomb says if you would know anything of a lady's meaning (always providing she has any) when she writes to you, look at her postscript. Now pray, dear sir, how came you ever to imagine what you are pleased to blazon to the world with all the confidence of self-belief, that you think farming the only thing worth manly attention? You, who, if taste rather than circumstance had been your guide, might have found wreaths and flowers almost any way you had turned, as fragrant as those of Ceres.[410]

Young's reply was fully as witty as Fanny's, thereby lending some weight to her last sentence:

You, "the willing advocate of every feeling I excommunicate from my bosom," knew you had thrown so bitter a potion into your letter that you could not (kind creature!) help a little sweetening in the postscript; but must there in your sweets be some alloy? Could you not conclude without falling foul of poor Ceres?

[410] Young, 1898: pp. 216–217.

Your letter, or rather your profession of faith, is one of the worst political creeds I remember to have read; you see no merit but beneath a diadem. In government a professed aristocrat, in political economy a monopolist, who commends manufactures, not as a market for the farmer, but for the much nobler purpose of contributing to adorn your *outside;* and who can attain not one better idea of the immortal plough than that of giving some sustenance to your *inside.* But, by the way, is not that inside of yours an equivocque? Do you mean your real or your metaphorical inside, your ribs or your feelings? If you allude to your brains, they are by your own account a *wool*-gathering. Do you mean your heart, and that the philosophical contemplation of so pure an engine as the plough is the sustenance of your best emotions? How will that agree with the panegyrist of a court, and the satirist of a farm? Or is it that this inside of yours is a mere bread and cheese cupboard, which, certes, the plough can furnish? Or is it a magic lanthorn full of gay delusions, lighted by tallow from the belly of a sheep? Till you have settled these doubts, I know not which you prefer, manufactures for improving your complection, or agriculture for farming your heart. Nor must you wonder at such questions arising while you use terms that leave one in doubt whether you mean your head or your tail. I know something of the one; the other is a metaphor. Though there is high treason against the plough in almost every line of your letter, yet the word *If I can I will* are not in the spirit that contains the Eleusinian mysteries; they bring balm to my wounded feelings.[411]

Fanny Burney actually went to Bradfield in October, but before she arrived an almost equally famous person, the Duc de Liancourt, had temporarily settled in Bury St. Edmunds as an emigré. Like so many enlightened nobles, Liancourt had favored the early stages of the Revolution, only to find the movement going far beyond his expectations and desires. Completely loyal to the person of Louis XVI he could not accept the events of August 10. Shortly after that tragic day he fled and after a romantic escape arrived in England about August 20, and later took a small house at Bury St. Edmunds.[412] Although Young's views had changed before Liancourt's arrival, the story of his friend's troubles, and the murder of Liancourt's cousin, the Duc

---

[411] *Ibid.*, pp. 217–218.
[412] *Cf.* Dreyfus, 1903: pp. 115–117, 200–208; de la Rochefoucauld, 1933: pp. xxii-xxxi.

de Rochefoucauld, whom Young knew well, must have contributed to his growing bitterness against the Revolution.

Fanny's account of her arrival at Bradfield was very graphic:

FRIDAY, OCTOBER 5th.—I left Halstead, and set off, alone, for Bradfield Hall, which was but one stage of nineteen miles distant.

Sarah, who was staying with her aunt, Mrs. Young, expected me, and came running out before the chaise stopped at the door, and Mr. Young following, with both hands full of French newspapers. He welcomed me with all his old spirit and impetuosity, exclaiming his house never had been so honoured since its foundation, nor ever could be again, unless I revisited it in my way back, even though all England came in the meantime!

Do you not know him well, my Susan, by this opening rhodomontade?

"But where," cried he, "is Hetty? O that Hetty! Why did you not bring her with you? That wonderful creature! I have half a mind to mount horse, and gallop to Halstead to claim her! What is there there to merit her? What kind of animals have you left her with? Any thing capable of understanding her?"

During this we mount up-stairs, into the dining-room. Here all looked cold and comfortless, and no Mrs. Young appeared. I inquired for her, and heard that her youngest daughter, Miss Patty, had just had a fall from her horse, which had bruised her face, and occasioned much alarm.

The rest of the day we spoke only of French politics. Mr. Young is a severe penitant of his democratic principles, and has lost even all pity for the *Constituant Revolutionnaires,* who had "taken him in" by their doctrines, but cured him by their practice. . . .

Even the Duc de Liancourt, who was then in a small house at Bury, merited, he said, all the personal misfortunes that had befallen him. "I have real obligations to him," he added, "and therefore I am anxious to show him any respect, and do him any service, in his present reverse of fortune; but he had brought it all on himself, and, what is worse, on his country."[413]

Several days after Fanny's arrival, Liancourt came to Bradfield to meet her:

The Duke accepted the invitation for to-day, and came early, on horseback. . . .

[413] d'Arblay, 1842: **2:** p. 361.

Mrs. Young was not able to appear; Mr. Young came to my room door to beg I would waste no time; Sarah and I, therefore, proceeded to the drawing-room. . . .

He is very tall, and, were his figure less, would be too fat, but all is in proportion. His face, which is very handsome . . . has rather a haughty expression when left to itself, but becomes soft and spirited in turn, according to whom he speaks, and has great play and variety. His deportment is quite noble, and in a style to announce conscious rank even to the most sedulous equalizer. His carriage is peculiarly upright, and his person uncommonly well made. His manners are such as only admit of comparison with what we have read, not what we have seen; for he has all the air of a man who would wish to lord over men, but to cast himself at the feet of women.

He was in mourning for his barbarously murdered cousin the Duc de la Rochefoucault. His first address was of the highest style. I shall not attempt to recollect his words, but they were most elegantly expressive of his satisfaction in a meeting he had long, he said, desired.

With Sarah he then shook hands. She had been his interpretess here on his arrival, and he seems to have conceived a real kindness for her; an honour of which she is extremely sensible, and with reason. . . .

After a little, the duke began a *tête-à-tête* with Fanny on her second great novel, *Cecilia*, and continued to draw her out until dinner:

Mr. Young listened with amaze, and all his ears, to the many particulars and elucidations which the Duke drew from me; he repeatedly called out he had heard nothing of them before, and rejoiced he was at least present when they were communicated. . . .

At length we were called to dinner, during which he spoke of general things.

The French of Mr. Young, at table, was very comic; he never hesitates for a word, but puts English wherever he is at a loss, with a mock French pronounciation. *Monsieur Duc,* as he calls him, laughed once or twice, but clapped him on the back, called him *un brave homme,* and gave him instructions as well as encouragement in all his blunders.[414]

[414] *Ibid.,* pp. 365–366.

In the last months of 1792 Young took two more farming trips, the first of a week in Norfolk,[415] and the second of more than three weeks into Essex, Kent, and Sussex.[416] These trips were chiefly visits to "spirited" farmers, men with large acreages and large flocks of sheep and who employed advanced methods. On the first trip he spent three nights at Holkham with Coke of Norfolk, who was abandoning Norfolk sheep in favor of the Bakewell breed, who had a total flock of 2,400, and who had planted more than a million trees in the decade, 1781–1791. Young left Holkham, "highly gratified, with that steady attention Mr. Coke is in the habit of paying to the plough. He truly loves husbandry, practices it with equal intelligence and success, and is always most liberally ready to make any experiments that promise to be of public benefit."[417] He also visited William Colhoun, member of Parliament for Bedford, who had an immense farm of 5,400 acres and a flock of 3,000 sheep. He was definitely a spirited farmer who engaged in experiments in marling, drilling, and irrigation. Young and Colhoun then went to Riddlesworth where Silvanus Bevin, on his farm of 2,000 acres, had built a new home and farm buildings, had carried on an irrigation project, had bought a flock of South Down sheep, and had even followed Young's advice to have a good piece of rouen for the spring feeding of sheep. In an attempt to explain such spirited exertions, Young concluded:

I cannot help answering the question, in a manner that will not be looked for;—I say then,—TO THE BRITISH CONSTITUTION; . . . The world offers no such spectacles as this,—the world then knows no such freedom. Here commerce flourishes to enrich,— and here agriculture is willing to be adorned: all is good, because all is under the protection of equal laws. Forbid it heaven, that this house should ever become the residence of political discontent!—of a reformer![418]

Such a statement shows the almost unbelievable change in

---

[415] *Annals* **19:** pp. 441–499.

[416] *Ibid*. **20:** pp. 220–297.

[417] *Ibid*. **19:** p. 457. While at Holkham Young may have discussed Coke's proposal in a letter of October 23, favoring a bill to "fix an assize of flour according to the average price of wheat," thus protecting the poor from "the shameful practices and combinations of the millers." *Cf*. Young, 1898: pp. 212–213. It is rather doubtful whether Young would have supported such a proposal at this time.

[418] *Ibid*., pp. 486–487.

Young's opinion of the English constitution from that he had expressed in the *Travels in France*.[419]

The second trip was more interesting, partly because he was visiting many old friends and more aristocrats than practical farmers. He started early enough on November 18 to breakfast at Clare with Thomas Ruggles who was building a new house. Whereupon Young composed a little essay on the *desiderata* in a home for men of moderate means:

. . . how many new houses, in which people have no more elbow room . . . than if they were in the stocks; how many in which comfort is sacrificed to shew; warmth to space; the sun's rays, in latitude 55, to the sight of a park or a lake; shelter to a prospect; and the convenience of a lumber room to the arrangement of an anti-chamber [*sic*]; and as to cupboards, closets, and stowage of many sorts, the fools in middling life allow their puppy architects to sweep them all away, because my lord, with forty servants, transfers such things to the offices. Where do we meet with a moderate house, well calculated for a small fortune? Where do we find one planned for a man who keeps the key of his wine cellar? Who has connected a kitchen and a dining-room in such manner, that the smell of the former should be excluded, without a long walk to the latter? Who has contrived a moving table, served through the wall, without any servants to wait in the room? There is not one apartment in a house, from the cellar to the garret, which has not been improved for men of large fortune; but for small incomes, I believe invention has either gone retrograde, or at least stood still.[420]

He spent three days with Montagu Burgoyne, who was employed in steaming potatoes for fatting cattle, and whom he accompanied to a meal with the Mutton Club, where the mutton was "so fat, that the only difficulty was to find lean." He was also present by chance at the "annual celebration of Mr. and Mrs. Burgoyne's wedding-day; fire-works, catches, glees, dancing, and a great entertainment. . . . Mr. Burgoyne is a good farmer, and Mrs. Burgoyne has the merit, the first a lady can possess, of making a good farmer a happy man."[421]

Among the practical farmers visited was William Dann, an

---

[419] *Cf. supra*, p. 286.
[420] *Annals* **20**: pp. 220–221.
[421] *Ibid*., p. 228. The contrast with Mrs. Young was all too obvious.

important contributor to the *Annals,* a friend of potatoes, lucerne, and South Down sheep. He also returned the visits of his two guests in the summer, John Boys and John Ellman. He spent four days with Boys who took him to visit a Mr. Wall, a famous Romney Marsh grazier who believed the Romney Marsh sheep to be superior to the Bakewell breed. On December 1 Young, Boys, and Wall had dinner with Edward Knatchbull, another patron of the Romney Marsh sheep. During dinner Wall agreed to allow Boys and Young to weigh alive, and then again after butchering, one Romney Marsh wether and one South Down. The day was just such a one as Young loved, "and thus closed a day, that has been, from the first moment to the last, a *farming* one." The test on December 2 resulted in showing that the Romney Marsh sheep was "beyond all comparison, the best."[422]

Between December 6 and 10, Young and Lord Sheffield were the guests of the Earl of Egremont at Petworth. Apparently this was Young's first visit there, but it was not his last, for Egremont and Young became intimate friends. Young had high praise for his host, "than whom no man wishes more cordially for the improvement of the agriculture of his country, nor would more readily contribute, in any way, to further and promote it, that appeared practicable and useful."[423] While at Petworth Young visited the Duke of Richmond's seat at Goodwood where he greatly admired a very fine barn, but was less impressed by "a most superb dog kennel. The dogs seem to rest here in most luxurious repose; some were in their dining, some in their drawing-rooms, others in bed chambers, some were stretched on a carpet of velvet lawn. These dogs are in luck not to be on the other side of the channel, for they would doubtless be hanged for aristocrats. . . ."[424]

Shortly after returning from this last trip, Young composed and printed a broadside addressed to *"such Persons in the Hundreds of Thedwastry and Thingoe as are desirous of testifying their content under the Constitution of this kingdom as* ESTABLISHED AT PRESENT, *and of Associating for those laudable Purposes which tend to secure the Blessings we derive from its Influence."* Dated December 18, the broadside was ap-

[422] *Ibid.*, pp. 269, 273.
[423] *Ibid.*, p. 291.
[424] *Ibid.*, pp. 228–229.

parently posted in public places, and called for a meeting at the Angel Hotel for December 29, where Young would state more fully his purposes and offer some resolutions.[425] Thus by the close of 1792 Young's position was clear, both nationally and locally. He had joined the conservative, anti-French, and anti-reform forces in England and was preparing to play a prominent role in all efforts to uphold the *status quo.*

[425] *An Address Proposing a Loyal Association to the Inhabitants of the Hundreds of Thedwastry and Thingoe.* The copy in the British Museum is addressed to the Rev. W. Godfrey, The Officiating Minister, Hawstead, with a note, "For the Chh. Door if you approve." A. Y.

VI. The Board of Agriculture, 1793-1796

IN MANY respects the middle 1790's mark the culmination of Arthur Young's career. Without question he was the most famous authority on agriculture in the English-speaking world. Although there were many more successful farmers, and several men with much greater technical knowledge, his was the name of greatest note. His agricultural writings went back now for nearly thirty years, and many had great merit. The *Annals of Agriculture,* the best-known agricultural periodical, now had a decade of successful operation behind it. His travels made him better known personally than any other authority. He was then the natural choice as secretary to the new Board of Agriculture in 1793, a post which he held until his death. In turn the new position added greatly to his prestige.

The *Travels in France* was recognized as a first-rate work, but unfortunately its tone had been somewhat too friendly to the French Revolution for the great majority of the landed class. Although the book had described the events of 1789 it had not appeared until 1792 when Burke's *Reflections,* and the very trend of events, had made most Englishmen hostile to the French Revolution. Late in 1792 Young completely switched to the conservative position and early in 1793 published his most influential pamphlet, *The Example of France a Warning to Britain.* Its rabidly anti-revolutionary position made Young again acceptable as spokesman of the agrarian interests. Two more pamphlets in 1795 and his part in proposing and advocating the Yeomanry Militia further identified him with the patriotic party.

Moreover his personal life was reasonably happy in this period. True he lost his second daughter Bessy in 1794, but this loss did not shake him very deeply. His son Arthur graduated from Cambridge in 1793 and began to write on agricultural subjects almost immediately. He had hosts of friends and was welcome in nearly every company. Never would he be as happy again, for in

306

1797 occurred Bobbin's death, a loss from which he never really recovered. There followed his conversion to a fanatical, evangelical pietism. He came to regard many of his former pleasures as sinful. Melancholia and fanaticism made him unwelcome and unhappy in many quarters.

The first really important event in Arthur Young's life in 1793 was the publication of *The Example of France a Warning to Britain,* which appeared in February. It was based upon three articles in the *Annals,* dating from August 20 to October 12, 1792, nearly every sentence of which was reproduced verbatim, although the material was completely rearranged and greatly enlarged. While the three articles totaled only 41 pages, the first edition of the pamphlet ran to 146 pages of about the same size. It first analyzed contemporary French conditions under the three headings of government, personal liberty, and security of property. Young supported his indictment by frequent quotations from the *Moniteur* of such prominent revolutionaries as Barbaroux, Marat, and St. Just. Government had broken down, he said, and France was actually in a state of anarchy. As for personal liberty, "THERE IS NO SUCH THING."[1] As noted in the previous chapter in analyzing the articles, the danger to private property was really the basis of Young's fear of the revolution and he documented this danger at some length. The second part of the pamphlet discussed the three causes for the evils described in the first part, namely personal representation, the rights of man, and equality. Throughout Young attacked all English reformers, moderates as well as radicals, and much space was filled quoting from Paine, Cartwright, and the resolutions of various pro-French and pro-revolutionary societies. He even defended the existing English government with all its abuses, even parliamentary corruption, and declared that the unrepresentative character of the House of Commons was a virtue:

If they are bribed in order to act wisely it . . . tends to prove that there is something on the verge of danger in all numerous assemblies, which, if not controuled by prerogative or influence, would hazard the public peace. . . . If the nature of such an assembly demands to be corrupted, in order to pursue the public good, who but a visionary can wish to remove corruption?[2]

[1] Young, 1793, 2nd ed.: p. 23. All citations are from 2nd ed.
[2] *Ibid.,* p. 76.

His conclusion on reform follows: "That the first lines of discontent are in fact the most dangerous; that moderate reform, or any reform at all, *on principle,* is a sure step to all that followed reform in France; jacobinism, anarchy, and blood."[3] He pointed out how the revolution had affected the landed, monied, commercial, and laboring classes, and contrasted their favorable condition in unreformed Britain. A short section described his proposed militia of property. Lastly the pamphlet urged the formation of loyal associations to combat the pro-reform and pro-revolutionary societies. At one point he even urged a boycott on all tradesmen of radical inclinations.

A second and enlarged edition of *The Example of France* appeared in April, which included additional quotations from French documents and English radicals, plus eleven pages justifying the war which had recently broken out between France and England. Young had been almost a pacifist since 1780, and had attacked every war of the century in which Britain had engaged, but the present conflict was entirely different from those earlier commercial and colonial struggles. It was "not only just, but absolutely and essentially necessary to the salvation of all that makes life desirable; the peace of families,—the surety of dwellings,—the safety of life,—the security of property."[4] The erstwhile pacifist even justified a preventive war: ". . . on the long account, every year of war, at this crisis, will probably secure ten years of peace in its train, and consequently . . . the policy of permanent peace is, of all others, that which most clearly calls for temporary war."[5] Young saw the Revolution as a challenge to the whole social and political fabric of eighteenth-century English society, a challenge which must be fought and defeated, both at home and abroad, and regardless of cost.

The pamphlet created a considerable stir. It passed through three editions in 1793 and a fourth in 1794. French editions appeared, at Brussels in 1793 and at Quebec in 1794. A German edition came out in 1793 and an Italian in 1794. Congratulatory letters poured in. At the end of the second edition Young in-

---

[3] *Ibid.,* p. 92.
[4] *Ibid.,* p. 162. 2nd ed. also had an appendix tracing English government back to Saxon times. A writer in *Gentleman's Magazine* **63** (1793): p. 450, claimed that this appendix and perhaps the re-editing was done by John Symonds.
[5] *Ibid.,* p. 154.

serted letters of thanks from John Reeves on behalf of the famous Crown and Anchor Committee which was a center of the whole loyal associations movement, and from a local association at Melford in Suffolk. One of the earliest responses came from Burke:

Mr. Burke thanks Mr. Young for his most able, useful, and reasonable pamphlet. He has not seen anything written in this controversy which stands better bottomed upon practical principle, or is more likely to produce an effect on the popular mind. It is, indeed, incomparably well done. We are all very much obliged to Mr. Young, and think the Committee ought to circulate his book.[6]

A little later his neighbor, the Countess of Bristol, wrote, "I think you may, without flattery, consider yourself as one of the means which has rescued this glorious country from the destruction which was preparing for it."[7] Lord Sheffield wrote "that every Body I have seen approves your Pamphlet very much. Those whose opinion you regard, think it excellent."[8] Dr. Burney reported high praise from many quarters. Horace Walpole had told him, "There it is; I read nothing else." A large party of bluestockings had agreed "that your book and Hannah More's 'Chip' were the best on the subject." Just recently Mrs. Crewe had written Burney: "Mr. Arthur Young's pamphlet makes a a great noise, and, I think, I never knew any book take more."[9] In April the *Gentleman's Magazine* published a six-page review with copious quotations, quite favorable to "this animated performance."[10]

Young himself wrote: "The pamphlet rendered the author exceedingly popular among all the friends of government and order, and as unpopular among the whole race of reformers and

[6] Young, 1898: p. 232. Dated March 5.

[7] *Ibid.*, p. 228. Dated March 20. The countess' earlier letter of January 4 (*ibid.*, p. 226) showed that she had encouraged him to write the pamphlet: "I . . . fairly confess that I did wish to set your pen a-going, because you had *experience* and *facts* to write upon, and . . . I knew your warm coloring would suit the picture—in short, I saw you were a convert. I wished you to make others, and if I have been the least instrumental by awakening the spark in you, I shall feel that I am not wholly useless to the community. . . ."

[8] Add. MSS. 35,127, f. 239. Dated March 31.

[9] Young, 1898: p. 233. Dated May 12, after 2nd. ed. had appeared. The countess of Bristol had also noted Walpole's approval.

[10] *Gentleman's Magazine* 63 (1793): pp. 345–351.

Jacobins."[11] As usual, the two-page review in the *Monthly Review* was caustic, being particularly disturbed by his rejection of even moderate reform, and regarding his proposal for a militia of property as one "to arm the rich against the poor."[12] Several correspondents in the *Annals* took him to task. Edward Harries declared his attack upon those who wished for peace with France was "unworthy your pen."[13] John Jenkinson treated him considerably less gently:

The zeal of your principles for the good of your country, were by many much admired; but now it is remarked, that you have received the golden ram from Spain, and that you preach accordingly against the proper reform for the good of your country. Now the burthen of taxes, tithes, &c. are nothing; all is right. Pray explain yourself on your ancient grounds; when I hope, for the honour of your old friends, they need not be ashamed of subscribing themselves your admirers. . . .[14]

In a long footnote Young felt called upon to repudiate such charges:

I am very much obliged to him for one of the handsomest compliments that could be paid, for he evidently thinks, that my pamphlet answers so directly the purpose of government, that he knows not how to conceive, that it was not written at the instigation of government, and paid for by a golden ram. Mr. Jenkinson is not very correct in his chronology; but no matter . . . I would also assure him, on the word of a man of honour, that I never received, directly or indirectly, one shilling from the government of England.[15]

Major John Cartwright, to whom Young had alluded rather contemptuously in an article, also made a dignified but very vigorous attack in a letter to the *Annals,* largely devoted to agriculture:

While perusing with delight and instruction your bold and manly reasonings against pernicious laws, I certainly did not foresee that I was so soon to behold you the apologist of an abuse which *poisons the very fountain of legislation.* . . . I little thought that I should

---

[11] Young, 1898: p. 205.
[12] *Monthly Review* 11 (1793): pp. 111–114.
[13] *Annals* 21: p. 368. Dated September.
[14] *Ibid.* 20: p. 185. Dated April 4.
[15] *Ibid.,* p. 186. In disputing Jenkinson's chronology, Young probably meant that he had received the Merino ram in 1791.

so soon have reason to blame you for inconsistency and want of candour, much less for becoming instrumental in promoting public
deception—a deception intended, as I suspect, for cheating the people
of their rights—yes, I say, *their rights;—*and as a means to so blessed
an end, to involve them in the calamities and hazards of an unnecessary war.[16]

On February 10, 1793, the Rev. Humphrey Smythies of Alpheton preached a sermon at Bradfield, which probably had been
written by Young,[17] from a text from Proverbs, xxiv, 21, "My
son, fear thou the Lord, and the King; and meddle not with
them that are given to change." Young defended Louis XVI's
benevolent intentions and pointed out how all classes had suffered
from the revolution. The English reformers were urging measures
quite similar to those which had inaugurated the revolution in
France, and Young hoped that the English lower classes, "when
warned of the evil by the dreadful examples of others, will be
too wise to listen to doctrines so insidious and so dangerous." He
emphasized the blessings of the poor in England, the Poor Laws,
and legal protection of their property and rights. "The same
sun of equal right shines here with equal beams on the cottage
and the palace."[18] The conclusion was almost obvious:

Let us all, with one mind, with one pious accord, easy with the station in which Providence has placed us—content with the lot which
the present order of social connection gives us—shew, by our peaceable and orderly demeanour, our resignation to the will of God—our
obedience to those authorities ordained by divine and human laws,
for our comfort and protection. . . .[19]

In the spring of 1793 Young was faced with a major decision
about his future. For nearly a decade and a half he had spent
the greater part of each year in his ancestral Bradfield, engaged
in farming, making many short and some longer trips, and
editing the *Annals.* Now he found two quite different courses
of action open to him, each involving a sharp break with the
routine at Bradfield. According to the *Autobiography,* Lord
Loughborough told Young about a fine and very cheap estate

[16] *Ibid.,* p. 38. Dated February 8.
[17] *Ibid.* 19: pp. 500–513. *Annals* does not state specifically that Young was the
author, but at the end appear the initials, A. Y.
[18] *Ibid.,* pp. 511–512.
[19] *Ibid.,* p. 513.

on the Yorkshire moors, consisting of about 4,500 acres in Knares-borough Forest at a price of £4000. There was a "handsome shooting-box, sufficient for the residence of a small family."[20] Most of the land was waste, but very capable of improvement. Ever since 1773 when he had published *Observations on the Present State of the Waste Lands,* Young had been interested in improving wastelands. His plan was to rent Bradfield and presumably to move with his family to the shooting-box. With the rents from his Suffolk farms he would gradually improve the estate through draining, paring and burning, liming and irrigating, cropping for a few years, and then laying down to grass. Each year would see a new farm established which could then be rented. Thus a small capital could be made gradually to go a long way. It was a seductive dream:

. . . . becoming the solitary lord of four thousand acres, in the keen atmosphere of lofty rocks, and mountain torrents, with a little creation rising gradually around me, making the black desert smile with cultivation, and grouse give way to industrious population, active and energetic, though remote and tranquil, and, every instant of my existence, making *two blades of grass to grow* where not one was found before. . . .

Even before Young had completed the final papers for purchase, he had been offered the secretaryship of the Board of Agriculture. As he put it, "The two situations were incompatible with each other." Consequently he put the estate up again for sale, but had to advertise for nearly a year "before I could sell it with much less profit than I had reason to expect."[21] Actually since Young was not a very good businessman and did not have a large capital to invest, it is far from certain that he would have made great profits. Moreover, his roots and those of his family were deep in East Anglia. Mrs. Young would never have been very happy anywhere, but the moors of Yorkshire were far from Bury, Lynn, or London.

The first intimation which Arthur Young had about a possible Board of Agriculture came from Sir John Sinclair late in 1792. Lord of Thurso Castle in the very most northern part of Scotland, Sir John was not quite forty years old, strikingly handsome, and

---

[20] Young, 1898: pp. 208–209.
[21] *Ibid.,* pp. 222–223.

almost unbelievably self-confident. In his propensity to go into print at the slightest provocation and in his tremendous capacity for work, Sinclair closely resembled Young. It is not clear when the two men first met, but Sinclair contributed his first article to the *Annals* in 1790. When Sinclair first approached Young about the establishment of a Board of Agriculture, the latter had remarked, "That it was perfectly unnecesary to take that trouble, as there was not the least chance of success."[22] This conversation resulted in a curious bet, Young wagering a set of the *Annals of Agriculture,* and Sinclair a set of his *Statistical Account of Scotland.* On January 5, 1793, Sinclair wrote Young:

I am to be with Mr. Pitt next week, respecting the proposed Board of Agriculture (you'll lose your wager) so send me *fully* your sentiments on the enclosed plan, which you may publish in your annals if you chuse it.[23]

Young's reply on January 10 still showed skepticism:

You are going TO Mr. Pitt, and I am to lose the wager. When you come FROM Mr. Pitt, I shall win the wager. Pray, don't give Ministers more credit than they deserve. In manufactures and commerce you may bet securely; but they never did, and never will do any thing for the plough. Your Board of Agriculture will be in the moon; if on earth, remember I am to be secretary.[24]

Apparently Young had suggested to Sinclair that he would like to be secretary when they had first talked about it. In the spring of 1793 Sinclair probably talked about the proposed Board to William Marshall who also had ambitions to be secretary:

He showed me his plan, and during my short stay in London repeatedly consulted me on the subject. At the time of my leaving town there did not appear to be the smallest probability of the measure being adopted: even its promoter assured me that he had no hope of its being then carried into effect.[25]

Thus early in 1793 the prospects of a board seemed slight to

---

[22] Sinclair, 1831: **1:** p. 406; Sinclair, 1896: pp. 1–21. For further details about the origins, early history, and organization of the Board of Agriculture, *cf.* the two recent very important studies by Rosalind Mitchison, 1959: pp. 41–69, and 1962: pp. 137–158. For a portrait of Sir John Sinclair, see fig. 21.

[23] Add. MSS. 35,127, f. 216.

[24] Sinclair, 1831: **1,** p. 407.

[25] Clarke, 1898: p. 3.

all those most interested. Young was so doubtful that he purchased the Yorkshire estate. Yet on May 15 the bill was introduced into Parliament and was passed by the House of Commons
two days later. According to Sinclair a major factor in changing
Pitt's mind was Sinclair's aid to the government in April in
regard to the issuance of exchequer bills which eased the government's financial difficulties in the early months of the war.
The story as told by Sinclair's son follows:

The value of my father's genius, in restoring commercial confidence,
in a great national emerency, was fully appreciated by Mr. Pitt.
He sent for the Baronet to Downing Street, and expressed, in em-
pathic terms, his sense of obligation, "There is no man," said he, "to
whom Government is more indebted for support, and for useful information on various occasions, than to yourself, and if you have any
object in view I shall attend to it with pleasure." . . . He replied
to the Minister . . . that the reward most gratifying to his feelings
would be support of the Minister to the institution by Parliament
of a great national corporation, to be called, "The Board of Agriculture."[26]

Sir John Sinclair's motion of May 15 took the form of an
"humble address" to his Majesty, couched in the vaguest language
and specific only on one point, that it should not cost more than
£3000 per annum.[27] Sinclair's speech made it clear that the
Board's chief purpose would be to encourage new agricultural
methods—to collect data, popularize new improvements, answer
questions, conduct foreign correspondence, and perhaps give
premiums. He mentioned specifically the proposed agricultural
county surveys. Members of the Board would receive no pay,
and the Board should be established only for a five-year period.[28]
Many members of Parliament, however, had probably read Sir
John's original proposal which he had sent to Young on January
5 and which had appeared in the *Annals*.[29] Sir John had proposed a board of twenty-four members, and had outlined how
£2,500 might be expended[30]—£1,500 for the county surveys, £500

---

[26] Sinclair, 1837: **1**: p. 252.

[27] For text of the motion *cf. Annals* **21**: pp. 139–140, and Clarke, 1898: pp.
5–6.

[28] For Sinclair's speech, *cf. Annals* **21**: pp. 129–139.

[29] *Ibid.* **20**: pp. 204–213.

[30] Sinclair's original estimate was £10,500, which he had reduced to £5,500,
and finally to £2,500. *Cf.* Sinclair, 1837: **2**: p. 49. The government finally set
amount at £3000.

for stationery and foreign correspondence, and the remaining £500 for "expense of the house, a secretary, and two clerks." Young had a footnote at this point: "A secretary ought to be able to write and read at least; but after paying house-rent and two clerks, the pay of a coal-heaver would not remain."[31]

Sir John's motion was seconded by Young's very good friend, Lord Sheffield, and supported by William Pitt, Dundas, and Wilberforce. On May 17 the opposition attacked the measure. Sheridan proposed an amendment to withdraw all public support, and Charles James Fox declared that "the measure was in itself objectionable, it being in his opinion a mere job, and likely to be converted into an instrument of influence."[32] After Sheridan's amendment was defeated, the motion was carried by the decisive vote of 101 to 26.

Sir John must have informed Young immediately of his success in Parliament, for Young wrote him on May 19:

Upon my word you are a very fine fellow, and I have drunk your health in bumpers more than once. You begin to tread on land; and what I conceived to be perfectly aerial, seems much less problematical than before. Premiums might be made to do much good; but they would demand another thousand to the sum you propose.

Let me have your speech fully and directly; and, if you establish a Secretary on a respectable footing, do not forget the farmer at Bradfield. I am, dear Sir, your faithful and obliged,

> A. Young

The Annals are preparing, and shall be bound and gilt handsomely.[33]

Young did not rely on Sinclair alone, but on the following day made his own application for the post of secretary directly to Pitt:

Bradfield Hall: May 20, 1793.

Sir,—I am informed by Lord Sheffield and Sir John Sinclair that the establishment of a Board of Agriculture is determined.

It has been the employment of the last thirty years of my life to make myself as much a master of the practice and the political encourage-

---

[31] *Annals* **20**: pp. 207–208.
[32] *Parliamentary History of England* . . . 1817: **30**: p. 952.
[33] Sinclair, 1831: **1**: p. 407.

ment of agriculture as my talents would allow. I have examined every part of the kingdom, and have farming correspondents in all the counties.

It is impossible I should know what is your intention in relation to the office of the secretary; but the same wisdom that established the Board will, without doubt, give such an appointment to that office as may fill it in a manner the best adapted to the business.

Should I be happy enough to appear in your eyes qualified for such a post, and you would have the goodness to name me to it, it might lessen the anxieties of a life that has been passed in the service of the national agriculture; and I should feel with unvarying gratitude the obligation of the favour.

I have the honour to be, sir, with the greatest respect,

> Your most humble and obedient servant,

> Arthur Young[34]

The government acted promptly, accepted Young's application, and asked George Rose to notify Young to that effect. On May 30 Young replied to Rose as follows:

> Bradfield Hall: May 30, 1793.

Sir,—It is with pleasure that I acknowledge the receipt of your letter, as it shows that, whatever may be the result of the present business, my exertions have met with the approbation of Government, whose public-spirited and laudable views I have long been solicitous to second.

The salary you mention is, I confess, less than I imagined would be assigned to the office, but its being adequate or not depends entirely on the circumstances of attendance, duty, residence, &c. If these be arranged on a footing any way liberal, the sum is equal to my desires; and I shall in that case accept the office with pleasure. If, on the contrary, these points be fixed as to overturn my present pursuits in life, they would render a larger salary less valuable to me than the sum you mention.

From the nature of the Board, intended to consist, as I understand, of members of the two Houses, with the objects in view, I take it for granted that the points above mentioned may, without the least impediment to the business, be easily arranged.

Trusting in this entirely to Mr. Pitt and yourself, I beg your good

---

[34] Young, 1898: pp. 220–221.

offices that if I should have improperly expressed my meaning, you will do me the justice to rely on the integrity of my views, and not imagine me eager in making a bargain for profit with a great and liberal benefactor.

I have the honour to remain &c.

Arthur Young[35]

Most unfortunately the missing link in the above correspondence, the letter from Rose to Young, has not survived. The last part of the first paragraph of Young's letter to Rose hints that Rose had praised the views expressed in *The Example of France*. George Rose was notoriously William Pitt's patronage-monger. The question is whether Young was bribed by the ministry to change his political views or rewarded for having done so by being made secretary to the new board. Hardly had the appointment become known than such accusations began. The point is very important, for it touches upon Young's essential integrity. Fox's warning that the new Board was a job seemed to be proved when the only important appointment under the Board which carried a respectable salary went to a man who had so recently and so completely changed his views on the French Revolution and upon reform proposals at home. To all the English reformers it seemed obvious that Young had sold himself and his pen to the government in return for a government appointment.

Charles Piggott, in his scurrilous book, *The Female Jockey Club* (1794), wrote: Mr. Arthur Young, author of the Farmer's Letters, who formerly devoted his literary talents to the service of the people, now secretary to the newly appointed board of agriculture, and notoriously a hireling apostate in the pay of administration."[36] Probably the most bitter attack was Major John Cartwright's in his pamphlet, *The Commonwealth in Danger* (1795), attempting to refute *The Example of France* where Cartwright had been attacked. He called Young "the disgraced disseminator of court delusions the most contemptible; the fabricator of false alarms, to serve the dangerous purposes of a domineering faction; and the very personification of political apostasy." He absolved Young from having taken a money bribe, but accused him of succumbing to "the silent, insinuating, serpent-

[35] *Ibid.*, pp. 221–222.
[36] Pigott, 1794: pp. 268–269.

like weazle of influence,"[37] and charged that he had received "a *salary, ex officio,* for his pains."[38] Much later William Marshall, who had proposed the county surveys as early as 1780, and had hoped to become secretary himself, wrote bitterly that the whole board was a "job," and declared that the only question was whether the government backed it "to avoid the importunities and quiet the still more ambitious cravings of the President, or to embrace a fair opportunity of rewarding a recent change of political sentiments in the Secretary."[39]

On the whole, it seems very unlikely that Arthur Young was bribed. Not even Cartwright could believe that. The present author has searched the Pitt papers at the Public Record Office in vain for any hint of a deal. Most conclusively, Young attacked the Revolution as early as August 20, 1792. It is almost certain that Sinclair had not even spoken to Young of a possible Board of Agriculture at that time.

On the other hand, it seems all too likely that the ministry regarded his appointment as a reward for services rendered. Young admitted almost as much in the *Autobiography:*

Mr. Le Blanc . . . informed me that this new board was established with a view of rewarding me for "Example of France." In a conversation with Lord Loughborough on the attendance required, he remarked, "You may do what suits yourself best, I conceive, for we all consider ourselves so much obliged to you that you cannot be rewarded in a manner too agreeably." If the appointment of secretary be considered, as it has been by many, a reward for what I had affected, it was not a magnificent one.[40]

Further on, in speaking of the friction at the Board between President and Secretary, Young remarked:

I was a capital idiot not to absent myself sufficiently to bring the matter to a question, and leave them to turn me out if they pleased. Mr. Pitt would probably have interfered and effected the object I wanted, and, if not, would have provided for me in a better way.[41]

[37] Cartwright, 1795: pp. iii, v.

[38] *Ibid.,* p. clxiii.

[39] Clarke, 1898: p. 4. Marshall's bitterness may have been enhanced because he thought that Young had worked to prevent him from getting financial aid from the Society of Arts to whom he had appealed in regard to a county survey. *Cf.* Gazley, 1941: pp. 144–145.

[40] Young, 1898: p. 219. *Cf.* also *ibid.,* pp. 223–224.

[41] *Ibid.,* p. 243.

The very men in the cabinet most closely associated with the scheme—Loughborough, Dundas, and Rose—were the most corrupt members of the Pitt administration. Is it likely that Young would have been appointed if the Board of Agriculture had been instituted in 1791 or 1792 when he was outspokenly hostile to the government? Sinclair, Sheffield, and Banks knew his attainments and might have backed him for the post regardless of his political ideas, but it seems very doubtful that Loughborough, Dundas, or Rose would have agreed, if he had not changed his views so strikingly and so completely. Thus, no matter how pure his motives, or how great his achievements in agriculture, he could never live down the charges leveled against him in 1793.

The details of organization of the new Board were still to be determined. The law officers decided that the Board should be given an organization similar to the Royal Society. Apparently the charter was finished early in August and Sinclair summoned a meeting of the Board for August 22. He then sent the Charter to Lord Chancellor Loughborough on August 21, with a note stating that he hoped that the formality of affixing the Great Seal could be gone through quickly, "as several gentlemen had come to town to attend the meeting tomorrow." Loughborough naturally resented such treatment and did not answer Sinclair's note until the evening of August 23 and then wrote a stinging letter of rebuke: "It must indeed be supposed that to affix the Great Seal is a mere form, if it is to be gone through so quickly."[42] As a result the meeting had to be postponed until September 4.

The Charter named the first officers, Sinclair as president, Sir John Call as treasurer, and "our trusty and well beloved Arthur Young, Esquire" as secretary. The Charter also established four classes of members of the Board. Sixteen members held their seats *ex officio,* including four ecclesiastics, several cabinet members, and such dignitaries as the president of the Royal Society, the surveyor-general of woods and forests, and the surveyor of crown lands. The ecclesiastics and cabinet members seldom attended. The thirty "ordinary" members were the real working members of the Board. The *ex officio* and ordinary members could elect any number of honorary members who could attend meetings, but could not vote. Finally there were corresponding

⁴² Clarke, 1898: p. 7.

members who could not attend meetings. All elections, of officers and members, had to be by ballot.[43]

The first regular meeting of the Board of Agriculture took place, then, on September 4, 1793. Among the *ex officio* members, Sir Joseph Banks, president of the Royal Society, was deeply interested in agriculture. Among the ordinary members were some of Young's close friends—the Duke of Grafton, the Earl of Egremont, the Bishop of Llandaff, Lord Sheffield, and Thomas W. Coke.[44]

In his inaugural speech, Sir John pointed out that naturally he had thought much about a program for the Board. The first thing was "to ascertain facts," and "to examine into the agricultural state of all the different counties in the kingdom." A "number of able men" should be employed, and their reports circulated "previous to their being published."[45] He was anxious to push through preliminary reports as quickly as possible, preferably within a year. They would then be corrected and enlarged, after which the Board could make recommendations to Parliament. The execution of the plan left much to be desired. In the first place, Sir John had acted pretty arbitrarily in actually appointing men to go into the counties before the Board had even met. Moreover, many of his appointments were unfortunate, and hence the reports were of very uneven merit. Sir John had also rushed into financial engagements far beyond the resources of the Board which thus by 1795 was in a very unsound financial condition. Young was very critical:

I was infinitely disgusted with the inconsiderate manner in which Sir John Sinclair appointed the persons who drew up the original reports, men being employed who scarcely knew the right end of a plough; and the President one day desired I would accompany him with one of these men, a half-pay officer out of employment, to call on Lord Moira to request his assistance in the Leicestershire Report, when this person told his Lordship that he was out of employment and should like a summer's excursion. To do him justice, he did not know anything of the matter. Still, however, he was appointed, and amused himself with his excursion to Leicester. But

---

[43] For text of charter *cf. Communications to the Board of Agriculture,* 1797: 1: pp. xxv-xxx.

[44] A list of the original ordinary members is in Clarke, 1898: p. 8, and *Annals* 21: pp. 141–142.

[45] For summary of Sinclair's speech *cf. Annals* 21: pp. 143–150.

the most curious circumstance of effrontery was, that the greater number of the reporters were appointed, and actually travelled upon the business before the first meeting of the Board took place, under the most preposterous of all ideas—that of surveying the whole Kingdom in a single year; by which manoeuvre Sir John thought he should establish a great reputation for himself.[46]

In addition to the outstanding events in Young's life in 1793—*The Example of France* and the Board of Agriculture—a considerable number of other points deserve mention. Both in the *Travels in France* and more specifically in *The Example of France* Young had urged the formation of a militia of property. The *Autobiography* certainly exaggerated the importance of his proposal: "A circumstance in the exploits of my public career which made, perhaps, a more general impression than any other event of my life, was the proposal in 1792 for arming the property of the Kingdom in a sort of horse militia."[47] He claimed that several such troops were formed "almost immediately," and that they later "multiplied rapidly through the Kingdom." Since he had originated these corps, Young declared they customarily drank his health immediately after the King's. Young himself joined the corps at Bury under the command of Lord Broome, son of Marquis Cornwallis, and as a private, after having learned "sword exercise" at London for that purpose.

Young was troubled with sheep rot in the winter of 1792–1793, apparently for the first time in his career. In January he had to kill six sheep and was still fighting the disease in May. He experimented with "verdigrease liquid" from Bakewell and with "red salve" from George Culley, a noted grazier in northern England. He even had leather boots made for some sheep.[48]

The Washington-Young correspondence in 1793 displays the basic differences between agriculture in England and America which made it impossible for Young to understand American conditions. He could imagine neither the richness nor the cheapnes of the virgin soil of America. His letter of January 15 questioned Jefferson's claims of his wheat crop which seemed inconceivable without keeping a greater number of livestock to manure the soil. "I do not want to come to America, to know that this

---

[46] Young, 1898: pp. 242–243. *Cf.* also *Elements* 1: ff. 3–4.
[47] *Ibid.*, p. 203.
[48] *Annals* **21**: pp. 58–69.

is simply impossible."[49] In his reply on September 1 Washington enclosed comments by Jefferson and Richard Peters. Jefferson met Young's criticism about the relations between grain crops and animal husbandry thus, "Manure does not enter into this, because we can buy an acre of new land cheaper than we can manure an old one."[50] He admitted, however, that Young had converted him to the desirability of keeping more sheep. Peters ridiculed many of Young's points because English assumptions were just not applicable to American conditions. There was no monarchy or privileged nobility or clergy in America, taxes were very low, most farmers owned their land, much farming was subsistence farming, and elaborate calculations of profit never entered the head of many American farmers. "We have here innumerable instances of farmers who get forward, without ever spending a thought on per centage, or other nice calculations. . . . Instead of calculating, he labours and enjoys."[51] Washington entrusted his letter, with its enclosures, to Tobias Lear and requested Young to give Lear information about British manufactures.[52]

In another letter on December 12, Washington stated that he wished to rent all the Mount Vernon estate except the home farm, and requested Young to sound out any prospective tenants. He was willing to rent the land at one dollar an acre if in large units, at a somewhat higher rate if in smaller units. If Young considered the plan unwise, Washington told him not to bother with it at all, but to destroy his letter.[53]

On April 25 Young probably attended the meeting of the Melford Agricultural Society started two years before by his friend Thomas Ruggles. In 1793 there were thirty-three mem-

[49] *Letters from . . . Washington to . . . Young,* 1802: p. 88.

[50] *Ibid.,* p. 94.

[51] *Ibid.,* p. 99.

[52] *Ibid.,* pp. 104–105. That Tobias Lear did meet Young, probably in 1794, is clear from Richard Parkinson's remark: ". . . Colonel Lear told me he had himself been in England, and had seen Arthur Young . . . and that Mr. Young, having learnt that he was in the mercantile line, and was possessed of much land, had said he thought he was a great fool to be a merchant and yet have so much land: the Colonel replied, that if Mr. Young had the same land to cultivate, it would make a great fool of him." *Cf.* Haworth, 1915: p. 279.

[53] *Letters from . . . Washington to . . . Young,* 1802: pp. 105–113. Two recent articles by American scholars have attempted to evaluate Young's influence on American agriculture, which was certainly real, even if impossible to measure. *Cf.* Loehr, 1969: pp. 43–56, and Woodward, 1969: pp. 57–67.

bers, drawn from Suffolk and Essex, many of them Young's old friends and neighbors. Besides Ruggles there were the Rev. Jonathan Carter of Flempton, the Rev. Mr. Fiske of Shimpling, Rev. Roger Kedington of Rougham, Lewis Majendie of Hedingham Castle, Robert Plampin of Chadacre, Rev. Mr. Ray of Tostock, and Rev. Mr. Symthies of Alpheton.[54] The meeting decided to give premiums to common laborers, rather than to farmers, for the best plowing, hoeing, and reaping. Several premiums displayed the landlords' desire to keep the laboring poor in their place. For instance, one was offered to "the servant in husbandry who had lived the longest, and behaved the best, in one service," another to the "labourer in husbandry whose children earn most by spinning and knitting," and a third to "the labourer in husbandry who has brought up the greatest number of children in wedlock without parish relief."[55]

On May 31, the day after Young accepted the secretaryship of the Board of Agriculture, he was the host to a distinguished party of farmers at Bradfield.

I wrote to Mr. Ellman . . . and to Mr. Boys . . . to inform them that South Down sheep would be the subject of conversation in a farming party, at Bradfield; that objections would be urged, and ought to be answered; they mounted their horses, and rode 130 miles for a single *batch* of farming.[56]

On the next day Young and some others probably attended a meeting at Colchester to discuss the establishment of a lamb fair at Horringer to supercede the traditional one at Ipswich.[57] On June 3, in company with Sir Joseph Banks, Young again

---

[54] *Annals* **20:** pp. 409-410.

[55] *Ibid.*, pp. 405-406.

[56] *Ibid.*, p. 509. When describing this meeting Boys told of Young's fat Bakewell lambs, his chicory plants too scattered to be of much value, his fine ploughing with two oxen, and his collection of grasses in his experiment ground. *Cf. ibid.* **21:** pp. 75-77.

[57] *Ibid.*, p. 78. Young, 1898: pp. 228-229 quotes a letter of January 17, 1793, from the Earl Bishop showing that Young attempted to get his aid in establishing this new fair, but that Bristol was not very sympathetic: "You are as great a quack in farming as I once was in politics, and therefore . . . I must be on my guard against you. . . . Ipswich has an old prescriptive right to our lambs—we have sold them well at that market. . . . Adieu! magnanimous Arthur. Reserve your prowess for a greater object than distressing poor Ipswich by bereaving it of its ancient patrimony."

visited the king's sheep and cattle at Windsor and Kew, and also attended a royal reception on Windsor Terrace:

I could see in every eye and hear from every tongue of numbers to whom Sir Joseph Banks introduced me on the Terrace at Windsor that I was considered as one to whom the nation was obliged. The King spoke to me, but not so graciously as some years before. . . . However, Sir J. Sinclair reported to me some days afterwards that his Majesty had expressed to him great satisfaction at my appointment to the secretaryship of the Board.[58]

Young probably spent much of August and early September in London. He almost certainly came down for the abortive Board meeting on August 22 and stayed for the actual meeting on September 4. He was almost certainly referring to this time when he wrote, "I dined out from twenty-five to thirty days in the month, and had, in that time, forty invitations from people of the highest rank and consequence." Still the farmer was not entirely happy! "Faith! I had need to be flattered to be kept in good humour—losing my time doing nothing in London in August." He was still in London on September 9: "Dined at Pinherring's, the American ambassador; he is a gentleman-like man; but for his company . . . they were so indelicate as to call for a war with England."[59] Such sentiments would hardly please Young's new born patriotism.

Arthur Young's only son, Arthur, received his bachelor's degree from Cambridge in 1793, must have been ordained shortly afterwards, and almost immediately was commissioned by the Board of Agriculture to make the survey for Sussex. As noted above, Young had criticised the way in which many county surveyors had been appointed, but he was certainly vulnerable when his son was given a commission just after completing his university course, and before the first meeting of the Board. Of course he had been brought up in the right atmosphere, and

[58] Young, 1898: p. 224. On April 8 John Symonds wrote Young that many of his friends admired the *Example of France:* "You should come to town and be presented, or, at least, take an opportunity to walk on the Terrace at Windsor, where you would not fail of being marked out." Part of this letter is identical with Add. MMS. 35,127, ff. 153–154, dated March 30, and attributed wrongly, I think, to 1792.

[59] Young, 1898: pp. 223, 225. He is referring to Thomas Pinckney, who had written him on June 20, enclosing a letter from Washington, which does not seem to be in the printed collection. *Cf.* Add. MSS. 35,127, f. 269.

had accompanied his father on at least one short trip in 1784. But he was certainly not well acquainted with Sussex, nor was he an agriculturist of considerable practical experience. He made an extended and very thorough journey through Sussex from August 5 to October 15, and his preliminary survey of 97 pages was printed in 1793, one of the first to be finished. The account of his trip in the *Annals* is much longer, almost 300 pages.[60] After the youthful surveyor had visited him, Lord Sheffield wrote Young:

He was very much approved here & was thought very like you in voice & other particulars. . . . I am not yet quite reconciled to your putting him in this line if you mean to push him in the Church—I revolted against it from the beginning, & I find others think as I do.[61]

"Young Arthur Young" never did quite make up his mind whether he was primarily a clergyman or a farmer. During this trip he probably first met Lord Egremont, the eccentric owner of Petworth who became his chief patron. Egremont was tremendously wealthy, kindly and easy going, sincerely interested in agriculture, and a most valuable member of the Board of Agriculture. But his private life and the character of his home were hardly the highest ethical models for a young clergyman.

Sinclair very naturally asked Young to prepare the preliminary report for Suffolk. In order better to prepare himself Young took "several journeys into different parts of the county at some expense,"[62] presumably in the autumn of 1793. The account of only one has survived, a fortnight's trip through east Suffolk.[63] Since the tour is very technical it has little material of general interest. He found much to admire, although some crop rotations were still bad, the farmers were too wedded to heavy wagons, and too few sheep were kept. He was gratified to find some farmers gradually accepting practices which he had advocated twenty-five years earlier.

This account of 1793 cannot be concluded without a brief reference to his review in the *Annals* of William Godwin's *Political Justice*. Young noted with fear and indignation Godwin's

[60] *Annals* **22:** pp. 171–334, 494–631.
[61] Add. MSS. 35,127, f. 289.
[62] Young, 1898: p. 247.
[63] *Annals* **23** pp. 18–52.

republicanism, anti-clericalism, anarchism, loose ideas on marriage, and sympathy with French developments. He believed that the attack upon private property was the real foundation for all Godwin's theories. "The object is palpable, and even avowed, from the first page to the last of this bulky emanation of 'mind,'— LEVELLING PROPERTY." A halt must be called to such expressions of radicalism or mankind would revert to barbarism.

I believe, in truth, that if such writings are allowed freely to be circulated, democracy will effectually abolish everything that has hitherto been respected in the world; all tangible property, and all moral good; it will abolish every possession, and eradicate from the heart and mind of man every feeling that does honour to his nature, and every ray of knowledge that raises him above a brute.[64]

At the very end of 1793 the Duc de Liancourt sent Young a pair of sleeve buttons made of the rock which served as a pedestal for the statue of Peter the Great, with the following inscription:

> this buttons are then an historical monument—I dare offer them
>   under this colour, to my friend Arthur Young—
> and although my heart are not so hard than this stone
>   I beg with Young to believe it is in friendship so solid like
>   a rock.[65]

When the Board of Agriculture really became established in 1794 an entirely new life opened for Arthur Young. Now most of the year had to be spent in London, and probably for some years apart from his family. There was a long vacation during August, September, and October, plus a month at Christmas and three weeks at Easter. "As I was determined to pass all the vacations at my farm in Suffolk, six journeys of myself and servants became necessary, and caused a considerable expense."[66] Until 1798 the Board had no separate quarters but only a room at Sir John Sinclair's home in Whitehall. As a result Young had to hire lodgings at 2½ guineas a week.

The president and treasurer of the Board of Agriculture received no salary, but the secretary was given £400 a year. As

[64] *Ibid.* **21**: pp. 181–183.
[65] Add. MSS. 35,127, f. 297.
[66] Young, 1898: p. 242.

noted above,[67] Young's letter of acceptance had expressed disappointment at the amount. The under-secretary, the famous traveler, John Talbot Dillon, received £200 a year. It is strange indeed that Young's *Autobiography* never once referred to Dillon when they were so closely associated for ten years. The first of the two clerks received £150 and the second £80.[68] The first clerk, William Cragg, became under-secretary on Dillon's death in 1805. Somewhat later an attorney was also in constant attendance on the Board. Twice the *Autobiography* expressed Young's dissatisfaction with conditions during the early years under Sinclair.

. . . the salary, 400 l. per annum, would have been desirable had it left me more time in Suffolk, but when I found a very strict attendance attached to it, with no house to assemble in except Sir John Sinclair's, and in a room common to the clerk and all comers, I was much disposed to throw it up and go back in disgust to my farm; but the advice of others and the apprehension of family reproaches kept me to the annoyance of a situation not ameliorated till Sir John was turned out of the Presidentship by Mr. Pitt, and the Board procured a house for itself.[69]

Lord Hawke had examined the rules and orders of many societies, and found that in all letters communications were addressed to the Secretaries, and answers given by them. Sir John Sinclair struck this out, and directed all such communications to be to the President (himself) , and for him also to sign all letters. This at once converted the Secretary into nothing more than a first clerk. . . . All letters were dictated by the Secretary and written in a book; this book was altered and corrected at the will of the President, and such alterations made as in respect of agriculture were absurd enough; the whole done in such a manner as not to be very pleasing.[70]

The Board of Agriculture started to meet regularly in February, 1794, and probably broke up on July 29 when Sir John outlined the progress already made and his future program.[71] Seventy-four county reports had already been submitted and he

[67] *Cf. supra,* p. 316.
[68] Clarke, 1898: p. 10. For salary figures *cf.* MSS. Agricultural Society Treasurers Cash Book, May 10, 1796. The manuscript records of the Board of Agriculture are in the Library of the Royal Agricultural Society, through the kindness of whose Librarian I was permitted to use them.
[69] Young, 1898: pp. 219–220.
[70] *Ibid.,* pp. 241–242.
[71] *Annals* **23:** pp. 200–217.

was confident that the preliminary surveys would be completed a year after the Board's establishment. He planned to print the preliminary reports with wide margins and to send them to competent farmers for further comment. Thus 80,000 individuals would be reached. After the comments had been received, final and corrected reports could be published. In the meantime a "General Report" along topical lines should be compiled from the preliminary surveys. This grandiose scheme, which was never completed, was to be finished during the Board's second year. Sinclair's letter to Young which accompanied a copy of the address, stated that he proposed to write the introduction, the first five chapters, and the five concluding chapters for the General Report. He then assigned twelve chapters to Young, those on agricultural labor, enclosures, rotation of crops, comparison between drill and broadcast husbandry, fallowing, culture of various grains, harvesting, culture of green crops, artificial grasses, two on wastelands, and manufacturers residing in the country. For the most part Sir John had picked Young for those subjects on which he was generally recognized as an expert. He certainly laid on the flattery pretty heavily:

I wish not to press on you too great a share of the burden; but as no man has a greater faculty for writing well, or more knowlelge of those subjects, I hope that you will undertake for as many of the other chapters as possible. . . .

You have already husbandry at your fingers end, and I hope therefore you will be able to complete the chapters above-mentioned before we meet in January.

Young's answer to the above has not survived, but it is easy to imagine his feelings at a request which would certainly consume most of his vacation. In the *Annals,* where he printed both Sinclair's address and letter, Young inserted the following footnote:

Our excellent president may depend on every exertion in my power to forward his patriotic designs. He estimates my ability for the work too highly. I can promise inclination and industry only. Too much praise can never be given to that unceasing activity of mind, which is ever promoting him to great undertakings.[72]

Young must have spent much of the spring of 1794 preparing

[72] *Ibid.,* pp. 199–200.

his preliminary report on Suffolk which was 75 pages long, about the average of these preliminary reports. Young's preface stated that it was purposely general in character, for he thought it unfair to mention individuals, since he knew certain regions much more intimately than others. Moreover, the lack of time prevented any treatment in detail. The twenty-one subdivisions covered extent, climate, soil, estates, tenures, farms, rent, wastelands, crop rotations, improvements, livestock, enclosures, implements, and labor. He found much to praise—the rich tenant farmers, the practices of hollow draining, the system of crop rotations, the excellent roads. On the other hand, paring and burning was only practiced in the fen district, the breeds of sheep left much to be desired, and irrigation was almost unknown. "Of all the improvements wanting in this county, there is not one so obvious, and of such importance, as watering meadows."[73] Arable lands were much better managed than grasslands. He insisted that fallow crops could clean the land just as thoroughly as a complete summer fallow and were also necessary to maintain many cattle and sheep. He hoped the Board of Agriculture could carry a measure through Parliament to simplify and cheapen the process of enclosure and improvement of wastelands. He declared that the necessity for discovering the best breeds of cattle and sheep "is perhaps the most important in the whole range of rural oeconomics."[74]

Young also wrote an important article in the *Annals* in the spring of 1794, "An Idea of the Present Agricultural State of France, and of the Consequences of the Events Passing in that Kingdom."[75] The article pointed out in some detail that France under the Convention displayed certain similarities with ancient Sparta. Under requisitions and the maximum system the peasants were becoming little better than helots, while the lack of liberty and the emphasis upon the military were likewise Spartan. Young summarized the results of the Revolution as follows:

Here then are the two great results of this new system which the French have established; the landlords murdered, the cultivators of every kind made beasts of burthen to the towns and armies; the

[73] Young, 1794: p. 57.
[74] *Ibid.*, p. 32.
[75] *Annals* **23**: pp. 274–311. This article was also published as a pamphlet in 1795.

trade and industry dashed to pieces!. . . . The IRON AGE of barbarism returned—and all that trade and industry, wealth and peace,
arts and science, civilization and elegance—all that the culture and
decoration of the human mind have done for man—levelled in the
dust—and, in their place, blood and rapine and horror triumphant.[76]

Young predicted that a French republic would, like ancient
Sparta, follow a warlike policy. He was very pessimistic about
an early peace, for a continued war would appeal, not only to
those who wielded power in France, but also to those who received wages as soldiers. Especially to be dreaded was a peace
which would permit the consolidation of the French system:

. . . but this must be clear to every neighbour of France, that if
a peace would enable them to consolidate and perfect any plan of a
republic that does or might tend to establish a system of the kind
I have described, there is no war that would in the end be more
fatal than such a peace. Such a republic is absolutely incompatible
with the safety of the property of Europe.[77]

With such a prospect his proposal for a militia of property
must be extended. Voluntary efforts were not sufficient, for a
force of 500,000 men must be raised. More formidable fortresses
must be built along the coast. Revolutionary France could only
be defeated by a strong coalition of all Europe. In case of a
French invasion of any of her neighbors, the country should be
"made a desert for them to march through." The allies must
make it absolutely clear that they were not fighting to gain territory, but only to destroy the Revolution and to reestablish a
liberal monarchy. Such a combined ideological and military offensive would encourage those Frenchmen who hated the Revolution and would make it possible to land large armies, including
those of Russia eventually, in the heart of France. Young admitted that it was hazardous for a civilian to make such proposals, but justified his temerity because only new measures could
hope to succeed in a revolutionary period:

. . . but the events of this fearful period are all *new;* the principles
on which every thing moves are new; they are novelties to the most
experienced statesman; and if none but old plans are pursued on

---

[76] *Ibid.,* p. 283.
[77] *Ibid.,* pp. 290–291.

one side, while the other is actuated by unheard-of principles and exertions, the event may be easily conjectured.[78]

Busy as Young must have been early in 1794, he seems to have had plenty of time for amusements. He had always enjoyed chess, and in June was elected to the Chess Club.[79] On finding that there was no agricultural club in London, he proceeded to establish one. He first approached the Duke of Bedford and the Earls of Egremont and Winchilsea, all of whom were favorable. Invitations were then sent out and meetings were held fortnightly at the Thatched House Tavern. The club was at first limited to fifty members, but later became very fashionable and the membership was greatly extended. The annual dues were two guineas, one of which went for running expenses and the other into a fund which Young and Sinclair tried unsuccessfully to tap for agricultural premiums.[80] Young also attended several conversaziones that spring. One was at the home of the Countess of Bristol, wife of his old friend, the earl-bishop. Occasionally he probably appeared at the large and rather ostentatious parties in Portman Square which Mrs. Elizabeth Montagu was giving about this time. His favorites were at the homes of Mr. and Mrs. Matthew Montagu and Mr. and Mrs Charles Cole, at both of which he met some of the bluestocking luminaries, Mrs. York, Mrs. Garrick, Mrs. Orde, and Hannah More. "The *petits soupers* at Mrs. Matthew Montagu's, and to which she asked a selection of eight or nine persons, were very pleasant, the conversation interesting, and this select number more agreeable than I ever found full rooms." Young also had great pleasure in the small parties at the Coles': "The conversation at these parties on the publications of the day, anecdotes of the time, with the conduct of many of the great men of the age, was usually very interesting."[81]

Although busy and enjoying himself in London, Young could

---

[78] *Ibid.*, pp. 304–305.
[79] Add. MSS. 35,127, f. 327.
[80] Young, 1898: pp. 244–245.
[81] *Ibid.* Matthew Montagu was Mrs. Elizabeth Montagu's nephew. He had legally taken her husband's name and became her heir. Charles N. Cole was a lawyer, the literary executor of Soame Jenyns, and a friend of John Symonds through whom Young probably met him. The Coles had been friends of the famous Mrs. Delany, who once described him as "a very entertaining man." *Cf.* Delany, 1862: **5:** p. 319.

not forget family troubles. His son's career was still unsettled as the following letter indicates:

>   Chandos Street
>   Cov. Gard. No. 4
>   Mch 26, 94
>
> My Lord,
> Some of your time having been unfortunately interrupted by illness, and all of it so occupied by business of importance that I have not ventured so often to your door as my inclination & gratitude would have led me—but may I without presuming too far, remind you of my son whose whole dependence, at present, except the assistance I can give him, is a curacy. Some kind expressions your Lordship had the goodness to use in relation to him gives me confidence that I shall not offend you by mentioning him.
> I can now assert that he executes the duties of his office in a manner that gives him some reputation, both as to reading & preaching; & his report of Sussex Husbandry to the Board of Agriculture as well as his Tour there now publishing in my Annals shew that he will make a good farmer.
> Have the goodness to excuse this liberty, and permit me to assure you that I have the Honour to Remain
>
>   With the greatest respect
>   My Lord
>   Your Lordships
>   Much obliged
>   & Humble St.
>   A. Y.

Yesterday the 4500 acres in Knaresboro forest became mine. It is as improveable a tract of land at an easy expence as ever I saw, but beyond my capital to do anything effectively with.[82]

Much more important was the death of Bessy Hoole on August 1, 1794.[83] She was already suffering from tuberculosis in September, 1793. In December her husband took her to Lynn in Norfolk and about February 1 on a long trip to Sidmouth on the south Devon coast where she stayed at least until early April. Three letters to her father from Sidmouth show, pitifully enough,

[82] Add. MSS. 35,127, f. 315. One can only guess to whom this letter was addressed. Egremont and Sheffield are good bets. Young's address was probably his boarding house in the spring of 1794.

[83] *Gentleman's Magazine* **64** (1794) : p. 769.

the sufferings which she endured, and the solicitous care of her husband and father. On February 22 she stated that the long trip had greatly tired and weakened her, "indeed I have had more fever and cough since I came here than ever I had." Since her husband must soon return to his parish, she asked her father's advice whether she should stay alone at Sidmouth or return to Abinger with him: "I do not think Abinger at all in fault; I have been well or *better* there than anywhere. But I am not unwilling to be left here, if it should still be thought advisable. Will you have the goodness to write as soon as you can, as we shall not determine till we hear?" A postscript from her husband showed his fears that she would never recover: "I fear this journey will be of no avail. I do not think our dear Bessy is in any immediate danger, but I much fear this cruel disease is gradually preying on her strength." The decision was to leave her there alone. When she wrote again on March 18, she was somewhat better: "I am now quite free from pain, and can sleep on one side as well as the other; I think the last blister was of use. I have been twice in the warm bath since Mr. Hoole went. My cough must have its course."[84] Her last letter on April 3 was still more favorable:

Many thanks for your last kind letter, there is nothing I am in want of, or that you could send me, but I am much obliged to you for the offer. I can see a newspaper whenever I please therefore would not think of your sending one. . . . Last week we had beautiful weather; I rode out double almost every day, I like the exercise exceedingly, and it is so gentle, as I ride, that I am not at all fatigued. . . . I think I am mending, indeed I am certain I am much better than some weeks ago when I cd hardly breathe, when I eat anything—and I have very little fever.

Young was much encouraged as shown in his note added in forwarding the letter to Hoole: "this is not bad news by any means."[85]

Bessy certainly returned home before her death. Young's own account of the sad event in the *Autobiography* follows:

She was of a most amiable, gentle temper, and in a resigned frame of mind, which gave me much satisfaction. The last visit I paid her at Abinger . . . she was very weak, yet not suspected to be so

---

[84] Young, 1898: pp. 250–252.
[85] Add. MSS. 35,127, f. 317.

near her end. But at the last parting with me, she did it in so feeling and affectionate a manner as seemed to imply that she thought she should see me no more. It made me, for a time, extremely melancholy, which was shaken off with great difficulty.[86]

On August 24 John Symonds wrote, attempting to comfort Young:

It is not to be wondered at that you have felt a deep concern for the loss of a daughter, who was always dutiful, and conducted herself with the utmost propriety; but it was an event for which you must have been, or ought to have been, long prepared: and a reflection on what she suffered should allay your grief. She is happy, and will have her reward.[87]

Bobbin's earliest letter to her father dates from 1794 when she was eleven years old. It indicates that even then her health may not have been too robust:

Bury, Nov. 15, 1794.

My dear Papa,

By your request I take the first opportunity of writing to inform you that my Cough is a great deal better—I have received the cakes you were so good as to send me they were very nice & for which accept my best thanks. I should be much obliged to you to send home Mr. Lloyd's history of England if you have not already it lays in the Room where I slept—I have found my *Odour of Roses*—Miss Macklin & I have just received the Money you were so kind as to send us, for which we are much obliged to you. . . . I shall have no objection to a Letter whenever you can make it convenient—I hope Mag & the Poney are quite well—I must now beg to conclude with Duty to yourself & remain, Dear Papa

Your dutiful Daughter

M. A. Young[88]

In the summer Young made a trip to Hampshire and Berkshire, the agricultural focus of which was his visit to W. P. Howlett (an honorary member of the Board and an M. P.) at Sombourne

---

[86] Young, 1898: p. 246.

[87] Add. MSS. 35,127, f. 336. An incidental remark in this letter showed how the professor of history regarded his field of study: "and of what use is history unless it be considered as a school for modern politicians?"

[88] *Ibid.*, f. 344.

in the North Down district.[89] Young enthused over the irrigation practices common in the district. Since he had always favored paring and burning, he was disturbed by the bad repute of the practice in this part of Hampshire, and inclined to attribute the failures to its use as preparation for arable instead of grass. Again he emphasized that satisfactory crop rotation was the foundation of all good farming:

The object of a right arrangement of the crops of a farm, cannot receive too much attention; almost every thing depends on it. If there is a circumstance to be named in Norfolk husbandry, which more peculiarly decides its merit than another, it is that of their course of crops, so well adapted to keep the land clean, to put in wheat at a small expence, and to support as many sheep as possible.[90]

After Sombourne, Young spent several days at Sandelford, Mrs. Elizabeth Montagu's Berkshire country estate. He found that she had been "as successful in the decoration of a villa, as in the more splendid exertions of Portman Square," and was especially enthusiastic about a picture window in the dining room, looking out into a grove of fine elm trees:

. . . one seems to dine in a wood; and the fancy is weak that does not give to the air a delicious coolness; the fanning breezes bring the perfumes of her shrubs and flowers; the eye becomes the avenue to every sense; and while every sense is in imagination gratified, the illusion of the scene deceives us into pleasure.[91]

Mrs. Montagu's comment on his visit, in a letter to a friend, indicates well his repute at this period:

The celebrated Arthur Young bestowed a few days of his company upon us, which he rendered very agreeable. He is a skillful Farmer, an ingenious philosopher, and a judicious Politician, as you will have perceived if you have read his book, The Example of France, a warning to England. He has travell'd much, seen Cities and men, as well as attended to the cultivation of land and produce of various Soils.[92]

While enjoying "a few very agreeable days, passed in this man-

---

[89] *Annals* **23:** pp. 163–179. After his visit to Powlett, Young wro e a letter to Sinclair which appeared as a postscript to the preliminary survey of Hampshire. *Cf. ibid.*, pp. 355–371.

[90] *Ibid.*, p. 364.

[91] *Ibid.*, p. 172.

[92] Blunt, no date, **2:** p. 307.

sion of taste and genius," Young visited Prosperous Farm, the former estate of the famous Jethro Tull of horse-hoeing and drilling fame, for whom he expressed an almost filial respect: ". . . I have a very great, though melancholy pleasure in viewing the residence of persons who rendered themselves celebrated by their actions or their writings, and particularly such as were noted for their exertions in husbandry."[93] The account concluded: "Every part of his works manifests strong talents, and no inconceivable learning; and he has left a name in the world, which will probably last as long as the globe we inhabit."[94]

Young's second trip in the autumn was made among old friends and scenes.[95] He again admired Mr. Ducket's threshing mill at Esher, Lord Egremont's cattle and sheep at Petworth, Mr. Boys's South Down sheep, Mr. Dann's clean fields and potato culture, and Mr. Majendie's cabbage and plantations. He also visited Mr. Taylor, whom he had last seen at Bologna in 1789, and who was a friend of the drill husbandry, but not of horse-hoeing. Young stopped at Dover where the views from the castle always thrilled him. This time his patriotic and anti-revolutionary emotions produced the following outburst:

The heights and cliffs are lofty enough to give an impression of grandeur, which reaches perhaps sublimity, and perpendicular enough to justify the epithet romantic. The commerce of the world may be said to pass before the eye and affords an endless variety to the sea view. But the circumstance most singular is the distinct sight of the coast of France, and even of the signals made upon their hills by the atrocious masters of that shore so hostile to humanity itself. To be seated amidst the smiling plenty, which freedom and national happiness can breathe around, and view the seat of rapine, tyranny, poverty, and blood!—weak must be the sensations of that bosom which from one mighty empire can thus contemplate another; where at one glance of the eye we can behold the contending regions that are to decide the fate of the human race;—that are to decide whether order, law, government, and property are to exist in the world; or all to be confusion, anarchy, and bloodshed. Cold is the heart that remains unmoved at such a spectacle![96]

[93] *Annals* **23:** pp. 172–173.
[94] *Ibid.*, pp. 177–178.
[95] *Ibid.*, pp. 374–385.
[96] *Ibid.*, pp. 379–380.

Among Young's miscellaneous articles in the *Annals* for 1794 one described a six-year experiment on crop rotations.[97] An old upland pasture was divided into thirty-six equal blocks upon which various rotations were tried. Among the conclusions reached were that potatoes were very exhausting, that beans were by far the most profitable fallow crop and the one that kept the land in best order, and that the regular alternation of beans and wheat, and beans and barley was very profitable. Another article maintained that the use of land for forests was definitely inferior to its use for food from every consideration except possibly the private profit of the landlord. For societies to give premiums for planting forests was absurd, for it was offering "to reward those who are the readiest to take these retrograde steps toward changing the corn, cattle, and sheep of Britain, into the savage robe of an American wilderness." He brushed aside the arguments that England needed more land devoted to raising fuel: "Coals are so inexhaustible in this island, that every man in Britain may be warmed by them for ten centuries to come.[98]

The only important letters of 1794, not already mentioned, are two from Jeremy Bentham in September. One requested information about the value of various kinds of property and the size of Britain's population, while the other revealed the friendly relations between the two men: "A thousand thanks for your kind letter—sorry you should fancy you have been bathing for health—hope it was not true—only idleness—we can't afford to have you otherwise than well."[99] Young's note of October 10 was not an answer to Bentham's letters, but referred to a project to establish a house of detention at Bury where Young had been for several years a justice of the peace:

I . . . yesterday attended a Committee of Justices to examine plans, when we reduced the candidates to two, . . . and the rest were rejected because they were not on the Panoptican principle which I write you as I think you should have the satisfaction of knowing how much your ideas are approved.[100]

In 1795 the burden of the French war really began to be felt. The poor harvest of 1794 resulted in a scarcity of bread and high

[97] *Ibid.*, pp. 471–507.
[98] *Ibid.*, pp. 394–395.
[99] Young, 1898: pp. 247–250.
[100] Add. MSS. 35,541, f. 609. Bentham Papers in the British Museum.

prices in 1795. Both as editor of the *Annals* and as secretary to
the Board of Agriculture, Young devoted much attention to the
food crisis. On January 23 he issued a questionnaire to his correspondents of the *Annals*,[101] asking about the reserves of bread
grains, the prospects of the 1795 crop, the methods used to relieve
the poor, the success of various substitutes for wheat, and the
changes in prices and wages. Seventy-nine replies were printed in
the *Annals*, some from such notables as the Earl of Egremont, the
Earl of Hardicke, Lord Sheffield, Sir Joseph Banks, and T. W.
Coke, others from practical farmers and old friends—William Pitt,
J. Boys, George Culley, Thomas Ruggles, S. Bevan, William
Dann. Dr. Richard Valpy, Lewis Majendie, and W. P. Powlett. In
his summary of the replies,[102] Young concluded that despite a
definite scarcity there was no real danger of famine. His prediction that prices would reach their peak in May and then decline
was mistaken, however, since the harvest of 1795 also proved bad,
and on October 27 he sent out a second questionnaire,[103] which
was answered by forty correspondents.

Young also published eight articles during 1795 in the *Annals*
bearing more or less directly upon the food shortage and the resultant high prices. One article pointed out to the poor that
rioting only raised prices and intensified the shortage: "That
exactly, in proportion to the internal tranquility of a country
being disturbed, is scarcity; that general riot and licentiousness
greatly increase scarcity, and that revolutionary confusion inevitably brings famine and all its horrors in the rear.[104]

On November 3 Parliament established a select committee to
deal with the crisis. Young reported the parliamentary debate in
the *Annals*, and followed up with a fairly long article, dated
November 9,[105] expressing his grave doubts about the wisdom of
such a committee, for parliamentary agitation only tended to
arouse public alarm and hence to raise not lower prices. In accord
with *laissez-faire* principles, Young believed that the lessening of
consumption through high prices during a scarcity was the best
way to make the short crop go the furthest. In the short run the
government could do very little to meet the crisis:

[101] *Annals* **24**: pp. 42–43.
[102] *Ibid.*, pp. 327–348.
[103] *Ibid.* **25**: pp. 344–345.
[104] *Ibid.* **24**: p. 541.
[105] *Ibid.* **25**: pp. 420–472.

The value of all you can do amounts but to a bagatelle; but the mischief by alarm is certain. By substitutes and other palliatives you may assist a few thousand people, but it will be at the expence of millions. Export prohibited; the ports open to importation; the distillery stopped; and consumption checking itself by a high price; what more is to be done by a thousand inquiries?[106]

Despite his general skepticism, Young proposed several short run measures. Thus he recommended the mixture of rice with wheat and made a number of experiments himself.[107] He also thought that much grain might be saved if fewer oats were fed to horses. He sent out a questionnaire on the subject.[108] "The fact is now well ascertained that you cannot, in the present situation of our agriculture, feed your people and your horses; one or the other must be lessened."[109] Although horses could not be forbidden, they could be heavily taxed.

Somewhat inconsistently Young supported a proposal to make wages proportionate to bread prices. On October 12 the Quarter Sessions at Bury passed a resolution, "That the members for this county be requested to bring a bill into parliament, so to regulate the price of labour, that it may fluctuate with the average price of bread corn."[110] Young's old antagonist, Capel Lofft, opposed the measure vigorously on strict *laissez-faire* principles. Young disagreed with such reasoning and expressed his own attitude on November 9:

The greatest evil attending the high price of corn is the singular instance of one commodity that has not risen by any means proportionably to it, and that is husbandry labour; this remains much too low; so low, that something must either speedily be done to raise it in some degree proportionably, or evils of the very worst aspect may be expected. This object is within the power of law and regulation, and though there are principles that would be hurt by interfering, yet the nature of the call is too pressing to be neglected.[111]

· Early in 1795 Young inserted in the *Annals* a very interesting proposal for a general enclosure bill, which would make private

---

[106] *Ibid.*, p. 470.
[107] *Ibid.*, pp. 535–537.
[108] *Ibid.*, pp. 25–26.
[109] *Ibid.*, p. 458.
[110] *Ibid.*, p. 316.
[111] *Ibid.*, p. 470.

enclosure acts of Parliament unnecessary by transferring the procedural arrangements to the quarter sessions. The poor should not only be safeguarded in their rights, but should pay none of the legal expenses or the cost of fencing their lands. Moreover some land for fuel should be kept in common. The poor man's allotment could not be sold separately from his cottage. Young admitted that the scheme might seem unduly favorable to the poor, but commented: "No measure, of this complexion, can be expected to pass, in which the interests of the lowest classes are not most pointedly attended to. This must be esteemed the *sine qua non* of such a bill. It must be popular amongst the poor, or it will not be at all."[112] His article of November 9 also urged "a general act for relieving all poor and other commoners, by generally dividing and allotting all the wastes in the kingdom, under such provisions and restrictions as shall secure to the poor cottagers their fair share, without expence, for ever and inalienably annexed to the cottage which gives the right."[113] Both his enclosure plan and his proposal to regulate wages proportionately to prices reveal a consideration for the poor in striking contrast to Young's earlier heartlessness. Now he appeared willing to aid the lower classes even if it meant a departure from strict *laissez-faire* ideas.

The Board of Agriculture also took an active interest in the food crisis of 1795. On February 20 the Board published a memorandum urging the growth of many more potatoes, and describing various methods of planting, preserving, and using them for bread.[114] On July 4 the Board issued a quite specific set of recommendations to meet the crisis—that all grain should be used except the coarsest part of the bran, that each individual eat less bread, that less grain be fed to horses and other livestock, that other grains be mixed with wheat to make bread, and that the poor be given more aid, either through private charity or increased poor relief.[115] On November 10 about eighty sorts of mixed bread were prepared, sent out to bakers, and sampled by the members of the Board.[116]

The Board of Agriculture, however, did not confine its work

---

[112] *Ibid.* **24**: p. 17.
[113] *Ibid.* **25**: pp. 456–457.
[114] *Ibid.* **24**: pp. 64–72.
[115] *Ibid.*, pp. 579–581.
[116] *Ibid.* **25**: pp. 574–598.

in 1795 to meeting the grain crisis. It supported one law to standardize weights and measures to prevent frauds upon the poor, and another to protect the funds of friendly societies.[117] By 1795 the preliminary county reports had been completed, and Sinclair was urging that the revised reports be finished. It was decided that the reports of Lancashire and Norfolk be published as models for the rest, and a uniform plan of organization for such reports was drawn up.[118] That Young was shouldered with considerable work in getting out the Lancashire report is evident from the minute of the Finance Committee of June 5:

This Committee being met to take into consideration the Reference from the Board of the 2d. inst. on the propriety of Publishing the Lancashire Report, with the additional Remarks and Observations that have been made thereon,—read the Chapter on Live Stock of the intended Publication, prepared by the Secretary, ordered that the Secretary do proceed and finish the said Report. . . .[119]

By 1795 the bad financial condition of the Board was revealed in the report of the Finance Committee.[120] The chief cause was of course the heavy printing costs of the county surveys and the fees to those who had prepared the reports. Several letters show haggling between the Board and the authors of the reports.[121]

Young published two political pamphlets in 1795. The first, *An Idea of the Present State of France,* had appeared in the *Annals* in 1794 and has already been analyzed.[122] Since at least six months had elapsed between the composition of the article and the time the pamphlet went to press, certain additions were made. Young had been shocked by the popular manifestations supporting the defendants in the state trials, since he was convinced that the accused favored violent revolution in Britain. He pleaded with the opposition to support the government even if they disliked it. He was especially disturbed by the demands for peace: "What can a peace produce, at such a moment, but imperi-

[117] *Ibid.,* pp. 610–611; Clarke, 1898: p. 19; Young, 1809: p. 17; Minute Books, Miscellaneous Committees, **2,** f. 17. The measure on weights and measures was passed, that on friendly societies was not.
[118] *Annals* **24:** pp. 548–559.
[119] Minute Books, Finance Committee, **3,** f. 17.
[120] *Ibid.,* f. 11. *Cf.* also Clarke, 1898: pp. 12–13.
[121] Copy Letter Book, ff. 9, 17, *re* W. Pomeroy and Charles Vancouver.
[122] *Cf. supra,* pp. 329–330.

ous superiority on one side, and disgraceful submission on the other? . . . Is all national dignity fled? . . . With an enemy so superior in the field, PEACE IS SUBMISSION."[123] There was, however, a reasonable outlook for ultimate victory, since the French would eventually tire of assignats and would overextend themselves.

The pamphlet was reviewed in the March issue of the *Monthly Review* and in the June issue of the *Gentleman's Magazine.* The latter review was almost entirely a summary, but the reviewer obviously thought the pamphlet important and sympathized with the views expressed.[124] Although the reviewer in the *Monthly Magazine* condemned Young's opposition to the English reform movement in no uncertain terms, the tone of the review was one of rather reluctant admiration:

Some of the author's principles we have condemned in the course of our review; but the pamphlet, on the whole, we must in justice allow to manifest much ability. The writer is unquestionably a party man, though acting with what he may consider as a majority of the country. . . . He appears to consult the interest of Great Britain with respect to foreign powers, and to feel on that head like a patriot. . . .[125]

Young's second pamphlet of 1795 was *The Constitution Safe Without Reform,* an answer to Major John Cartwright's attack upon him in *The Commonwealth in Danger.* Young had attacked Cartwright pretty severely as a reformer in his *Example of France.*[126] Hence Cartwright's *The Commonwealth in Danger* was his revenge. He quoted at length from Young's *Travels in France* to prove Young's apostasy and to point out that before the publication of *The Example of France* Young had attacked the British constitution just as violently as he. As noted above, Cartwright accused Young not of accepting a money bribe, but of having succumbed to flattery:

I cannot—I will not, attribute a change so extraordinary, to the corruption of the heart. . . . As gold resists aqua-*fortis* but dissolves in aqua-*regis,* so there are minds which resist gold, but dis-

---

[123] Young, 1795: p. 48.
[124] *Gentleman's Magazine* **65**,1: *pp.* 497–500.
[125] *Monthly Review* **16**: p. 294.
[126] *Cf. supra.* pp. 296, 307.

solve in the fumes of incense. The sturdy farmer of Bradfield, I doubt not, held fast the portal of integrity, but I suspect that his other and dearer self, the F.R.S. the author of a library of agriculture and politics; the honorary member of more than half the philosophical, literary, oeconomical, physical and agricultural societies of Europe; the darling child of his own brain . . . unguardedly suffered the wicket of vanity to be opened. . . .[127]

One further example of Cartwright's attack is too good to be omitted:

When a man condescends, for the gratification of new friends, to eat his own words. . . . If a farmer will depart from his old approved practice, to try new experiments; take infected seed from the government's granaries; sow at a wrong season and on an improper soil; and totally neglect to clean his crop; what can he expect but lean corn and luxuriant weeds, disappointment and disgrace.[128]

Certainly Young's *Constitution Safe Without Reform* was much more temperate in language than Cartwright's pamphlet. He denied that he had accused Cartwright of supporting massacres and declared that his attack upon Cartwright had been purely political, not personal. Since Cartwright had been unable to meet his arguments, he had attacked Young's character. "He could not refute the book, and therefore he reviles the author."[129] On several points Young's position deserves quotation. His refutation of his apparent inconsistency follows:

Because I thought liberty, before the 10th of August, a blessing to France, *therefore* I was to think French liberty, after that period, a blessing also! Because I thought reform in the constitution of England wholesome, before the French revolution, *therefore* I am to think it safe and expedient now![130]

Young's most complete defense of his integrity in regard to the appointment to the secretaryship is also found in this pamphlet:

In regard to his various insinuations, and in some passages direct accusations, of my having been influenced to write by public money,

[127] Cartwright, 1795: pp. iv-v.
[128] *Ibid.*, p. xxxiii.
[129] *Annals* **25**: p. 264. Since the pamphlet has not been available, I have used the article with the same title in *ibid.*, pp. 246–293.
[130] *Ibid.*, p. 278.

it is a calumny that every man connected with government . . . knows to be utterly unfounded. Mr. C. himself informs his reader that the first publication of the essay, which afterwards became the *Example of France,* was in autumn 1792. Several editions of the work were printed before my appointment at the Board of Agriculture. Living retired on my farm, without the most distant connection or correspondence with administration—without instigation—or the smallest idea of reward,—my feelings of the public danger alone produced the work. I have never been urged to take pen in hand by any of their friends—never received one shilling of public money, nor any other advantage direct or indirect from government, except the salary openly and publicly, as secretary to the Board of Agriculture. Mr. C. must have strange ideas of the bribes with which ministers influence men to become parties in conspiracies, to think that such a salary, by annual election, with an expensive attendance . . . to a man occupying a considerable farm . . . could be considered as a bribe to turn conspirator—or as a reward for having been one!!!—NO—such were not the ideas that made me secretary to the Board. My agricultural pursuits for nearly thirty years—the works which I had published . . .—and the minute surveys I had taken of three kingdoms—pointed me out as a person . . . proper for the office.[131]

At another point he explained his ideas on reform in England in greater detail:

I am as sensible, I believe, as others are, of the abuses which time has brought into our government . . .; I trust that they will be gradually corrected. Before the revolution of France degenerated into anarchy. . . . There was nothing then dangerous in innovations or the spirit of reform. Since that period such numbers of men have associated . . . with views so dangerous in my estimation, and we have been deluged with publications, so thoroughly in the revolutionary spirit—that I feel all reform on such principles— all tampering with, or altering our constitution, to be in such perilous times the very height of insanity. As to the French revolution, I detest and abhor it, as a change that has been utterly destructive of human happiness. . . .[132]

A study of the few reviews available makes it unlikely that the Cartwright-Young pamphlet war changed the views of many. The *Gentleman's Magazine,* favorable to Young and his views, only

---

[131] *Ibid.,* pp. 268–270.
[132] *Ibid.,* p. 286.

reviewed Cartwright's pamphlet in June, 1796, and then used Young's words to condemn it.[133] On the other hand, the pro-reform *New Annual Register* wrote that Cartwright "completely refutes Mr. Arthur Young's pamphlet . . . and clearly convicts the author of the grossest inconsistency, and the most contemptible apostasy of principle." It also declared that Young's answer "leaves his opponent's material statements uncontradicted, and his arguments unanswered."[134] The *Monthly Review* for August published a long review of Cartwright's pamphlet and completely accepted his attack on Young:

> . . . nothing can be more certain than that . . . he has gained a most complete victory over Mr. Arthur Young, whose pretensions to consistency, to respect for the constitution, and to fair dealing in stating the opinions and arguments of his adversaries, the Major proves to be totally unfounded. Some of our readers will readily believe us, when we say that he has convicted the Secretary to the Board of Agriculture of a complete dereliction of his own most favorite principles . . .[135]

The *Monthly Review* for March, 1796, in a very perfunctory notice of Young's reply, stated that the public was tired of the whole controversy, but it did admit: "Mr. Arthur Young has repelled the attacks of Major Cartwright with vigour and ingenuity . . . those who have perused the severe strictures of Mr. Cartwright . . . ought,—in common justice—to peruse this defence of Mr. Y's principles and political conduct . . ."[136]

Late in 1795 Young was disturbed by an attack upon his integrity as revealed in a letter from Robert Fraser, one of the Board's county surveyors: "This morning I received your Letter of the 25th of Nov. in which you state that Mr. Sheridan asserted in the House of Commons that I was employed while in Cornwall and Devonshire to circulate your Pamphlet called *the example of*

---

[133] *Gentleman's Magazine* 66, 1: p. 502.

[134] *New Annual Register*, 1795: p. 267.

[135] *Monthly Review* 17: pp. 429–430. Cf. *ibid.* 19: p. 78 for a brief notice of a pamphlet entitled, *A Reply to a Pamphlet entitled, "An Idea of the Present State of France &c." By Arthur Young, Esq.* (Owen, 1795). The anonymous author had composed a dialogue between Young and a Whig, using Young's own words. The reviewer thought that "he comes off with flying colours, and leaves Mr. Young, like Braddock's forces, to lament a defeat by an enemy whom he can scarcely see, and with whom he had not a fair opportunity of grappling."

[136] *Monthly Review* 19: pp. 347–348.

France a warning to Britain." Fraser denied the report and declared that he did not even own a copy of the pamphlet.[137]

Two of Young's most interesting letters to Thomas Ruggles
date from 1795, and both deserve extensive quotations.[138] The
first was written on February 26:

> I recd. your favour in which I am ready to confess you com-
> plain with some degree of reason for I have been but a bad corres-
> pondent; I have however to plead much ill-health—much anxiety—
> and a great deal of business, for the Board has been so anxious
> on the subject of Potatoes and the apprehended Scarcity that I have
> worked double tides; and not only have had to dictate Lrs. for
> hours, but to write memoirs in the bargain. I believe from the
> returns which I have recd. from above thirty counties that the
> apprehension is not well founded and that there is corn enough
> in the kingdom to last with the spare consumption that always
> attends a high price till the next crop. . . .
>
> Why do you complain of having spent yr. money, seeing that
> it is to make a comfortable house in wch. you will live happily
> *in the family way;* the only way a man can be happy at all; and
> I am one instance of the want of it. I own I wish you had laid
> out the same money in a neighbourhood where you would have
> had more of what, however, you do not want, Society; by the bye,
> that is one of the objections to a man confining himself too much
> to a systematic life at home; the world becomes too indifferent to
> him, and he gradually contracts his circle till at last it becomes
> composed of nothing more than his wife, his cat, his dog and his
> child. That is bad, and I hope Spains Hall will see you a better
> animal.
>
> I have not had good spirits, or too good health; otherwise I
> should pass this winter agreeably enough. I have been much sought
> after and had a good round of dinners with evenings well enough
> in circles that never see a card table, but I like the country and
> more quiet scenes. I have established a farmers club that meets
> for the first time next saturday at the thatch'd house. . . . There
> is no such club, and I hardly named it but it took like Wildfire.
>
> What do you say to politicks? To the thriving, growing, rising,
> increasing resources—the inexhaustible resources of this country
> yt. will bear in one year 16 hund. thous. pounds new taxes & laugh
> at them like a feather. . . .

[137] Add. MSS. 35,127, f. 373. A search of the *Parliamentary History* has not
revealed any such attack by Sheridan.

[138] These are letters II and III in the valuable collection of letters from Young
to Ruggles. Letter I in the series has no year date and is not very important.

Young's second letter in 1795 to Ruggles was written exactly four months later on June 26:

I recd. yr. very kind & friendly Lr. for wch, accept my best thanks. You are happy in ye country amidst a noble property and a fertile farm, a good wife and good children—you have all good things—therefore being forgotten by what is called ye world, were it the case is not a very heart breaking malady. I cannot but break a commandment & envy yo. tho' not by coveting—I am here the end of June in smoak, stink & noise; and shall not quit it till 5th July when I go to Woburn . . . for two or three days, & thence to Suffolk . . . My very good fd. Betsy is in my cottage; I shall live with her and Oakes[139] for some time, comfortably after the racket of London. If I can find time & prosy [*sic*] to come to Spains Hall I will—I shall do it with pleasure, that yo. know; & as to feasting yr. tench & carp beat our Turbots. Your schemes for the poor are excellent. . . . I see another very dear year, whatever ye crop, wch. in general is good, though bad in some districts.

News you request, but we have none only skirmishes by sea, but in our favour. We have no hope but in assignats which are now so low that their end must be near. We have corn certainly coming, 125,000 qrs from Dantzick, wch. is something—but ye price rises terribly; farmers now hoard, they expect L 5 a qr, There is large sples. [supplies] in Wilts, Hants, Berks, etc. God send us free from riots, but ye aspect is not pleasant. I recomd. to Govt. long ago to prohibit by act oats being given to horses; but in vain.

Among the Chatham papers at the Public Record Office, there is a curious letter to William Pitt from Mrs. Arthur Young, which must have embarrassed Young considerably. Since it is rather typical of Mrs. Young's letters, although somewhat less confused than was often the case, it is being quoted in full:

Sir

By the death of our late very worthy friend Captain Barrett the *small house in Middle* Scotland Yard is become vacant, he had an appointment in the Kings Household/gentleman of the Pantry/but was of course generally absent with the West Kent Militia by his Majesty' permission—I have heard a report that those houses are soon to be down, if not, I should be very glad if I might be allow'd

---

[139] "My very good fd Betsy" was Elizabeth F. Plampin who had married Orbell Ray Oakes on April 23, 1795. She had been a county belle since the days when Lazowski had visited Bury. She was Young's closest woman friend. Orbell Ray Oakes was the son of James Oakes, chief banker of Bury. *Cf.* Gage, 1838: p. 494.

with my daughter to inhabit it a year or two untill such alterations may take place, one reason is to save rent & another as a centinal [*sic*] always guards it, which is truly a very great charm to a small circle, where gentleman [*sic*] are not stationary.

I hope you will have the goodness to pardon the liberty I take in making this request

I have the Honour to be
Sir your obedient Servant

M. Young

Bradfield Hall Novbr. 29th 1795[140]

Young spent the spring of 1796 at London working on Board of Agriculture business. As papers and communications to the Board multiplied, a procedure for weeding out the less valuable had to be made. In February it was decided that no papers should be laid before the Board without prior examination by the president or secretary, while in May the Committee on Papers passed the following vote:

That an abstract of each Paper be previously prepared by the Secretary, before it is laid before the Committee.

That the Secretary to prepare a List of all Papers to be laid before the Committee, to be sent to each Member of the Committee four Days previous to meeting, & that such list to not contain more than ten Papers.[141]

Thus the responsibility for determining what papers should be read now rested primarily upon the Secretary, although he may have delegated some preliminary weeding out to the undersecretary and clerks.

In 1795 the Board had procured a parliamentary grant of £1,000 to Joseph Elkington "for the purpose of inducing him to instruct others in the important discoveries he had made" in the field of draining.[142] In 1796 the Board had to work out the details, but Elkington seems to have been singularly elusive and unbusinesslike. On April 29 Sir John reported to the special committee on draining that Young "from his zeal to promote the objects of the Board, would dedicate the months of June & July, to the sole

---

[140] Public Record Office, Chatham Papers G. D. 8: f. 193.

[141] Minute Books, Finance Committee, 3, ff. 47, 73.

[142] *Annals* 24: pp. 525–529, 563, 611; Clarke, 1898: pp. 19–20; Young, 1809: pp. 23–24.

purpose of ascertaining the Principles of Mr. Elkington's system, without putting the Board to any expence. . . ."[143] Naturally this proposal was adopted, and on May 10 Elkington accepted the plan. But when Young tried to work out the details, he was appalled by Elkington's vagueness:

The Secretary having consulted Mr. Elkington . . . was informed by him that his motions were so very uncertain that he could lay down no Plan before hand. . . . That he would be in London between the 4th & 10th of June, when he would inform the Secretary where he was going & when they met would there fix on the next place of meeting, but it was out of his Power to be more explicit. The Secretary wished to procure a List of the places Mr. Elkington had drained, that he might also view these: but Mr. E. could give him very few. He keeps no Books and makes no memoranda; & he had very little Recollection of what he has done, even whether he had worked in any particular county, or not. Such uncertainty both of his motions and also of his works, will necessarily throw many difficulties in the way of procuring that satisfactory information which the Secretary wishes, & which he presumes is the object of the Board.[144]

The committee then voted that Elkington would not get his money unless he gave up two months to accompany Young and a land surveyor, John Johnston. The following letter indicates that Young had further difficulties with the unsystematic drainer:

On coming to London on the 13th I was very much surprised to receive your Letters. You must have forgotten when you say a months notice is necessary that it is six weeks since you positively engaged by word of mouth to the Board to go the Journey and you were informed by me, that I would meet you at any place where your business called you. You named Dorchester and said you would there fix on some other place of meeting; thus the scheme was to be made subservient to your private convenience, yet you now say you must be at Warrington tomorrow, too soon for it to be possible to meet you, and you name no other place where you go to from thence—as to my hearing from you again, and again coming to London to be a second time disappointed it would be idle to suppose you would keep one engagement with me if you break one openly come into with the whole Board.[145]

[143] Minute Books, Miscellaneous Committees, **2**, f. 68.
[144] *Ibid.*, ff. 71–72.
[145] Copy Letter Book, **1**, ff. 166–167.

It is not clear exactly what happened next. Young made a two-month trip in the summer of 1796, but Elkington is not mentioned in the account of it. On the other hand, Johnston wrote an account of Elkington's system which appeared in 1797 and went through five editions.[146] Sinclair reported to the Board praising Elkington for his generosity,[147] and on April 13, 1797, Young wrote: "I am very anxious that Elkington, who has behaved so well should get his money."[148]

Some time during the spring Young was asked to advise the Prime Minister about parliamentary regulation of labor, which he had advocated in 1795. Young was impressed with Pitt:

I answered all his enquiries, and could not but admire the wonderful quickness of his apprehension of all those collateral difficulties which I started, and of which he seemed in a moment to comprehend the full extent. I found him hostile to the idea.[149]

Young was one of the guests at a rather interesting dinner given by Lord Somerville. The occasion was to complete the Duke of Bedford's experiment upon the fatting propensities of four breeds of sheep. Lord Somerville bought one of each breed, and had them butchered and cooked exactly alike:

The legs of mutton were placed on the table at the same time, for the company to pronounce which was the best; and the result was, that by much the larger part of the company declared in favour of the Leicester—thinking, while so doing, that it must be the South-Down. The difference in fact, however, was very small; the flesh of the Leicester was of a closer texture, but rather of a paler colour; the gravy of the South-Down of a higher flavour.[150]

Late in April the newspapers carried an advertisement of an intended pamphlet, to be written by Edmund Burke in the form of letters to Young "on some projects talked of in Parliament, for regulating the price of labour." A doctrinaire believer in *laissez-faire* principles on labor questions, Burke disagreed completely with Young on the subject. After Sir John Sinclair had requested

[146] Clarke, 1898: pp. 19–20.
[147] *Annals* 28: pp. 77–93.
[148] Copy Letter Book, 1, f. 192.
[149] Young, 1898: pp. 255–256.
[150] *Annals* 26: pp. 433–434.

Burke for a memorandum on the subject for the Board, and had received no satisfactory reply, he suggested that Young pay a visit to Burke "in order that I might discover whether that celebrated character continued his intention of throwing his thoughts upon paper."[151] Young arrived at Beaconsfield before breakfast, remained there all day, and stayed over night. He must have been pleased by Burke's opening comment:

Why, Mr. Young, it is many years since I saw you, and, to the best of my recollection, you have not suffered the smallest change; you look as young as you did sixteen years ago. You must be very strong; you have no belly; your form shows lightness; you have an elastic mind.

After breakfast Burke took Young on "a sauntering walk for five hours over his farm. Young has left a moving picture of Burke in decline in the year before his death:

. . . I was shocked to see him so broken, so low, and with such expressions of melancholy. I almost thought that I was come to see the greatest genius of the age in ruin.

And I had every reason to think, from all that passed on this visit, that the powers of his mind had suffered considerably. . . .

His conversation was remarkably desultory, a broken mixture of agricultural observations, French madness, price of provisions, the death of his son, the absurdity of regulating labour, the mischief of our Poor-laws, and the difficulty of cottagers keeping cows. An argumentative discussion of any opinion seemed to distress him, and I, therefore, avoided it. . . . it hurt me to see the languid manner in which he lounged rather than sat at table, his dress entirely neglected, and his manner quite dejected. . . .

But to behold so great a genius so depressed with melancholy, stooping with infirmity of body, . . . and sinking to the grave under accumulated misery; to see all this in a character I venerate, . . .

---

[151] Young, 1898: p. 256. Cf. Burke, 1800: pp. viii-xi. The editor of this posthumous work stated that he had fused Burke's memorial to Pitt with a fragment of his letter to Young, and wrote (p. x): "The principal alteration has been the necessary change of the second for the third person, and the consequent suppression of the common form of affectionate address, where Mr. Young is named. That gentleman alone can . . . complain of this liberty, . . . as it may . . . have deprived him of that, which in some sort was his property, and which no man would have known better how to value." In the only direct reference to Young in the pamphlet itself (p. 22) he accuses "my friend, Mr. Arthur Young," of exaggerating the profits of farmers in his "most useful works."

wounded every feeling of my soul, and I left him the next day almost as low-spirited as himself.[152]

Young made one very important and interesting trip in 1796 which lasted about two months. He went down along the south coast as far as Devonshire, then up through Somerset, and back through Bristol, Bath, and Reading. The entire tour covered more than thirteen hundred miles.[153] Unlike his other journeys reported in the *Annals*, this one was printed in sixteen driblets, running through four volumes. Young left Bradfield on June 13 and stopped at Clare to see Thomas Ruggles and where William Shrive gave him a fine tench dinner which made Young envious. "Why cannot I have this at Bradfield pond, as well as he at Clare river? Not quite so well, to be sure—mais n'importe—I might still have tench. Laziness is the orgin of half our wants." He found poor rates very high in the manufacturing towns of Essex, and advocated compulsory friendly societies as a remedy: "... to ordain that in towns of a certain description, and in country parishes, every master manufacturer should either limit his work-men to those who subscribed six-pence a week, or pay that sum for them. . . ."[154]

In Sussex he visted Petworth, "where one is ever sure of finding some new effort to improve the agriculture of the country; nor is that princely residence often without some person whose exertions have been beneficial to society."[155] He then stopped at Arundel for a casual visit, but being recognized, was asked to dinner and to stay the night. The 11th Duke of Norfolk proved a cordial host and the wine was excellent:

. . . our noble and festive host was in spirits, and gave the seasoning of wit and good humour to every topic of conversation, even to politics:—and though his Grace has some ideas which I think too democratical for the premier Duke . . . of England, . . . yet there was such liberality and pleasantness in all that passed, that the subject caused not a moment's regret. . . .[156]

[152] Young, 1898: pp. 257–261.
[153] *Annals* **30**: p. 94.
[154] *Ibid*. **28**: pp. 100–101.
[155] *Ibid*., p. 109.
[156] *Ibid*., p. 113. The 11th Duke of Norfolk was a Whig, a very heavy drinker, and in general an eccentric. He was elected president of the Royal Society of Arts in 1794 and spent vast sums in remodeling Arundel Castle.

The sight of the ocean at Bognor led Young into more patriotic expressions:

> . . . that ocean, which, while it enriches this island with the commerce of the world, gives it security. Europe over-run by the new Goths; but England, triumphant on the bosom of the waves, sees the hostile shore of France with no emotions of fear. Thanks to the glorious energy of British seamen, who prove themselves irresistible; and worthy of the fame which has in every age attended the flag of real liberty.[157]

Young visited several capital farmers in Wiltshire and Dorset. The first was Mr. Dyke near Amesbury, "the greatest farmer in Wiltshire," who kept 5,000 sheep, and was shifting from Wiltshires to South Downs like many of his neighbors, a fact which Young attributed to the information in the *Annals.* Young was also pleased to learn from Dyke that the yeomanry cavalry corps were flourishing in Wiltshire, that some men traveled as far as twenty miles to drill, and that the mere presence of five troops had sufficed to quell a riot at Devizes.[158] Young called upon another famous farmer, Mr. Bridge, south of Sherbourn, whose farm had been honored by a visit from the king, who had built a fine cow barn, was a great cultivator of cabbage, and had an elaborate apparatus for steaming potatoes. Young praised Bridge's farm highly and was sure that the king's visit had acted as an incentive: *The King's attention* in matters of husbandry *is as dew upon the grass.*"[159]

Young spent three days at Exeter and was greatly pleased with his reception. Having heard of the "known liberality" of Mr. "Francis" Baring, he took the liberty of applying to him for information. Baring invited him to dine with the Quarter Sessions, where he met thirty magistrates and received more invitations than he could accept. During dinner Young supported the proposed general enclosure bill very warmly, not realizing that the chairman of the session, Mr. Leigh, clerk of the House of Commons, was very strongly opposed to the measure. In the *Annals* Young wrote of the dispute: "There was some good-humoured prancing on the impropriety of *delegating powers constitutionally*

[157] *Ibid.,* pp. 116–117.
[158] *Ibid.,* pp. 361, 365.
[159] *Ibid.,* p. 481.

lodged, at present, in safe hands."[160] The *Autobiography* gives the impression that the dispute was quite acrimonious until he realized who his opponent was. He attributed Leigh's opposition to his vested interest in fees from private enclosure bills.[161] Shortly after leaving Exeter, Young spent several very pleasant days with the Taylor family near Newton Abbot, who saw that he was given all possible information about the country. The father, T. Taylor, Esq., was one of the justices whom Young had met at Exeter. His son, Colonel Taylor, also proved most hospitable and one day rode with Young as far as Torquay. Young also met the colonel's lady: "Interesting—What place would not be rendered so by Mrs. Taylor? We began a conversation which I wished to have pursued. . . ."[162]

Young was shocked at the bad conditions among the poor weavers in this area, even though their wages were good:

There was a dark, dirty, even blackness, in their cottages, offensive; the people ill-dressed; and their habitations carrying, in almost every circumstance, a countenance of wretchedness. Their gardens or rather patches of ground, miserably cultivated. These are indications against the landed proprietors. It is much to be lamented when the poor of a country are not instigated to personal and domestic cleanliness, by every decent provision being made for them in respect to habitation and garden. They have a fair claim to this attention; and a country ought to be esteemed backward, barbarous, and unjust, when the habitations of the cottagers do not partake of those improvements which every thing else indicates. To hear of a great rise in rents, to see the houses of landlords and farmers improving, and cottages no better than in the last century, is a horrid spectacle.[163]

Here Young sounded almost like a social reformer. His position was certainly very different from what it had been twenty-five or thirty years earlier.

On leaving the Taylors Young went to Totnes where he took boat for Dartmouth and was enchanted: "Of all the beautiful

---

[160] *Ibid.*, p. 630. Almost certainly the Mr. Baring mentioned was John, 1730–1816, M. P. for Exeter and brother to Sir Francis, founder of the family's fortune.

[161] Young, 1898: pp. 261–262.

[162] *Annals* 29: p. 199.

[163] *Ibid.*, pp. 195–196. This new sympathy with the lower classes is also reflected in a comment favorable to Lord Rumford's soup kitchens, etc. *Ibid.* 26, pp. 170–172.

scenery that I have viewed in England, I recollect nothing equal to this ten miles by water. . . . There are some views, that in beauty almost rival Killarney."[164] From Dartmouth he went overland to Plymouth, stopping on the way to see Richard Hawkins whom Sinclair had engaged to make the survey of Devon. A little later he reached the seat of Paul Treby Treby, one of his correspondents at Goodamore on the edge of Dartmoor. One of the most interesting days of his whole trip was spent on the moors. "We were on our horses at day-break."[165] The party consisted of Mr. Treby and Mr. Richard Hawkins. They visited an extensive improvement on the moor by Justice Francis Buller around Princetown, where they were hospitably entertained for dinner. Contrary to Treby's belief, Young was convinced that most of Dartmoor was capable of being laid down to grass. His summary was: "Dartmoor, at present, is a disgrace to the kingdom; it might be what Prince Town is—a garden."[166] At night they returned to Goodamore, "after a day as interesting as any I had ever passed."[167]

After leaving Goodamore, Young's route lay across Devonshire to Barnstaple on the north Devon coast, where he was entertained by Mr. Exter, a great advocate of paring and burning, a writer on potato cultivation, and almost a fanatic on the drill husbandry. Exter accompanied him on a day's excursion over Exmoor, which Young felt was even more capable of improvement than Dartmoor. His reactions were natural for a life long advocate of reclaiming wastelands:

I know not a more melancholy spectacle than to see two such enormous deserts, within sight of each other . . . And both these wastes situated in a part of the kingdom where land products of all sorts sell much higher than in the Eastern counties. Periodical complaints of a want of bread, and of dearness of all provisions, and such tracts left to the wretchedness of common-rights![168]

A chief subject of his enquiries in Devonshire was the fashionable breed of "North Devon" cattle. He interviewed some eminent breeders, especially the Quarterly brothers. The strong

[164] *Ibid.* **29:** pp. 207–208.
[165] *Ibid.*, p. 562.
[166] *Ibid.*, p. 578.
[167] *Ibid.*, p. 587.
[168] *Ibid.* **30:** p. 192.

points of the North Devons were their strength as oxen and their line beef, but Young displayed less enthusiasm for them than for Bakewell or Sussex cattle. The breed was described, however, at great length, including one entire driblet in the *Annals*.

Near Taunton he visted his old friend, Robert P. Anderton of Henlade who had made great improvements since Young had visited him twenty-nine years earlier, and was a complete contrast to the unprogressive type of landlord stigmatized below:

. . . he who in such a period of advance, has consumed all his income in eating and drinking, or in hounds and horses, or in the dissipation of public places of resort, and has done nothing to smooth and brighten the face of his patrimony and residence, has lived more like an ox than a man.[169]

Young's account of the last lap of his trip through Bristol, Bath, and Reading is very sketchy. "A couple of agreeable days" were spent at Reading "with my good friend Dr. Valpy."[170] The two men had always been very companionable. Valpy introduced him to several spirited farmers in the neighborhood, including his father-in-law, Mr. Benwell, a strong believer in the merits of rouen for the spring feeding of sheep, and a Mr. Carter who did all his farm work with she-asses and raised a very large number of rabbits which manured fifty-two acres a year.

Young's most important writing during 1796 consisted of three letters to the Yeomanry of England, which formed the core of a major pamphlet published in 1797 as *National Danger and the Means of Safety*. The first letter[171] started off very mildly, declaring that conditions were less threatening than earlier but that the roots of danger were still present. Since the regular troops would soon be concentrated on the coast, the interior of the country would lie defenseless against riots over high prices unless more yeomanry companies were formed.

The second letter,[172] written under the shadow of Bonaparte's first astonishing victories in northern Italy and the dictated peace with Sardinia, was much more alarmist. Italy had relied on a

[169] *Ibid.*, p. 309.
[170] *Ibid.* 31: p. 86.
[171] *Ibid.* 26: pp. 516–521. The *Annals*, where these letters first appeared, does not date them, but in the pamphlet the dates are July 1, August 1, and December 1. The first letter was addressed to the Yeomanry of Suffolk, the later ones to the Yeomanry of England.
[172] *Ibid.* 27: pp. 49–54.

regular army and when it was defeated she was overrun. There-fore England should not rely upon regular armies for her pro-tection. True she had the navy, but national security required more. England could never be successfully conquered if each county had a force of ten to thirty thousand men ready for any emergency. He pointed out in lurid phrases what the conquest of the "new Goths and Vandals" might mean: "The plunderers of the world demand pictures and statues; they seize equally the liv-ing Venus's and the dead Apollo's." What were Englishmen doing to provide security?

But when I see an English county, full of wealth, of fabrics, of seats, of parks, and gardens, but without one troop of yeomanry established, I see nothing more than Italian imbecility—that heaps up native riches without a thought for their security.[173]

Before his third letter[174] was written the king had requested 20,000 cavalry to defend England against French invasion. Such a proposal was "without doubt a wise and salutary measure,"[175] but hardly adequate. The yeomanry companies offered the cheap-est method to raise a large enough force "as should deter the most daring foe."[176] It is not clear just how large a force he had in mind, for at one point he mentioned 200,000 and at another 500,000. He made strong appeals for volunteers, especially from the fox hunters, whose habits, horsemanship and knowledge of the country, made them ideal members of such corps. At one point he lashed out against "supine apathy and sloth" of the rank and file of landowners and farmers. While urging volun-teers to join the corps in one part of the letter, in another part he demanded that every man in the age limit fifteen to sixty, who kept a horse, should be forced to enroll and to arm himself.

Much attention in the third letter was devoted to the problem of financing the corps. Apparently thus far the national govern-ment had furnished pistol and sabre, while voluntary subscrip-tions had paid other necessary expenses, including a small fee for the weekly or monthly drill. Voluntary subscriptions would prob-ably not be adequate if the number of corps were greatly in-creased. Should each trooper bear his own costs, should parlia-

[173] *Ibid.* pp. 49, 51.
[174] *Ibid.*, pp. 528–538.
[175] *Ibid.*, p. 528.
[176] *Ibid.*, p. 534.

mentary appropriations be made, or local taxes imposed? In the following letter to William Pitt Young explored the various possibilities and related his own activities:

The important business you are at present engaged in of adding to the internal defence of the Kingdom, induces me to inform you of a fact which possibly may not have been laid before you.

There are five troops of yeomanry cavalry in Suffolk, wch. if compleat would amount to about 300: They were raised, cloathed & hitherto supported by subscription; but on closing the second call for contributions it was found that money recd, would support the establishment no more than three months, & no hope remained of a better voluntary assistance. As a subscriber and enrolled in the ranks of Lord Broome's troop, I was present at a late meeting and moved this resolution.

"That from this account of the funds subscribed & their appropriation, it appears that a general meeting of the county should be called for the express purpose of taking the same into consideration, in order that the inhabitants of this county may be informed, either that this measure of arming the yeomanry for its internal defence must be abandoned at the very period when His Majesty has communicated to his Parliat. the intention of the enemy to invade these Kingdoms, or, that some new and more efficacious mode of support must be adopted."

This was voted & ordered to be advertized; and the new meeting appointed for the 5th of Jany. in expectations that before that time you would bring forward some proposition in Parlt. relative to these corps.

On the same day it was calculated that to keep up and support these five troops the expence of L 2500 per ano. would be necessary. 300 horse that have been long trained will be lost in that county for want of so small a revenue. I then mentioned my intention of moving on ye 5th Jany. a petition to the House of Commons for power to raise the sum by addition to the county rate.

Probably something similar to this necessity may have occurred in other counties; if so, I submit to your attention whether there is any way of raising cavalry so cheaply (100,000 costing less a million sterl.) and whether it would not be expedient to adopt some measure that shall increase these corps rather than allow them to diminish?[177]

Among Young's miscellaneous writings in the *Annals* in 1796 was the rather smug "Sermon on the Scarcity of Corn" preached

[177] Public Record Office, Chatham Papers G. D. **8** Dated November 2. I have been unable to discover any follow up.

at Bradfield and some other Suffolk churches. The general thesis was that the scarcity was the act of God, that the poor should bear it without complaint, and that the rich and the government would give every possible aid if the poor were submissive. He congratulated the poor in Suffolk for their "exemplary and highly meritorious conduct during the crisis. He referred to "this happy nation blessed with the best constitution ever yet framed by the weak efforts of mankind."[178] He declared:

No where are the properties and the rights of the poor so carefully guarded. The sun of diffusive prosperity, that gilds the palace, illumines the humblest cottage. In no other country in the world is there any general and legal provision for the poor; and such is the benignant spirit of our laws, that no man is great or rich enough to dare to oppress them.[179]

His own journal of his summer's trip bore eloquent witness that things were not quite as fine as this, and that in another mood he fully recognized the evils that existed.

On May 1 he summarized the replies to his second questionnaire on food prices and then added some more smug observations. He completely exonerated the farmers from any blame for the high prices, for they accepted "no more than that offered." Rather government actions had raised an alarm which accounted in large measure for the prices so much higher than the actual scarcity warranted. He maintained, however, that high prices were much better than low ones, which led to unemployment which was a much worse evil. And he ridiculed those who preached that England was facing ruin: "... if it is ruined, it has the most marvelous symptoms of health that ever indicated the vital principle of strength and vigour."[180]

Young also published in 1796 a long article on his experiments on various grasses.[181] He had experimented with oat-grass, trefoil, ray grass, chicory, burnet, timothy, yarrow, crested dog's tail, meadow fox-tail, and meadow fescue. All served admirably for several years, but then tended to die out with continual feeding. Very interesting in this article is the enumeration of his fields with such quaint names as Pond Lay, Barnfield, Adam's Lay, Butt Lay, Little Ardera, Three Corner, Little Pakes, Great Pakes, Burd's Bit, Strawberry Lay, Jermyn, Bess Reed, and Churchfield.

---

[178] *Annals* **26:** pp. 198–199.
[179] *Ibid.,* p. 207.
[180] *Ibid.,* p. 471.
[181] *Ibid.* **27:** pp. 372–407.

VII. Domestic Tragedy and Religious Conversion, 1797-1799

THE YEAR 1797 certainly marked one of the most significant turning points in Arthur Young's life. In that year religion became the dominant force of his thoughts and activities. A new circle of friends, with similar religious ideas, eventually became more important than his old friends. His amusements, his daily habits, his reading, all were changed by his new religious outlook. There is absolutely no doubt about the immediate occasion for the change. In the summer of 1797 his darling Bobbin, a lovely girl of fourteen and the very apple of his eye, died from consumption. It was by far the most severe emotional shock which he ever suffered and it is doubtful whether he ever really recovered. The only consolation in his bereavement was the hope that, "by dedicating the rest of my life here to God, to join my dear child hereafter...."[1]

"This year, so fatal to every worldly hope, which overturned every prospect I had in life, and changed me almost as much as a new creation, opened in the common manner by my going to London to attend the meeting of the Board."[2] Thus began the *Autobiography* for the tragic year 1797. The work of the Board was not very interesting or important, however, during the spring of 1797, and Young seems to have found plenty of time for writing. Several fairly important short essays appeared in the *Annals*. One discussed the necessity and the difficulty of keeping detailed agricultural accounts.[3] In order to make comparisons separate accounts should be kept for each field, but such comparisons were worth little when the grass borders in the different fields were of differing widths. Furthermore no proper comparison of the profits of arable and grassland was possible without care-

[1] Young, 1898: p. 293.
[2] *Ibid.*, p. 263.
[3] *Annals* 28: pp. 47–64.

fully estimating the costs of the teams of horses and oxen. To calculate the costs and value of manure made from straw, or of the turnips or grass which sheep or cattle consumed, was very difficult. A careful annual inventory should be made of cattle and implements. It was also very hard to draw up an annual balance, since many operations lasted over several years. Young was rather humble in the face of these very real difficulties. He wanted agriculture to be a science just as completely as chemistry or physics, but at the same time realized that it was a business which must be measured in terms of profit or loss.

A brief article attempted to answer the queries of a Durham agricultural society whether it was best to expand money for publications, premiums, or an experimental farm. He pointed out that many foreign societies had been ruined by the indiscriminate publishing of mediocre material: "They printed, and printed, till nobody would buy their productions, and so their works sank into well-merited oblivion; and their institutions were forgotten."[4] Was he possibly thinking of Sir John Sinclair's propensity to publish too much too rapidly? In line with his earlier criticisms of the premiums of the Society of Arts, he urged societies to give few and large premiums and only for really practical and important subjects. Any experimental farm must contain at least 400 acres and must be stocked at the rate of £6 an acre. Some money might be spent to send out a practical farmer on a tour for some specific project. He summarized:

Thus . . . any society who could give 1000 l. a year in premiums, and assign as much more to a farm, would be able to effect more for the promotion of agriculture, than ever was effected yet. To such a society, printing and publishing would be a SURE FUND of profit, not of expense; for they would have merely to register what they had effected —MEMOIRS OF THE GOOD THEY HAD DONE.—Nothing questionable—nothing theoretical—nothing trifling.[5]

He also prepared an elaborate index or *table raisonée* of all the articles in the *Annals* on potatoes, under such headings as varieties, cultivation, time of planting, expenses, comparison with turnips, and use as human and animal food.[6]

[4] *Ibid.*, p. 283.
[5] *Ibid.*, p. 289.
[6] *Ibid.* **29:** pp. 38–62.

His most important short article in the spring of 1797 was entitled, "A Word in Season, at a Critical Moment, to Landlords, Yeomen, and Farmers."[7] It was occasioned by a financial crisis in February, a run on the banks, and the suspension of specie payments. He attempted to restore confidence in the government, the Bank, and paper money, which must be the objective of all enlightened landlords and farmers. England was basically sound, and her industry, commerce, and agriculture were all prosperous. Paper money was a necessary basis of exchange in an advanced country, and probably more reliance on paper was necessary, not less. But should not the issuance of all paper money be by government, rather than by private banks, many of them unsound country banks with inadequate backing? If the landed interest helped to prolong the crisis by playing politics because of their dislike of the ministry, the results might be low prices, unemployment, and increased poor rates.

There remain his two most important writings completed in the spring of 1797. The first was his pamphlet, *National Danger and the Means of Safety,* not quite half of which reprinted almost without change four articles in the *Annals.* The first three, published in 1796, have already been analyzed.[8] The fourth was dated March 1, 1797,[9] and predicted repeated French attempts at invasion since they must realize that the continent could never be completely subdued as long as Britain resisted. Again he appealed to the gentry, yeomen, and farmers to enroll in the Yeomanry Companies. If this whole class enrolled voluntarily, the French might realize that invasion was hopeless. He also urged the women to bring pressure upon their men folks and friends, and even to ostracize those who had not enrolled:

> . . . tell him that he cannot have a real taste for literature, or relish fine arts, if he is not ready to repel the Goths and Vandals of modern France; that he can feel neither friendship, love, nor respect for the sex, if he is not in a situation to defend it.[10]

[7] *Ibid.* **28**: pp. 426–443. Although numbered Letter IV in the series "To the Yeomanry," no part of this article is included in *National Danger,* except the postscript, pp. 442–443. The letter is undated, but it was written before the naval mutinies.

[8] *Cf. supra,* pp. 356–357.

[9] *Annals* **28**: pp. 177–187.

[10] *Ibid.,* p. 184.

To these articles was added a new section entitled, "On the Necessity of Men of Property being armed," longer than the four articles together. It was probably not written until after the naval mutiny, which completely shocked Young: "A convention of delegates in the wooden walls of the Royal Charlotte!!! . . . it is ON OURSELVES THAT WE MUST RELY AT LAST."[11] Since the navy was in revolt, only the men of property could now defend England. He predicted that, if defeated, England might expect to become one or more departments of France. "Such is the alternative: LIVE by new measures or DIE by your old ones."[12]

In case conscription was necessary, the cost of arms and equipment must be borne by the individual, not by the government. Moreover, conscription would apply only to men of property, for Young clearly feared revolution at home fully as much as foreign invasion. Since the French always appealed to the underprivileged, Young foresaw England governed by the scum of the cities and by criminals emptied from the prisons. He recognized, however, the inconsistency of demanding large numbers of troops and yet excluding the masses, and felt it necessary to make the following unconvincing explanation:

As I have raised my imaginary edifice entirely on the basis of property, it is necessary to declare that I have the most perfect conviction of the loyalty . . . of a vast majority of his majesty's subjects. . . . It is not any doubts of that nature that have induced me to avoid the supposition of calling them out, but merely the idea, that, when a measure is not to be supported by taxation, no burthen ought to be laid on men of that description.[13]

The remainder of the pamphlet makes it all too clear that Young felt that only men of property could be trusted to safeguard property and the constitution against any revolutionary attempts in England.[14]

The second important work completed during the spring of 1797 was the *General View of the Agriculture of Suffolk,* the

[11] Young, 1797: pp. 31–32.
[12] *Ibid.,* p. 42.
[13] *Ibid.,* pp. 66–67, note.
[14] *Gentleman's Magazine* **68:** (1798): **2:** p. 693 had only a very perfunctory review of this pamphlet. The *Monthly Review* **25:** pp. 461–462, was hardly more satisfactory.

revised and greatly enlarged survey for the Board of Agriculture.[15]
It should certainly have been the model report, for here was a
highly trained reporter covering the county which he knew best.
Actually it seems sloppy and several times the same paragraph
has been inserted at two different points without substantial
change.[16] It is deadly dull, but perhaps that is not a fair criticism of such a work. It is full of long quotations from his correspondents, but there is almost no attempt to integrate the material.
The ten-page review in the *Monthly Review* is quite favorable:

> His present work contains, indeed, a great variety of interesting infor-
> mation, delivered with much apparent accuracy, and combined with
> the reflections and observations of a philosophic mind. No man knew
> better than Mr. Y. what his subject required, or was more able to
> execute it to the satisfaction of the Board of Agriculture and of
> the public.[17]

Young followed closely the uniform organizational framework
prescribed by the Board. The introduction stated that the county
reviewers should concentrate on the unique features of the county.
His aim then was not to pile up details of Suffolk husbandry
which might well be repeated for most other counties. Rather
he would emphasize carrots, cabbage, hemp, and the pollard
breed of cows. For those familiar with Young's agricultural ideas
for the past twenty years, the views expressed were just what would
have been expected. Blame and praise for the practices of his
home county were meted out impartially.

During the spring of 1797 Young was increasingly worried
about Bobbin's health, as her illness failed to respond to treatment. She had returned with him from Bradfield in January and
was sent to a fashionable girls' school at Campden House in London, which cost £80 a year. Young wrote "she abhorred school."
The food was poor and Bobbin had to share a bed with a girl who
was deaf and hence only slept on one side, "because the vile beds
are so small that they must both lie the same way." At home she
had been accustomed to an out-of-doors life, but at school "no air

---

[15] The preliminary survey of 1794 had been 92 pages long; that of 1797 was 329
pages. The edition available to me was that of 1813, which was 432 pages.

[16] For instance, *cf.* pp. 126 and 129 on the use of carrots for horse food; also
pp. 121 and 203 about making butter from cows fed on cabbages instead of hay.

[17] *Monthly Review* **28:** (1799): p. 70. I could find no review in the *Gentleman's
Magazine*.

but in a measured, formal walk, and all running and quick motion prohibited." Young blamed her sickness partly upon the school, and naturally was pretty bitter: "Oh! how I regret ever putting her there, or to any other, for they are all theatres of knavery, illiberality, and infamy."[18]

When she became ill in March, Young removed her from school to his own lodgings in Jermyn Street. Here she was treated by the famous Dr. John Turton who completely misunderstood the case and purged and physicked her until she was "little more than skin and bone." On April 12 Young took her back to Bradfield where the same treatment was continued by a Dr. Smith acting on Dr. Turton's general orders.[19] Fortunately there have survived from April, May, and early June eleven letters between Bobbin at Bradfield and her father at London, which deserve very generous quotation, even though they are already in the *Autobiography*.[20]

I. *Bobbin to her Father*

My dear Papa,—I received your letter this morning. Thank you for it. My strength is much the same as when I saw you; my appetite is getting better a good deal. Mr. Smith saw me yesterday, and said it was a running pulse, but that he thought me better. I think if anything I am better than when I saw you. Thank you for the wine. . . . As for the bad news, I am tired of it. I want, and should very much like, a nice writing-box to hold pens, ink, paper, all my letters, &c. . . . this is just the thing for a birthday present. As for sweet things, I do not wish for them particularly; any little thing that you think wholesome I should be glad of. The weather is as yet so bad that I cannot stir out. . . .

> Believe me, dear Papa,
> Your dutiful daughter,
> M. Young[21]

II. *His Reply*

My dear Bobbin,—I am much obliged to you for the description you gave me of your health, but I beg you will repeat it directly, and do

---

[18] Young, 1898: pp. 263–264. For miniature portrait of Bobbin, see fig. 3.

[19] *Ibid.*, p. 264.

[20] *Ibid.*, pp. 264–276.

[21] I should date this letter shortly after April 12 when Young brought her back to Bradfield. Mr. Smith is the Bury doctor, of course.

not forget appetite, pulse, sleep, pains, swelled legs, fever, exercise on change of weather, thirst, &c., for I am extremely anxious to know how you go on. I have looked at a great many writing-boxes, but find none yet under 1*l.* 5*s.* and 1*l.* 11*s.* 6*d.*; but I hear there are good ones to be had at 15*s.* The moment I can find one I will buy and send it packed full of seals, or something else.

Politics are melancholy, for the fleet is satisfied, the army is not, and the same spirit there would be dreadful. . . . The French will make no peace with us, but bring all their force to the coast and ruin us, if they can, by invasion expense.[22]

I cannot read half your mother's letter; but enough to see that she is very angry, for I know not what. I am not paid, and have nothing to send.

Dear Bobbin,
Yours affectionately,
A. Y.

III. *Bobbin to her Father*

My dear Papa,—I received your letter this morning, for which I thank you. My appetite is a great deal better, pulse rather too quick, sleep *very well,* no pains, no swelled legs, no fever. We have had Sunday, Monday, and yesterday fine, and only those since you went. I walked in the Stone Walk. . . . I saw Mr. and Mrs. O. Oakes in Bury yesterday, they have made but a short stay in town. I am much obliged to you for the wine and porter, which I have received safe. . . . write soon. . . .

N. B. By the time M. Kedington comes his strawberries will be ripe; ask him if he would give me a few if I send for some. Pray remember a patent lock, so it will be a guinea besides that.[23]

IV. *His Reply*

I was very glad to receive my dear Bobbin's letter, for it gives me the best account I have yet received of your health. As you wish a guinea box you shall have one, though I can very ill afford it at present. As to your mother's idea of my being paid, nothing can be said to it, if she knows better than I do. . . .

---

[22] The reference here is probably to the first naval mutiny at Spithead which occurred on April 16 and was settled fairly quickly.

[23] If Bobbin mentioned meeting Mr. and Mrs. Oakes in Bury, they presumably no longer occupied a cottage at Bradfield. There were several Kedingtons who were neighbors and close friends.

Lady Hawkesbury, Lady Hervey and Lady Erne have given the Macklins and me some opera tickets several times—last night for the benefit of the sailors' widows and orphans . . . but sailors are not in fashion; the pit was not more than two-thirds full. . . . Every person that comes from France asserts the same, that their whole force will be brought against this country. I have been sent for, and had an interview with a cabinet minister on arming the landed interest; but I fear nothing will be done effectually, though they seemed determined that *something* shall. Don't mention this out of the family. . . . Continue to write me. Tell your M. I have no money; therefore why worry me? She might as well ask blood from a post. . . .[24]

V. *Bobbin to her Father*

My dear Papa,—I received your letter this morning and am extremely obliged to you for your attention about the writing-box, but if it be not purchased, I have seen one at Rackham's which suits me exactly in every respect, therefore, if you will send a patent lock, I can have it put on. The price of Rackham's is a guinea, but if you have bought it I shall like it as well. My chief complaints are weakness, and a very bad cough, nothing else that I mind. I dare say you were entertained at the opera.[25]

I have just got six of the most beautiful little rabbits you ever saw, they skip about so prettily, you can't think, and I shall have some more in a few weeks. Having so much physic I am right down tired of it. I take it still twice a day; my appetite is better. What can you mind politics so for? I don't think about them. Well, good bye, and believe me, dear papa. . . .

VI. *His Reply*

My dear Bobbin,—I received your letter this morning, and am sorry it did not come in time to stop my buying this box, which is twenty-five shillings besides carriage; but I hope you will like it; the lock is good, and not common. I cannot afford a patent one, which is fifteen shillings alone.

I am sorry to hear you have a bad cough and are weak. God send

---

[24] Young was telling the truth about the money. Minute Books, Finance Committee, 3, f. 142, makes it clear that the secretaries were not paid their salaries due March 1 until June 23. Lady Hawkesbury and Lady Erne were daughters of Lady Hervey and the Earl-Bishop, neighbors from Ickworth.

[25] J. Rackham was the printer of the *Annals* at Bury. Bobbin's letter implies that he also kept a kind of stationery and bookstore.

this fine weather may make you well soon. Continue to let me know
how you do, particularly your cough. You do not say if you are
upon the whole better, nor whether you have got on horseback yet,
which I must have you do, . . . or you will not get well at all. Be
more particular—what physic do you take?

You are right not to trouble yourself about politics. . . .

The Directory of France has ordered all my works on Agriculture to be
translated in twenty volumes, and their friends here would guillotine
the author. The "Travels" sell greatly there in France, the third
edition coming. . .[26]

Note from A. Y. to Bonnet (Farm Bailiff)

Miss Patty is to ride out in the chaise or whisky, or on double
horse, whenever Bonnet is not obliged to be absent from the farm.
If he is at market when the days are long and Miss Patty rises
early, she can have a ride before breakfast.

Bonnet to pay Miss Patty a shilling a week.

VII. *Bobbin from her Father*

My dear Bobbin,—I know your understanding, and therefore shall
not write to you, young as you are, as a child. Mrs. Oakes writes me
from Smith that Dr. T. ordered you physic which you have not taken,
at the same time that he does not at all like your case. Now this is
a very serious business for your health, and consequently it makes
me very uneasy. You are extremely weak by your own account, and
steel is to strengthen. I gave it with my own hand to my father for a
year, and with great effect; why you should doubt the efficacy of anything prescribed by so great a physician is more than I can understand; as to ill tastes, it is beneath common sense to listen to anything of the sort.[27]

But, my dear Bobbin, you ought to bring some circumstances to
your recollection; the expense I have been at is more than I can
afford, and I am now paying your school the same as if present.
It is surely incumbent on you to consider, that when a father is
doing everything upon earth for your good, yet you ought from feelings of gratitude and generosity to do all you can for yourself.
I ask nothing but what another would positively insist on, and would
order violent means of securing obedience; I, on the contrary, rely
on your own feelings and your good sense, and so relying, I do beg

---

[26] For the letter from the French government, *cf. Annals* **29**: pp. 473–477.
Dated June 2.

[27] The friendly relations between Young and Mrs. Oakes are here apparent.

that you will take everything ordered without murmur or hesitation, for I assure you it is with astonishment I hear that you have omitted this some time. Call your understanding to your aid, and ask yourself what you can think must be my surprise at hearing that while all around you are anxious for your health, that you alone will be careless of it. It is a much worse thing than ill health, for I had rather hear you were worse in body than that you had a malady in your heart or head. Think seriously of such conduct, and I am confident it will cease, for I know your disposition, and that makes me the more surprised. . . . I am sure I shall hear . . . that you are acting worthy of yourself; and having so much patience in your illness, you will show it in this, as in so many other things. (The first cheap lobsters I shall send you three by mail, the weather being hot.) I think you are not strong enough to ride a dicky alone. Surely double-horse would be better, but if you have tried you must be able to judge. . . .

God bless you, my dear girl. I talk of your physician and your physic, but God forbid you trusted to either without asking His blessing regularly. You tell me that you always say your prayers; you cannot deceive God, and I hope you have a reliance on His blessing, which you cannot have if you do not ask it, and gain the habit of asking it.

VIII. *Bobbin to her Father*

My dear Papa,—I received your letter yesterday. Thank you for your advice; I had taken the *steel* and *draughts* long before I received it, besides which I take some more stuff . . . and ask him likewise how long the steel, &c., must be taken before you feel any effect from it, for one might take physic for ever without receiving any benefit. Let not my giving you my opinion make you think that I do not take mine regularly; I assure you I do. My dear papa, how can you imagine that I should ever neglect my prayers? No! believe me, I know my duty too well for that. I believe *once,* the last time I was at the cottage, when I was too weak to say them out of bed, I then said them when Betty brought the asses' milk. One morning I fell asleep and forgot them; I thought of it at night, and told her to remind me of them, which she did—this she can tell you. I thank you for some fine cod and lobsters, which came very fresh and good.

I am very much the same as when I wrote last, the cough very troublesome still. . . . Adieu, my dear papa. . . .

IX. *His Reply*

My dearest Bobbin,—The moment your letter came I went to Dr. Turton, but he was, as I feared he would be, out. However, I shall call on him this evening and send you his answer by to-morrow night's mail, if he is not gone to Kent. . . . If so, I cannot see him till Tuesday, as I shall be Sunday and Monday at the Duke of Bedford's. . . . I am of opinion that you should leave off steel till you have my answer. . . . It is to give you strength. . . . You are a very good girl for having taken it, and equally so for saying your prayers. Always preserve the habit of doing so. God protect and bless you!

. . . . Great expectations of a peace. Tell Mary, St. Paul's would come to B [Bradfield] as soon as Dr. T., but her thought was a good one. . . .

X. *Bobbin from her Father*

My dear Bobbin,—As I desired you to write to me twice a week I expected a letter yesterday, but hope when I go by and by to the Board that I shall find one from you, for I am very anxious to hear how you do, and what Dr. Turton has ordered Smith to do for you. I had a very disagreeable journey to town, and did not sleep a minute, but, thank God, did not take cold; went to bed next night at nine o'clock and recovered the fatigue. The Macklins and Kedington were in a good deal of rain. They have been to two plays; I go to none. K. lives with us, and I am to charge him what he costs. They will go somewhere every night. I order everything just as usual before I go to the Board, and though I am to pay no more than my common expenses when alone, yet, as I necessarily live much better, I think it but fair to be quite economical, and we have a great deal of pleasant laugh at my pinching them, and not permitting their being extravagant. I allow no scrap of a supper which they make a rout about, for they come hungry as hounds from the play, and drink porter when they can get it; the wine I lock up, and have been twice in bed when they return. If you saw them devour at breakfast you would laugh; K. who is to be with the Oakes' when they come . . . threatens that he will give us nothing but potatoes.[28]

The news you see. It is said that there will be mutinies in the army as soon as the camps are formed; if so, and no immense army of property to awe them, the very worst of consequences may be expected. Ireland is in a dreadful state of alarm and apprehension—in a word, everything wears a threatening appearance, and nothing but the greatest wisdom and prudence can save us.

[28] The Macklins were also Young's neighbors. He was a clergyman, she a sister of Mrs. Oakes, née Sophia Plampin. Apparently Young's quarters in Jermyn Street were large enough to accommodate both Kedington and the Macklins.

I hope you have got rid of your lameness, and use your legs much more than you did when I was with you. Pray, my dear Bobbin, exert yourself, and take much air and exercise on the Stone Walk, which will do in all weathers except rain. . . .

XI. *Bobbin's Reply*

My dear Papa,—I received your letter this morning. I am sorry you had not a pleasant journey. Every day since you went we have had nothing but rain all day . . . long, so I have not been able to stir out, only in the chaise. I am much the same as when I saw you, but hope that when we have fine weather I shall get better. My leg *is* a great deal better. Mr. Smith advises porter, the beer is so new. . . . If you like to send a quarter cask my mother will pay the carriage; she has no opinion of Bury porter. . . . What terrible news you write me in your letter. I really have nothing more to tell you; write soon. . . .[29]

N. B. Papa, you said you would send me some red wine, as there is none drank here; he speaks very much against my drinking so much water without red wine in it, because my ankles swell so much.

On June 11 Young went to the Duke of Bedford's sheepshearing at Woburn and did not get back to London until June 16. The next day he received such an alarming letter that he left London on June 19 and went directly to Capel Lofft's seat at Troston Hall where Bobbin had been taken some days earlier for a change of air.

Good God! what a situation I found her in, worse than I had conceived possible in so short a time, so helpless and immovable as to be carried from a chair to her bed, evidently in one of those cruel consumptive cases which flatter by some favorable symptoms, yet with fatal ones that almost deprive hope.[30]

In less than four weeks Bobbin was dead. Certainly her father did all that he could to save her. The eminent Dr. W. H. Wollaston of Bury was consulted, who gave little hope, but recommended a milk and vegetable diet. Another doctor advised exactly the contrary, an egg and meat diet. The famous Cambridge scientist, Dr. Martyn, suggested the inhaling of ether. For more than two weeks Young remained with Bobbin at Troston, and it is pleasant

---

[29] This letter and the preceding imply that Young visited Bradfield early in June.

[30] Young, 1898: p. 277.

to record that Capel Lofft thus opened his very lovely Elizabethan home to the sick girl and her father, although Young and Lofft disagreed upon almost every public question.

After about two weeks at Troston Young decided on another change of air and engaged a house near Boston near the sea. They left Troston on July 5 and reached Boston on July 11. The poor child, worn out by this trip which should never have been attempted, died three days after reaching Boston. Her father, mother, and sister Mary were all with her at the last. Young's touching account of her death and his evaluation of her character must be quoted very nearly in full.

My poor child breathed her last at twelve minutes past one o'clock on Friday morning, the 14th. I was on my knees at her bedside in great agony of mind. She looked at me and said, *"Pray for me."* I assured her that I did. She replied, *"Do it now, papa,"* on which I poured aloud ejaculations to the Almighty, that He would have compassion and heal the affliction of my child. She clasped her hands together in the attitude of praying, and when I had done said, "Amen" —her last words. Thank God of His infinite mercy she expired without a groan, or her face being the least agitated; her inspirations were gradually changed from being very distressing, till they became lost in gentleness, and at the last she went off like a bird.

Thus fled one of the sweetest tempers and, for her years, one of the best understandings that I ever met with. She was a companion for mature years, for there was in her none of the childish stuff of most girls. . . . Her disposition was most affectionate, gentle, and humane; to her inferiors, full of humility, and always ready to perform acts of beneficence, thus attaching the poor by her charity, whilst she was equally courted by those above her for that fascination of manners possessing the attraction of the loadstone. Her countenance in health beamed with animation, and her dimpled cheeks smiled with the beauty of a Hebe. Dear interesting creature! Endowed, too. with a sensibility that shrunk from the gaze, her appearance in society produced.[31]

[31] *Ibid.,* pp. 279–280. About two weeks after her death Young wrote bitterly about Bobbin's medical care: "On reviewing her last illness I am filled with nothing but the most poignant regret . . . for putting so much reliance in the medical tribe, for she had the personal attendance or correspondence of five physicians and none agreed. I did for the best and spared nothing, but had she been a pauper in a village she would, I verily think, have been alive and hearty. Such are the blessings of money; it has cost me 100 1. to destroy my child. . . ." *Ibid.,* p. 287.

After a few days at Boston, Young returned to Bradfield to arrange for Bobbin's funeral. Presumably Mrs. Young and Mary followed with the remains. At midnight on July 18, Bobbin's brother, the Rev. Arthur Young, read the services for her "in a most impressive manner." Arthur Young hugged his grief to him, and so, "I buried her in my pew, fixing the coffin so that when I kneel it will be between her head and her dear heart." Besides the date of her birth and death, the inscription on her tombstone included her last words, "Pray for me papa—Now! Amen." Young seemed fearful that he might recover from the shock of his loss. He hoed her little garden "in which I have so many times seen her happy,"[32] and examined each plant and tree. He had a copy of his wife's miniature of Bobbin made for himself, "twill serve . . . as a melancholy remembrance, and, I hope, recall my mind should it ever wander from the lamented original."[33] On Sunday he went to church, "sitting over the remains of my child."[34] He determined that her room should "never be altered, but everything continued as she left it,"[35] and when her books and letters arrived, it was he who unpacked and arranged them in her room.

The only consolations for the bereaved father were the thought that she was happy in heaven and the hope that at some future time he might rejoin her there. Within a week of her death he read the Gospel of St. John, the Acts, and all the Epistles. He "devoured" all the sermons and other religious works upon which he could lay his hands. He was especially concerned with the "hideous doctrine" of the sleep of the soul until the Last Judgment, for he wanted to believe that his darling was already happy. He wrote his friend, Charles Cole, asking him to consult the famous Dr. Jacob Bryant on the question.[36] When his old friends, John Symonds and Jonathan Carter, spent the day with him, he discussed this question with them.

He was desperately anxious that the remainder of his life might atone for his past sins and religious negligence, so that he might join Bobbin at his death, and he wished to lead such a life as

---

[32] *Ibid.*, p. 281. In 1938 Bobbin's stone and those to Mary Young, the Rev. Arthur Young, and Mrs. Martha Young were all on the wall in the very small vestry to the left of the nave and just behind the pulpit.

[33] *Ibid.*, p. 286. For reproduction of this miniature, *cf. ibid.*, opp. p. 265.

[34] *Ibid.*, p. 284.

[35] *Ibid.*, p. 282.

[36] Add. MSS. 35,127, f. 428, with Bryant's answer to Cole enclosed.

might not pain her. His views were expressed somewhat amusingly in his diary for August 2:

Can I then hope, by dedicating the rest of my life here to God, to join
my dear child hereafter, my mother, my other daughter, and my
sister; and should it so please the Almighty in His mercy, my father
and brother? Of the females I can have little doubt, or rather none.
I know too little of the lives of the others to venture to pronounce.

A strange commentary on the clergymen in his family! His ideas
were continued in the diary for the following day:

No! she lives, and as there is reason to believe, the departed spirits
have some knowledge of what passes here. What a call is it to conduct
myself so as to give no pain to her! Let me imagine myself for ever
seen by the spirits of my mother and my child. Let me have a keen
feeling of the pain any unworthy action or impure thought would
give to them, and of the pleasure they would reap from seeing
the reverse. . . .[37]

Since Young had given very little serious attention to religion, he found it difficult even to understand much of what
he read. Dr. Henry More "gets so high in the region of fancy,
and is so full of jargon and supposition . . . that I am disgusted with his farrago; . . . so much on witches, apparitions,
&c., as to be mere rubbish." Likewise his patience was exhausted
by Sherlock on immortality, "the verbiage is such that it sickens
one."[38] Unquestionably the book which contributed the most
powerfully to his conversion was Wilberforce's *Practical View*.
After reading it, he wrote Symonds for his opinion, but Symonds
was far from enthusiastic:

I have not read Wilberforce's "Practical View of Christianity," nor
am I indeed much solicitous about it. . . . W. is a strict Calvinist,
and is therefore orthodox, for he is supported by our Articles of
Religion. I who think that the Articles on this head are not founded
on Scripture, am a heretic, as I take you to be also.[39]

Young's impressions of Wilberforce were quite different:

I read it coldly at first, but advanced with more attention. It brought

[37] Young, 1898: pp. 293–294.
[38] *Ibid.*, pp. 291–292.
[39] *Ibid.*, p. 307. Symonds's long letter is dated June 8, a month before Bobbin's
death.

me to a better sense of my dangerous state; but I was much involved in hesitation and doubt, and was very far from understanding the doctrinal part of the book. This was well, for it induced me to read it again and again, and it made so much impression upon me that I scarcely knew how to lay it aside. It excited a very insufficient degree of repentance, and a still more insufficient view of my interest in the Great Physician of souls.[40]

By chance Young received a letter from Wilberforce on some agricultural subject, while he was reading Wilberforce's book for the fourth time. Young's reply raised certain religious questions with Wilberforce who answered very cordially:

I must have been utterly devoid of feeling, if I could have perused your obliging letter without lively emotions of mind. . . . I hope however that I can truly declare that the feeling which *abides* in my mind, is a feeling of joy to hear, that a Person, such as you are in all respects, has . . . been turned to seek for happiness where alone it is to be found. . . . But it may justly be productive of more than ordinary satisfaction, to perceive this change in anyone, who by talents, or energy of mind, is qualified to be eminently useful in the cause of God—My dear Sir the freedom with which you have spoken to me, encourages me . . . to address you with the same unreserved frankness. I beg you to ask me any questions you please, & to make use of me in any way you can.[41]

There followed an invitation to visit him, and thus began a most intimate friendship. There can be little doubt that the kind of Christian which Arthur Young became resulted first and foremost from Wilberforce's influence.

Young's friends naturally rallied to him in his bereavement. Samuel Hoole, Bessy's widower, argued that those who die young escape many of the miseries of maturity.[42] Richard Valpy urged Young to believe that Bobbin's death was the will of Divine Providence, best for her and best for him.[43] Two beautiful letters from Charles N. Cole were marked by deep religious feeling and a tender friendliness. In the second letter, replying to Young's expressed fear that Bobbin's death might leave no permanent imprint upon his own life, Cole wrote one very signifi-

[40] *Ibid.*, pp. 287–288.
[41] Add. MSS. 35,127, f. 438.
[42] *Ibid.*, f. 424. This letter is undated.
[43] *Ibid.*, f. 434. Dated August 5.

cant sentence which can hardly be dismissed as mere flattery: "But consider—you are a moral man, and have always been so—you have no favourite vice—you have not been used to live with those that have."[44]

His neighbors called in friendly fashion. Capel Lofft spent July 30 at Bradfield, and they talked religion: ". . . his conversation is ready on any subject, and mine led to serious ones, which he seems to like. We had much that was metaphysical on the soul . . . and a future state, etc."[45] An anonymous Mrs. —— called, perhaps Betsy Oakes, "to persuade me into company for regaining cheerfulness."[46] He refused all invitations to dine with his neighbors: "I could associate with nobody with comfort but those whose religious acquirements could tend to strengthen my present habits."[47] His rather unhealthy morbidity is shown by his comments on a visitor on August 1:

At night a Dane came, recommended by Sir J. Sinclair. Unfortunate to all my feelings. I refuse dining with all my friends, and to be tormented with a trifler who can speak neither French nor English. My mind is in a state that cannot bear interruption. I love to mope alone, and reflect on my misery.[48]

Young spent August, September, and October, 1797, in Lincolnshire. The preliminary report of that county had been made by Thomas Stone, who had made himself very obnoxious to the Board by his demands for compensation. Consequently on June 2 the Finance Committee had commissioned Young to make the final report and had voted him £50 for that purpose.[49] After Bobbin's death, the impending trip was anything but attractive:

This tour hangs on my mind; nothing would suit my feelings so well as to stay here in my present melancholy gloom, reading divinity, and endeavouring so steadily to fix my mind on eternity and the hope of joining my dear child, as to work a change in my habits, my life, my conversation, and pursuits. . . . These thoughts, however, I shall try to preserve in spite of a journey. I will take the New Testa-

[44] *Ibid.*, ff. 428, 436. Dates August 4 and 13.
[45] Young, 1898: p. 291.
[46] *Ibid.*, p. 285.
[47] *Ibid.*, p. 286.
[48] *Ibid.*, pp. 292–293.
[49] Minute Books, Finance Committee, **3**, f. 132.

ment and Wilberforce with me, and read a portion every day, and spend the Sundays in a manner I have never done yet in traveling.[50]

In his introduction to the Lincolnshire survey Young also used his state of mind during the tour to excuse its shortcomings:

In such a state of mind, to act as if present evils were forgotten, to give an undivided attention to business, to prosecute inquiries with keenness, and even ardour, and to dress one's face in smiles, that the anguish at heart might not cool or turn aside the readiness of communication I every where found; this, though certainly necessary, is not an easy task.[51]

That his sorrow and his determination to lead a new life still obsessed him appears in the scanty quotations from his diary in the *Autobiography*:

August 7. To Ely. . . . View the Minster . . . and venerated the piety of former ages that raised such noble edifices in honour of God Almighty. . . . I once thought such buildings the efforts of superstition, perhaps folly! . . .

August 8 . . . To Peterborough . . . it is singular that even while I am depressed with deep melancholy at the loss I have sustained, yet unholy ideas and imaginations will intrude. Is this the devil and his powers of darkness which buffet and beset us? . . .

October 15. I have torn my heart to pieces with looking at my dear child's hair! Melancholy remains, but how precious when their owner is no more! I am to see her no more in this world. Gone for ever![52]

The journal for the tour, the basis for his Lincolnshire survey, has not survived. The survey itself is so impersonal that it is impossible to trace his route. Sir Joseph Banks had almost certainly urged his appointment and had promised to accompany him for part of the trip and to aid him with introductions. On August 19 Banks wrote to Young, in care of Sir Cecil Wray at Summer Castle, complaining of a gout so severe that "I fear also that I have no chance to undertake the Tour I wished to have taken in your company," and stating that he really could not give Young all the letters he desired, but listing half a dozen men

[50] Young, 1898: p. 296.
[51] Young, 1799: pp. ii-iii.
[52] Young, 1898: pp. 297–298.

on whom Young might call. He did promise to meet Young at Revesby on September 11.[53]

Young's disappointment at Sir Joseph's lack of aid appears very clearly in two letters written during the trip to the president of the Royal Society.[54] The first, written from Lincoln on August 22, must have crossed with Banks's letter of August 19 from which quotations have just been made:

From not finding a Lr. fro. you either here or at Grantham I am afraid my last to you miscarried. I have done pretty well . . . thus far but now I know nobody in the North nor Axholm. . . . I am sorry to be so troublesome . . . but wishing to do the best possible for the Board I am desirous of Lrs. I had none to Mr. Chaplin, but made known my object I wished more than I got there, & should have had it perhaps if I had had Lrs. tho he was very communicative by dragging for half a day; but it demands time to pick out every thing a man has in him—I go tomorrow to Tunnels (no Lr.) Reesby, after bulls—Moody of Riswin oil cake (no Lr.) Mr. Pell of Tupham cattle (no Lr.) Mr. Slater of Carlton (no Lr.) then to Gainsbro & so to Axholm there is Mr. Lyster & Mr. Johnson but I have Lrs. to neither & know not a soul. . . . I hope ys. Lr. will catch you in time for some—for you were my main dependence for ye whole Co.

Young's second letter to Banks was dated September 4 from Barton:

I this morn recd. yr. favour in which you liken me to what I have no resemblance to, a comet—I have made but one rule & yt. is to stay every where as long as I can procure intelligence, & no longer I have been above a month in ye county & have a great deal to do to finish only a part of it. . . . upon ye whole I have found people very communicative very civil & very intelligent & I have collected much more important information than I expected; or than I ever did before in an equal space of country. The Isle of Axholm is very important, I called on Mr. Johnsons but he was at Buxton Mr. Lyster, a bad accident in ye family & Mr. Stovin not at home Dr. P. a physn. so costive I cd. get 0 out of him—but I found a Suffolk parson at Haxey of whom by his means or his farmers I got much—& I followed warping to ye source & got a thorough view of it & accounts satisfactory. . . .

---

[53] Banks Correspondence, Yale University Library. Banks's handwriting is so bad that frequently one must guess at the exact word. Obviously he wrote in great haste with little attention to clarity or even grammar. For a portrait of Sir Joseph Banks, see fig. 22.
[54] *Ibid.*

I will call on all yo. have named and hope to reach Revesby after sweeping the wolds & marsh before me. . . .

I have gone only fro. Gainsbro to Barton in 9 days & you talk of a comet. Depend on it I neglect nothing but the county is so large yt. my time is much too short—and here is a wet harvest for my corn at Bradfield on 300 acres & ye farmer rambling in Lincolnshire!

Fortunately Young found Sir Joseph at Revesby. His bailiff, Mr. Parkinson, was as valuable as any one whom Young met on the trip. Sir Joseph's plantations were admirably conducted, but Young was still skeptical about how profitable they were:

There is not a clearer head in Great Britain than that of the Right Honourable Possessor of these woods; and whenever the immense extent of his respectable pursuits will permit attention to such questions of his private interest, he will doubtless reflect on the vast capital he has thus employed at an interest, to speak in the mildest terms, rather inadequate.[55]

He was tremendously impressed with Sir Joseph's system of keeping records. An iron-plated door separated an outer work room provided with desks, tables, and bookcases from an inner room with 156 filing drawers.

There is a catalogue of names and subjects, and a list of every paper in every drawer; so that whether the inquiry concerned a man, or a drainage, or an inclosure, or a farm, or a wood, the request was scarcely named before a mass of information was in a moment before me.[56]

In spite of his gout, Sir Joseph took a boat trip with Young through the great East Fen. Young had heard it said that Sir Joseph opposed all draining, enclosing, and improving of this area, "like a great bull at Revesby, ready with his horns to but [sic] at any one that meddled." Young repeated this statement to his host who replied:

Very true Sir Joseph is that bull, to repulse those who would pretend to carry the measure upon wild and ill concerted plans in spite of him; but let them come forward in the right way, and with any prospect of success, and they shall find that Revesby bull a lamb.[57]

[55] Young, 1799: p. 221.
[56] *Ibid.*, p. 20.
[57] *Ibid.*, p. 234.

Indeed Sir Joseph completely convinced Young that he had all the documents prepared for such a move, but that he would not take the initiative until a well-ordered plan had been drawn up and the money raised. Young also noted that Banks was determined not to distress the poor by throwing together his lands into larger units. Although recognizing that Sir Joseph was motivated by humanitarian considerations, Young believed that small holdings actually contributed more to human misery than happiness.

One of the high spots in Young's Lincolnshire trip was his visit to Major John Cartwright at Brothertoft Farm, about five miles west of Boston. Two years earlier the two men had been bitter antagonists over revolution abroad and reform at home. Probably Capel Lofft helped to bring them together, as evidenced by Lofft's letter to Cartwright after Young's visit:

I am glad to hear of the intercourse which has taken place between you and Mr. Arthur Young, and of your conferences and correspondence on agricultural subjects. He is greatly to be·pitied for his late loss of a daughter of every amiable and virtuous promise.

I believe his spirits were soothed by conversing with you on his favourite topic of agriculture, and it is pleasant to reflect, that two men at the widest distance in political matters, yet amicably confer and act together on this other great subject of public interest.[58]

Cartwright's biographer believed that politics were never mentioned during Young's visit,[59] and certainly Young's references to Cartwright's practices and improvements are completely friendly in tone. Since neither man had changed his political views, the whole episode does credit to both.

Thoroughly alive to the common improvements, John Cartwright was a driller and had fine herds of sheep and hogs. But he must have had some of the mechanical genius of his more famous brother, Edmund Cartwright, inventor of the power loom. He had a very fine collection of agricultural implements, some invented or improved by him and others by his bailiff, Mr. Amos. He was experimenting with a reaping machine, had invented a contrivance for covering corn stacks, and a special boat for carrying sheep on the canal between Boston and Lincoln

[58] Cartwright, 1826: 1: pp. 242–243.
[59] *Ibid.* 1: p. 230.

which passed by Brothertoft.[60] He had constructed very elaborate apparatus for the growth and manufacture of woad.[61] He wrote Young long accounts of many of these improvements and most of the plates in Young's Lincolnshire survey were contributed by Cartwright. He was also interested in social questions and advocated friendly societies to provide regular sickness benefits.[62]

Some knowledge of Young's Lincolnshire trip can be gained from articles in the *Annals*. Probably early in September, Young crossed the Humber River and spent a few days in the Holdernesse country near Beverley and at Hull,[63] where he was much interested in the vast improvement resulting from a large drainage scheme. Young philosophized about the great changes since his visit there thirty years earlier:

It is in these amazing exertions, which have added so immensely to the national territory, changing pestiferous marshes into well cultivated districts, that we are to seek the causes of that matchless prosperity which renders this country the envy of the world. Imagined, undertaken, and executed, in that confidence which every rational man feels in the GLORIOUS CONSTITUTION of this kingdom, by which property is safe, and equal protection given to all—from the peasant to the prince. *Esto perpetua.*[64]

Nearby, however, there were still large districts just as unimproved as ever, and which some of the proprietors, including the Earl of Egremont, had tried in vain to enclose. Young was indignant at such stupid opposition, for to him the remedy was obvious: "Every day's experience tells us, that a general bill of enclosure, attended with far more facility of operation than the present laws, is necessary, and ought to pass."[65]

On his way to or from Lincolnshire Young visited the Earl of Winchilsea at Burley-on-the-Hill in Rutlandshire.[66] Lord Winchilsea had long experimented on various breeds of sheep, had fine herds of North Devon cattle and Suffolk hogs, and used oxen and one-ox carts. He had the finest field of rutabaga which Young had ever seen, and experimented with potatoes and cabbages.

[60] Young, 1799: pp. 69–72, 405–406, 442–445.
[61] *Ibid.,* pp. 149–157.
[62] *Ibid.,* pp. 409–410.
[63] *Annals* **31:** pp. 113–164.
[64] *Ibid.,* p. 115.
[65] *Ibid.,* p. 120.
[66] *Ibid.* **32:** pp. 351–382.

Young found there a new type of horse-hoe which he prevailed upon Miss Finch to draw for the *Annals*. "That lady will permit me to dedicate it to her; a very small return for the pleasure I have had in her conversation."[67]

Young returned to Bradfield on October 30 and a week later went to London, accompanied by Mrs. Young and Mary, who had not gone with him in previous years. He confided to his diary: "I knew it would be a very uncomfortable plan; but to do as I would be done by made it proper."[68] The application of Christian principles to his relations with Mrs. Young shows how much in earnest he was!

Both Arthur Young, Esq., and the Rev. Arthur Young attended the Petworth cattle show in November. On November 20 Young served on a committee of five, along with Sir John Sinclair and William Marshall, to award the silver cup for the best bull. On the following day Young and his plowman from Suffolk, Thomas Bailey, each won the first prize of three guineas "for ploughing an acre, in the best manner, with least assistance, and with the fewest oxen." The five other competitors used four oxen, a plowman and a driver, while Young's plowman used only two oxen and had no driver. The plow was a swing-plow improved by Young on the principles of his old friend, John Arbuthnot. The award of the four judges, including Sir John and William Marshall, was unanimous.[69]

Young ended his account of the Petworth meeting in the *Annals* by noting that there were now three such annual meetings, at Petworth, Woburn, and Holkham, and urging other great noblemen to spread the good work in other regions. Eloquently he pointed out the great benefits from such meetings:

Men philosophically inquisitive, men of theory, and men of practice, assembled at the same table; the breeders of one sort of stock contending with the breeders of other sorts, meditating on old facts, strenuous to ascertain new ones; . . . this collision of interest, this contest of opinions . . . tend strongly to excite emulation, to awaken the torpid, to inspirit the active, and to infuse new energies into the minds of all. . . .

[67] *Ibid.*, p. 365.
[68] Young, 1898: p. 298.
[69] *Annals* **29**: pp. 505–520.

And here let me for a moment contemplate the vast effect of such annual meetings, . . . that every part of the kingdom, animated by the same principle, and inspirited by the same attentions, might see the great lord, the farmer, the breeder, the mechanician, and the philosopher, at the same table, each striving in the sphere of *their* own ideas to throw out hints, to debate propositions, to elucidate difficulties, and to bring all to the test of experiment. To slaughter —to weigh—to plough—to cart. To do this in the eye of the country; and gild the whole with the splendour of hospitality of a great fortune, a great residence, and the energy of an ample mind. This we have seen; and those who best know the human heart, will least doubt the admirable effect. Were I a poet, I would sing the merit— Were I a minister, I would reward it.[70]

If the above wasn't poetry, it was at least eloquent prose, but in his diary composed after the meeting, the new religious convert gave quite a different emphasis:

I have been a week at Petworth, an interesting, splendid, gay and cheerful week, and, as too often the case, a vain, frivolous, and impious one. Sir John Sinclair would have me on the Sunday go to Goodwood. Never a serious word, never a soul to church from that house to thank God for the numerous blessings showered down upon it, and the means of good which 60,000 l. a year confers. Yet Lord Egremont does all that could be wished as far as humanity, charity, and doing moral benefits can—but no religion. In the chapel, no worship, no hats off but my own—dreadful example to a great family and to 2,500 people in the town. I talked to Arthur, and strongly recommended to him to attend constantly and to keep himself clear from such a want of piety. He disapproved of it much; and I pray to God that yet he may not be corrupted by such evil examples, but imbibe a dislike to such want of gratitude.[71]

On Friday, November 24, William Wilberforce wrote in his diary: "A. Y. with us alone. Interesting talk, eager and vehement. I wish he were walking more softly."[72] Such a comment from a man of Wilberforce's unquestioned piety and sincerity seems to indicate that Young was already moving towards fanaticism. An undated letter from Wilberforce to Young, which may well come from about this time, shows the nature of the relationship be-

[70] *Ibid.*, pp. 519–520.
[71] Young, 1898: p. 299.
[72] Wilberforce, 1838: **2**, p. 252. For a portrait of William Wilberforce, see fig. 23.

tween the two men which was to prevail for more than twenty
years:

It has often been in my mouth & still more often in my thots for
many Days past, it is a long time since I saw Mr. A. Young, & my
dear Mrs. to whom the Remark has been made, confirmed the justice
of it. . . . Your friendly letter just received, was written about the
same time that I was saying to Mrs. W. that I had been thinking of
walking around by Hanover Square to persuade you to walk down
& dine with us—My dear Sir, we have already seen enough of each
other to know that we are neither of us people of form; & being
both of us much occupied ceremonious visits are out of the question.
But if you will knock at my door any morning & ask if I dine at
home, 4 times out of 5 the question will be answered in the affirmative & I shall at any time be glad to see or hear you will favour me
with your company. . . .[73]

Several letters in 1797 reflect Young's solicitude to obtain an
adequate living for his clergyman son. At the height of Bobbin's
illness, Young received the following very friendly letter from
Sir John Sinclair:

The season is still so favourable, that I hope your daughter's recovery
will soon take place. In the mean while, I thought it right in a friend,
to avail himself of your distress and misfortunes, in one respect, for
the purpose of endeavouring to obtain another object, you have
naturally much at heart, namely a settlement for your son, for the
attainment of which object, I have taken two chances, the first
was an application to the Chancellor, through the Secretary at War,
who expressed himself much pleased with your last Publication; and,
I hope, through that channel, that something may be expected—The
other application I made was to the Bishop of Durham, whose answer
I enclose. I am sure that he would take a pride in doing it, and that
you are sure enough of hearing from that quarter, *when an opportunity offers*. I have also hinted it to the Duke of Bedford—I hope that
you keep up your spirits, and propose soon commencing your Lincolnshire Tour. There is money reserved at the Bankers, for the payment
of your Salary on the 4th of September.

Actually the bishop's reply to Sinclair was anything but re-
assuring, for he said that his invariable rule was "never to make
a promise, or to raise an expectation," a wise rule but not one

[73] Add. MSS. 35,127, f. 451.

calculated to make Young very optimistic about aid from that quarter.[74]

After an interval Young followed up Sinclair's application to William Windham, Secretary at War. Young's letter has not survived, but the following from Windham to Young almost certainly refers to a requested living for the Rev. Arthur Young:

I did not fail immediately upon the receipt of your letter to make the application, which you wished, and shall be extremely happy, if . . . it should prove successful. I urged what I could say with great truth; that I thought your publick services (I mean particularly in respect to Political writings) as fairly to entitle you to the consideration of those who, being charged with the publick interests, are the natural patrons & guardians of persons, by whom those interests have been especially served: & certainly according to my opinions & views of things, there are few by whom that task has been undertaken, on better principles, & with better effect, than by yourself. . . .[75]

Since nothing came of these applications to government, the Youngs, both father and son, put their chief reliance upon the Earl of Egremont. The praise of the earl by the Rev. Arthur Young was so fulsome as to be disgusting. On January 27, 1797, he wrote from Glynd an open letter to the *Lewes Journal*, expatiating at considerable length upon Egremont's great benefits to British agriculture:

That active encouragement which animates the noble Lord to whatever is conducive to the improvement of mankind, is become the theme in every circle. . . . His residence at Petworth is the levee of whatever is . . . connected with the cause of the plough. . . . A character of this complexion is above all praise, superior to all merit. . . . There is nothing in the farming world which has the chance of being useful, but its merit or demerit is ascertained at Petworth, the repository of science. . . . The father of the county, the protector of the plough, the noble Lord has acquired a celebrity, as durable as it is distinguished; and whilst the guardian angel of our island speeds the British plough, and patrons like his Lordship direct it, we may laugh at the impotent fury of our foes. . . .[76]

[74] *Ibid.*, ff. 415, 417.
[75] *Ibid.*, f. 455. Windham, 1913: **2**: pp. 63–64, assumed that Young had applied for a pension for himself, but Sinclair's letter makes it clear that application was for his son.
[76] *Annals* **28**: pp. 204–206.

On February 6 he wrote an article for the *Annals*, praising Egremont for his charity to the poor: "In whatever light so truly, liberal, and disinterested a conduct is considered, we cannot help acknowledging, that it flows from the purest and most genuine principles of humanity, and Christian beneficence. . . ."[77] On March 16 he wrote an article from Bradfield on the desirability of using draught oxen instead of horses, which included the following: "If landlords held out to their tenants an encouragement of a similar nature to that which has been adopted by the Earl of Egremont in another county, we might then expect to see scarcity fly before plenty and abundance in our markets. . . ."[78] In a note to an advertisement for the plowing contest at Petworth in November, the Rev. Arthur Young added, "We find Lord Egremont pursuing his meritorious and incessant labors, in forwarding improvement throughout every branch of rural economy."[79] That Lord Egremont was not unsympathetic with the aspirations of the two Arthur Youngs is shown in an undated letter which may be of 1797:

I was in hopes to have been of some use in the business & I may still, but I have engaged to a friend of mine (whom by the bye I do not much care about but this must be secret) to ask for another thing held by Dr. Buckner which I have done but in such a manner that I expect & hope for a refusal & then I may interfere for your Son, but I have a letter from Rose to say that Mr. Pitt is always at Walmer Castle shooting and drinking (pretty employment in these times) & that he can not yet learn the destination of Dr. Buckner's preferment but shall soon & will write me word. . . . You must keep all this secret & let me know when anything fresh happens about it.[80]

There was also quite an extensive correspondence between Young and Jeremy Bentham in 1797. In the spring Bentham requested certain information about the poor laws which Young furnished readily, although apparently not very satisfactorily.[81] Early in September Bentham wrote a long letter to Young, en-

[77] *Ibid.*, p. 257.
[78] *Ibid.*, pp. 580–581.
[79] *Ibid.* **29:** p. 307, note.
[80] Add. MSS. 35,127, f. 456. This was Charles Buckner, who became bishop of Chicester in 1798.
[81] There are three brief and quite unimportant notes from Young to Bentham, dated March 21, April 12, and July, in the Bentham Papers, Add. MSS. 33,542, ff. 462, 471, 485.

closing two complicated blank tables relative to poor law relief, which he hoped the readers of the *Annals* would fill out. A few paragraphs from Bentham's letter will show its friendliness and humor:

It was but t'other day that I became master of a complete series of your Annals: accept my confession, and record my penitence. Having . . . lent to a friend—who had lent to another friend, whom we neither of us could recollect—the twenty-five or thirty numbers I had taken in . . . I postponed from time to time the completion of the series. . . . When at last shame and necessity got the better of procrastination, what a treasure of information burst upon me! No—so long as power . . . shall have left me an annual guinea in my pocket . . . not a number of the Annals shall ever be wanting to my shelves.

This waits upon you with a proof of a blank Pauper Population Table: being a Table framed for the purpose of collecting an account of the Pauper Population in as many parishes. &c. as I may be able to obtain it from. . . .

Is it worth while to give the Table this indiscriminate kind of circulation? At any rate, your Editorial Majesty will I hope be pleased graciously to grant unto me your Royal Letters—*patent* or *close,* or both, addressed to *all* your loving subjects, my fellow-correspondents; —charging and exhorting them, each in his parish— . . . to fill my Tables, and send in their contributions. . . .

You will not easily conceive—few heads, at least, but your's are qualified to conceive—the labour it has cost me to bring the two Tables to this state. As to the work at large, it will occupy two independent, though connected volumes. *Pauper Systems compared:—Pauper Management improved;*—the last the Romance, the Utopia, to which I had once occasion to allude.—Romance?—how should it be any thing *less?*—I mean to an Author's partial eyes. . . .[82]

This letter arrived at Bradfield apparently when all the Youngs were away. On October 17 Mrs. Young wrote a long and typically confused and irrelevant acknowledgment. Poor soul, she tried to help, but Jeremy Bentham must have smiled sympathetically and yet not without irritation:

Sir,

I have the satisfaction on enquiry to see all your papers appear to have come quite safe hither, at the time stated in the letter I had

[82] *Annals* **29**: pp. 393–426. For text of letter *cf.* also Young, 1898: pp. 308–311.

the honor of receiving this day from you! I & my (now) only daughter being returned last night from a *ten weeks* Tour in order to prevent if possible the mind corroding itself by an irreparable loss we have for ever to deplore! ! ! and which I fear may in some measure, have weakened Mr. Youngs efforts in a present degree, for he never sustained such deep affliction before! ! ! I, who am pretty nearly his age, find my memory much impaired, & hope dead! because the two children I have the blessing still to keep, are only painful witnesses of that sorrow which their dutiful behaviour does so much to alleviate; therefore I am highly culpable for dwelling on the past who when alive appeared to rank only as a third in affection, which all who saw her granted our plea in almost adoring her! ! pardon my egotism—my mind &c. are relaxed beyond cure!

The bailiff has opened your parcel, & acted according to the orders *therein* contained gave it to me this morning and I write this post to Mr. Y. at *Stamford Post Office Lincolnshire* to mention the business of the Poor being ye subject of yr papers. I am hurt the delay has so long happened—my son is at Brighton (or perhaps at Petworth) his father was on his way hither last Sunday, at Stamford, but had a cross call back to Hull again, he writes me word yesterday. It cannot be the least trouble for me at all times to have the pleasure of answering the enquiry of Mr. Young's friends, having only to lament my present & future inability from heart felt grief to do it every way more satisfactorily.

> I have the honour to remain
> Sir yr obedt. &c.
> M: Young

Bradfield Hall Oct. 17th, 1797
the bailiff says he forwarded your letter under cover to Mr. Fydell of Boston M.P.[83]

The confusion incident to forwarding mail prevented Young from receiving Bentham's letter until he returned to Bradfield at the end of October. His letter of apology follows:

Bradfield Oct. 31, 97

My Dear Sir
I have rarely been so provoked as ys. morn, I returned home last night and found this morning the Lr. my bailiff sent to me with above 30 others that had travelled from post to post after me & at last sent home by orders to a post-master at a Town I did not go to. I

[83] Add. MSS. 33,542, f. 488.

have been out 12 weeks and passed 1200 miles. What must you think of me—and horrid for the Annals too—the last No. only half a No. for want of Ms. I find a supply here but your valuable papers stronly [*sic*] seconded by me shall have precedence of all—and the proof sent you; you will have one in a week & all in next No.—I have sent them to press. This is all my time & harrassed mind can permit at present. A thousand thanks for yr. good opinion of a Work wch. you will contribute to make so much better—send me botany bay;—send me the other, send all yo. can.

The beginning of next week you will see me

Yrs faithfully

A. Young[84]

Bentham's letter of September 8 had referred to a manuscript entitled *Pauper Management improved,* which eventually also appeared in instalments in the *Annals,* totaling nearly three hundred pages,[85] and which applied his famous Panoptican plan to poor houses. In Bentham's own account of this work, he connected it with Pitt's poor law scheme in 1797:

I took in hand this bill. I dissected it, I proposed a succedaneum to it; this succedaneum I couched in the form of letters, addressed to Arthur Young, for proposed insertion into the Annals of Agriculture . . . Arthur Young was in a state of rapture: he presented me with 250 copies of those Nos. of his Annals in which the matter was contained.[86]

In December, 1797, the Pitt ministry introduced the so-called "Triple Assessment" tax upon houses, windows, and especially upon such luxury items as male servants, horses, carriages, dogs, and watches.[87] On December 20 Young wrote the prime minister protesting parts of the plan and urging substitute levies:

From a knowledge of country gentlemen & resident clergy, more extensive perhaps than is possessed by any other man in England, I am confident either yt. yr new tax will not be paid, or if paid do more mischief to government than double ye amount raised on differ-

[84] *Ibid.,* f. 498.
[85] *Annals* **30:** pp. 89–176, 241–296, 393–424, 457–504; **31:** pp. 33–64, 169–200, 273–288. On p. 288 further instalments were promised, but apparently this was the last.
[86] Bowring, 1843: **11:** p. 102.
[87] *Cf.* Fay, 1928: pp. 40–41; *Annual Register* 1798: pp. 186–192.

ent principles—taking it for granted therefore that it must either be reduced very greatly or that you will give it up when the collection becomes too oppressive, permit me to suggest another to make up the deficiency.

This is an excise of ¾*d*. per lb on meal, with exemption to the poor mans hog, which upon ye lowest calculation wd. produce L 1,500,000 & perhaps L 2,000,000. . . .

A stamp upon all places of public diversion, public dinners, clubs &c &c not forgetting debating societies & jacobin meetings would produce more than commonly supposed & tend to restrain that violent emigration to towns wch. the measure dreadfully threatens. At Bath they are in high spirits, their town will be crouded [*sic*]

Licenses to every sort of trader

Do. to shopmen

If tripling & quadrupling &c be persisted in, much time must be given before the first payment, or it cannot be collected. . . .

I know numbers who on an income of L 600 a yr. keep a four wheeled carriage footman & postillion; it is self evident yt. all must go except perhaps the footman, & many will for him substitute a maid.

I avoid expressions of respect and attachment because your time is precious; none feels them stronger than
            Sr.
                Your most faithful
                & obliged sert.

                    Arthur Young

If there is one principle in taxation clearer than another it is that the weight should bear proportionably light on an infinite number of points—heavily on none. I cannot calculate ye income of the country at less than 200 millions, the mass of taxation is therefore on ye whole light; but it is the multitude *of points* hitherto yt. makes it so.[88]

In comparison with 1797 the year 1798 was unimportant in Arthur Young's life, and the documentation is much less complete than for most of that decade. The first six months, however, were quite hectic. He had spent the Board's long Christmas vacation at Petworth: "A good deal of rabble, but some better."[89]

[88] Public Record Office, Chatham Papers G D 8.
[89] Young, 1898: p. 313.

Still disturbed by the proposed Triple Assessment tax, he published a brief article in the *Annals*, "On Certain Principles of Taxation," reiterating many arguments in his letter to Pitt. Most important, taxes should be apportioned "equally on all the classes of the people, proportionably to their capacity of bearing them."[90] Although in principle he believed in an income tax and denied "that there would be great oppression in forcing people to declare and register their income,"[91] he feared that merchants and manufacturers could conceal part of their income. Hence a direct income tax would in practice prove more burdensome to the landed than to the urban interests. A tax upon theaters, concerts, and all public meetings would help to equalize the burden by falling more heavily upon cities. Young's opposition to the new proposed taxes was also personal, for he estimated they would add £100 to his own tax bill. His financial embarrassment was already great for he closed the year 1797 with debts greater than his total income for the year.[92]

Young's only publication for 1798 was *An Enquiry into the State of the Public Mind amongst the Lower Classes*, a brief pamphlet which was in form a letter to William Wilberforce, dated February 25. Wilberforce's diary for March 25 reads: "Arthur Young's manuscript."[93] Wilberforce's *Practical View* had been aimed at the lack of true piety among the middle and upper classes, and Young was interested in the same phenomenon among the lower classes. Hence his pamphlet was a sort of companion piece to Wilberforce's much more influential work. The first of Young's writings to reflect his religious conversion, it combined his alarm at the danger of popular revolution with his newly born belief that both private and national salvation could be attained only through religion. Young was convinced, after covering three thousand miles in England during the past three years, and after living in London during the past three winters, that revolutionary sentiments were much more widespread among the lower classes than most upper-class Englishmen realized. Revolutionary propaganda had been carried on tirelessly and unscrupulously. "The attack is insidious, reiterated, incessant, perpetual. The best in-

[90] *Annals* **30**: p. 178.
[91] *Ibid.*, p. 181.
[92] Young, 1898: p. 313.
[93] Wilberforce, 1838: **2**: p. 273.

tentions are vilified; the best actions traduced; the wisest systems ridiculed; . . . and the freest government that ever blessed mankind, represented as the vilest of tyrannies."[94] The defenders of order had been very much less effective: "Blasphemy, sedition, treason distributed for a penny; their antidotes for a shilling, or half a crown."[95]

Young argued that revolutionary propaganda could best be countered by converting the masses to vital Christianity, "the doctrines of that truly excellent religion which exhorts to content, and to submission to the higher powers."[96] Again he wrote: "Genuine christianity is inconsistent with revolt, or with discontent in the midst of plenty. The true christian will never be a leveller; will never listen to French politics, or to French philosophy."[97] The conservatism of his social philosophy appears in another striking passage:

No society can, nor ever did exist, without the distinctions of rich and poor. Equality is a romantic phantom of the imagination, never realized, and where most talked of, as in France under the new order of things, it is least found.

Every where, while the world endures, the rich will be purchasers of the labour of the poor. To be forced to labour for life, may appear, speculatively considered, to be a moral evil; but an evil no more to be banished from political communities than sickness or death.[98]

However, Young severely criticized the upper classes and the church itself for failure to provide the poorer classes with incentives to become good Christians. The churches were built to make the rich comfortable, but contained no seats for the poor and no mats for kneeling. Many more churches should be constructed for the poor who could not build churches themselves. They should be built in the form of theaters, with the lower classes sitting in the pit, and the upper classes in the gallery or in a row of boxes. Young even declared that to build two churches for the cost of one naval vessel would be just as conducive to national security. The rural poor were neglected because the

---

[94] Young, 1798: p. 14.
[95] *Ibid.*, p. 10, note.
[96] *Ibid.*, p. 19.
[97] *Ibid.*, p. 25.
[98] *Ibid.*, p. 6.

clergy were absentees and pluralists, who too often spent the mornings in fox hunting and the evenings in dissipation:

. . . several parishes are often served by the same person, who, in order to double or treble his curacy, hurries through the service in a manner perfectly indecent; strides from the pulpit to his horse, and gallops away as if pursuing a fox. Burials and other solemnities are performed in the same manner; and reluctant will be the attendance on the sick, when six to eight miles are to be travelled. But where, under such circumstances, can be found that attention to the morals of the parishioners; that friendly advice; that adaptation of discourses to the circumstances of his people, of whom he knows nothing. . . .[99]

The year 1798 saw the fall of Sir John Sinclair from the presidency of the Board of Agriculture and the establishment of a permanent and separate home for the Board. The year began dully enough. Four meetings of the Finance Committee in January and February were adjourned for lack of a quorum.[100] Two meetings of the full board in February debated how to spend £10 which George Washington had paid for various publications and decided to bind a handsome gift set of all Board publications for him.[101]

The election of officers for the ensuing year took place on March 27. Lord Somerville received thirteen votes for president and Sir John Sinclair twelve. There exist two somewhat different accounts of this event. The first by Archdeacon William M. Sinclair was based upon Sir John's *Memoirs*:

. . . after working day and night in its interests, for five years, most laboriously and without emolument, he was suddenly displaced, through the influence of Pitt. Sinclair was exceedingly independent; and it seems that Pitt became jealous of the general popularity and influence with the landed and farming interest acquired by the indefatigable President and Founder. . . . At any rate, at the instance of Lord Chancellor Rosslyn, the bitter opponent of the General Enclosure Bill in the legal interest, Pitt set up Lord Somerville in opposition to Sinclair at the election in 1798. Somerville had, at his own request, been made a member of the Board by Sinclair, and came forward with great reluctance, and only when assured that if

[99] *Ibid.*, p. 27.
[100] Minute Books, Finance Committee, **3**, ff. 145, 147.
[101] *Ibid.* **4**, ff. 4, 5, 9.

he refused another competitor would be found, or the Board extinguished. On the day of election, the official members, who had never attended before, crowded the Board, and Sinclair was defeated by one vote. The active members of the Board, with one exception (Sir Joseph Banks, who afterwards wrote to regret the step) were staunch in their support, and some of the official members declined ministerial dictation. . . .[102]

The second account is by Young:

While Sir John Sinclair was engaged in this pursuit he . . . imagined that his indefatigable exertions, misplaced as they were, gave him a claim to the attention of Government, and, it is said, induced him to ask a peerage. But Mr. Pitt not acceding to the proposition, he next desired to be a Privy Councillor. When this second gentle request failed, he set hard to work to form a party of his own in the House of Commons in opposition to Government, which by degrees completely estranged Mr. Pitt from him; and he was, by the votes of the official members, turned out of the chair. Lord Carrington taking me to Holwood, we walked about the place . . . before Mr. Pitt came down. When he arrived, ordering a luncheon, he said he had desired Lord C. to bring me, that he might understand what members of the Board of Agriculture were proper to fill the chair.

I named Lord Egremont. "He has been applied to," rejoined Mr. Pitt, "and declined it." I then mentioned Lord Winchilsea; the same answer was returned. I named one or two more, but the minister seemed not to relish their appointment. I next said Lord Somerville, who was famous for the attention he had paid to some branches of husbandry. Mr. Pitt's reply was, "He is not quite the thing, but I doubt we must have him," and the conversation concluded with an apparent determination that Lord S. should be the man. He was accordingly elected.

Under a later date Young added:

The world gives it all to politics, but it was not caused solely by that motive; his management of the 3,000 a year was next to throwing it away, and gradually created much disgust; had his industry been under the direction of a better judgment he would have been an admirable president.[103]

---

[102] Sinclair, 1896: pp. 13–14. For further details about Sinclair's defeat for the presidency in 1798, *cf.* Mitchison 1959: pp. 41–69, and 1962: pp. 159–174.

[103] Young, 1898: pp. 315–316. Robert Smith, Lord Carrington, was a very close friend to Pitt and completely in the latter's confidence. That Egremont was Young's first choice is interesting.

Without doubt Sir John had grounds for feeling aggrieved, but Pitt's action was hardly surprising. Sir John was largely to blame for the bad state of the Board's finances. He had also been very importunate in his demands upon the Prime Minister. After being defeated for Parliament in 1796, he wrote on July 6:

I certainly would feel some concern, in relinquishing a career, to which I have dedicated so much attention. . . . But if the Government of this country think otherwise, the exertions of a single individual can be of little avail, more especially without a seat in either House of Parliament. I still flatter myself, however, that you will see the object in such a light that you will give the President of the Board of Agriculture a seat in the upper or the lower House. . . .[104]

Sinclair had always advocated parliamentary reform, in 1797 he had urged peace negotiations, and his support of the General Enclosure Bill had alienated Chancellor Rosslyn. The following letter to Pitt was hardly tactful:

It was my intention, on my arrival in London, to have subscribed to the Books opened at the Bank, for voluntary subscriptions, to have promoted the Plan as much as possible, and indeed to have suggested some hints by which it might be expanded. But since I came to Town I am informed that it is your intention, to oppose any additional parliamentary grant to the Board of Agriculture, in which case, I must devote what I should otherwise subscribe at the Bank, to carrying on the objects of that institution.[105]

Finally Pitt was tired of Sinclair's omniscience. J. H. Rose has put it aptly:

In truth the thane of Thurso had become a bore. His letters to Pitt teem with advice on foreign politics and the distillation of whisky, on new taxes and high farming, on increasing the silver coinage and checking smuggling, on manning the navy and raising corps of Fencibles. Wisdom flashing forth in these diverse forms begets distrust. Sinclair the omniscient correspondent injured Sinclair the agrarian reformer.[106]

It is difficult to evaluate Young's part in Sinclair's fall. Apparently he did not defend his superior officer when consulted

---

[104] Public Record Office, Chatham Papers, 178.
[105] *Ibid.*
[106] Rose, 1911: p. 296.

about a successor. Sinclair's kind efforts to secure preferment for
Arthur in 1797 should not be forgotten. Of course Young had
reasons for irritation with Sinclair. He was the older man and
probably knew more about agriculture, yet Sinclair had treated
him almost like a clerk rather than a secretary. It may be that
Young feared to lose his place if he attempted to defend Sinclair.
As it was, he was reelected unanimously.

At the same meeting where Sinclair was defeated, the secretary
was ordered to "look out for & provide a house for the meetings
of the Board & residence of the Secretary."[107] Young acted
promptly, and on April 6 wrote to Pitt from his new address,
32 Sackville Street.[108] There were still certain questions which
were bothering him. Would an adequate allowance be granted
for running the house? Who would pay for furnishings? Would
the chief clerk, William Cragg, also be lodged in the house? All
three questions were answered completely to Young's satisfaction.
He was given an annual allowance of £90 to run the house, "for
a Porter Housekeeper, Maid, Coals, Candles, Turnery, Glaziery,
and other Incidents. . . ."[109] and also given £300 for furnishings,
to which the Board added another £100 in 1799.[110]

Young was really worried over Cragg. His diary on April 8
read:

Crag [*sic*], the clerk, wants an apartment, and I have befriended him
with Lord Somerville . . . much against my own convenience, for the
house is not large enough; but, do as we would be done by, must be
a rule far more obeyed by me in future than formerly, and it is
more a convenience to him than an evil to me. It would be easy for
me to prevent it and time has been that I should have taken that
part; but God send me the power to follow better dictates.[111]

Still Cragg did not get the apartment, as the diary for June 23
shows:

It is an admirable house, and Mrs. Young's only apprehension was
the plan of Cragg . . . having apartments in it; but when it came
to be debated, Lord Carrington procured it entirely to me. . . .

---

[107] Minute Books, **4,** f. 36.
[108] Public Record Office, Chatham Papers G D 8.
[109] Minute Books, Finance Committee, **3,** f. 159. In 1800 this allowance was
raised to £ 105. *Cf. ibid.,* f. 245.
[110] *Ibid.,* ff. 172, 179.
[111] Young, 1898: p. 316.

Upon the whole it is an arrangement which is equal in all to 100 1. a year to me, and in comfort, saving me the trouble of thrice a year seeking lodgings, as good as a hundred more.[112]

Cragg's request that his wife be made housekeeper was also refused. Since Sinclair had apparently given him reason to expect such an arrangement, the Board added £40 to his annual salary as a recompense.[113]

The house really was "admirable," located conveniently near fashionable Piccadilly. Much of the London which mattered to Young could easily be reached on foot. The building was typically Georgian with a severe brick façade with stone facings. It consisted of a basement, ground floor, and three upper stories. The front door and vestibule were "elegant," the former with a fine fanlight, the latter with beautiful medallions which look like Wedgwood. Each of the upper floors had four windows across the front. On the first floor was the Board Room, which has been immortalized by Rowlandson's print. Although not as large as the print indicates, the room is clearly identifiable with its deep set windows and handsome ceiling decorations. There was a similar room on the second floor. The decoration throughout was in the style of Robert Adam and must have pleased Young's tastes.[114]

The new president of the Board, the fifteenth Lord Somerville, was still a young man of thirty-two and was noted for his fine Devon cattle and merino sheep.[115] Young had visited his large estates in Somersetshire in 1796. As president, Somerville's first concern was to restore order in the finances. His initial address on May 8, given at the first meeting in the new house, was both a financial statement and a program for future action. The Board was clearly in debt for £400 and in addition Sir John had conditionally promised £1600, which had not been sanctioned by the Board. Somerville claimed that the debt could be liquidated in one year, and Sir John's commitments in four additional years,

---

[112] *Ibid.*, p. 319.

[113] Minute Books, Finance Committee, **3**, f. 161.

[114] I was kindly permitted to go through the house in 1938. It was unquestionably the same house of the Rowlandson print, although many changes had been made in the arrangement of the rooms, etc. It was in the occupation of several business firms. For a copy of the Rowlandson print, see fig.14.

[115] *Cf.* Clarke, 1897: pp. 1–20.

if the Board decided to honor them. Printing costs must be drastically reduced and limited to one annual volume of communications. Most of the work on the amended county reports should be at least temporarily halted. Since most practical farmers could not read, much of the Board's printing had been wasted. Moreover the inquiries incidental to the county surveys had been somewhat resented by the farmers and had made the Board unpopular, "for, unwillingly, I must say, it has not in the country been a popular institution."[116] He suggested that the £1000 saved in publishing expenses should be spent chiefly in premiums. Once the Board's debts had been liquidated, the amount spent in paying them might well be devoted to an experimental farm. At the same meeting on May 8, thanks were voted to Sir John Sinclair for his zeal as president. Whether this motion was proposed by the new president or by some of Sir John's friends is not clear. Another vote reflected the widespread fear of French invasion: "Ordered that the Board do adjourn sine die in case of ye enemy landing or the danger of invasion being such as to induce Government to call out the volunteer corps, many members of the board, and the officers being engaged in these corps. . . ."[117] On May 18 the Committee on Expenditures established a definite budget for office expenses as follows: incidentals £25, postage £42, stationary £40, totaling £107.[118] On May 29 the Board approved in principle the president's proposals for premiums and authorized him to enquire about an experimental farm.[119]

During May and June, Arthur Young was greatly embarrassed and worried by an escapade of his son. On May 1 the Rev. Arthur, nearly thirty years old but still lacking in preferment, wrote to Gamaliel Lloyd of Bury St. Edmunds, describing a fictitious conversation with some of the prospective jurors for the famous trial of Arthur O'Connor. The damning part of the letter is in the first paragraph:

I dined yesterday with three of the jurymen of the Blackbourn hundred, who have been summoned to Maidstone to the trial of O'Connor and Co. . . . These three men are wealthy yeomen, and partizans

[116] *Annals* 31: p. 68.
[117] Minute Books, 4, f. 40.
[118] *Ibid.*, Finance Committee, 3, f. 161.
[119] *Ibid.* 4, no folio number. *Cf.* also *Annals* 31: pp. 77–78.

of the "High Court party." Now this is as it ought to be, and as they are good farmers and much in my interests, to be sure I exerted all my eloquence to convince them how absolutely necessary it is, at the present moment, for the security of the realm, that the felons should swing. . . . I urged them, by all possible means, in my power, to hang them through mercy, a memento to others. . . . These with many other arguments, I pressed, with a view that they should go into court avowedly determined in their verdict, no matter what the evidence. . . .[120]

If what he had written was true, the Rev. Arthur had bragged that he had been tampering with the jury before the trial. Lloyd was a political Radical and sent the letter to Capel Lofft, an even stronger Radical, who in turn, without consulting any of the Youngs, forwarded the letter to Radical leaders in London. Consequently the defendant's counsel read the letter in court at the beginning of the trial on May 21. Lloyd testified to the court that the handwriting was Young's. Both Justice Buller and the attorney general promised to punish Young if it was clearly proven that he was guilty of trying to influence the jurymen, as the letter implied.

The whole nasty mess broke just when Young was fully occupied with the business of the Board, its new house and its new president. It was reported in the papers, and was noted by Wilberforce in his journal, May 23, "Strange letter from Arthur Young to Gamaliel Lloyd."[121] Young's own account of the incident showed how deeply he was disturbed and to what great trouble he was put:

Last Sunday se'nnight a new scene of sorrow and vexation. Arthur sent me a foolish letter of his written to Lloyd from Dover, by way of a stupid joke, describing an ideal conversation with some of O'Connor's jurymen, to frighten Lloyd, who sent it to Lofft, and he to Walker, to Erskine, &c. It was read in court at Maidstone, and Lord Egremont told me it had an immense effect, exciting universal indignation.

The Attorney-General pledged himself to punish it. The Jacobin papers kindly assigned it to *Arthur Young*, so all believe it to be me. I had a letter contradicting sent to four papers, and have been in incessant worry ever since . . . employing Gotobed and Garrow,

---

[120] Howell, 1819: **26**: pp. 1224–1225.
[121] Wilberforce, 1838: **2**: p. 280.

and seeing Lord Egremont often on it. I sent an express to the Attorney-General . . . and another to O'Connor's counsel. All agree Lofft to be a base villain, pretending so much friendship for all the family . . . and never asking any explanation or naming it to Mrs. Y., Mary, or A. I have fretted about this affair and worried myself terribly and with reasons, for it will be the utter ruin of my son. Possibly a fine of 500 l. and two years imprisonment if he is not able to prove it to be a jest. To avoid being punished as a rascal, he must prove himself to be the greatest fool in Christendom, which he certainly is, for the letter was unquestionably a humbug.

On June 23 Young wrote further:

Arthur has been in Kent and procured nine or ten affidavits of the jurymen; those who refuse he never set eyes on till he made the application, so he has cleared himself to me, but whether it will do for the Attorney-General is another question. It is a sad business, and will be very expensive, when I can ill afford it.[122]

Reflections upon this affair appear in letters from two friends. On June 8 John Symonds wrote from Bury St. Edmunds:

. . . I am sure the imprudence of your son must have caused you great uneasiness. . . . There can be doubt that he meant to speak ludicrously; but I am afraid that this will be allowed by few except the narrow circle of his acquaintance. The letter could not possibly have fallen into so bad hands, as those of Capel Loft [sic], who is so much warped by party; and . . . often betrays a considerable want of judgment. But nothing can excuse him for sending away this letter, before he had conversed with your son about the design of it. Mrs. Young (who, as well as your son, called upon me) told me, that she heard Mrs. Loft [sic] say her husband would have acted toward a child of his own precisely as he did toward Arthur.[123]

Everything we know about Capel Lofft indicates that he would have been as good as his word. Poor Mrs. Young must have had quite a time of it also, with her husband in London. The other letter from Wilberforce on June 10 replied to Young's complaint that Wilberforce seemed indifferent to his troubles:

. . . & I actually, on first seeing your Letter to Mr. Editor, took up the defense of your son of the charge which was attempted to be

[122] Young, 1898: pp. 317–318. John Gotobed was a lawyer whom Young frequently consulted.
[123] Add. MSS. 35,128, f. 31.

fastened on him & tho' certainly I regretted so gross & misplaced an act of impudent merriment, I believed & therefore argued strongly & that with many, that it was no more.[124]

Apparently no legal action was actually taken against the Rev. Arthur, and his protestation that it was meant only as a joke was probably accepted by his father's friends. Where the joke came in is difficult to discover. At best it was a stupid one. This was the first specific incident showing the somewhat blunt if not perverted moral sense of the young man who certainly was never suited for the clergy.

Young's trip to William Pitt's estate at Holwood to discuss Sir John Sinclair's successor was not his only visit there in the spring of 1798. Early in the year Lord Carrington applied to him to get someone to drain Pitt's land:

I told him none to be had but from a distance, and at a considerable expense; that perhaps it was an easy job, and if so his own people could do it if the drains were marked out for them, and I would go and look when nobody there. Next day he came again from Pitt with thanks, and desiring me to go when he was there.

I went, and examined the land. A hill wet from springs, the cure obvious. So I am to do it for him. He and Lord Auckland and Lord Carrington walked round the place with me, and then returned to a cold dinner. . . .[125]

On April 6 Young wrote Pitt that he had marked out some drains at Holwood the previous day and that he was enclosing a diagram for further drains. The letter ended: ". . . after ye 8th of May I shall be able to attend them so often as to preclude any error and I beg you will permit me to give such attention, merely & solely as a mark of gratitude for the goodness I have already experienced at yr. hands."[126]

During the spring Young began to give dinners to poor children in memory of Bobbin, a practice continued for the remainder of his life. The following notes from a memorandum book are reproduced in the *Autobiography*:

[124] *Ibid.*, f. 33.
[125] Young, 1898: p. 314.
[126] Public Record Office, Chatham Papers, G D 8. There is also a brief note from Young to Pitt dated December 1, regarding his estimate of the number of acres of land in England and the gross produce of the land.

March.—A dinner for fifteen poor children, 11*s*. 10*d*.
Another dinner for thirty-seven children, 16*s*. 6*d*.
Another dinner for forty-seven children, 1*l*. 6*s*. 6*d*.
April.—This month seven dinners to about forty-eight children each
time.
May.—Four dinners to about forty-eight children each time.[127]

In June Young was at Woburn again for four days at the
famous sheepshearing: ". . . a very great meeting. The duke
desired me to preside at the lower end of the table, he told me
to keep Stone from it."[128] He was back at Bradfield by July 1, and
presumably spent most of the summer there, for there is no hint
of any trips. His diary notation for July 14 was a sad one: "This
day twelvemonth it pleased God to take to Himself my ever dear
and beloved child. In the evening at the Hall my wife, and self,
children, and Miss Griffith joined in prayer."[129] With Lord
Somerville's help, he had cleverly escaped from Sir John Sinclair's
proposal that he should spend the summer recess studying grasslands.[130] Young was one of the charter members when the Smithfield Cattle and Sheep Society was founded on December 17. The
Duke of Bedford presided and Lords Winchilsea and Somerville,
and John Ellman, were also present. Young was appointed to
receive subscriptions of 10*s*. 6*d*. each, to be used for annual
premiums for the best cattle and sheep, and was also made a
member of the executive committee.[131]

The year 1799 was not particularly notable for Arthur Young,
although it is quite well documented—a few pages in the *Autobiography*, some records of the Board of Agriculture, considerable
correspondence, and many short articles in the *Annals*. There
were no long trips, no great new books. The most important family event was the marriage of his son, recorded laconically in the
Autobiography: "My son this year married Miss Jane Berry,
daughter of Edward Berry, Esq. The connection arose from her
being at school with my dear Bobbin at Campden House, and

---

[127] Young, 1898: pp. 319–320.
[128] *Ibid.*, p. 318. The reference is probably to Thomas Stone, one of Young's
worst professional enemies.
[129] *Ibid.*, p. 319.
[130] Copy Letter Book, f. 282. The proposal indicates that Sir John did not sulk
after his defeat.
[131] *Annals* **32**: pp. 208–210.

afterwards visiting us at Bradfield."[132] Arthur Young was very devoted to his daughter-in-law, who became for many years a sort of substitute for Bobbin.

The *Autobiography* makes clear that religion was becoming the dominant force in his life:

April 1799.—In London. I am alone, therefore at peace. I rise at four or five o'clock and go to bed at nine to ten P.M.

I have no pleasures, and wish for none, saving that comfort which religion gives me. . . . I go to no amusements, and read some Scripture every day; never lay aside my good books but for business. I have dined out but little, and wish for no more than I have. . . .

May 4 . . . I went to the opera with Mrs. Oakes; and that amusement which had for so many years been my delight, I met so coldly as to be almost asleep through much of the performance. What a change had taken place in my mind! This was the last public diversion at which I have been present.[133]

Such a change from former tastes and manner of life was more than merely growing old. John Symonds could not forbear taking a dig at Young's trip to the opera: "Yesterday I saw Mrs. O. . . . So you went to the Play and Opera! Alas! are you no longer Wilberforced? But it seems you made some amends for the impropriety of your conduct by falling asleep at the Play."[134]

Religion also brought Young in 1799 quite close to his neighbor, the Duke of Grafton at Euston. Although the Duke had earlier been a notorious profligate, he was now deeply religious. Early in 1799 he had lost a daughter, and Young's letter of condolence opened a correspondence along theological lines. Grafton was a Unitarian, while Young, under Wilberforce's in-

---

[132] Young, 1898: p. 323. *Cf. Gentleman's Magazine* **69** (1799) p. 620: "Rev. Arthur Young, son of the Secretary to the Board of Agriculture, to Miss Griffiths, niece of Edward Berry, esq. of Elsworth park, co. Gloucester." Young, 1898, makes two references to a Miss Griffiths. On p. 312 he stated: "Miss Griffiths, the friend and mother of my ever dear Bobbin at school, coming to Board with Mrs. Y." On p. 319 Miss Griffith [*sic*] joined the family prayers at the first anniversary of Bobbin's death. There is a mystery here, but almost certainly Miss Griffiths and Miss Berry are one and the same person. The most likely explanation is that Jane was illegitimate, going under her mother's name, but that in the statement above quoted in the *Autobiography* her father-in-law discarded the fiction that she was Berry's niece.

[133] Young, 1898: p. 320.

[134] Add. MSS. 35,128, f. 119.

fluence, was a strict Calvinist. Moreover, Young defended the doctrine of original sin which the Duke rejected. On March 29 the Duke commented upon a letter from Wilberforce which Young had passed on to him:

I wish Mr. Wilberforce and myself were agreed upon all points as we are on the (I fear) hopeless attempts to abolish totally the slave trade. Depend upon it that no one who knows that gentleman so little honours him more than myself. . . . I believe him to be an upright, sincerely pious and beneficent character, treading a road that leads to future happiness, even though he be under great but involuntary errors. . . .[135]

Although Young felt that the Duke was in grave error himself, he could only respect Grafton's character in his old age.

Quantitatively the documentation for 1799 is predominantly on Young the agriculturist. The Board of Agriculture went along more circumspectly under the new president. When Sir John asked the Board to print the Dumfries report, the Finance Committee recommended "that no more printing be undertaken, till the Board is out of Debt, with the Exception of the Annual Volume of Communications."[136] At one meeting two members were thanked for a handsome mirror for the Board room,[137] and the Board for the first time awarded its gold medal to the famous Mr. Ducket, "as a mark of the esteem of the Board for his general merit as a cultivator; and also for having presented his skim-coulter plough to the Board, and offering to fix that coulter to ploughs sent to him from any part of the kingdom free of expence." Young called the skim-coulter "perhaps the most important mechanical invention in agriculture for the last fifty years,"[138] and also reprinted in the *Annals* the two letters on Ducket's husbandry earlier contributed by George III under the pseudonym Ralph Robinson. The chief clerk, William Cragg, was suspended at another meeting on the president's complaint

---

[135] Young, 1898: pp. 325–326. For this letter *cf.* also Add. MSS. 35,128, f. 90. *Cf.* also *ibid.*, ff. 69–74, for two letters from Young and one from Grafton on theological issues. John Symonds was the Duke's very close friend and spent much time at Euston.

[136] Minute Books, Finance Committee, **3,** f. 177.

[137] *Ibid.* **5,** f. 46. This mirror may be the one in Rowlandson's print at the far end of the Board Room.

[138] *Annals* **33:** p. 60.

that "he had behaved in a disrespectful and improper manner to the President." Two weeks later Cragg made his submission and apology and was reinstated.[139]

The year 1799 also witnessed the Board's negotiations with the eccentric James Powell of Herefordshire. The whole business was unimportant, but it illustrates the sort of vexatious detail to which the secretary was subject. It is hard to follow because Powell was so emotional and his handwriting so bad. He was apparently a small farmer with a small flock of very good Ryeland sheep. He was in financial difficulties and requested the Board's aid. As early as February 23 he wrote Young:

Your kind & *Friendly* letter arrived at a moment when My Heart was making its last struggle, & I was in every sense *humbled to the dust.* . . . May the blessing of Heaven ever attend the man, who shall enable me (tis all I wish) to procure bread for him [his son].[140]

In the same letter he had offered the Board some samples of his wool, but had requested that his name be kept secret. Early in March, Young brought Powell's case before the Board, and Lord Somerville wrote him that the Board was interested in his attempt to preserve the pure Ryeland blood.[141] On April 3 Powell complained hysterically that the Board had made public his distress by implying that he would sell his sheep:

. . . *Disgrace* was in my mind, worse than *Death*—All my hopes depended on secrecy. . . . Now . . . two or three of my neighbours have this day come to my house fully Informed that I *must* part with my sheep, the news will soon be propagated—my lease . . . probably stop'd, & myself perhaps torn from my child & hurried to Jail.[142]

On May 14 Lord Somerville had flattered Powell in a speech before the Board, coupling him with the great John Ellman. Nevertheless on May 22 Powell wrote his wildest letter to Young:

I consider myself so exceedingly ill used by you & Lord Somerville in extorting from me under the mask of *Friendship* & *Humanity* my private circumstances & distress & then endeavouring in the most *ungentleman* like manner to take advantage of that distress. . . . I am determined the publick shall be made acquainted with the use

[139] Minute Books, **5,** ff. 48, 66–67.
[140] Add. MSS. 35,128, f. 80.
[141] Copy Letter Book, f. 317.
[142] Add. MSS. 35,128, f. 98.

of the Board of Agriculture, & they will best determine whether such institutions are for the *Good of the Community,* or the amusement & emolument of Lord Somerville & the *King.* . . .[143]

Since there was already considerable public and partisan criticism of the Board, Somerville and Young wished to avoid any such publication and hence Young wrote a very careful reply:

I received your Letter which I think upon the whole is a lamentable instance of error and delusion.

As to your threat of publishing the correspondence . . . you would not think of such a thing if you did not suppose it a means of making known your situation and expecting to derive some ideal advantage from it as an ill treated injured man. . . . As to Lord Somerville's humane feelings; he offered to subscribe 5 guineas—afterwards he even named 30, but the idea of a subscription was postponed till the result of the wool enquiries in the Board were finished & the various specimens compared, which business at the present moment is not compleated. . . . My own opinion was (and I think I wrote you so) that your farm would be a burthen to you unless you had a much larger capital and therefore buoyed you up with no hopes. . . .

But now Sir permit me for yr own sake in real friendship to a gentleman in distress . . . to persuade you not to think of taking so rash a step as to publish a correspondence by way of attack against persons who have done & were to the present moment doing all they could to serve you and who will continue to do it if you will permit them. You write evidently under the influence of passion . . . So many persons have been applied to here: & so many others have been witnesses to Lord Somerville's attempts to serve you with the Board that you should reflect on the appearance which it will have with the world for you to break every tie to private confidence against the very persons that have proved to so many their wish to assist you.[144]

The letter worked and Powell's reply of June 7 stated that he would "certainly not do anything to offend Lord Somerville or you."[145] The episode finished amicably enough as seen by Lord Somerville's letter to Young on August 17:

I enclose for you Mr. Powell's treatise on Ryeland sheep—you will have the goodness to ask him in what form he would have it published

[143] *Ibid.,* f. 109.
[144] *Ibid.,* ff. 111–115.
[145] *Ibid.,* f. 132.

in the annual volumes of Committee: and say that next June I shall be glad of his first choice of a ram at 5 guineas. He will inform you what he would have omitted, what inserted. I would not have any controversy with him.[146]

Whether a subscription was actually raised is not clear. But Powell had evidently saved his flock, had an article published in the Communications, and received some pecuniary aid from Somerville.

On March 15 the Finance Committee of the Board requested the secretary to collect "under respective heads, all the information that has from time to time been communicated to the Board, in the old Reports, as well as in those reprinted."[147] As a result Young published articles in the *Annals* in 1799 on the following topics: courses of crops; fallowing and extirpation of weeds; manufactures mixed with agriculture; artificial grasses; culture of grain; green crops; wastelands.[148] These articles were generally dull compilations of materials from the county reports, and only the few pages of comment by Young were interesting. The general impression given is that less progress had been made than Young had expected to find, and that the Board was therefore necessary to spread more advanced methods.

In the article on crop rotations, Young observed that the Norfolk rotation had "probably done more general good in the kingdom than almost any other circumstance of modern husbandry." On the contrary he remarked of Sussex: "The courses noted throughout this country are in general so bad that they are undeserving any other attention but to condemn them."[149] He found "that in many parts of the kingdom, the farmers are in a state of barbarous ignorance; of an ignorance so dark as in this age to be truly surprising."[150] The main objective of a proper crop rotation was to banish fallows by substituting a crop which would not exhaust the soil.

But by rejecting the heavy, useless, and barren expence of a fallow, and substituting crops for the support of cattle and sheep, the soil receives as much manuring, all given (it ought always to be so) to

[146] *Ibid.,* f. 140.
[147] Minute Books, Finance Committee, **3**, f. 181.
[148] *Annals* **32:** pp. 1–23, 217–223, 329–339, 489–510, 585–609; **33:** pp. 12–59.
[149] *Ibid.* **32:** pp. 5,7.
[150] *Ibid.,* p. 17.

the hoeing crops, that the live stock may be in a state of perpetual improvements and this I take to be the ultimate perfection of the right arrangement of the courses of crops;—by far the greatest and most important improvement of all modern husbandry.[151]

In the short article on artificial grasses—clover, ray grass, sainfoin, burnet, and tares—he commented on clover:

It is one of the most extraordinary circumstances of the backward state of agriculture in which many districts of this kingdom yet remain, that so valuable an acquisition as clover has not yet been well established in them; and it proves clearly the necessity to disseminate, in every part of the territory, the means of applying the knowledge of one county to the improvement of another.[152]

In the article on green crops Young found that turnips, cabbages, and carrots were grown only in limited areas. Such backwardness seemed to prove the necessity and utility of the Board of Agriculture, "of such an institution as HIS MAJESTY has been graciously pleased to found, for inquiring into the agriculture of Britain."[153]

The longest article was devoted to the somewhat over six million acres of wastelands in England and Wales.[154] If such lands were opened, more population could of course be supported, there would be greater demand for manufactured goods, and hence greater taxes could be collected. Young was certain that the method of improvement should be paring and burning rather than fallowing, and that such lands were unsuitable for arable crops, but could provide good permanent grasslands.

Young also wrote other agricultural articles for the *Annals* in 1799. One, accompanied by four plates, developed elaborate plans for farm buildings centered around a threshing mill. Corn stacks could be moved on iron tracks in a circle to the mill. Beyond these tracks were sheds for cattle, beyond them receptacles for manure. Within the circle of the tracks were hay stacks. Thus all the necesary buildings for feeding cattle and for collecting their manure were conveniently grouped with the threshing mill as the focus. Young complained that often farm build-

---

[151] *Ibid.*, pp. 19–20.
[152] *Ibid.*, pp. 329–330.
[153] *Ibid.*, p. 586.
[154] *Ibid.* 33: pp. 12–59.

ings seemed to have no plan at all. "On the contrary, there ought not to be the smallest convenience on a farm . . . that is not placed so precisely in the right spot, that to move it any where else would be a loss of labour or manure."[155]

Young also published on account of gross cruelty to a woman and young child in a workhouse in Framfield, Sussex. He commented very unfavorably upon workhouses in general, and the system of farming out the poor in them, and urged more generous outdoor relief for families with many children. The following quotation shows how far Young had moved from his earlier unfeeling attitude towards the poor:

The management of workhouses is, in too many cases, a horrible abuse; and when they are farmed, it is likely to become a still greater abuse. . . . The price of labour should be raised, or the weekly allowance to large families should be stated by act of parliament. . . . Something should be done to give relief that would keep the poor out of workhouses. Of all evils there is none more abominable than that of families of seven, eight, nine and ten people, without the least relief from the parish, because they will not go into a (perhaps) crowded workhouse.[156]

Some of Young's correspondence in 1799 reflected his ties with continental agriculture. Albrecht Thaer, the leading German agriculturist, wrote complimenting Young and noting that he had recently sent a German work full of references to Young.[157] Herbert Marsh, later Bishop of Peterborough and resident at this time at Leipsig, wrote Young about the invention of beet sugar, enclosed some German beet seeds, and offered to correspond with Young on German agricultural developments.[158] A later letter from Marsh revealed that Young had considered a possible tour in Germany.[159] And Count Rumford wrote that a letter from Young "has induced me to add several paragraphs to an Essay on Kitchen fire-places now in the Press. . . ."[160]

Young's only book in 1799 was his survey of Lincolnshire based

---

[155] *Ibid.*, pp. 497–498.
[156] *Ibid.* **32:** pp. 386–387.
[157] Add. MSS. 35,128, ff. 78–79.
[158] *Ibid.*, f. 204.
[159] *Annals* **33:** p. 375.
[160] Add. MSS. 35,128, f. 161. Rumford's essay on kitchen fireplaces was quite an important work, but does not mention any ideas gained from Young. *Cf.* Young, 1898: pp. 323–325, for Rumford's letter of January 8.

upon his trip of 1797. In some ways it is more satisfactory than the one on Suffolk. Lincolnshire was comparatively unknown to Young, and for that very reason perhaps he took more pains. Two reviews appeared in 1799. The *Gentleman's Magazine* was quite critical of the somewhat disjointed, ancedotal character of the survey. The reviewer admitted that there were valuable parts of the volume and concluded "that the *manner,* and not the *matter,* is the most exceptionable part of the work."[161] The more substantial and longer review in the *Monthly Review* was much more favorable: "Even into the fens of Lincolnshire we have no objection to accompany Mr. Arthur Young who, to a clear judgment, unites great perseverance, and treats even dry subjects in a manner which renders them interesting."[162] Many highlights of the survey connected with Sir Joseph Banks and Major John Cartwright have already been discussed.[163] Perhaps the most interesting and novel additional information, was Young's description of the practice of "warping," by which tidal mud was used to create new and very fertile lands.[164] It is curious that Young made absolutely no reference to Thomas Stone's preliminary survey of Lincolnshire in 1794. True, Young and other Board members had been dissatisfied with Stone's survey, else he would have been chosen to make the finished one. But there must have been some material which Young could have used, and common courtesy demanded that Young should have mentioned his predecessor. That Stone took his revenge in the following year will be noted below.[165]

June was a busy month for Young. On June 4 he attended George III's birthday review of the volunteer companies of London in Kensington Gardens. Although he presumably enjoyed this event, his diary entry was gloomy enough:

These corps owe their origin I may, without presumption, say to me, and I should in a former part of my life have been full of mortification and envy at the gay and brilliant situation of others, whilst I was a humble spectator lost in the crowd . . . but, thank God, I had no such ideas, and am more free of sin of such thoughts than I am

[161] *Gentleman's Magazine* **69** (1799): p. 322.
[162] *Monthly Review* **30** (1799): p. 55.
[163] *Cf. supra,* pp. 379–381.
[164] Young, 1799: pp. 276–288.
[165] *Cf. infra,* pp. 421–424.

from that of entering this note of it. My mind was much occupied in thinking of such multitudes of people of all ranks, all ages, from infancy to decrepitude, gay, lively, and running at the tilt of pleasure, . . . all to be in a few years in their graves, their souls in their eternal doom. . . .[166]

Of course Young was present at the Woburn sheepshearing, June 17–22.[167] Those who sat down to dinner varied from 100 to 190. The Smithfield Club held an adjourned meeting on June 17, at which Young collected £36 in dues from seventy additional subscribers. Plans were made to hold their first show in December, and Young was requested to plan for the Club dinner at the Crown and Anchor. Considerable controversy arose at Woburn over the respective merits of Leicester and South Down sheep. T. W. Coke proposed a trial of the two breeds which was refused by the Leicester men who in turn made offers refused by Coke. Young seemed to favor Coke and the South Down men, pointing out the fairly common barrenness of the Leicestershire sheep, presumably resulting from too much in-breeding, and he felt that many trials should be made, all of which would have some value and would keep the spirit of emulation alive.

Sometime late in June, Young spent a very thrilling day at Windsor. He had written to Sir Joseph Banks for permission to see the royal farm. The permission was readily granted and the royal bailiff, Frost, ordered to show Young everything:

. . . as Frost knew the King would like to see me, he went when likely to meet him on his return from a review. The Queen, Prince of Wales, the Princesses, &c., were in two sociables, and the King on horseback, with his train of lords, aides-de-camp, &c. He inquired who I was, and called me to him; rode up to the Queen, &c., and introduced me. The Queen said it was long since I was at Windsor, &c., not recollecting me at first: they passed on, and then the King rode with me over his farm for two and a half hours, talking farming, asking questions without number, and waiting for answers, and reasoning upon points he differed in. Explained his system of crops, . . . enquired about the Board, the publications of it, the "Annals," and asked if I continued to work on my "Elements," which I have been many years about; recommended me to compress the sense of

---

[166] Young, 1898: p. 321.
[167] *Annals* **33:** pp. 306–327. For a picture of the Woburn sheepshearing, see fig. 15.

quotations in short paragraphs . . . said the work would be highly useful. . . . He enquired about my farm, grasses, sheep, &c; . . . His strong land farm is in admirable order, and the crops all clean and fine. He was very desirous that I should see all. . . . I found fault with his hogs. He said I must not find fault with a present to him; the Queen was so kind as to give them from Germany, and while the intention was pleasing, we must not examine the object too critically. *"The value of the intention,* Mr. Young, is greater than a better breed." He told me he learned the principles of his farming from my books, and found them very just. Quoted particularly the "Rural Economy:" Cattle give manure, and manure corn. *"Well understood now, Sir, but not so well before you wrote."* When I said anything that struck him he turned about to tell it to the nobles that followed. He is the politest of men, keeps his hat off till every one is covered. . . .[168]

No wonder that Young wrote an account of such a flattering visit to Mrs. Oakes and to Sir Joseph Banks. One part of the Banks letter is interesting:

I little thought that he had a very strong land farm managed without an aborn [sic] of fallow, wch. highly delighted me, and I think its so interesting, and the course of crops so excellent (1 Cabs. 2 Oats 3 Beans. 4 Wheat) that I wish to have an article in the Annals on it provided his Majesty gives permission—that perhaps you could have the goodness to mention it to him.[169]

Arthur Young was much worried in 1799 over the bad crops. On April 25 he sent out a questionnaire, asking his correspondents for the temperature during the frosts, the effects of the frosts upon various crops including cattle food, the rise in prices, and the effects upon the poor.[170] In summarizing twenty-four replies he emphasized that many sheep had starved. He pointed out that the *Annals* had repeatedly urged some spring provision for sheep in addition to turnips which were apt to freeze in severe weather, and sarcastically commented, ". . . the general mass of farmers continue to pursue the practice of their grandmothers, and are much too wise to follow book husbandry."[171] He recommended a plentiful supply of rouen, the introduction of the

---

[168] Young, 1898: pp. 322–323.
[169] Banks Correspondence, Yale University Library, f. 491.
[170] *Annals* **33**: pp. 129–131.
[171] *Ibid.,* p. 401.

much hardier Swedish turnips, and the harrowing in of turnip seed upon wheat stubble.

By October, Young feared a real scarcity in wheat with resulting high prices and great suffering among the poor. Although he was quite cautious to avoid undue alarm, he stated: "In truth, these and various other causes, offer too much reason to expect the following winter to be marked by as high a price of corn as we have ever known."[172] He urged the upper classes to feed their horses on hay instead of oats: "Their horses may subsist on hay, but the poor cannot."[173] He pleaded with the rich to abandon some luxuries and to give more generously to the poor. Again he strongly supported a poor law amendment to give large families more relief in proportion to their children without having to go to the workhouse. Conditions among the poor were much worse in many regions than commonly supposed, food prices were terribly high and spinning wages much too low.

And I well know many cottages, in which the bedding is miserably bad, the father, mother, and two or three children, in one bed, and four children in another, which does not deserve the name; straw instead of a bed:—no blankets, no sheets, or only rags, and nothing to cover them but the clothes they wear in the day, and these so bad as scarcely to hide their nakedness.[174]

On November 28 Young addressed a remarkable letter to William Pitt:

In the situation I hold at the Board of Agriculture in the absence of the President & . . . a vacation of the Board, I consider it as my duty to give you such information as I am able relative to the most alarming prospect of scarcity I have witnessed in forty years.

The intelligence I have received from various parts of the Kingdom makes the crop deficient a third below a mean produce, & ye quality of every sort of grain worse than ever known. The stock in hand from the lateness of harvest, very little; and the seed now put in & putting in so late that an early harvest next summer impossible. The assistance parochially given to the poor in Suffolk (whence I am lately come, & have had as an acting magistrate means of extensive knowledge) has been by selling flour at half price. . . . ensuring a

---

[172] *Ibid.*, p. 622. This article, pp. 621–629, is dated October 11 from Bradfield.
[173] *Ibid.*, p. 624. For similar views on horse food in 1795, *cf. supra*, p. 339.
[174] *Ibid.*, p. 627.

large consumption of wheat and I find that this is practised in other parts of the Kingdom. These circumstances will have an effect wch. is much to be dreaded.

At the same time that wheat is so dear lean cattle & sheep are so cheap as to be unsaleable; such make good soup and could 2 or 3 million of women & children be taken from the consumption of flour & turned to that of meat with rice & other vegetables, not at all impracticable the difference it would make would be material.

By the first clause page 2 of your poor bill (as amended by the Coms) you gave an allowance of 1s./per week per child beyond 2. I know hundreds of poor families with 6, 7, & even 8 & 9 children, young & at home who have not more than 6d a head for the children now that wheat is £5 a qr, yet we (the magistrates) have found the consumption very generally to be ½ a stone (14 lb ye stone) of wheat flour or 2s. per head all round for parents & children of all ages per week. Judge then the inconceivable distress they suffer; but where there are workhouses we have no power to interfere. . . .
I forbear to take up your time on points wch. you are probably well informed, but if I had the honour of a few minutes conference with you I could further explain the hardships the poor suffer & your talents would doubtless find a remedy. . . .[175]

Only Young's religious conversion can explain a letter so different from his earlier views that the poor were the sole authors of their own misery.

---

[175] Public Record Office, Chatham Papers G D 8. Young's reference to the lack of Board meetings is borne out by Clarke, 1897: pp. 8–9. "Soon after this his health gave way, and his attendance at the meetings of the Board ceased for a considerable period, his last appearance . . . being on June 25, 1799. Presumably from this cause, the meetings of the Board ceased too, and it was not until January 21, 1800, that a quorum was again formed." The reference to "your poor bill" refers to the poor law reform advanced by Pitt in 1796 which proposed that the poor be given an allowance for all children more than two, but which never came to a vote. *Cf.* Webb, 1929: Part II, 1: pp. 34–39. Young was now backing a similar proposal.

VIII. The Friend of the Poor, 1800-1804

THE EARLY YEARS of the new century witnessed no decided change in Arthur Young's life. They were very busy and are well documented. He continued to edit the *Annals* and still contributed a large proportion from his own pen. The Board of Agriculture consumed an increasing amount of Young's time and energy. During three of these summers he made extended journeys for the Board which resulted in an important pamphlet in 1801 and the corrected surveys of Hertfordshire and Norfolk in 1804. The Board was under considerable criticism during this period, and its secretary came in for more than his share. Friction between Young and Lord Carrington, president of the Board from 1800 to 1803, constituted another of Young's worries.

Neither were the developments in Young's private life revolutionary. An extensive but not very lucrative Irish living was finally found for Arthur in 1801 through Lord Egremont's aid, but Arthur did not reside in Ireland. During these years Young's faith was both deepened and narrowed as he became more closely associated with the Evangelicals. He became a regular attendant at the Lock Chapel under Thomas Scott and his successor Thomas Fry, and ardently admired the great Cambridge Evangelical, Charles Simeon. As he grew more unhappy in company not deeply religious, even the Woburn sheep-shearings became increasingly distasteful.

Perhaps the most distinguishing feature of these years in Young's life was his increasing interest in the welfare of the poor. The change of heart certainly resulted primarily from his conversion. Arthur Young the Evangelical was also Arthur Young the humanitarian. Although the change began before 1800, nevertheless in these early years of the new century his writing seemed to focus, as at no other time in his career, upon the problems of the poor, with many detailed proposals to ameliorate their conditions. The change was especially notable in his attitudes towards enclosures, which he had always advocated because they opened wastelands

415

and led to improved techniques. He continued to favor enclosures, but now he insisted that the interests of the poor *must* be safeguarded in the process.

The last few pages of the preceding chapter revealed Young's concern with the bad harvest of 1799. The year 1800 was one of unparalleled scarcity and resulting high prices. Young wrote at least ten articles in the *Annals* on the crisis and prepared two questionnaires for his correspondents whose replies also appeared in the *Annals*. Moreover, the *Annals* reprinted many reports of parliamentary committees studying the scarcity. The activities of the Board of Agriculture reveal its natural interest in the problem. Early in the year Young published *The Question of Scarcity Plainly Stated*, a pamphlet severely criticized in some quarters. In the summer he went on a long trip at the Board's expense to study the closely related subjects of wastelands and enclosures. This trip resulted in 1801 in another important pamphlet, *An Inquiry into the Propriety of Applying Wastes to the Better Maintenance and Support of the Poor.*

Young was thrown directly into controversy by his first pamphlet, the complete title of which was *The Question of Scarcity Plainly Stated, and Remedies considered. With Observations on Permanent Measures to keep Wheat at a more Regular Price.* The first part of the pamphlet attempted to analyze the situation, based upon reports from his many correspondents and upon information coming to the Board of Agriculture. He maintained that the crop of 1799 was deficient by about a third while the supply on hand was certainly not greater than usual, and thus disagreed with those who estimated the deficiency at only a quarter which was pretty well compensated for by an unusually large stock in hand. Young claimed that such a view denied the reality of the scarcity and hence blamed the high prices chiefly upon artificial manipulation by farmers and those engaged in the corn trade. He summarized the two positions:

Mark the difference of our language . . . mine has been, *the deficiency is great and real; the stock in hand small; a very high price the necessary consequence—the evil is from God.* . . . What has their's been? *No; the deficiency was small; the stock in hand great; there is plenty of corn, but they will not let you have it.*[1]

[1] Young, 1800: p. 55. This pamphlet contained 100 pages and sold for 2 *s.*

The second half of the pamphlet discussed both short-term and long-term remedies. As short-term remedies, Young urged premiums to extend potato culture, prohibitions upon feeding oats to horses kept for pleasure, and small allotments to the poor from the wastelands for potatoes and a cow. He devoted much more attention to the long-term remedies. The government should ascertain corn prices more exactly and take a census every five or ten years both of population and the acreage devoted to wheat and rye. He pointed out that he had urged a census nearly thirty years before and stated that such a numbering of the people "becomes more and more necessary."[2] Poor relief consisting only of potatoes, rice, and soup would gradually wean away the poor from an excessive reliance on bread. In future crises Parliament should obtain its information from the Board of Agriculture, instead of from public hearings of parliamentary committees, which were bound to intensify the alarm.

Most important of all permanent remedies were a general enclosure act and allotments to the poor. A general enclosure act would greatly reduce the unreasonably high costs of enclosure under private acts. Under past enclosures "the heavy rich deep soils that have been constantly yielding wheat under a low rent, are inclosed and converted to grazing land under double or treble the rent."[3] Enclosure to date had thus resulted in growing less grain, since only the rich lands suitable for grass could afford the costs of private bills, while at the same time the poorer dry lands well adapted to corn had been left waste. If such wastelands could be brought into grain cultivation through a cheap process of enclosure, England could still feed herself without corn imports.

Equally important was a measure by which "every country labourer in the kingdom, that had three children and upwards" should be provided with "half an acre of land for potatoes, and grass enough to feed one or two cows." The land was to be taken from the commons and the cow provided by the parish. The laborer was to repay the value of both land and cow in gradual easy payments. The plan could be administered by parish officers, county justices, and a county inspector appointed by the London government and perhaps responsible to the Board of Agriculture.

[2] *Ibid.*, p. 83.
[3] *Ibid.*, p. 74.

The great object is, by means of milk and potatoes to take the mass of the country poor from the consumption of wheat, and to give them substitutes equally wholesome and nourishing, and as independent of scarcities natural or artificial, as the providence of the Almighty will permit.[4]

Young ended his pamphlet on a strongly religious note:

Proud in wealth, and gorged with prosperity; flourishing amidst a general misery, and rearing aloft the banners of victory and security; . . . with peace at home, while the nations of Europe lie in ruin around us; have we been grateful to Heaven for such mighty blessings? Irreligion, luxury, extravagance, and perpetual dissipation, mark too many in the higher walks of life; while profligacy, idleness, immorality, vice, and depredation, the sure effects of neglected education, prey amongst the lower classes. Are our poor provided for in the manner they ought to be, in a kingdom that expends forty millions per annum? . . . It may not be in the councils of the Almighty, that this nation should be conquered by foreign arms, or destroyed by internal commotion; but it evidently is HIS will that it should be chastized; or the punishment we feel at present would not have taken place.[5]

In an article in the *Annals* he painted a dark picture of the sufferings of the poor and again appealed for their relief on religious grounds:

It is disgraceful to a Christian country to have our poor in the situation I have described. With the commerce and wealth of the world in our hands, our cottagers are miserable; their wives and children half starved and naked, without bedding, without fuel, unless stolen, and in many places inhabiting buildings, or rather ruins, which keep out neither wind nor rain. Can this meet the approbation of that Divine Being who has hitherto showered down such plenteous blessings on this favoured and for every class except the lowest, most happy country?[6]

Late in January or early in February, Young testified before a Commons committee, attacking the practice of selling the poor grain at half price, which only tended to increase the consumption of wheat and hence aggravated the crisis. He also described his own experiments in providing the poor with soup, probably in 1799:

[4] *Ibid.*, pp. 77, 79.
[5] *Ibid.*, pp. 85–86.
[6] *Annals* 34: pp. 190–191.

I made for some days a copper full of soup every day; and in order to try the effect of the leanest meat that could be procured, I killed ten of the very leanest sheep there were in a flock of 500; and that effect was exceeding good. To each copper, containing 30 gallons, I put one sheep of from 25 lbs. to 30 lbs. a peck of potatoes, half a peck of onions, a peck of carrots, a peck of turnips, half a peck of pease, and 6 lbs of rice, and it made most excellent soup, which the poor relished exceedingly. . . .[7]

Of twenty-one premiums offered by the Board of Agriculture in 1800, seven reflect either the scarcity or Young's new interest in ameliorating the conditions among the poor.[8] One was offered for the best plan to amend the poor laws, a second for the most cottages for the poor with land enough for a cow, a hog, and a garden, a third for the best demonstration that the poor could keep a cow on poor land, a fourth for the best answer to the objections to a general enclosure bill, a fifth for the best proposal for preventing future scarcities, a sixth for the best cheap and comfortable cottage for the poor, and a seventh for the best substitute for leather shoes for the poor.[9] On May 27 the Board of Agriculture approved the resolutions of the Grand Jury of York, favoring the opening of wastelands and the passage of a general enclosure act, and ordered that these resolutions be printed and submitted to other Grand Juries.[10]

Young used the *Annals* to propagandize for a general enclosure act which would benefit the poor:

If the direct amelioration of their condition be not a leading object in such a measure, I confess, for one, that I do not wish to see any enclosure bill pass. . . . But the light in which this subject ought to be viewed, is, that the labouring poor have as much right to subsistence as any man can have to his land. . . . Go ahead, do something: TRY TO FEED THE PEOPLE.[11]

A later article urged the landed classes to pass resolutions favorable to a general enclosure bill, so that "the table of the House of Commons will be covered with petitions, and the

[7] *Ibid.*, p. 497.
[8] Minute Books, **5**, ff. 84, 86. The rough minute book vol. 4, contains posted sheets in his handwriting stating the premiums.
[9] *Annals* **34**: pp. 405–414.
[10] Copy Letter Book, ff. 422–426.
[11] *Annals* **35**: *pp.* 87, 88, 90.

Speaker in his chair smothered with parchment."[12] Young also
reprinted in the *Annals* George III's proclamation of December
3, urging all to reduce the consumption of bread by one third, to
abstain from the use of flour for pastry, and to restrict the use
of oats for horses. In his comment Young was especially severe
upon the feeding of oats to pleasure horses:

500,000 fat horses, crammed on the food of man, move about the
kingdom spectacles of envy to the starving poor. Abominable and
scandalous spectacles, which in times like these ought to be removed
from the view of those whose miserable children might be fed on
the corn thus saved.[13]

All Young's writing on the scarcity of 1800 naturally provoked
considerable comment and criticism. His pamphlet, *The Question
of Scarcity*, was reviewed by the *Monthly Review* and the *Gentleman's Magazine*.[14] The former sympathized with Young's proposal to give land to cottagers, but queried whether such writing
might not tend to raise prices still more. The latter was cautiously
skeptical of Young's statistics and maintained that no prohibition
upon feeding pleasure horses with oats could be enforced. Moreover the poor should receive a cow only after the full purchase
price had been paid in advance.

Young was also severely criticized by the Earl of Darnley, both
in the House of Lords and in repeated letters, some of which
Young published in the *Annals*. Basically Darnley accused Young
of grossly exaggerating the grain deficiency. Hence he had increased public hysteria, and contributed to hoarding, speculation,
and the rise of prices.[15] In his article of May 12 Young referred
to his opponents, "with my ingenious correspondent Lord Darnly
[*sic*] at their head,"[16] to which Darnley replied on August 14:

I have endeavoured to support the cause of truth, of reason, and
of common sense, in opposition to a general and mischievous panic
and clamour of scarcity, in raising of which so many eminent per-

---

[12] *Ibid.*, pp. 141–142.
[13] *Ibid.* 36: p. 197.
[14] *Monthly Review* 31: pp. 440–442; *Gentleman's Magazine* 70, 2: p. 758.
[15] *Parliamentary History* 34: pp. 1497–1498, 1503–1504; *Annals* 34: pp. 622–
635; 35: pp. 588–599. John Bligh (1767–1831), fourth Earl of Darnley, was a
member of the Smithfield Club.
[16] *Annals* 35: p. 85.

sons, with the Secretary of the Board of Agriculture at their head, so liberally contributed. . . .[17]

By autumn, however, Darnley seems to have accepted much of Young's position and even some of his remedies.[18]

Much more virulent was an anonymous pamphlet, *A Letter to the Right Honourable Lord Somerville*, by "A Society of Practical Farmers."[19] It attacked the Board of Agriculture in general, its *Communications*, and especially the secretary and his recent pamphlet, *The Question of Scarcity*. It impugned Young's evidence about the extent of the scarcity, blamed him for exonerating the speculators, and ridiculed his impractical and ineffectual remedies. It attacked the Board of Agriculture as visionary and especially singled out Young's survey of Lincolnshire and the *Communications* which was hardly more than an additional volume of the *Annals*. The *Gentleman's Magazine's* review of *A Letter to Lord Somerville* started out: "The good sense and candour of this letter entitle it, in the highest degree, to public attention." And the review ended: "Those who are convinced by his reasoning, as we confess we are, will thank him for detecting the sophisms of an heterogeneous Board, and rescuing them from the imputation of supporting a *jobb*."[20] On the other hand, while admitting that the pamphlet was able, the *Monthly Review* defended the Board from the attacks made, and declared, ". . . there appears a degree of acrimony against the Board, and particularly against its Secretary, which will probably lead to the suspicion that the pamphlet . . . partly originated in narrow and interested views."[21]

Late in 1800 there appeared an even more scurrilous attack upon Young by the surveyor, Thomas Stone, in his review of Young's corrected survey of Lincolnshire. Since he had prepared the preliminary survey of that county, Stone had undoubtedly been deeply offended that he had not been asked to write the amended report. Moreover, it must be remembered that Young had completely snubbed Stone by not once referring to his

[17] *Ibid.*, p. 589.
[18] *Ibid.* **36**: pp. 92–103.
[19] Some have attributed this pamphlet to Thomas Stone, others to Stone and Thomas Marshall jointly.
[20] *Gentleman's Magazine* **70**, 2: pp. 763–767.
[21] *Monthly Review* **33** (1800): p. 371.

preliminary report. In reviewing Stone's new book, the *Monthly Review* judiciously summarized Stone's grievances:

Mr. Stone may perhaps have been slighted; his abilities, which are certainly considerable, may have been undervalued; and his communications may not have met with the handsome treatment which they deserved: Mr. Young's appointment to the Lincolnshire Survey may have been a great mortification to Mr. Stone, considering the pains which he had taken, and the service which he had already rendered in that department; and there may have been a want of delicacy in the Secretary in entering, without sufficient apology and explanation, on ground which he knew to have been previously occupied; yet, after we have made every allowance for Mr. Stone on the score of mortification and disappointment, we cannot approve the manner in which he records his feelings.[22]

Stone started with an address to the Board of Agriculture, followed by a letter to Young. The body of the work consisted of a point by point criticism of Young's Lincolnshire survey. Stone also included a vicious attack on Lord Somerville and some of his recent writings. His attack upon Young's *The Question of Scarcity* was very brief and only a part of his general onslaught against the Board:

The world would not have believed that you were in possession of so little accurate information concerning the causes and the true state of the present scarcity and dearth of provisions; if your secretary had not been suffered to divulge your *official* ignorance and his own, in a pamphlet which respect for you, if not concern for his own ill-earned literary reputation, should have forbidden him to make public.[23]

Stone concentrated his venom in the introductory letter to Young, which well illustrates how his adversaries regarded the great agriculturist:

I do not consider it a duty incumbent upon me to begin this address with a prolix apology. You, of all men, have the least reason

----

[22] *Ibid.* 34: p. 368. The full title of Stone's work was *A Review of the corrected Agricultural Survey of Lincolnshire, by Arthur Young, Esq. published in 1799, by Authority of the Board of Agriculture; together with an Address to the Board, a Letter to its Secretary, and Remarks on the recent Publication of John Lord Somerville, and on the subject of Inclosures.* For earlier references to this affair, *cf. supra,* pp. 376, 410.

[23] Stone, 1800: p. vii.

to expect a ceremony of this kind, since you have invariably commented upon the agricultural productions of others with the most unrestrained freedom. You have, Sir, long assumed to yourself the office of a Reformer, and if your talents as an agriculturist had been equal to your industry in travelling and writing, you would long since have ranked amongst those to whom the present and future ages of the world are and will be ever indebted. . . . But, Sir, unfortunately for the country and yourself, you commenced a Reformer before you had acquired any practical knowledge of the subject upon which you had undertaken to treat. . . .

I feel no sentiments of a vindictive nature, although I suspect myself to have long been the object of your dislike. This is in some degree proved by your review of my essay on agriculture, published in 1785, as it stands recorded in the fourth volume of your Annals. . . .

If we turn our observations to Bradfield Hall Farm in Suffolk, the *seat of your improvements,* the most useful lesson of practical husbandry may be instantly learnt by adopting the contrary of your example. Were the practical husbandmen, within the circle of your fame, questioned by an inquisitive traveller respecting who was the worst farmer in the county, they would immediately answer, Arthur Young.

If you had travelled ten times more than you have done, and written ten thousand volumes upon agriculture, not having been grounded in the knowledge of practical husbandry before you set out, you would have continued only to heap theory upon theory, without adding a mite to the general stock of practical knowledge . . . . you became the dupe of every sly, artful knave who proposed to himself either interest or pleasure in misleading you. Hence the innumerable exaggerated accounts in all your tours and travels of crops that never were reaped, and improvements in cattle that never existed; and whilst I give you credit for the purity of your intentions, I cannot help remarking that, by all the practical husbandmen who have traced you in your eccentricities, and compared your writings with the true statement of facts, you have been regarded as the Munchhausen of the age. . . .

Nevertheless I must pay a just tribute of respect to a periodical work called the "Annals of Agriculture," compiled by you.

Considered as a repository of the reports and opinions of experimental farmers, it is a work of great public benefit; but merely of entertainment, where your own theories are introduced.[24]

[24] *Ibid.,* pp. 1–26.

Stone had certainly hit upon Young's weak points. His virulence must have hurt severely, but Young makes absolutely no mention of Stone in the *Autobiography*.

James Powell was still a nuisance in 1800. Late in March he wrote that Lord Somerville had promised to buy some of his ram calves for the Board. By this time Somerville was ill and had ceased attending the Board. Perhaps Young had forgotten the transaction. At any rate Cragg replied that he could discover no action committing the Board to any such expenditure.[25] There followed a blustering and almost blackmailing letter from Powell to the new president, Lord Carrington.

I have rec'd your Lordship's Favor of the 31st ultia [*sic*] and must acknowledge that your Lordship's Answer *is not* such as I think I had a *right* to expect.

I thought the words of such people as Lord Somerville & Mr. A. Young . . . were far superior to the Entries of a Book Keeper—I had long ago too much reason to believe that Lord Somerville & Mr. Young (to say no worse of it) were Triffling [*sic*] with me. . . . I had an Intention of publishing a little Treatise upon the subject . . . and being at hand to procure the assistance of a Literary Friend, who is editor of one of the morning papers, I am now determined, as soon as possible, to put my former Intentions into execution. . . .

Because I have it *officially* under the Secretary's hand, by *direction* of Lord Somerville as *Pres.* & saying, "The *Board of Agriculture will* take what rams you have left of the Hereford Breed at five Guineas each . . . and *Further,* That his *Lordship* had *no doubt* but that if you breed very fine rams in future the Board will also be ready to take eight or ten annually, till they have sufficiently dispersed the race.

Facts, My Lord, are stubborn things—and people cannot *easily* deny their handwriting.

I won't at *present* say the circumstance is actionable—but had I not relied upon the *Honor* of the *Board,* the lambs in question would have been cut long since—They are now too old.[26]

An undated manuscript note in Young's handwriting in the Young Manuscripts proves that Powell had some justice on his

[25] Copy Letter Book, f. 393.
[26] Add. MSS. 35,128, ff. 215–216.

side: "The President engaged to take for the use of the Board 5 or 10 Ram Lambs of Mr. Powell . . . & wrote to him last year to that purpose; the price was to be 5gs. each."[27] The upshot was that the Committee on Finance voted to buy six of the calves at five guineas each and pacified Powell.[28]

With Lord Somerville's illness a new president became necessary. In the balloting on March 25 Lord Carrington received 11 votes, Lord Somerville 5, and Sir John Sinclair 4.[29] Not yet fifty years old, Carrington had only recently been ennobled. A business man, he had long been bound closely to Pitt and had given him financial aid. He was certainly not an outstanding agriculturist, but he seems to have been a fairly good administrator and he had the ear of the ministry. Young's relations with the new president were not cordial, but there is no evidence of any friction in 1800. Shortly after the election Young visited Lord Carrington's estate at Wycombe in Buckinghamshire. Significantly Young gave no description of Carrington's husbandry and had to limit himself to telling how the new president had destroyed moss in his lawn by wood ashes and that he kept a flock of South Down sheep for his table. Nearby Young found a common of 4,000 acres and exclaimed:

Are these four thousand acres to be under the eye of a President of the Board of Agriculture, and remain in this state? No: I will hope for better things; his ideas are perfectly correct on this subject, and there is energy enough in his character to render ideas active and efficient. I think he will not sleep quietly in his bed till something is done. . . .[30]

Arthur Young spent the summer and much of the fall of 1800 in a long journey, the origin and purpose of which he described as follows:

In the summer, in consequence of much conversation with Lord Carrington on the importance of enclosures, I proposed to him that I should take a tour expressly for the purpose of ascertaining what the effect had really been in practice. He approved of the idea,

[27] *Ibid.*, f. 142. This memorandum is arranged with the MSS. of 1799, but the spring of 1800 seems more likely, after Powell had raised the question again.
[28] Minute Books, Finance Committee, **3**, f. 221.
[29] *Ibid.* **5**, f. 104.
[30] *Annals* **35:** *p.* 170.

and desired me to execute it; and, in regard to the expense, I told him that if he would allow 100*l.*, I would expend it in travelling, and report to him the country travelled, and the enclosures examined, and then he might extend or not the undertaking at his pleasure. He approved the plan, and I accordingly employed twenty weeks on the journey.[31]

Our knowledge of this tour from June to November, certainly one of the most important and interesting that he ever conducted, is tantalizingly scanty. The *Autobiography* includes a few excerpts from his diary. The trip resulted in 1801 in his pamphlet *An Inquiry into the Propriety of Applying Wastes to the Better Maintenance and Support of the Poor,* which gave details of some of the areas covered. Lengthy articles on enclosures in the *Annals* in 1804, 1805, and 1806 are probably based on notes taken during the trip.[32]

The fragmentary account in the *Autobiography* begins with a paean of praise to his beloved Bradfield:

Bradfield.—I never come to this place without reaping all the pleasure which any place can give me now. It is beautiful and healthy, and is endeared to me by so many recollections, melancholy ones now, alas! that I feel more here than anywhere else. Here have I lived from my infancy, here my dear mother breathed her last, here was all I knew of a sister, and the church contains the remains of my father, mother, and ever beloved child. . . . All that locality can give an interest to in this world is here—sweet Bradfield, to use an epithet of my dear mother fifty years ago. . . . the scene also of many and great sins; and of none perhaps greater than the black ingratitude of never thanking God with fervency for the blessing of such a spot till misery turned my heart to Him . . .[33]

From Bradfield he went to Woburn for the Duke of Bedford's sheepshearing. Although expected on the Sunday before the meeting, Young purposely postponed his arrival for religious reasons:

He expected me to-day, but I have more pleasure in resting, going

---

[31] Young, 1898: p. 333, footnote.
[32] *Annals* **42:** pp. 318–326, 471–502; **43:** pp. 42–59, 111–118; **44:** pp. 39–62, 174–201, 288–307; 426–432. Miss Betham-Edwards's note, *Autobiography,* p. 339: "full accounts of these tours are given in the *Annals of Agriculture,*" is unfortunately incorrect.
[33] Young, 1898: p. 329.

twice to church and eating a morsel of cold lamb at a very humble inn, than partaking of gaiety and dissipation at a great table which might as well be spread for a company of heathens as English lords and men of fashion.[34]

Young also expressed a kind of morbid satisfaction in two tragic scenes in cottages on his way to Woburn. Between Royston and Baldock he found the married daughter dead in a poor cottage where he stopped. Voicing his hope that the young woman had died a Christian, he declared that her mother ". . . seemed to feel very little. And it is the blessing of God that they do not—they cannot afford to grieve like their betters." Unfortunately upper-class Evangelicals were too often guilty of such smugness. On passing through Millbrook the following day he found a dead child at another cottage:

On a bed, which was hardly good enough for a hog, was the woman very ill and moaning; she had been lately brought to bed, and her infant was dead in a cradle by the bedside.

What a spectacle! She had four children living; one, a little girl, was at home, and putting together a few embers on the hearth. My heart sank within me at the sight of so much misery, and so dark, cold, tattered and wretched a room. Merciful God, to take the little child to Himself, rather than leave it existing in such a place.

Inquiries of several neighboring cottagers revealed a shocking insecurity.

These poor people know not by what tenure they hold their land; they say they once belonged to the duke, but that the duke had swopped them away to my lord (Lord Ossory). How little do the great know what they swop and what they receive! . . . How very trifling the repairs to render these poor families warm and comfortable! Above their gardens on one side there is a waste fern tract now enclosed, from which small additions might be given them, yet would enable them to live from their ground at least much better than at present. What have not great and rich people to answer, for not examining into the situation of their poor neighbours?[35]

From the squalor of Millbrook, Young passed on to the "wealth

[34] *Ibid.*, p. 330.
[35] *Ibid.*, pp. 331–333.

and grandeur and worldly greatness" of Woburn, "but I am sick of it as soon as I enter these splendid walls."[36] The detailed account in the *Annals* expressed none of this disillusionment.[37] The meeting was very well attended by more than eight hundred, ranging from a prince of the blood to common farmers and breeders. Many close friends were there, but also his worst enemy, Thomas Stone. On the last day both Young and Stone, along with several others, attended a meeting of the Committee of the Smithfield Society, where Young's accounts for the preceding year were duly accepted and plans made for the show in December.[38] Young's final paragraph in the *Annals* follows:

May the new century open auspiciously to the plough! May the spirit of this sheep shearing improve the flocks of Britain! May her fields smile again with ample harvests, her wastes by a GENERAL ENCLOSURE covered with cultivation, her farmers rich, her poor well fed and happy; and may we all, by reverence of THAT BEING from whom all blessings flow, endeavour to deserve them![39]

About two weeks later Young reached Kimbolton Castle in Huntingtonshire, the seat of the Duke of Manchester, who had hired a Northumberland bailiff who was introducing turnips and making many improvements. The duke was all right, but the duchess captured Young's heart: "The duchess pleases me as much or more than any woman I have met these many years. Her character in every worldly respect is most amiable. There is a native ease, simplicity, and naïveté of character in her which delights me."[40] She told him her whole story, how she played barefoot as a girl in the Scottish hills, and had never had any formal education. The only trouble was that she was not truly religious, although she went often to church and took her children there. In his notes on Kimbolton in the *Annals*, Young wrote:

If a farming traveller comes to Kimbolton, and forgets its mistress, may his sheep rot and crops blight. . . . A young duchess, ever in the

---

[36] *Ibid.*, p. 333.
[37] *Annals* **35:** pp. 225–257.
[38] *Ibid.*, pp. 561–564.
[39] *Ibid.*, pp. 256–257.
[40] Young, 1898: p. 334.

country, loving it, and free from a wish for London, a character, that if I was to give my pen scope, it would run wild. . . . Happy, happy would it be for England if such instances were to abound; if the great found their pleasures amongst the people that depend on them. . . . These are the paths where reciprocal utility, affection, and respect are found. Rents expended in the country that pays them; the farmer rich and the poor comfortable. . . . There is a natural connection between brilliant parties in town and uncultivated deserts in the country: when I see in the Morning Post long lists of London assemblies, the Dutchess of ——— routs, and Mrs. ——— parties, I conclude the country residence is well ventilated from extensive heaths of desert mountains; and that moors, bogs, and wastes spread an ample range for every animal except those that feed mankind.[41]

During the remainder of July, Young examined the effects of enclosures on the poor in Huntingtonshire, Cambridgeshire, and Norfolk. When he saw four good-looking but unmarried parson's daughters, he reflected upon the evils of fornication and whoremongers which deprived these girls of husbands. At St. Ives he met a drunken doctor of divinity who "introduced himself to me, and breathed like a puncheon of rum in my face."[42] At Thorney Abbey he gladly forgave a party of ladies who attacked him on his Lincolnshire report when he found them "good Christians" and devotees of Wilberforce and Hannah More. At Downham Market he attended church twice on the Sunday. In the morning he approved the rector's strictures against sitting rather than kneeling in prayer, but in the afternoon he was indignant that there was no sermon for a large congregation, and remarked that the growth of Methodism was no wonder.

Late in July he was at the Earl of Hardwicke's seat at Wimpole where he "beat the country well for enclosures." On July 31 a lord lieutenant's gala took place with seventy-three persons to dinner, and with "turtle, venison, and everything that could be." Young found Lord Hardwicke "very clever, has very good parts and a clear head, a man of business." In the future Young was

---

[41] *Annals* **35:** pp. 432–434. William Montagu (1767–1843), fifth Duke of Manchester, was later famous as Governor of Jamaica. The duchess was Susan Gordon, daughter of the famous Duchess of Gordon, noted as a political hostess. In spite of the duchess's character, as painted by Young, her marriage was not happy and the ducal couple not long afterwards separated.

[42] Young, 1898: p. 336.

to see much of him as president of the Board of Agriculture from 1813 to 1816, and again from 1819 to 1821. He was pleased that Hardwicke attended church twice on Sunday and held family prayers on Sunday evening. "I am glad to find a great Lord who is not ashamed of praying to God."[43]

Young was back at Bradfield for over a week in August, managing two Sundays at home. Nevertheless the interlude was not especially happy:

Mrs. Y. in great health, and when that is the case in too much irritation—God forgive her—life is a scene of worrying, time trifled with, a book never looked in, quarrels and irritation never subsiding. My daughter and daughter-in-law reading cart loads of novels.

A little later he commented on the compensations for his hard work on the road:

This first week of my second journey I have laboured very hard in my enquiries, and travelled many miles on bad roads, not finishing the day till six in the evening, and then dining and having much writing. Such a life I should earnestly wish to avoid if I had a home tolerably comfortable, but mine is so far from that description in almost every respect that I submit the better to being ever in harness.[44]

It is almost impossible to follow Young after he left Bradfield. When near Norwich he wrote very favorably of the cottagers on Mousehold Heath.[45] Among other counties certainly covered were Bedfordshire, Essex, Surrey, Sussex, and Hampshire.[46] Early in October he was near London and at Enfield was taken sick "with a violent purging colic and vomiting," attributed to drinking coffee "out of copper." Before he had regained his strength he was greatly disturbed by two letters from a lawyer of Sir Cecil Wray, which threatened to hold "me to bail on my bond to them on buying Knaresboro Forest, as Abbey of Northampton has not paid one shilling rent or interest." The business is far from clear, but the threat greatly worried Young for he did not know where to turn to get bail. He saw the hand of God in the episode. "I

[43] *Ibid.*, pp. 338–339. Philip Yorke (1757–1834), third Earl of Hardwicke.
[44] *Ibid.*, p. 339.
[45] *Annals* 36: pp. 550–566.
[46] *Cf.* Young, 1801: *passim.*

shall pray earnestly to be spared, but if it is His will, be it done, and may He grant me resignation, patience, and submission to His correction. . . . May God be appeased and spare me the affliction."[47]

Young was certainly back in London by the middle of November when the Board of Agriculture resumed its meetings. In December he was busy with the annual cattle show of the Smithfield Club, of which he was secretary, treasurer, and a member of the executive committee. Young also had to arrange for a dinner for one hundred guests at the Crown and Anchor on December 15, the charge to be 9s. 6d. to include a bottle of wine. At the banquet the Duke of Bedford, the president, made a long speech and proposed the new constitution which was duly accepted, after which "the president's health was then called for, and drank with *three times three*."[48]

In June, Young received a long letter from Jeremy Bentham,[49] inquiring about price increases since 1793. Disappointed that the Board had adjourned until November, he suggested that a subcommittee might furnish the desired information, or if such a committee meeting was impossible, that Young would make an estimate. He was very complimentary and asked, "for where else could any other equally competent opinion be obtained?" And again at the end of the letter, "But these matters are so perfectly A. B. C. to Mr. Young that I am sanguine enough to hope, if not for a definitive solution, at least for an answer with an approximation."

The documentation for Young's life in 1801 is very uneven, very complete for the spring, very scanty for the second half of the year. He spent most of the spring in London, except for the long Easter vacation at Bradfield. As usual he attended the Woburn sheepshearing in June. He was at Bradfield again in early July. In September he was gathering material for the corrected Hertfordshire survey. He published two separate works in 1801, *Letters from his Excellency George Washington to Arthur Young*,[50] and *An Inquiry into the Propriety of Applying Wastes*

[47] Young, 1898: p. 340.
[48] *Annals* **36**: p. 260.
[49] *Young*, 1898: pp. 341–344.
[50] These letters from Washington contain very little that is important for Young's biography, and have already been analysed for the years in which they were written.

to the Better Maintenance and Support of the Poor, the first a slim volume, the second a rather long pamphlet.

When Young dined with the Duke of Grafton on May 26, 1801, he was informed of some of the current criticisms of the Board in general and of its president and secretary. Grafton had told the grumblers that he could understand the criticisms of Lord Carrington, but had asked what could be the basis of the attacks upon Young. The answer was "Oh, he is careless, and does nothing." Young was naturally upset and penned a memorandum which he gave to Symonds to pass on to Grafton. On one side of the card he wrote, "Some use in rising at 4 A. M.," and on the other a summary of his work for the Board:

From January 20 to May 23 are 90 days, Sundays and vacation excluded, 50 Boards and Committees; 340 essays read, and every one commented on. Report to the House of Commons on Potatoes. Report to the Lords on Grass Lands. Enquiry into cottagers' land published, but drawn up for the Board. Memoir of Salt, from more than fifty authors. . . . Memoirs on wastes, paring, burning, and arable land.[51]

Actually the Board's records for 1801 show that most enquiries were referred to the Secretary for a report. On February 14 when the president and vice-presidents of the Board were empowered to lay out £100 for books for the new library, it was also ordered "that the Secretary do prepare a List of Books for the Consideration of the President and Vice Presidents."[52] On March 6 the Committee on Papers requested Young to make a study of more humane methods of slaughtering cattle.[53] When Sir Christopher Willoughby offered on March 3 some lands for the experimental growth of grass seeds, the Board "Ordered that the Secretary do with Sir C. Willoughby's permission, view his farm at Croyden, and fix on a field for the culture of separated Grasses, pursuant to his obliging offer."[54] On March 31 the Board voted to send a new paring plow to Bradfield for Young to test.[55] On May 22 the secretary read to the Committee on Papers "Extracts from various authors relating to salt. Ordered that he do continue to make

[51] Young, 1898: pp. 361–362.
[52] Minute Books, **5,** ff. 174–176.
[53] *Ibid.,* f. 180. *Cf.* also *ibid.,* ff. 126, 177; Copy Letter Book, f. 433.
[54] Minute Books, **5,** ff. 178–179.
[55] *Ibid.,* f. 203.

further Extracts, and report the same to the Committee on a future occasion."[56]

The Board's most time-consuming work of the spring was the consideration of the essays on grasslands. In December, 1800, Lord Carrington's committee in the House of Lords had referred to the Board the question how grasslands might best be converted to arable and then reconverted to grass without damage to the land. In turn the Board offered premiums for the best essays on the subject. Since essays were encouraged from people familiar with only one kind of land, whether clay, loam, sand, chalk, or peat, the Board was swamped with 340 essays. No wonder that Young had to rise at 4 A.M. and could find no time for his *Elements of Agriculture*. Fourteen Board members were appointed to read the essays, and the Duke of Bedford read forty during his Easter vacation. Young apparently read them all. Four major premiums ranging from £200 to £40, plus thirty-six smaller premiums were awarded. The essays had been due on February 1, and the awards were made on June 12. When the third prize of £60 went to the Rev. Arthur Young, his father must have been pleased, but some eyebrows were probably raised.[57]

The Board held a special meeting on March 27 to test Captain Hoar's claim that he could discover springs and water by means of the *Virgula divina*. After assembling at 32 Sackville Street, the Board adjourned to Hyde Park, among them Lords Carrington, Winchilsea, Egremont, and Sheffield, and Sir Joseph Banks. After receiving information from the director of the water works about the water mains, "Mr. Hoar was requested to follow the secretary who led him in the direction described across a part of the Park; across a small Enclosure at the end of the Riding House into Grosvenor Street. . . ."[58] Young and the Board were completely satisfied with the results.

There was considerable friction between Carrington and Young at this time. Late in March, Young requested Carrington to employ his son as a clerk during the vacation at Bradfield. Young commented in the *Autobiography*: ". . . he said it was mean to

---

[56] *Ibid.*, f. 223.

[57] *Ibid.*, ff. 148, 152, 239. Young, 1898: p. 353, stated number of essays as 360. *Annals* **36:** pp. 207–210, gave Board's instructions for the essays. *Cf.* also Clarke, 1898: pp. 26–29; Young, 1809: pp. 34–39.

[58] Minute Books, **5,** f. 199: Young, 1898: p. 350.

make him a clerk—but everything is wrong that is proposed to this man, even the things which, let alone, he would propose himself."[59] Despite his objections, Carrington apparently relented, for on May 28 the Committee on Papers voted £50 to the Rev. Arthur Young "for the adbridgement which he made of the Returns sent to the Committee of the House of Commons of the quantity of corn &c. grown in Parishes where Enclosures have been made. . . ."[60] Young was both surprised and pleased:

I thanked them very awkwardly, and talked of gratitude, for it came on unexpectedly; that readiness which is never at a loss I have not an atom of. My heart always speaks at a sudden; whereas in many cases the head is most wanted. But the fault was on the right side, it was more than I expected.[61]

Young was also greatly irritated by Carrington's letter which necessitated an earlier return to London from Bradfield than planned:

A nonsensical letter from Lord Carrington requiring me to go to the Treasury for the 800*l.* for the Essays, which is entirely the treasurer's business. He is as unfeeling as a log; this is a return for my being at work from 4 A. M. in the morning for ten weeks. I should once have been full of indignation and abhorrence; thank God, I am more calm. . . . But I wish he had let me alone. I am vexed, but the world is full of nothing but great miseries or teasing vexations, the more the better; they wean us from it effectually.[62]

Young's unfriendliness to his chief is shown by his comment on a piece of gossip retailed to him by his former son-in-law, Samuel Hoole:

He has heard at a great table . . . a very so-so account of Lord Carrington—fidgeting, restless, dissatisfied, ambitious, avaricious, with a mere show of parts and knowledge. He has made immensely by the loan; and the richer he grows, so much the worse. The eldest girl said to Mr. H. when he called: "My papa used to have prayers in his family; but none since he has been a peer." What a motive for neglecting God! Also he is a dissenter and a democrat. A Unitarian he may be, but certainly no democrat. The Lord show

[59] Young, 1898: p. 352.
[60] Minute Books, **5,** f. 226.
[61] Young, 1898: p. 364.
[62] *Ibid.*, pp. 356–357. Dated April 12. The £800 was the especial appropriation for additional premiums on grasslands.

mercy to him, and by interrupting his prosperity or lowering his health, bring him to repentance![63]

A few days later Young made another outburst in his diary, just when the public criticism of the Board seems to have reached its height:

Lord Carrington fretting and worrying, and upon the full fidget about the newspapers' abuse, and the criticisms in the House of Lords upon the publication of the Board; swearing that he will allow no nonsense to be published. . . . Oh! Mr. Pitt, Mr. Pitt, that thou shouldest have formed such a Board as this, and then permit it to frame such a constitution as should render it absolutely dependent on the folly and caprice of a president! What might it not have done had its laws been what they ought to have been.[64]

A major cause for Young's irritation was disappointment with the Board's refusal to support his proposal for allotments to the poor from the wastelands. There may well have been a misunderstanding between Young and the Board about the real purpose of his trip in 1800. Young stated that it was undertaken "expressly for the purpose of ascertaining what the effect [of enclosures] had really been in practice."[65] Although his report included many evidences of these effects, his emphasis had rather been upon the need of allotments for the poor, a measure which he had already advocated. To opponents of such allotments it seemed that Young had used the Board's money to propagandize for his panacea. Since the Board was under fire for other reasons, it hesitated to support a measure which seemed visionary to many. Young's own account of the reception of his report follows:

March 28 . . . Tomorrow will be published in the "Annals" the first parts of my essay on applying waste lands to the better support of the poor. I prepared it some time ago for the Board, as it was collected in my last summer's journey; I read it to a committee—Lord Carrington, Sir C. Willoughby, and Mr. Millington—who condemned it, and, after waiting a month, Lord C. told me I might do what I pleased with it for myself, but not print it as a work for the Board; so I altered the expressions which referred to the body, and sent it to the "Annals."[66]

[63] *Ibid.*, p. 361.
[64] *Ibid.*, p. 363. Dated May 28.
[65] *Ibid.*, p. 333, note. *Cf.* also *supra*, pp. 425–426.
[66] *Ibid.*, pp. 350–351. Young had read part of his paper to the Committee on February 14. *Cf.* Minute Books, **5**, f. 168.

Naturally Young was greatly disappointed. A private publication even by the famous Arthur Young would have much less effect than a report by the secretary of the Board stamped with that body's approval. In effect the secretary had been rebuked by the Board and it hurt his pride.

Nevertheless, *An Inquiry into the Propriety of Applying Wastes to the Better Maintenance and Support of the Poor*, deserves to rank among Young's most important pamphlets. In the *Annals* it runs to about 160 pages and was also printed separately as a pamphlet by Rackham at Bury. The form is rather awkward. The pamphlet proper consists of about fifty pages and summarizes very concisely his findings, both the effects of enclosures and how the possession of cottages and cows contributed to the spirit of independence among the poor. The appendix is more than 100 pages long and is almost a transcript of his notes, referring to individuals by name and going into minute detail.

This work is by far the best expression of Arthur Young the humanitarian and Christian friend of the poor. It is the most complete recantation of his earlier views about the poor. It is interestingly written and full of good quotations, some of which have been used again and again to describe the evils among the poor about 1800. Young showed himself still a strong friend of enclosures and an ardent believer in the necessity of a general enclosure act. Nevertheless he had become convinced that enclosures as hitherto carried out had resulted in injustice to the poor, and his observations in 1800 confirmed this conviction. One of the most often quoted statements from this pamphlet declared of the poor, ". . . the fact is, that by nineteen enclosure bills in twenty they are injured, in some grossly injured."[67] Again he wrote:

What is it to the poor man to be told that the Houses of Parliament are extremely tender of property, while the father of the family is forced to sell his cow and his land. . . . and being deprived of the only motive to industry, squanders the money, contracts bad habits, enlists for a soldier, and leaves the wife and children to the parish? If enclosures were beneficial to the poor, rates would not rise . . . after an act to enclose. The poor in these parishes may say, and

[67] *Annals* 36: p. 528. My citations are from the text in the *Annals* 36: pp. 497–658.

with truth, *Parliament may be tender of property: all I know is, I had a cow, and an act of Parliament has taken it from me.* And thousands may make this speech with truth.[68]

Like all landlords and farmers of his day, Young was deeply disturbed by the alarming rise in poor rates which, he believed, might result in total ruin for the landed interest. To make matters worse, the misery of the poor seemed to increase proportionately with the rise in poor rates. In one place he predicted that poor rates would "in no long period, absorb the rents of the kingdom—not to give ease and comfort to the lower classes, but to leave them, if possible, in a worse situation."[69] At Bocking in Essex he found a badly administered workhouse, very high poor rates, and miserable conditions:

The misery among the poor, notwithstanding the immense sum expended for them, is terrible. In lodging, bedding, clothes, and food the account, confirmed by what I saw, is all distressing. The whole town an object which calls loudly for some effective remedy to be applied to a wretched system which beggars the inhabitants of one class without removing the misery of the others.[70]

On the other hand, Young saw many cottagers with cows who were reasonably prosperous and who received no poor relief. If the poor could hope for property of their own, they would "exert every nerve to earn, call into life and vigour every principle of industry, and exert all the powers of frugality to save."[71] Unfortunately, few of the poor could look forward to landownership: "To acquire land enough to build a cottage is a hopeless aim in 99 parishes out of 100." Another often-quoted passage expressed this point even more strongly:

Go to an alehouse kitchen of an old enclosed country, and there you will see the origin of poverty and poor rates. For whom are they to be sober? For whom are they to save? . . . For the parish? If I am diligent, shall I have leave to build a cottage? If I am sober, shall I have land for a cow? If I am frugal, shall I have half an acre for potatoes? You offer no motives; you have nothing but a parish officer and a workhouse!—Bring me another pot—[72]

[68] *Ibid.,* pp. 538–539.
[69] *Ibid.,* p. 504.
[70] *Ibid.,* pp. 621–622.
[71] *Ibid.,* p. 508.
[72] *Ibid.,* pp. 505, 508–509.

Of course Young visited many miserable cottagers. Richard Binfield's at Farnham in Surrey was "a most wretched, miserable hovel, much worse than any thing I have seen in the remotest parts of Ireland." Still worse was that of John Binfield:

He is in the workhouse; but his wife and two girls grown, one of them 23, at home, in a hovel worse than the preceding. Imagination can hardly conceive any thing so miserable. It is quite open to the weather on one side; no bedstead, only straw and some rags on the ground; filth and vermin. 4s6d. a week from the parish. The spot a rood of uncultivated waste and a dunghill (the cabin) in the middle.[73]

Such conditions were extreme, but he saw plenty of poverty among cottagers, many of whom were on the rates. Sometimes the holding was too small or the cottager lacked a cow and hence the manure to grow good crops. One of his favorite questions to such cottagers was whether they would give up all claims to poor relief in return for a cow. The answer was almost always in the affirmative. Young even showed some sympathy with that much abused class, those who encroached on the commons, and in the appendix described in detail the cottagers who had encroached on Mousehold Heath near Norwich. Although the parish officers were very suspicious of them, and although in some cases their fences had been torn down by irate farmers, Young showed that they had cost the parish far less per head than the poor who had not taken lands from the commons. The following passage about the "Mousehold folks" would certainly have alarmed some conservative members of the Board of Agriculture:

I viewed their little farms with singular pleasure, yet with a sinking heart at the thought of the evils an ill-framed enclosure act might bring upon them. Suppose the commons divided in proportion to value of the lands already enclosed throughout it, the regulation so common in enclosures, the whole mass of these people are ruined at one stroke. A cottager here who keeps three cows might get half an acre. To set fire to his house would be an equal favour.[74]

Young's solutions were much the same as those he had proposed in 1800 in *The Question of Scarcity*. The parish authorities

[73] *Ibid.,* p. 597.
[74] *Ibid.,* p. 566.

should establish the poor on the wastes, giving each family a cottage, a cow, and enough land to support the cow. The parish should borrow the necessary money and the recipient of the holding should re-pay the parish in annual instalments. He calculated that the amount spent annually on poor relief would suffice to care for all the poor in such a manner. The family should retain the holding so long as it did not come to the parish for relief. Such a plan would "attach the people to their king and country by the closest ties, and give every man such a stake in it's prosperity as would ensure the last drop of his blood to defend that which was the parental source of all his comforts."[75]

Many of Young's short articles in the *Annals* in 1801 are related in one way or another to the ideas expressed in his great pamphlet. He prepared a brief "Cottager's Garden Calendar."[76] He addressed a letter to Lord Carrington pointing out in considerable detail how a cottager might get along on a chalky soil where grasslands were scarce.[77] More important was a brief article entitled "Price and Consumption of Corn." After commenting on the continued high prices and poor rates, Young declared that the efforts of statesmen and Parliament had not "been adequate to the demands of the moment." Although human wisdom could not avert famine, it could plan so that future scarcities would be less cruel to the poor:

What I mean is, that all *country* poor, should have land enough, in property or rented, to keep them entirely from the parish; and all *town* poor so to contribute, by themselves, or their employers, . . . to a friendly society fund, as to secure them all necessary relief.[78]

Later in 1801 Young submitted a draft of an act to establish friendly societies in all parishes which had no houses of industry.[79] Contributions should be made in a rough proportion of three by the members themselves to one from the poor rates. Benefits would include sickness and accident payments, funeral and survivors' payments, incapacity and old-age pensions, and family allowances for large numbers of children.

[75] *Ibid.*, p. 528.
[76] *Ibid.* **37**: pp. 145–147.
[77] *Ibid.*, pp. 352–360. This letter was read to the Board which suggested it be published in the *Annals*.
[78] *Ibid.* **36**: p. 486.
[79] *Ibid.* **37**: pp. 562–571.

The *Annals* also contains Young's "Sermon to a Country Congregation," especially interesting in linking his religious beliefs with his attitudes towards public affairs. If England was to remain prosperous and secure, she must show gratitude and piety to the God from whom all blessings are derived. People guilty of common sins were not only in danger of eternal damnation, but might also be "throwing obstacles in the way to a secure and lasting peace." Furthermore the true Christian will not complain of the terms of the peace: "Peace itself is one of the greatest blessings which the Almighty gives to man; and to complain of it's establishment because you dislike certain of the conditions, is direct rebellion against Providence."[80] Young also emphasized that the farmers and gentry must see to it that the children of the poor were instructed more carefully in religious principles. All such efforts would repay them "in a thousand ways":

. . . you will find those obedient friends who without it would be, perhaps, the rancorous children of *the rights of man*. The blessing of God will be sure to attend such endeavours, and you will have the satisfaction of being surrounded by peaceable, honest, humble, and pious Christians for your poor neighbours, instead of those worthless, profligate, swearing thieves and plunderers you have so often complained of.[81]

Here is some of that hypocritical cant which gave the Evangelicals such a bad name in later years. Although Young realized that the preliminary terms of peace left France with predominant power on the continent, he had no fear of Bonaparte if a spirit of piety pervaded England: "Bonaparte may be the Assyrian to our sins as well as to those of France and other nations; but it is sin alone that gives him power. . . . Let us tremble, not at France, but for our own offences. . . ." What worried Young was the complacency, the vice, the very wealth of England:

Look around, and what do you behold but mingled vice and wealth; profligacy and success; the riot of luxury and the misery of poverty; God forgotten; Christ denied; . . . . I cannot view the perpetual wars abroad and the impiety at home without surprise that the whole human race has not long since been swept from a globe they inhabit only to disgrace.[82]

[80] *Ibid.*, p. 624.
[81] *Ibid.*, pp. 627–628.
[82] *Ibid.*, pp. 630–631.

Like most busy people, Arthur Young kept a diary only fit-fully. Often the most interesting, because most busy, periods are entirely omitted. In March, 1801, he resolved to keep his journal more regularly in order "to use it as a memento of the progress I make in the only business worth real attention—my salvation through the merits of the Blessed Saviour."[83] He set aside a period when he would surely be uninterrupted, well be-fore six A.M., just after he had said his morning prayers. Conse-quently the diary is quite complete for the later spring of 1801.

Besides the Board, Young had other activities connected with agriculture. The Farmers' Club usually met on Saturdays, and on April 19 he had an argument there with his old and dear friend, Lord Egremont, on land for the poor. "Everybody is against it. What infatuation."[84] There were also committee meetings of the Smithfield Society, unfortunately some held on Sunday, drawing up lists of premiums and preparing for the fair in December.[85]

Young was also going a little into society, but not very happily. Although invited to the Monday evening conversaziones of Mr. and Mrs. Matthew Montagu, he seldom attended, but on April 20 he talked there with the Bishop of Durham who agreed with him on land for the poor, with Lady Harcourt who wished some restrictions on farmers, and with Lord Somers about "table turnings" in Norfolk. "But all company of the sort is flat to me." What Young really wanted was more religious society. There were but few to whom he could open his heart and mind on religious subjects. Wilberforce "was so full of business that I might nearly as well be unknown to him." About this time he began to attend quite regularly the services of the Rev. Thomas Scott at the Lock Hospital. "Scott is now my favourite preacher, and I have heard him ever since I came to Town with great pleasure and attention in spite of a very bad manner. His matter is most excellent."[86] On May 3 Young took Mrs. Oakes to hear Scott at the Lock Hospital in the morning and Rowland Hill at the Surrey Chapel in the evening. "Neither of them pleased, though she admitted Scott's matter was excellent. She was most struck with the extreme fervency of Mr. Wilberforce's devotion, who, sitting in the reading

---

[83] Young, 1898: p. 347.
[84] *Ibid.*, p. 358.
[85] *Ibid.*, pp. 348, 351.
[86] *Ibid.*, pp. 348–349.

desk for the convenience of hearing better, she saw him clearly."[87]

Early in April, Young was at Bradfield for the three-weeks vacation. Apparently he was alone, attended only by one old woman, and truly happy, for "I had nobody to wrangle and quarrel with me."[88]

The pleasure of coming into the country from such a place as London is great and pure. The freshness and sweetness of the air, the quiet and stillness, the sunshine unclouded by smoke, the singing of the birds, the verdure of the fields, the budding out of vegetation, altogether is charming.[89]

He lived most simply. Up at 4 A.M., he took his morning dip up to his neck in the garden pond, even when there was snow on the ground as on Sunday morning, April 12. "I do not mind it at all, and sometimes stand in the wind till dry; it is, however, sharp work."[90] Young was nearly sixty years old! The day was spent in reading, walking, and farming, "and no moment hangs heavily on my hands." He also saw something of his neighbors. On April 6 he walked the six miles to Bury, for he had no horse or chaise with him, to dine with Betsy Oakes. "I had much talk, and tried hard to impress her with good religious notions, but I fear in vain; she will not be converted but by misfortune and misery. . . . I can only pray for her."[91]

One day he dined with a neighboring parson, the Rev. Mr. Balgrave, who had no vices, but "neglects the duty of his church, idle, indolent, drinks his bottle of port and reads his newspaper. . . ."[92] He was shocked to find only two couples at Bradfield church, one the clerk and his wife, but the rector, the Rev. Mr. Sharpe, "is past everything, preaches and reads worse than any human being."[93] On that point London was certainly preferable. His reading consisted of sermons and tracts, and he conceived of the possibility of publishing a book of excerpts from various religious works: "I think I could produce a very useful

[87] *Ibid.*, p. 359.
[88] *Ibid.*
[89] *Ibid.*, p. 352.
[90] *Ibid.*, p. 356.
[91] *Ibid.*, p. 354.
[92] *Ibid.*, p. 355.
[93] *Ibid.*, p. 352. According to the Bradfield Parish Register, Sharpe was rector, 1790–1803.

work without presuming to compose any part of it myself."[94]
His conscience was bothered by his failure to visit the poor while
at Bradfield. "I can read, speculate, write, and meditate, but in
doing good am negligent and slothful." True the poor came
to him in large numbers, both from Bradfield and from neighboring villages. He tried to give them all something, but wondered
whether it was wrong to give to charity when still in debt. Yet
he had always been ready enough to spend on himself, and felt
that he must give to "miserably poor people in such times as
these."[95] One Sunday, when there was no church service, Young
invited eleven poor village women to Bradfield Hall and berated
them for lax church attendance. He read from a Bible commentary and then preached to them. "One made a defence, and was
inclined to prate. I took it coolly, and presently brought her to
better reason. I doubt they liked a sixpence better than my
sermon, yet three of them cried. . . ."[96]

While at Bradfield, Young sent the following to Thomas Ruggles at Clare:

We have seen the moments of no king, no minister, a famine and
seven wars—we have now got ½ a king, and ¼ of a minister, with
addition of riots, rebellion and Jacobinism rearing its head—but
we, yt. is, you & I, have more than all this, got a bone to pick—
Land for ye poor—so much to talk on & never meet. Where are
you, & what about? If at Clare I will spend a day with yo. . . .

If you have not seen No. 207 of Annals we shall want the basis of
our edifice the thread of our discourse.[97]

After Easter Young spent two months in London, but by
early June he was restless:

Charming weather for the country, now in its full beauty, and I am
stoved up in this horrid place. . . . it will be the 20th before it is
possible for me to see Bradfield, and hardly then; the longest day
before a man gets into the country!!![98]

[94] *Ibid.,* p. 355.
[95] *Ibid.,* pp. 357–358.
[96] *Ibid.,* p. 360.
[97] No. 4 in the Ruggles Collection, dated April 7. Young is referring to the
king's insanity, the fall of the Pitt ministry, and the appointment of Addington.
The footnote refers to the issue which contained *An Inquiry.*
[98] Young, 1898: p. 364.

From June 15 to 18 Young attended the Woburn sheep shearing, along with nearly 900 others. There were more bets than usual and the business done in selling sheep and cattle and letting tups amounted to the considerable sum of £1687. Young admirably summarized the benefits of such gatherings:

Whatever some persons may be inclined to think of exhibitions of cattle fattened to a very extraordinary degree, . . . none can question the propriety of comparing different races of cattle and sheep in various particulars, exclusive of excessive fatness; none can doubt the utility of premiums for promoting a more correct tillage, for bringing into use new and improved implements of husbandry; none can hesitate in admitting the importance of that extensive communication of ideas and emulation of excelling which necessarily flow from bringing the farmers of the remotest parts of the kingdom into contact with each other. . . . The drillers of Norfolk describe their system to the adherents of broadcasting from Cornwall and Kerry, the enemies of paring and burning are enlightened by the practice of Kent and Cambridge, and every effort in tillage may be expected when the bets are in decision that shall decide the merit of the most important of all machines.[99]

During the summer England feared actual invasion, and the Duke of Grafton's son, Lord Euston, took a census in Suffolk of carts, wagons, horses, mills, and ovens. The trumpeter of the Bury Corps of Yeomanry visited Young, and asked him to pay a forfeit if he did not attend the drills.

I was always exempted on account of my necessary absence; however, as they expect to be called into actual service, I would not now retire when an invasion is expected, so I signed; but when the alarm is quite blown over . . . I shall withdraw, for I am too old and too weak, and my pursuits too far off and too numerous to permit attendance.

Many of Young's friends feared that, because of very high food prices, half the country would welcome the invader, and he commented, "I must freely confess I dread the result." Naturally he added that the poor would not be as disaffected if they had been given land. "At present they see nothing done or doing for them, and have their hearts almost broken by penury—without resource—without hope."[100]

[99] *Annals* **37**: pp. 193–194.
[100] Young, 1898: pp. 365–367.

During September Young toured Hertfordshire for a corrected survey of that county. On September 2 he wrote to Ruggles:

. . . I am on my survey of the Co. of Hertford & shall not be home till end of ys. Mo. If I can get over from Macklins to Spanes H: I will, but question time. I am glad you are awake at last. If you expect water & flax to thrive don't travel on a Sunday but appropriate the whole day to serving God in reading prayer & contemplation & yt you can't do at folkes houses, therefore be always on sunday at an Inn. . . .

Pull all the parsons' ears if they dont return thanksgiving.[101]

On September 7 Young stopped at Dunstable where he arranged to send a Bradfield girl to learn straw plaiting, which she could then teach his Bradfield villagers. He estimated the total cost at only £10. His complete acceptance of child labor involved in the plan is rather shocking:

. . . so for 10*l*. I shall be able to introduce this most excellent fabric among our poor. The children begin at four years old, and by six earn 2*s*. or 3*s*. a week; by seven 1*s*. a day; and at eight and nine, &c., 10*s*. or 12*s*. a week. This will be of immense use to them.

In the following year he reported that about twenty-five of the Bradfield poor had learned the trade, that "my splitting machines are all distributed," and that he had sent off their work to Dunstable, "and hope I shall have a good sale for the poor children."[102]

He spent October at Bradfield busy in writing up his survey of Hertfordshire. Unfortunately it is almost useless to his biographer. It is extremely impersonal, like that of Suffolk, and very different from his Lincolnshire. The preface lacks any acknowledgments to the individuals visited, and there are no high spots comparable to his descriptions in the Lincolnshire survey of the improvements of Sir Joseph Banks or Major John Cartwright.

Two more excerpts from Young's diary reflect some of his religious views in 1801. That on March 22 is as nearly obscurantist as anything Young ever wrote:

So near the expiration of the 1,260 years of Daniel and St. John;

[101] Letter No. 5 in the Ruggles Collection.
[102] Young, 1898: pp. 367, 383.

the Turkish Empire on the point of destruction; a strange and un-
thought-of establishment in Egypt, a country that is to have much
to do in the return of the Jews—ourselves in India, they may have
some unknown relation to that phial to be poured out on the
Euphrates to make way for the Kings of the East—altogether combine strangely to give suspicion that we are on the eve of some
great events which are to usher in the final consummation of all
things, and consequently the fall of the ten Kings of Europe. The
times are truly awful, and demand such piety and resignation as
no other period of modern history even approached to.[103]

On June 6 he discursed on the tranquil joys which accompany
a truly religious life:

It is a comfort which exceeds all others, that as age advances the
end of life is viewed as a mere change of residence. . . . I would not
give this conviction for the wealth of the Indies, for the empire
of the world. And what does one lose by religion? I enjoy all such
pleasures of life as are unattended by remorse, just as much, or
more indeed, far more, than I did while I was a dissipated character. Reading, composition, serious conversation on any topic worth
discussing, the rural beauties of Nature, and the pleasures of agriculture, friendship, affection, not love, as it is called, the whole
of which I fear is founded in lust, and proves nineteen times in
twenty the tyrant of the breast, and the fertile source of ten thous-
and miseries! Happy those in whom it terminates in a settled,
quiet, tranquil friendship, sufficient to satisfy without the wanderings of the heart that lead to so much misery.[104]

It is difficult not to relate the last part of the above to Young's
friendship with Betsy Oakes. The *Autobiography* states that he
kept no diary during November and December, but that he
"wrote journal letters paged to my friend."[105] What a pity that
his numerous letters to her, which would have been a mine of
information to his biographer, have not survived.

The year 1802 was one of the few in which Young published
no book or pamphlet. Moreover, the *Annals of Agriculture* contained no very significant articles from his pen. The documentation in the *Autobiography* is reasonably complete, but so confused that there is considerable doubt what material belongs to

[103] *Ibid.*, p. 349.
[104] *Ibid.*, pp. 364–365.
[105] *Ibid.*, p. 370.

1802 and what to 1803. After very careful study it seems probable that most of the material belongs to 1802, but if so, the documentation for 1803 is almost totally missing.[106]

As usual Young was very busy in the winter and spring working for the Board of Agriculture. The General Committee seldom met without instructing the secretary to investigate something. On February 8 he was requested to suggest premiums for the use of salt as a fertilizer, to inquire of a practical mechanic about a new drill plow, to propose terms for a premium contest about a method to restore the effectiveness of lime in the soil by the addition of some alkaline salts. Four days later he was asked to prepare some queries respecting dairies. On February 15 he reported on salt and the drill plow, and was instructed to furnish material on the smut in wheat. Three days later he brought in material on the smut, but was asked for further extracts.[107] Such were the secretary's routine tasks.

Young was still greatly dissatisfied with the president, Lord Carrington, who spent most of the session, 1801–1802, preparing for the press a volume of *Communications,* including some of the Essays on Grass Lands for which prizes had been awarded. Various county reports, including Young's Hertfordshire, were ready for printing, but everything stood still, while Lord Carrington was giving the *Communications* literary polish. Young himself had always written quickly and had never bothered with the refinements of style. Hence his impatience:

He corrected the proofs and made them dance up and down to Wycombe, and wait as if time was of no consequence, and a whole Session will pass with this for its only employment. . . . He is as fit to be President of the Board as Grand Llama of Thibet. . . . The whole of this flows from the most fastidious coxcombical pretensions to purity of language: the time is spent in making phrases . . . which ought to be employed in devising and executing plans of improvement and pushing on the county surveys. Lamentable![108]

Further in the spring of 1802 Lord Carrington rented six acres of land for experiments at Brompton, a mile and a half

---

[106] Without much doubt, Young 1898: pp. 370–384, concern 1802, and probably also most of pp. 384–389. Only the paragraph on p. 389 beginning "War!" is certainly about 1803.

[107] Minute Books, **5,** ff. 263, 266–267, 271.

[108] Young, 1898: p. 378.

from Hyde Park Corner. Young was never consulted until after the lease had been signed, then was asked to view the land, and directed to draw up a list of experiments, some of which were accepted and some rejected. "The whole idea is stark, staring folly; it will cost 250 l. a year, and the harvest well deserved ridicule."[109]

Since Lord Carrington was not a trained agriculturist he aroused Young's contempt. As a self-made man, he was apparently rather tactless of the secretary's feelings. In their general outlook, the two men were poles apart. That Carrington meant to be friendly is apparent, however, from a curious incident. On April 7 he informed Young that as his brother was in charge of a public loan, he, Lord Carrington, had invested £5000 in Young's name, that the loan had advanced 4½ per cent since it had been floated, and that he had sold Young's bonds at the increased rate. Therefore he presented Young with a draft for £221 17s. 6d. The action was perhaps a little highhanded and patronizing, but it indicated Carrington's desire to help Young. Young's reaction is interesting:

I thanked him much. Such a thing never entered my thoughts, and consequently surprised me much. It was very kind and considerate, and I am certainly much obliged to him for it. . . . I was thankful to God for this, and meditated much on it. If God had not been willing it would not have entered his head, and I find it comfortable to attribute everything to God, as, indeed, everything ought certainly to be attributed, and the more we trust entirely to Him the better I am persuaded it is for us. . . . If God pleases to give me money He has a thousand ways of doing it, and in these reflections I have had hard work to guard my mind against the temptation to consider it in the light of a reward which would be vile where is no merit, no desert. . . . But for two years past of His infinite goodness He has made all money matters very favourable to me, and I thank Him for an uninterrupted stream of His bounty without let or hindrance, and this notwithstanding my sensual mind and many offences.[110]

It certainly required considerable rationalizing to believe that the completely worldly Lord Carrington was God's chosen instrument to give Arthur Young, himself a sinner, over two

---

[109] *Ibid.*, p. 379. The land was that of "Salisbury's botanical garden."
[110] *Ibid.*, pp. 379–380.

hundred pounds based upon speculation in the public funds! Young's improved financial condition is borne out by the statement in the diary that he hoped to retire £700 of his debts.[111]

One of the greatest patrons of English agriculture, Francis, Fifth Duke of Bedford, died prematurely in March, 1802. Young's relations had been very close with the duke for nearly a decade. Bedford had been a charter member of the Board, the founder of the famous Woburn sheepshearings, and president of the Smithfield Club. The Board of Agriculture voted unanimously to strike a medal in his honor, to inscribe a forthcoming volume of *Communications* to his memory, and to order a bust prepared for the Board Room.[112] Young's graceful and completely respectful obituary in the *Annals* began: "The agricultural world never, perhaps, sustained a greater individual loss than the husbandry of this empire has suffered by the death of the Duke of Bedford." He also commented most favorably on the Duke's character:

Affable and engaging in his manners—mild, serene, and beneficent in his temper; none ever approached him but with pleasure, or quitted him but with regret: the firmness of his mind would have kept all in order; and an unvarying good humour would have given the wings of inclination to the feelings of respect.[113]

Shortly after his brother's death, John, the sixth duke, requested Young's aid in carrying out the agricultural improvements at Woburn. Young of course replied that he would be only too willing to help, and paid another moving tribute to the dead duke: ". . . to me . . . he was a kind, most amiable, and indulgent friend, nor shall I ever cease to lament the loss of the best temper I ever met with; good humour seemed to spring from a perennial source in his bosom."[114] On the other hand, Young's narrowness of mind come out unpleasantly in his diary:

The Board has been busy in voting testimonials to the memory of the Duke of Bedford. . . . These people are carnal and worldly, except, however, Mr. Wilberforce, who much promoted it. . . . His example is authority, or I should have considered the whole

[111] *Ibid.*, p. 384.
[112] Minute Books, **5,** ff. 283–284. *Cf.* also *Annals* **38:** pp. 374–377.
[113] *Annals* **38:** pp. 369, 372.
[114] Young, 1898: p. 374.

as a worldly-minded business, and bad. This Duke, with vast powers and immense influence, set an example . . . to a great circle of friends and dependents, of an utter neglect, if not contempt, of religion: all was worldly in his views; . . . his example mischievous to religion and the souls of men. All this praise and veneration is therefore very questionable. . . . Of what consequence is religion to the world if farming and beneficence and good temper, and a life highly useful in a worldly view, is to outweigh the evils of irreligion, and so very bad an example in morals and want of piety? I cannot approve of it, much as I liked the man in all worldly respects.[115]

Young also wrote another agricultural obituary for the *Annals* in 1802, on William Ducket. Young had known Ducket ever since 1769 and had repeatedly visited his famous farms at Petersham and Esher. In spite of the patronage of George III who, under the pseudonym of Ralph Robinson, had written several articles for the *Annals* on Ducket's farm, Ducket himself always remained a simple farmer, noted especially for his drilling methods and his invention of several agricultural machines. Linking Ducket's name with those of Bakewell and Arbuthnot, as he did later in his special publication on their work, Young declared:

He was certainly one of the great agricultural genius's produced in the present reign, to carry the husbandry of the kingdom to something of that perfection which manufactures have so rapidly attained in the same brilliant period. It was a great and interesting spectacle to see the admirable exertions of an ARBUTHNOT on strong land keeping pace with those of a DUCKET on sand; and both rivalling in merit, though not in fame, the astonishing efforts of a BAKEWELL. . . .

The obituary ended by expressing a view sharply different from his diary on the Duke of Bedford: "We hear of erecting statues to men who have deserved greatly of their country:—we cannot hear too much of paying that, or any tribute to the memory of those who have done honour to the British name."[116]

The *Annals* reveal that Young was still carrying on agricultural experiments in 1802. He had greatly increased the number of hogs at Bradfield and was trying to determine whether they produced enough dung to convert his straw into good manure.

[115] *Ibid.*, pp. 375–376.
[116] *Annals* **38**: pp. 629–630.

As usual, he drew up elaborate calculations of profit and loss. He made strenuous efforts to keep his hogs well and clean:

To keep them in health, I ordered every stie, as soon as empty, to be well washed, and all whitewashed within and without twice a year, and never to have too many hogs together; provided three roomy yards, well littered, with sheds under which they retire at pleasure.[117]

He also made a new experiment on the effects of various manures on soil. He put some soil from the poorest part of Bagehot Heath into thirty-two pots, to each of which was added some substance, everything from salt, charcoal, soda, lime, and chalk, to magnesia, cream of tartar, and oil of vitriol. Oats were planted in March, and turnips in June. The condition of the plants was observed at various intervals in April, June, and August.[118]

As usual Young attended many agricultural meetings in 1802. In March it was the Farmers' Club in London to set premiums for the ensuing year: "Carried with some difficulty a premium of fifty guineas for the best plough; several voted against it, because impossible to decide which of several should be the best! These folks can hardly know the right end of a plough."[119] In June he was at Woburn for the sheepshearing, but although the attendance was good, the atmosphere was quite different. He wrote in the *Annals*:

Those who attended this meeting, hitherto so gay and cheerful, animated as it had been by the enlivening presence of a nobleman so greatly beloved and respected, looked around in every scene with heavy eyes, and sorrow in their hearts. There hung a gloom over this whole business, which would not dissipate.[120]

From Woburn he accompanied Mr. Coke to Holkham and attended, for the first time, the meetings there. The business was much the same, inspection of crops and cattle, letting of rams,

---

[117] *Ibid.* **39:** p. 377.
[118] *Ibid.* **40:** pp. 97–104. There is also a note on this experiment in *Elements* **4,** f. 120.
[119] Young, 1898: p. 376. *Cf. Annals* **39:** p. 188, for text of the premium offer.
[120] *Annals* **39:** p. 42. *Cf. ibid.* pp. 385–458, for a long and very technical account of the fifth duke's agricultural changes, illustrated by several plates on various farm buildings at Woburn. There is also an account of his husbandry in Clarke, 1891: pp. 123–145.

sale of ewes, and exhibition of agricultural machinery. Young's own reactions were mixed:

He does it handsomely; 200 dined on plate.

The dinner better than at Woburn, I think from vicinity to the sea, which gives plenty of fish.

At the Holkham meeting, had I entertained my former feelings of pride and discontent, I should not have been too well pleased, for Mr. C. was personally civil and attentive; and yet he took not the smallest public opportunity of mentioning me, the Board, my report, or anything about it, although the occasion certainly called in reason for it. Once this would have mortified me, but now I value such matters not a straw.[121]

In December, Young had a busy week in London as secretary of the Smithfield Society. The show was held at Wooton's yard as usual, and premiums awarded. On the last day Young was duly thanked for "his zeal and attention in executing the office of secretary" and was reelected for the ensuing year. At the annual banquet at the Crown and Anchor Inn, the new president (the sixth Duke of Bedford) in his inaugural speech admitted that the club's purposes had been attacked and misrepresented, but concluded, "At all events our object is, and ought to be, to increase the quantity of animal food brought to market; and while we steadily endeavour to attain it, we must be promoting the publick service, unquestionably and laudably."[122]

The narrowness of Young's religious views was shown when he commented in his diary, after a dinner with the Duke of Grafton: "I never fail to combat his Unitarianism, but do no good; yet his arguments are weak as water."[123] He was also attacked by his old friend, John Symonds: "I am sorry to hear your declaration 'that you are apt to thank God that you cannot read the Scripture in Greek.' This, which is the common cant of the Sect to which you are so much inclined, is totally unworthy of you. . . ."[124]

Young's diary and letters in 1802 contain several interesting points about Betsy Oakes and her family. One day during the

[121]Young 1898: pp. 385–386. *Cf.* also *Annals* **39**: pp. 61–66.
[122] *Annals* **39**: pp. 598, 600–601.
[123] Young, 1898: p. 376.
[124] Add. MSS. 35,128, f. 435.

Easter vacation Betsy and her husband dined with him at Brad-field, "but the day so bad I could not show her the Round garden, which was got in very neat order."[125] A letter in the spring from Betsy's husband requested Young to write for him to Lord Broome on Yeomanry Cavalry affairs, because his little boy was so ill with convulsions that he had no time to do it himself. The same sheet contained a note from Betsy: 'O my friend it is for my unworthiness I fear that this dear Babe suffers—pray God that this trial may bring me to a sense of my sins—Adieu—a thousand thanks for all yr kindness yrs. ever. . . ."[126] On his way to Woburn in June he took with him part of the way a new chaise costing 170 guineas which he had bought for Betsy, presumably not as a gift, but on commission.[127] There is also a rather amusing letter from Betsy's sister, Sophia Macklin, also a friend of long standing:

You are certainly a mighty comical fellow. . . . you complain bitterly of my want of faith in you, & Betsys want of faith in you, but you say nothing of Orbell who I suppose has a soul to be saved as well as us women, so I conclude *he* has faith in you, & if you have converted one in the family, you ought & I dare say do rejoice. I cannot agree with you "that the things esteem'd amongst men are all abomination in the eyes of God," for what think you for instance of Betsy & myself, who are certainly in high estimation with *the men* deny it who can. You say right unless a man be born again he cannot see the kingdom of heaven, now as I do not see the women mentioned in the same way I conclude they are born with more purity & suffer more in this world than the men, by producing these very animals that are of so mischievous a nature, & which no doub[t] deserve all the punishment that many of them will meet with & when they are born again it is to be hoped for our sakes it will be in a different manner. . . . you may be right in some matters tho' you are a sad jumble on the whole & begging your pardon a little crack'd, but so much for you. . . . I wrote Betsy a very *agreeable sensible* letter last Monday & I am now expecting one in return, you are somehow a kind of necessary beast of burthen between us, as the blows fall upon you sometimes very conveniently, & then if you kick, we give you a little more to quiet you. . . . our garden is now in high beauty & I wish to

---

[125] Young, 1898: p. 382.
[126] Add. MSS. 35,128, f. 450.
[127] Young, 1898: p. 385.

know when you mean to visit us again, that I may arrange things accordingly, but I hope never to hear of your traveling again of a Sunday tho' a Duke should invite you. . . .[128]

A few items of family affairs in 1802 remain to be mentioned. Early in the year he was thinking about preparing a will. The letter from his lawyers seemed so confused to him and his marriage settlement especially was so complex that he drew up his own will without legal aid:

I am under such complex settlements that I do not understand what power I have; and Gotobed's draft was so full of law jargon, that I understand nothing of it. I wait no longer, but have made one plain and simple, and such as I hope, with the blessing of God, will not nor can be misunderstood. I have disposed of what I have to the best of my conscience, that is, if I was to die at Christmas.[129]

His diary while at Bradfield at Easter contains the following very melancholy passage:

I have been whitewashing the house, cleaning about it, and keeping all things in pretty good order to do justice to the place as well as I am able; but my dear child's recollection brings forcibly to my heart the impression that it is the will of God I should have hardly any chance of this property being kept in my family. My son has no children, nor likely to have any. Mary, no chance of marrying, so that my posterity ends with the next generation. The will of God be done, but human vanity and feelings will arise in the bosom. . . . Bradfield has been ours 200 years, and I should have liked that my name and family might here have continued. But God has punished me for my sins: I can have nothing at his hands that I do not deserve.[130]

The above seems perhaps unduly pessimistic. Mary, it is true, was thirty-five years old, but Arthur had been married only three years. Nevertheless his apprehensions were correct in that Mary remained a spinster, and that the Rev. Arthur Young had no children by his wife, Jane Young. The Evangelical Young would

---

[128] Add. MSS. 35,129, ff. 53–54. The collection attributes this letter to 1803, but the mention of Sunday travel, combined with the *Autobiography* account, p. 384, makes 1802 seem more likely.

[129] Young, 1898: p. 388. This quotation may refer to 1803, but I have attributed it to 1802 because of the lawyer's letter of February 6, Add. MSS. 35,128, f. 394.

[130] Young, 1898: p. 382.

hardly have been greatly comforted had he known that the Rev. Arthur Young would, after his father's death, have two illegimate sons, one of whom did hold the estate in the family until 1896.

After several years pressure upon Lord Egremont, the Rev. Arthur late in 1802 secured a very large Irish living, the Union of Agassin in County Clare in the Killaloe diocese, covering six parishes and 42,000 acres. He reached Killaloe on October 23, and was hospitably entertained by the bishop, who examined his papers, questioned him on the New Testament, and ordained him in the beautiful old cathedral. But the Rev. Arthur was contemptuous, both of the cathedral and the town of Killaloe: "this cathedral is rather better than our parish church of Bradfield; tho not much, & the city is the filthiest, nastiest, stinking little place that can be conceived. . . ."[131] Three weeks later he was at Ennis, the county town of Clare, engaged in conducting services in the separate churches in his living. His letter revealed his attitude towards his new responsibilities:

As I have not finished reading the assent & consent & the whole service of the church, which it is necessary should be gone thro' in each parish on a Sunday (reading both morning & evening service in each of them) I shall not be able to leave this part of Ireland before the end of the present month: after that I shall return to England as fast as I possibly can. . . .

There is no evidence that he ever returned to Ireland, although he held the living during the rest of his life. Thus he became a typical absentee Irish clergyman, chiefly interested in increasing the possible revenue from the living. Twice he warned his father not to paint things too brightly to Lord Egremont, for he hoped to get another living out of that nobleman, contiguous to Agassin:

But let me beg of you not to magnify but to diminish the value (or rather to speak the truth) if ever you have any conversation with Lord Egremont about it—because, as his Lordship had in fact promised me another & a more valuable union adjoining mine & tenable with it, if he find this that I now have can ever hereafter be raised, he may possibly think it sufficient without the other. . . .[132]

[131] Add. MSS. 35,128, f. 491.
[132] *Ibid.*, ff. 500–501. For further detail *cf. ibid.*, ff. 506–507; Gazley, 1956: pp. 362–363.

Immediately, however, he had to borrow £10 from his father. Actually he had endless trouble trying to collect his income from the living.

One last pathetic sentence in the *Autobiography* presumably dates from late in 1802: "I have never lived so well with Mrs. Young as for five weeks past."[133]

The *Autobiography* is almost completely silent on Young's activities during 1803. Only one brief paragraph is certainly for this year, a reflection upon the renewal of war with France. A number of letters have survived, some of considerable interest. There are the usual records of the Board of Agriculture. Although Young published no book or pamphlet in 1803, the *Annals* show that his pen was very busy.

Young was certainly back in London by February 15 for the first meeting of the Board, and was busy as usual on Board affairs during the spring months. Late in February he received a rather pathetic letter from his Bradfield bailiff, stating that the sheep lacked food except an insufficient supply of hay. The ewes had started to lamb and the first two lambs were born dead and the next two were just barely alive:

Independly [*sic*] of cruelty, &c. tis shocking to starve animals that are pregnant so that their mothers are unable to afford one drop of milk to their young. . . . Cabbage we must have another year. . . . No dry food; no wet food! The hay we are so sparing of, because so little, & yet no turnips. . . .[134]

For many years the *Annals* had contained an article telling how Young had fed his sheep over the winter, but this letter makes one doubtful about his success. Obviously he had failed to provide adequately for the early months of 1803.

At the very first meeting of the Board Young read letters from the Duke of Bedford, stating that the bust of his deceased brother was ready, and also from John Nollekens, the sculptor. Young was requested to ask Nollekens to attend a committee meeting to consider where the bust should be placed. Eventually it was installed in a niche near the window of the meeting room, on the wall behind the president's seat.[135]

[133] Young, 1898: p. 389.
[134] Add. MSS. 35,129, ff. 17–18.
[135] It is shown in this position in Rowlandson's famous prints. *Cf.* Minute Books, **5,** f. 351.

In 1803 Lord Carrington resigned as president of the Board, and on March 15 reported on the Board's failures and successes during his presidency. On the one hand were the debacle on the importation of rice to relieve the scarcity of 1800, the failure to attain a satisfactory General Enclosure Act in 1801, the attacks on the Board by the Church and the House of Lords; on the other hand the prestige gained through the essays on grasslands, and the progress of the county surveys.[136] On the whole it was a dignified and judicious accounting. Young might have smiled ironically when Lord Carrington referred to "the uninterrupted harmony and cordiality which had on all occasions prevailed"[137] among the Board members. Shortly after Carrington's retirement the Board passed an unanimous vote of thanks for "his judicious management . . . particularly of the funds of the Board."[138] And rightly so, for there was a favorable balance of £3300.[139] The new president was Lord Sheffield who held the post until 1806.[140] Young must have been greatly pleased at the election of a very old and intimate friend. There is no evidence of any friction between the new president and the secretary.

Unquestionably the outstanding development at the Board in 1803 was the inauguration of the lectures on agricultural chemistry by Humphrey Davy. Two lectures a week were to be given and at the Royal Institution because laboratory equipment was lacking in Sackville Street. Professor Davy was paid 10 guineas a lecture, and was also offered an annual stipend of £100 to serve as "Professor of Chymical Agriculture to the Board," in the hope that he and his assistants would analyze soil and manures for individuals wishing such a service.[141] Two general lectures were followed by one each on plants, soils, and the atmosphere, two on manures, and a final one on various applications of his principles to some of the major contemporary agricultural issues. Without doubt Davy's lectures, which were repeated annually through 1812 and published in 1813, were

[136] This speech is reprinted in *Annals* **40**: pp. 289–312.
[137] *Ibid.*, p. 309.
[138] *Ibid.*, p. 314.
[139] Clarke, 1898: p. 25.
[140] Minute Books, **5**, f. 376. Lord Sheffield was elected unanimously.
[141] *Ibid.*, ff. 413, 428, 430. *Cf.* also Clarke, 1898: pp. 31–32; Thorpe, 1896: pp. 94–99. Text of the lectures is in Davy, 1840: **7**: pp. 169–391; **8**: pp. 1–88. The decision to give these lectures had been made in 1802 under Carrington.

pioneer efforts in agricultural chemistry. As Young stated later in his lecture in 1809 on the advantages resulting from the Board of Agriculture:

The lovers of Science will rejoice to see, that the exertions of the Board directly tend to give the same foundation to agricultural knowledge, which so many other efforts of the human mind have long rested upon; by turning the attention of an undoubted genius, who has enriched Science with many splendid discoveries, to this neglected sphere of chymical research, the happiest consequences may be expected to flow.[142]

On June 3 it was voted to pay Young £100 for his Norfolk report, and in response to a letter from him to authorize him to survey Suffolk again for a new edition of his report of that county, at an expense not to exceed £50.[143]

In June, Young attended the annual sheepshearings, first at Woburn and then at Holkham. Just as the Woburn meetings were starting the new Duke of Bedford addressed Young: "Mr. Y., I beg you will take your old seat, and preside at one end of the table, for which purpose I have ordered a servant to keep your chair."[144] The *Annals* portray a very full and brilliant meeting.[145] There were present as guests at the Abbey seven peers of the realm, two former presidents of the Board, a foreign grandee, Prince Esterhazy, the American chargé d'affaires, Christopher Gore, and many of the great practical farmers.

Although nearly three hundred attended at Holkham, it was not quite as brilliant a gathering as at Woburn.[146] Young praised highly Coke's own "capitally cultivated farm," and enthused over three of his tenants' farms, one famous for the drill husbandry, a second for draining and irrigation, and the third for Devon cattle and South Down sheep. There were sweepstakes for the best guess on a wether's weight. Young and Edmund Cartwright gave the premium for the best agricultural implement to a Mr. Burrell for his machine for drilling seed and cake-dust at the same time. They also greatly admired a scuffler and a turnip drill submitted by the Rev. St. John Priest, the friend of that famous Norfolk parson and diarist, James Woodforde.[147]

[142] Young, 1809: pp. 68–69.
[143] Minute Books, **5,** ff. 433, 435.
[144] Young, 1898: pp. 384–385.
[145] *Annals* **40:** pp. 481–514.
[146] *Ibid.,* pp. 604–622.
[147] Woodforde's diary contains many references to St. John Priest.

The *Annals* for 1803 contain many interesting articles by Young. He devoted one long article to his crop rotations at Bradfield.[148] Twenty-nine fields were minuted, the largest one of twenty acres, the smallest of one. There were two kinds of land on the estate, one a wet sandy loam on a clayey marl bottom, the other a dry sound loam on a gravelly bottom, the latter worth twice the former in rent. He had gradually worked out a twenty-one year rotation. Since the land seemed to be "tired of clover" he did not use it at all. Young's rotation was peculiar because it included nine crops of grasses, an arrangement permitting large numbers of sheep to be kept. His rotation also provided within the twenty-one years four crops of wheat, one of oats or barley, four of beans, one each of turnips, cabbages, and tares. His plan also required less plowing than the Norfolk rotation of turnips, barley, clover, and wheat.

Several articles reflect the international situation, the peace of 1802, the renewal of war in 1803, and the danger of invasion. One was originally written in 1801, probably during the peace negotiations, but not published at that time. He admitted the complete continental predominance of France with "every idea of a balance of power gone for ever,"[149] and expressed doubts that the ensuing peace would prove permanent, but at the same time pointed out the tremendous growth in Britain's wealth and power. Since one of England's greatest dangers was a scarcity of foodstuffs, the Corn Laws should be revised and the wastelands brought into cultivation. All his old arguments against colonies were repeated, in the light of the temptations offered by the acquisition of Trinidad from Spain. He calculated what could have been done with the money spent on Gibraltar if used to bring English wastelands into cultivation:

In the way it has been expended, it gave us a rock, and that rock we possess at present—and upon that rock we shall, in the next century, expend fifty millions more to an equally beneficial purpose, provided the Almighty shall have patience to behold a world thus employed for a century more![150]

This same article voiced Young's disappointment with Parliament's refusal to pass a general enclosure act, and with the unjust criticisms leveled against the Board of Agriculture.

[148]*Annals* **41:** pp. 97–158.
[149] *Ibid.*, p. 81.
[150] *Ibid.*, p. 89.

In another article Young strenuously opposed the government proposal to increase war revenue by raising taxes upon tea and especially on malt. He made it very clear, however, that he was not opposed to the war itself:

At the opening of such a just and absolutely necessary war as the present, that great sacrifices must be made by the people, cannot for a moment be doubted. The whole energy of the country must be called for in contributions and in arms; taxes to an extraordinary amount must be levied.[151]

But these proposed increases would fall especially upon the two drinks of the working classes, tea and beer. In regard to beer Young wrote: "If Bonaparte could have dictated a tax for this country, he would have said—*Lay it on malt.*" And in respect to tea:

But I shall only remark, that beer and tea, and spirits (a great proportion of the last smuggled) are the only beverages drank by the poor, who can afford more than water; to raise greatly the duties on the two former is to drive them to the last—and the exchequer is to be filled by a double premium on immorality; first, by forcing a vile consumption at all, and secondly, by a direct encouragement to smuggling. Were milk or beer to be procured by *the poor,* tea would not class as a necessity of life; but for many years past I personally know it to be, with a small quantity of butter (with bread) the dinner of thousands. Ministers it is to be presumed, know nothing of this circumstance; but what must be the feelings of poor creatures, who have been driven to such food for their main support, to see it taxed again for funds to carry on a war which ought to be rendered (if possible) popular among them? What must they think of almost a prohibition on the use of beer—to thousands it will be an absolute prohibition—and at the same time find their tea raised in price?[152]

Instead of taxes upon tea and malt, Young proposed taxes on horses used for luxury, servants, amusements of all kinds, and meat. He explained in some detail that a tax on meat would bear much less on the poor than one on malt and tea, for the "poorest classes of society consume none, or next to none."[153] He also painted a dark picture of the effects of the new proposed

[151] *Ibid.,* p. 40.
[152] *Ibid.,* pp. 42–43.
[153] *Ibid.,* p. 45.

taxes upon the agrarian interests. Labor would demand higher wages, while its work would be less efficient because they could not get good beer. Moreover, the malt tax would lessen the demand and hence lower the price of barley.

A brief note on the evils of hoarding expressed Young's belief that Englishmen must resist invasion to the last ditch: "There is but one idea that should animate every bosom: to fight to the last drop of blood in the veins of every man capable of bearing arms: let the French seize the country, and it will be worth no man's living in."[154]

Young's most important article in the *Annals* in 1803 was an attempted reply to Thomas R. Malthus, the second edition of whose *Essay on Population* (1803) spent about ten pages criticizing Young's pamphlet *The Question of Scarcity* and its proposal to give the poor land for potatoes and a cow.[155] Malthus accused Young of inconsistency, since his *Travels in France* had claimed that much French poverty resulted from the smallness of properties which encouraged superfluous population. Malthus claimed that Young's proposal would encourage early marriages and large families. He especially feared that potatoes might become the staple diet of the English poor, and argued that Irish misery resulted largely from a potato diet.

Young could not permit an attack upon his favorite panacea, by such a distinguished man, in such an important work, to go unanswered. First of all, he denied that a writer should strive primarily for consistency, using essentially the same arguments made in 1793 in connection with his changed attitudes on France. This statement was an excellent summation of Young's aims as a publicist:

. . . a writer like myself, who has employed not a short life in the acquisition of facts, which he has been in the progressive habit of laying before the public, is not bound to reconcile such facts, or to withhold any, because they militate with others, that he had before communicated. He is rather bound in candour, to the directly contrary conduct: his business is to search for important facts for public use; and though he may adduce his own conclusions and reflections, as they strike him at the moment, it ought not to be expected from him, that his ideas are to remain stationary,

---

[154] *Ibid.*, p. 239.
[155] Malthus, 1803: pp. 570–581.

while he is advancing in inquiries. He knows that the main object is facts; to comment on them is an inferior business, and it would be a pernicious one, more futile even then [*sic*] mere theory, were his attention first given to former opinions, drawn from very different premises, and thus become solicitous, not to illustrate and apply his new information, without a caution first given to reconciling it with former circumstances.[156]

Young nevertheless denied that he had been inconsistent in this case. What he had objected to in France was the absolute ownership of land in very small units with the custom of subdividing it still further among all the heirs: ". . . the subdivision goes to such an extent, that a single cherry-tree, with the spot it covers, has been the whole of the property, yet a possessor kept at home by the charms of this property."[157] On the contrary in his proposal the cottager would rent the land from the parish, and would hold it only so long as he did not receive poor relief. He complained with some justice that Malthus largely overlooked this part of his proposal under which the allotments went only to those *not* on poor relief.

Young argued strongly that in England "*cottage* and *family* [should] be accepted as synonymous terms." Each cottage housed a family, each family had a cottage. Young people married as soon as they could get a cottage. To allot land with the cottage would not mean that population incentives would increase. On the contrary, those families with allotments might be more prudent because they had an incentive towards self-respect. If they had too many children they could not avoid poor relief and hence would lose their allotment. Malthus seemed "to assume the false supposition that if they are *not* comfortable they will not increase; though the proofs to the contrary are to be seen by millions."[158] Young was also very skeptical about Malthus's belief in the efficacy of moral restraint to lessen population. "And on what is the success of this revolution made to depend? why on young men and women avoiding matrimony and keeping themselves chaste without it!!!" Actually the only result would be "a general and promiscuous intercourse of the sexes."[159]

[156] *Annals* **41**: pp. 209–210.
[157] *Ibid.*, pp. 211–212.
[158] *Ibid.*, pp. 218, 220.
[159] *Ibid.*, pp. 221, 223.

Malthus had declared that the poor were the authors of their own misery by marrying and having children, that society had no obligation to support these superfluous children, that the poor individual "had no claim of right on society, for the smallest portion, beyond that which his labour would fairly purchase."[160] Young reacted indignantly: "Were the providence of the Almighty, no better dependence for such a man, than the speculations of such philosophical politicians, sad, indeed, would it be for the human race; better be a *young Raven,* than a man!"[161] He continued in one of the most radical passages that he ever penned:

To me it appears that the poor fellow is justified in every one of these complaints, that of Providence alone excepted. The price of labour IS insufficient to maintain a large family; the parish in all probability, IS tardy and sparing; the rich land owner IS (relative to him) avaricious of the land which would support him: he HAS AN INADEQUATE SHARE of the produce of the earth: his complaints ARE founded: he had NOT done wrong, and he ought NOT to accuse himself for following the dictates of God, of nature, and of revelation. He may see other cottagers with 3 or 4 acres of land, for which perhaps, they pay as large a rent as any farmer . . . living comfortably and independent of the parish, and has he not cause to accuse institutions which deny him that which the rich could well spare, and which would give him all that he wants? to tell such a man while in health and vigor, that, *he is not to marry but to burn:* . . . remaining chaste is a cruel insult:— to be chaste until what happens: until he has saved enough to support a family without a house to put them into, or land to feed them![162]

Young's article ended by summarizing the benefits from his own plan:

Land thus annexed to cottages, places the inhabitants in a superior state of comfort; encourages industry, increases the quantity of labour from a given number of people; promotes sobriety, and creates frugality. And the children thus bred make better labourers, servants, workmen, and soldiers, than others educated in poverty and vice.[163]

[160] Malthus, 1803: p. 540.
[161] *Annals* **41**: p. 222 note.
[162] *Ibid.,* pp. 225–226.
[163] *Ibid.,* p. 231.

Such expressions seem to make Young almost a forerunner of later Tory and Christian Socialism, of Shaftesbury and Kingsley.

Ever since 1786 John Rackham of Bury had printed the *Annals*, but late in 1803 the business was transferred to the famous Richard Phillips of London, who began the work with volume 41 and finished the few remaining volumes. In examining the *Annals* about this time, one senses that Young was finding it difficult to get enough material and that indeed he had lost interest. The Board took most of his time during the first half of the year. He had become increasingly engrossed with religion and was now over sixty. In the last two volumes printed by Rackham, Young had to scrape pretty hard at the bottom of the barrel, and they included about 250 pages, almost certainly consisting of his reading notes on various subjects—sheep, manures, climates, sainfoin, cabbages, mountains, and poultry.[164] They were the raw materials of his great unpublished manuscript work, the *Elements of Agriculture,* but for the most part they went back twenty years and were not up to date. On December 19, 1803, John Rackham wrote three brief notes to Young, complaining that he still did not have enough material for the forthcoming issue of the *Annals,* that his men were standing idle, and that this would be the last issue that he would print.[165]

In recent years an increasing proportion of the *Annals* had also been filled with agricultural news—of the Board of Agriculture, the Woburn and Holkham sheepshearings, the Smithfield Society, and various provincial societies. About the time he was changing printers, Young determined to fill up the *Annals* as painlessly as possible by making them the vehicle for the various agricultural societies, and hence on September 1, 1803, addressed a circular letter to them:

This circumstance, combined with the unquestioned utility of making known the transactions of all these public bodies, in some regular publication which might be easily and with certainty consulted, has induced him in the new form under which the Annals

---

[164] *Ibid.* **39:** pp. 164–184, 198–216, 280–286, 289–296, 309–321, 331–342, 481–522, 522–532, 633–658; **40:** pp. 6–50, 142–150, 172–185, 195–213, 579–604. They consist of excerpts from various authors from classical times on, and include many from his own works.

[165] Add. MSS. 35,129, ff. 88–93.

will in future be arranged, to assign a considerable portion of the work for the register of the British Societies, in which their rules and orders;—their premiums proposed;—their rewards conferred;—the journal of their meetings;—the names of the attending members, and all other particulars with which he may be favoured, shall be duly and immediately inserted.[166]

Two letters late in 1803 reflect the prevalent fear of invasion. One thanked him for a basket of game, and then discussed the prospect of invasion, but commented: "The Bury Ball, I imagine, would have prospered under every circumstance except that of actual invasion."[167] The other gave Young instructions in case of invasion to slaughter all horses and draft cattle and to destroy all carriages which could not be driven out of reach of the enemy.[168]

By far the most interesting and important of Young's letters in 1803 were two in December to his old friend, Thomas Ruggles.[169] Though full of unexplained allusions and hence difficult to understand, they are probably worthy of extensive quotation.

Dear Friend (Not Sir)

You are a whimsical fellow—to accuse me of not coming, as if it had been in my power. We went from Rogers at Ardleigh to Macklin's at Chesterford without information hungering & thirsting for Spains Hall & you gadding abroad at Freeland's. I was sorely sorry at your absence, I wanted twenty things; but have no great notion of your way of laying to grass, unless the soil is exquisite. You say 0 of irrigation.

When do you come to town to your benchership—& when to Clare —But what matters enquiry when we are so soon to be swallowed up by ye Corsican? I do not like the aspect of things. I fear that God has a controversy with us in which swaggering will not prove a satisfactory answer in mitigation of ye Divine wrath. He preserved us marvellously before—& I fear ye national heart was far enough from gratitude, and that he will now give us a bit of his mind, as ye farmers say. All things will work for the good of his

[166] *Annals* **41**: p. 26.
[167] Add. MSS. 35,129, ff. 67–68.
[168] Young, 1898: pp. 389–390.
[169] Letters VI and VII in the Ruggles-Brise collection. I think there is a mistake in dating the second letter which obviously follows the first, but is dated before it. The references to "My Smithfield book" would indicate that the letter was written on December 13, not 3, for the first meeting of the Smithfield Club that year took place on the 13th. *Cf. Annals* **41**: p. 446.

true & real Church, but being put into the furnace of afflection [*sic*] may be one of the means of his providence.

Your acct. of Phillips is only excusable fro. his not knowing who you were; a gentleman from Essex might be an Essex calf & not T. R.—I wrote him what you said.

I want to know how you rate ye tithegatherer—and by what rule they are or can be rated so as to be made to pay fairly.

God bless you & yrs.

My good friend

& believe me

Yrs.

faithfully

A. Young

London

6 Decr. 1803

Read Fletcher's appeal to matters of fact & comn sense.

The second letter is fully as interesting.

Dear Friend,

Sixty acres above L 1000! and for you to send no account for the Annals! Take the pen directly—and naked barley—a new subject. . . .

I know not how I mixed religion with farming to remind you of ye rascal Sterne—not that I am not as bad as him, but in a different way; and thought I was at best a passable character till I became a convert to Xtianity, but from that moment I saw myself in a different light and became sensible of a mass of depravity yt. I never dreamt had been in me. Now I am perfectly convinced that except in the blood of my Saviour I deserve to be damned for every thought of my heart, every picture in my imagination, & even for every prayer. I say but thanks be to God, Christ died for me: I know not at all what your Xtianity is, but I hope not that of the world, but ye true vital doctrine of our Incomparable Chh, yt. is; do all wch. ye world does not; and do not whatsoever the world does. . . .

You do not convince me that you could not furnish matter relative to the poor, but I suppose Rose has you. Nothing will do but land. Have you read Malthus on Population? Buy it directly.

I wrote you all abt. the picture to Phillips—he will do as he likes. It is for his public characters. . . .

Fletcher's book is on original sin and excellent. I shall be at Bradfield the 21st and if we do not meet I think it will [be] a strange business. My Smithfield book begins today and I should have no time to write a line had I not seized ye pen at present (5 in the morn)

Adieu. Remember me kindly to all yours

Faithfully

A Y

Dec: 3, 1803

The year 1804 was interesting and important in Arthur Young's life. The *Autobiography* is especially good on his private and religious life. The *Annals* are full of articles from his pen, and three of his books were published in 1804. The voluminous material will be organized, first under his professional and public life, and second under his religious and private life.

As usual the spring months were filled with the business of the Board of Agriculture. Nearly every meeting resulted in more work for the secretary and his staff. At one he was ordered to prepare a report on gypsum as a manure, at another to decide what machines in the Board's possession should be engraved.[170] On Young's initiative four new premiums were established about the process of laying down arable to grass.[171] Early in May Young gave evidence to a special Commons committee, chaired by Sir John Sinclair, on some questions about barley.[172] At the request of the Corn Committee of the Commons the Board distributed a circular letter throughout the country on the comparative expenses of arable land in 1790 and 1803. At the last meeting of the Board that spring it was:

Resolved that the Secretary do take the answers to the Circular Letter with him into the country, & in the ensuing recess arrange, with proper Extracts the answers received and report the same at the first meeting after Christmas next: and that he employ an amanuensis at the expense of the Board[173]

So here was a lot of work for the summer, nor was it all. On November 2 he wrote in his diary: "I have every post a packet of

---

[170] Minute Books, **5**, ff. 457, 459.

[171] *Ibid.*, ff. 460, 465. In this Minute Book the folio numbers 450–460 are given twice.

[172] *Ibid.*, ff. 471–472.

[173] *Ibid.*, f. 487. Young, 1804 (Norfolk): pp. 504–528 discuss the results of this circular letter. *Cf.* also Young 1809: p. 39. The amanuensis was the clerk, Mr. Alfred Vigne, who presumably was at Bradfield that summer, and was paid 30 guineas for the work. *Cf.* Minute Book, **6**, f. 10.

letters from Lord Sheffield, filling the tables and for many hours
employment every day."[174] For some unknown reason Young
feared that Lord Carrington would succeed Lord Sheffield as
president: "I have little doubt but Lord Carrington will be again
President of the Board. He likes me not, and I shall be much
more uncomfortable than I have been with Lord Sheffield."[175]
Actually Sheffield was reelected in 1804 by a vote of 13 to 3 for
the Duke of Bedford.[176]

In March, Young made a short trip down to Cricklade in
Wiltshire to observe the operation of a famous mole plow, used
chiefly to cut drains, and drawn not by horses or oxen, but by a
mechanical device of windlasses operated by eight women.[177] He
was quite favorably impressed and felt that a similar device could
be used for other agricultural purposes, including the operation
of a threshing mill.

When Young accompanied Lord Sheffield to Woburn on June
17, he was disturbed because he had to travel on Sunday. "I
detest this profanation of the Sabbath, but he urged me so to
accompany him that I yielded like a fool." He would have liked
to refuse to go to Woburn but did not feel quite free to do so,
in light of his regular attendance in the past, and because he was
still secretary to the Smithfield Society which always met during
the Woburn sheepshearings.[178] At Woburn he served as one of
three judges to determine the best plow out of eight entrants.[179]
His diary contains several pages of meditation while at Woburn.
The new duke was spending too lavishly and "immense debts
will prove a canker in all the rosebuds of his garden of life."
He was very critical of a new steam engine built by Edmund
Cartwright for threshing and grinding corn at an expense of
£700, "and yet a one-horse mill, price 50*l.*, would thresh all
the corn that will ever be brought to this yard."[180]

Some time in 1804 Young received a gift which must have

---

[174] Young, 1898: p. 402.
[175] *Ibid.*, p. 393.
[176] Minute Books, **5**, f. 460.
[177] *Annals* **42**: pp. 413–422.
[178] Young, 1898: p. 395. He did refuse the invitation to Holkham for the follow
ing week, for "a whole fortnight is horrible."
[179] *Annals* **42**: pp. 425–438.
[180] Young, 1898: pp. 396–397.

pleased him, in spite of his rather sour comment in the *Auto-biography*:

Mr. Smirenove came last night to dinner, and brought Count Rostop-chin's snuff-box. It is turned in his own oak, lined with gold, and has a tablet containing the representation of a building dedicated to me. The inscription in Russian, *A Pupil to his Master,* set around with sixty-six diamonds. Query—Should not all such toys be turned into money and given to the poor?[181]

Eventually he did sell the diamond chips for the anti-slavery cause, but kept the snuff-box.

Two of Young's books in 1804 were corrected county surveys. First to be published was the Hertfordshire, the trip for which had been taken in 1801. As noted above,[182] the work has almost no value for his biographer because it is so completely impersonal. The article in the *Monthly Review* only pointed out that Young's long residence in Hertfordshire added to his qualifications for making the survey, and concluded that the work "will not be found to lessen Mr. Young's character as an agriculturist or a writer."[183]

The tour in Norfolk had taken place in 1802. Although Young's knowledge of the county was very extensive, he apparently thoroughly covered the ground again. Many great Norfolk farmers who were old friends—Coke of Holkham, Stylman of Snettisham, and Bevan of Riddlesworth—figured largely in the survey. Much attention was also given to Purdis of Egmore with 1,900 acres; to Money Hill was a fine herd of South Down sheep; to Overman of Burnham, Coke's most famous tenant, who carried his grain to the London market in his own ship; and to the Rev. St. John Priest who also kept fine South Down sheep, improved many agricultural implements, and became secretary of the Norfolk Society. There were also many references to his interesting family connection, Martin ffolkes Rishton, whose wife was Maria Allen, a niece of Young's wife, and who was a driller, had South Down sheep, and used seaweed for manure.

---

[181] *Ibid.,* pp. 400–401. Smirnove was chaplain at the Russian Embassy and was active in engaging Englishmen to go to Russia to spread the new agriculture. Both he and Rostoptschin had earlier been Young's pupils. In 1931 the snuff box was owned by Mrs. Rose Willson who kindly had it photographed for me. For copy see fig. 8.
[182] *Cf. supra,* p. 445.
[183] *Monthly Review* **45**: p. 157.

Young's introduction apologized for a new survey of Norfolk by a new author, when the earlier survey had been made by Nathaniel Kent whose competence was generally recognized:

A SECOND REPORT for the County of Norfolk, by a different Writer from the Gentleman who executed the first, demands a short explanation, to obviate any idea tending to lower the estimation in which the Original Report is justly held. . . . The introduction of a new breed of Sheep, and the rapidity with which the practice of Drilling spread in the County, had effected so great a change in the State of Norfolk Husbandry, that all former works on the Agriculture of that celebrated County must necessarily be deficient, however excellent in other respects. The present Report does not appear to the exclusion of the former, but merely in assistance of it. . . .[184]

To what extent does Young's survey correspond to the above statement? Kent had not mentioned drilling, while Young devoted about fourteen pages to its rapid spread especially in western Norfolk. Kent favored the old native breed of Norfolk sheep, and had warned against introducing new breeds. Young, an enthusiast for South Downs, delighted in pointing out how common they had become with many of the best farmers. The two reports differ greatly in character. Young's was twice as long. Kent usually described things in general, with comparatively little specific detail. Young's work was much more like minutes, or notes. Kent's is therefore much better reading. For instance, Young devoted more than one hundred pages to minutes on various enclosures, while Kent discussed enclosures in less than ten pages. One reason why Kent was not chosen for the amended report may have been his disregard for the form which the Board had prescribed. He wrote in his introduction:

As to the arrangement of the matter . . . , it will not follow in the exact form of the preceding general Plan, as the greatest part of my scheme was digested and settled prior to my being acquainted with it; but I trust that under the following heads, I shall embrace all the material objects which the Board has pointed out.[185]

A few comments in Young's Report deserve attention. He concluded his section on the farm-yard:

[184] Young, 1804: pp. xv-xvi. The date of introduction was July 14, 1804.
[185] Kent, 1796: p. xv.

I wish I had it in my power to add, that I saw a good farm-yard in the county, manifesting contrivance, and in which no building could be moved to any other scite [*sic*] without doing mischief. Where is such an one to be seen?[186]

His praise for Coke was very high, but probably not excessive:

Mr. Coke . . . has done more for the husbandry of this country than any man since the turnip Lord Townshend, or any other man in any other county. . . . Those who have visited Holkham as farmers, will not accuse me of flattery, if I assert of Mr. Coke, that he is *fairest where many are fair.*[187]

While enthusiastic about the famous turnip husbandry of Norfolk, he was critical of its use on wet clay lands. While William Marshall had regarded the husbandry of East Norfolk as the only true Norfolk husbandry, Young felt that praise should also be given to the much poorer lands of West Norfolk, where correspondingly greater exertions were necessary to gain prosperity.[188]

Young's third book in 1804 was a "greatly enlarged and improved" edition of the *Farmer's Kalendar*. The *Autobiography* contains the following note:

An application from Phillips for another edition of the "Farmer's Calendar." He printed 2,000 of the fifth, and 1,200 sold in a month; they will all be gone before it can be reprinted. I had 100 *l.* for that edition, 40*l.* more for this six months after publication, and in future 25*l.* for each succeeding one. . . . The sale is an extraordinary one.[189]

It was the fifth edition which appeared in 1804 and was exhausted so rapidly that Phillips was calling for a new one, which duly appeared in 1805. Four more editions were published before his death, thus making the *Kalendar* by far Young's most popular writing.[190]

---

[186] Young, 1804: p. 25.
[187] *Ibid.*, pp. 31–32.
[188] *Monthly Review* **48:** pp. 40–47, reviewed Young's *Norfolk* favorably.
[189] Young, 1898: p. 392.
[190] Amery, 1925: p. 18, lists ten editions during Young's lifetime. The edition of 1771 had 399 pp., that of 1815, 658 pp. I have not had access to the editions of 1804 or 1805, but that of 1815 is probably not very different, for the preface, dated Bradfield, October 20, 1814, is almost the same as the preface of the 1804 edition quoted in *Monthly Review* **48:** p. 284.

Young's most important technical writing in 1804 was probably the *Essay on Manures,* for which, in competition with four other candidates, he was awarded the first Bedford Medal from the Bath and West of England Society. The Bath Society's advertisement for the premium had specified that the essay should be "founded on practical experience, on the nature and properties of manures, and the mode of preparing and applying them to various soils. . . ."[191] Young's essay was slightly more than one hundred pages long. He divided manures into two great categories, those which are dug or made on the farm, of fifteen different kinds, and those which are bought, of twenty-two kinds. As might have been expected, he spent much time on marling and liming, and paring and burning. He stated his views, in picturesque language, on the advantage of using dung, "fresh and long," instead of allowing it to ferment on dunghills:

He who is within the sphere of the scent of a dunghill, smells that which his crop would have eaten if he would have permitted it. Instead of manuring his land, he manures the atmosphere; and before his dunghill is finished turning, he has manured another parish, perhaps another county.[192]

His conclusion foreshadowed the future close alliance between agriculture and chemistry:

The more carefully these substances are analized [*sic*] and the more intimately we examine the effect of combining them with different soils, the better will this branch of agriculture be understood. This is to be effected only by the application of chymistry, going hand in hand with the vegetating process. Very few chymists have been farmers; we can therefore do no more than combine the facts discovered by one set of men, with the result of the observations made by another set.[193]

It is interesting to note that Humphrey Davy in his lectures on the chemistry of manures referred to Young's essay with great respect.[194]

Young's very numerous articles in the *Annals* in 1804 may be subdivided roughly into those essentially technical, and those

[191] *Annals* **41**: p. 444.
[192] Young, 1804: p. 152.
[193] *Ibid.*, p. 198.
[194] Davy, 1840: **8**: pp. 16, 32.

more economic and social. Two of his agricultural implements at Bradfield were described and illustrated by plates, a horse hoe and a potato harrow.[195] There was a detailed plan, with elaborate explanations, of a farmery without a threshing mill, quite different from his earlier circular plans.[196] Inspired by Davy's lectures in 1803 Young renewed and enlarged his experiments of 1802 on various manures.[197] Sixty-four small flower pots were used this time. The article shows that Young was familiar with the works of Ingenhousz, Darwin, Kirwan, and Senebier. He was especially impressed that he had completely dissolved charcoal in water by means of pearlash, while most chemists had declared that it was soluble only to a very limited degree. He also found that the mixture of charcoal and pearlash, chalk, salt, nitrate of potash, sulphate of lime, and spirits of wine were beneficial fertilizers.

Another important technical article was entitled "New Information on Paring and Burning." Divided into four installments, it totaled somewhat more than 100 pages.[198] The material was drawn almost exclusively from his 1800 trip to study enclosures which had also furnished the basis for his great pamphlet, *An Inquiry*. At the very beginning Young stated his purpose as "the improvement of the wastes of the kingdom." Such an improvement could be made in many cases only by this process of paring and burning, by which a special plow sliced off a thin portion of the soil, after which the sliced portions were gathered in piles and burned, and the ashes spread over the whole area. The practice was not universally accepted by the great agriculturists, however, and Nathaniel Kent listed it among the "reprehensible practices," and even attacked Young for his devotion: ". . . and I was not a little surprised at Mr. Arthur Young's coming forward, . . . not only with a sanguine recommendation of this reprobated system, but with a sort of censure upon such of the reporters as are of a different opinion."[199] The critics claimed that paring and burning "reduced" the land or lowered "the

---

[195] *Annals* **42:** pp. 84, 264–265.
[196] *Ibid.* **43:** pp. 473–478.
[197] *Ibid.,* pp. 433–455. For the 1802 experiments *cf. supra,* p. 451.
[198] *Ibid.,* pp. 133–152, 198–231, 300–321, 539–573.
[199] Kent, 1796: p. 186. Was this attack upon Young possibly another reason why Young conducted the corrected Survey of Norfolk?

staple" of the land, meaning that part of the valuable top-soil was destroyed by the process. Young, however, exclaimed, "When is conviction to be universal upon this question? It has been ascertained again and again, till doubt is folly."[200] He believed that the practice was absolutely necessary for any improvements on bogs and ferns, and satisfactory for all lands except perhaps sands. Young devoted most of the last installment to the serious situation in the fen country, where recurrent floods in recent years had driven many valuable lands out of cultivation. He was convinced that the government must act promptly to improve the drainage in the area, to deepen the rivers and control their flow. To permit such loss when food was scarce was little less than criminal.

Although the article was predominantly technical, in many places Young discussed related social questions. He was shocked by the lack of available churches for the poor, and urged the landlords to erect chapels:

. . . this woman had not been in a church for some years; and told me that very few fen people thought of going four miles for that purpose. This is not a way to make good subjects, or honest workmen; they are nearer being hottentots than christians.[201]

He was also wrathful to find that the poor, who taken land from the common and had built themselves cottages, were now being threatened by the parish. He referred to "another meritorious encroacher," and continued:

Wherever we go, instances are to be found of what the poor would do were they suffered; but rates at 12*s.* in the pound are preferred, rather than enable the poor people to support themselves without any rates at all. I met with no such instances without being astonished at the infatuation of mankind, to see so much misery among the poor, and such vast means of removing the whole, and establishing comfort and happiness in lieu of it, but resting torpid amidst their wastes, thoughtless of improvement.[202]

The sight of Epping Forest, so near to the capital, when food was so scarce, overcame his conservatism:

Such a space of land in a waste so near such a capital, filled with

---

[200] *Annals* **43:** p. 138.
[201] *Ibid.*, p. 546.
[202] *Ibid.*, p. 306.

complaints periodically for want of provisions, forms a spectacle abhorrent from every idea of policy and good government. Were it possible for any assemblage of the people to be justifiable, it would be one in every avenue to parliament, for shouting in the ears of the members—CULTIVATE THE WASTES![203]

From the same quarry of notes upon his trip in 1800 Young filled another 100 pages in the *Annals* with accounts of enclosures in Bedfordshire and Cambridgeshire.[204] In general these enclosures had benefited agriculture and were not prejudicial to the poor, but in several the poor had suffered needlessly. All too often the cottagers completely lost their rights on the commons or were inadequately compensated for their losses. Hence they had come to dread all enclosures. He concluded: "When an evil could be so easily prevented, and inclosures converted to their advantage as well as to that of every other class, it is to be lamented that measures are not taken with this view."[205]

The *Annals* in 1804 contain many articles connected more or less directly with a proposed new Corn Law. At one point Young briefly summarized the English Corn Laws since 1670. He still defended the bounty system of 1670 and 1688 and felt that the Acts of 1773 and 1791 constituted "unfortunate deviations from the old corn laws." The results of the later laws had been "almost a constant importation of foreign corn, three years of dreadful scarcity, such a fluctuation of price as to throw all calculation into confusion. . . ." Under such conditions certainly a revision of the laws "seems to be demanded by the most imperious necessity."[206]

Young's most important economic article in 1804 was stimulated by George Chalmers's bitter criticism of the agricultural interests and the "modern system of agriculture" in his revised *An Historical View of the Domestic Economy of Great Britain and Ireland*.[207] Chalmers claimed that the new agriculture had resulted in the consolidation of farms and consequent rural depopulation, and that its product was "a worse commodity at a

---

[203] *Ibid.*, p. 319.
[204] *Ibid.* **42**: pp. 22–57, 318–326, 471–502; **43**: pp. 42–59, 111–118.
[205] *Ibid.* **42**: p. 497.
[206] *Ibid.*, pp. 181–182.
[207] *Ibid.*, pp. 299–318. Chalmers's original work was entitled *A Comparative Estimate* . . . (1782).

dearer rate." Young challenged Chalmers's statistics to prove rural depopulation and denied that consolidation of farms was an evil. On the contrary, farms were still too small. He interpreted Chalmers's remark that agricultural commodities were worse *and* dearer to mean that meat was too fat and not lean enough. He thus associated Chalmers with those who attacked sheepshearings, cattle shows with premiums for fat cattle, and the Board of Agriculture which encouraged the same ends and wasted the public money in general. Young quoted Bakewell that the poor wanted fat meat, but carefully pointed out that the Board had never given a premium for fat cattle. Chalmers had also hinted that the scarcities and high prices for the poor had been caused by the nefarious practices of the great farmers and graziers, backed by the Board of Agriculture. Young answered that the rise in prices and the scarcities themselves were really caused by the Corn Laws of 1773 and 1791 which kept normal prices so low as to discourage production. Indeed if Mr. Chalmers wanted the real explanation for the scarcities he should have looked at "an agriculture so far depressed as corn laws can effect depression," and if he wished to assess blame, he "should have looked at home, and sought the cause where he might have found it;—IN HIS OWN OFFICE,"[208] a reference to Chalmers's position as chief clerk of the Committee of the Privy Council for Trade and Foreign Plantations.

Later in 1804 Young wrote a short article inspired by the Board's circular letter about the increased costs of arable agriculture. He had just spent most of the summer tabulating and studying the replies to the circular letter and he concluded that the tremendous forty per cent increase, in the decade from 1793 to 1803, was caused chiefly by higher wages for agricultural labor. On the other hand, grain prices had not increased proportionately, indeed not at all, except in years of scarcity. No wonder farmers were putting more and more land into pasturage:

Under such a state of prices, it is evident that the profits of husbandry must greatly depend on a due proportion of the lands of a farm being under grass, natural or artificial, and such other crops as feed cattle and sheep; and that such courses of crops as depend altogether, or nearly so, on white corn, must be very hazardous.[209]

[208] *Ibid.*, p. 317.
[209] *Ibid.* **43**: p. 42.

By November, Young was worried about the wheat crop which had been spoiled by a very wet season and the mildew, and which he rated the poorest in his whole farming experience. The poor should depend less upon wheat and more upon what they could produce themselves, presumably potatoes and milk. The shortage proved again the necessity to cultivate the wastelands. He reiterated his firm belief that the inadequacies of the corn laws constituted the most important cause for the scarcities. Population had greatly increased in the past thirty years, but the production of food grains had not, because grain prices had been kept artificially low by the Corn Laws. Naturally the farmers under such conditions had converted arable lands to pasture which of course increased the danger of scarcities: "give your farmers a steady adequate price for wheat *at all times,* and scarcities, as much as they depend, or can depend, on human exertions, would be at an end. . . ."[210] Of course scarcities were also acts of God as shown in the seasons and Young deplored the neglect of God "as if His was the only hand unacknowledged and Omnipotence the only resource forgotten."[211]

Early in November, Young also sent out a questionnaire on the mildew.[212] Twelve questions were asked, about the degree of the crop shortage, what soils were most subject to mildew, and what types of tillage, manure, and seed had escaped it the most. Replies from over fifty correspondents filled up about one hundred and twenty-five pages in the *Annals.*

Ever since 1800 Young had become increasingly interested in the housing of the poor. Very interesting was his plan for "A Cottage Cheap to Build and Warm to Inhabit," with a plate showing both an elevation and a ground plan.[213] His introductory paragraph constituted an indictment of existing conditions:

Cottages are, perhaps, one of the greatest disgraces to this country that remain to be found in it. Great numbers of them continue wretched hovels, that do not protect the inhabitants from the inclemency of the weather; ill built with mouldering walls of clay, windows broken, and so ill made as to admit the wind and rain even when in repair: doors out of repair, and opening immediately

---

[210] *Ibid.,* p. 251.
[211] *Ibid.* p. 248.
[212] *Ibid.,* pp. 321–323.
[213] *Ibid.,* opp. p. 262 for the plate, and pp. 284–286 for the explanation.

to the fire-place; pavements, if any, full of holes; and no circumstance of comfort to the wretched tenants, who pay 40s. or 50s.
for what is hardly a protection from wind and rain.

Young's cottage was strictly utilitarian and practical and would
have lacked the charm of many English cottages about which
romanticists rhapsodized. It would be very simple with one
large "keeping room" or living room, bedroom, and kitchen
combined, 15 by 16 feet in size. Above there would be a loft with
one or two additional beds which would be reached by a ladder
on hinges which could be hooked up to the ceiling when not
in use. There were also two smaller rooms, one a vestibule so
that the heat of the keeping-room would not be lost whenever
anyone entered or left the house, and which could also hold
boxes and casks and be fitted up with shelves. The second small
room would be a dairy in case a cow were kept, or an extra
bedroom where there was no cow. One large window faced the
south in the keeping-room, and one small window the north
in the dairy. Young also substituted for the ordinary fireplace
a sort of Franklin stove which would be cheaper and much less
wasteful of heat. Situated in the middle of the room, the whole
family could sit around it, and a pipe would carry the smoke
out of doors. One is impressed as much by the Spartan simplicity
and unadorned plainness of the plans as by the convenience and
comfort which it undoubtedly offered.

Nearly everything that is known about Young's private life
in 1804 was connected with his religious interests, directly or indirectly. On February 23 he attended for the first time a religious
conversazione held fortnightly by Thomas Fry, Scott's successor
at the Lock Hospital Chapel. About thirty were present, including Wilberforce, Scott, and Zachary Macaulay. The subject
was Providence and the following one was to be Temptation.
Young was much pleased, but thought his own remarks rather out
of place:

Scott, Macaulay, and Fry were the only speakers except myself, who
threw in a word or two in a bad manner and not in unison; but
I went without preparing the temper of my mind, and it proved
to be a mere temptation to sin. . . . I wished to touch on the state
of the King's health, where the hand of God is so evident; but
they would attend only to little and private things, and probably

were right. They made every possible event, the most trivial, providential.[214]

Young's attempt to connect religion with public affairs reflected his breadth of view, but apparently resulted in a little snubbing by the others.

Sometime in the late spring he was the guest of Mr. Anson, son-in-law of Coke of Norfolk, at a brilliant agricultural dinner, attended also by Coke himself, the Duke of Bedford, and Lord Somerville. Although greatly impressed with the house and its fittings, Young was disquieted, both on agricultural and religious grounds:

Nothing but a farming conversation makes the company of these people proper for me to have anything to do with; and that it is not to be conceived how little they know on the subject, considering it's a favourite pursuit . . . I have—I think I have—no envy of these doings. . . . I hate parties, and my heart condemns me whenever I go to them, in which not a word ever occurs to give God the glory due to His Holy Name, which is sometimes profaned, but never honoured, where Grace before and after meals is discarded.[215]

One reason why Young did not go on to Holkham after attending the Woburn sheepshearings in 1804 was his hope to spend a quiet Sunday at Cambridge and to hear a very famous Evangelical preacher, the Rev. Charles Simeon of Holy Trinity Church and fellow at Queen's College. On reaching Cambridge, Young immediately sought out Simeon who accepted his invitation to spend a few days at Bradfield later in the summer. Young, of course, knew Cambridge very well, but he could still enjoy its beauties: "I walked behind Trinity and John's, &c., twice—a delightful day." His diary for Sunday follows:

Night—I have been at Trinity Church thrice today. In the morning a very good sermon by Simeon, a decent one by Thomason, and in the evening to a crowded congregation a superlative discourse by Simeon. . . . Vital, evangelical, powerful, and impressive in his animated manner.

Two weeks later Simeon was at Bradfield for several days. At

[214] Young, 1898: pp. 391–392.
[215] *Ibid.*, pp. 394–395.

this time he was probably at the peak of his powers, about forty-five years old.

10th.—Yesterday, Simeon came. His character singular. His piety—his strong expressions—his fervency in prayer—a powerful mind!

13th.—Simeon went this morning. I have been horribly negligent in not writing down many of his conversations. What he thinks of me I know not, but he spoke to Jane with great freedom and candour, and as became a good Christian.

His abilities are considerable, his parts strong, his ardour and animation uncommonly great. His eloquence great, and his manner impressive. . . . He came with a servant and two very fine horses. . . . From his life and expenses must have a considerable income. . . . He is remarkably cheerful and has much wit, or something nearly allied to it.[216]

Thomas Fry also visited Young at Bradfield late in the summer. While he was there Young requested the use of the Bradfield church for his guest. The request was refused, but the rector apologized, saying that he was "talked to." Young burst out furiously:

Oh! for the dumb dogs of our clergy who will neither preach the Gospel themselves nor let others do it. I told him that my request was to him a safe one, for I asked, of course, only for a regularly bred clergyman, and who possessed preferment in the national church. Very unlucky!

During Fry's visit John Symonds dined at Bradfield and met him. Young commented: "Symonds dined here, and his conversation never does any good. He explained his chance for salvation in the merits alone of Jesus Christ, but denies original sin. This seems a contradiction; I would not think so for a thousand worlds!"[217] Fry's thank you letter to Young is interesting:

I must not forget to thank you for your very kind hospitality at Bradfield Hall; where I can most truly say I spent not only a most pleasant but a most profitable season; and should be sorry if I did not say I learnt many lessons from your own lips & example. . . . I meet with very few whose hearts are sufficiently impressed to enter into that. . . . experimental conversation which I own is my greatest satisfaction & which I found in you.

[216] *Ibid.*, pp. 398–400.
[217] *Ibid.*, p. 400.

Mrs. A. Young was at Chappel yesterday. I wished very much to have gotten an hours conversation with her & inquired about best things; but no opportunity occurred. I pray God to keep her from the dangers to which she is still exposed.

Mrs. Young and your daughter called on us some time ago and attend the Chappel pretty constantly. Sometimes I fancy they feel a good deal, especially under a view of judgment to come. But God only knows. . . .[218]

On November 5 Wilberforce replied, in very friendly fashion, to Young's complaint that he no longer seemed interested in continuing their friendship:

My private & public correspondence form together so immense a tax on my time, & I cannot alas work as hard as you do . . . Your letter really grieved me . . . for it shewed that my having seen so little of you & not having written to you, which had arisen solely from my many occupations & engagements had appeared to you to proceed at least to some degree from want of friendly regard. But believe me, that cause was *in no degree whatever* instrumental in producing the effect which I sincerely assure you I myself have always regretted—So let me once for all assure you, that I can never cease to feel a friendly regard for you. . . .[219]

The *Autobiography* also reveals Young's relations with his Suffolk friends and neighbors. The following quotation shows that Orbell and Betsy Oakes had rented a farm at Bradfield: "May 22.—My dear friend at Bradfield writes me in a most melancholy strain, on the ill success of her husband's farming. I doubt I shall lose largely by a scheme which was executed merely to keep him out of greater mischief."[220]

The account of an unidentified summer day has its points of interest:

Tuesday morning, C. Coke, of Holkham, Allen, and Moore called on me to see the farm, and would have me dine with them at Moore's. Haunch of venison, &c. Mrs. M. young, and in the world-

[218] Add. MSS. 35,129, ff. 156–157.
[219] *Ibid.*, ff. 228–229.
[220] Young, 1898: p. 392. What Young meant by "greater mischief" is not clear. He had no very high regard at this time for Orbell Ray Oakes, but in later life Oakes was in general highly respected. He marked the transition in his family from bourgeois to landed gentry, a change which came largely through his marriage to Betsy Plampin.

ly sphere very agreeable; her dress horrid. She contrives to force out her prominent bust in a manner that must take no small attention in dressing, is very big with child, and thinly clad; such a figure is common in these times, but the fashion is contrived purposely.[221]

Twenty years before Young had frequently visited at Ickworth. At last in 1803 his friend of those days, the earl-bishop, had died abroad, full of years and dishonor. His successor, Frederick William Hervey, 5th Earl Bristol, was just as friendly with the great agriculturist as his father had been. On July 30 he invited Young to dinner on August 1, and to stay over night: "If he will take a bed also & go round the grounds on Thursday morning & give some lessons to a young farmer, it will be particularly acceptable."[222] During one week Young spent two nights at Ickworth:

On Monday I breakfasted, dined, and slept at Lord Bristol's; Lady B. and her sister, Miss Upton, sung Italian airs till twelve o'clock at night. They were many years ago a horrible temptation, now a frivolous waste of time, but ever a bad tendency on the heart. Pressed me greatly to stay. . . . yesterday I dined at Betsy's. All this visiting is very bad for the soul.

Friday I dined again at Lord Bristol's, by desire of the Oakes . . . Music in the evening, slept there . . .

The following morning he met a fellow justice of the peace, John Benjafield, in Bury. Young's humanitarian attitude contrasted clearly with Benjafield's:

He wanted to commit a woman for being a lewd woman, on the statute of King James, by which it can only be for a year; if this was executed, all the prisons in the country would not hold them, and the time is far too severe; nor do I conceive that the case comes within it: I declined. . . .[223]

---

[221] *Ibid.,* p. 400. This passage is full of mysteries. Who was C. Coke, of Holkham? Did Young make a mistake in the initials? Stirling, 1907, has no reference to any such member of the family. The husband of the attractive Mrs. Moore was probably Mr. Moore of Long Melford.

[222] Add. MSS. 35,129, f. 150. Later he was created the first Marquess of Bristol. He had already served in the diplomatic services for some years and was still under forty years of age.

[223] Young, 1898: p. 401. The Countess of Bristol had been Elizabeth, daughter of Clotworthy Upton, later created Baron Templetown in the Irish peerage. It had been a love match, and apparently the countess brought no wealth to her husband.

The following letter from Bristol indicates that Young was at Ickworth at least once if not twice more that fall:

I am very desirous of knowing that you are better than the day you left us, & well enough to take one more evening of music. . . . you will find us quite a family party on Wednesday if you are disengaged—you are the person the most worthy of the true Italian musick that I can met [*sic*] with. . . . You need not send any answer about dining—do as you feel when Wednesday comes; but I should be glad to hear that you are better.[224]

Apparently Young did not let his conscience spoil the pleasure of others in Italian music, else he would not have been invited back again and again to listen.

Sometime in the summer or fall the Rev. Arthur Young and his wife, Jane, were in Wales, visiting Jane's father at Court St. Lawrence, as revealed in Arthur's letter to his father. As usual he was discontented:

We are all much obliged to you for the newspapers which come very regularly . . . but I have not seen the face of one single soul, nor an acre of land beyond these two farms. We shall get off the first moment we can after Bayleys arrival for we are both deadly sick of the place. . . .[225]

At the very end of the year an opportunity opened for the Rev. Arthur Young which completely changed his life, and which had nothing whatever to do with his vocation as a clergyman. In December, Arthur Young was asked to dine, probably at the Russian Embassy, with Novosiltzoff, who was in London "on a political mission of great importance," with Davidson who ran Alexander I's farm, and with Smirnove. At the dinner the Russians revealed their hope to prepare a series of reports on conditions in their country, comparable to those of the Board of Agriculture:

. . . and Mr. de N. talked much of the great advantage of my going. I would not offer myself, but said that the whole would depend on the sort of man employed. Since I came down [to Bradfield] I

---

[224] Add. MSS. 35,129, f. 168.
[225] *Ibid.*, ff. 162–163. This letter is neither signed nor dated, except "Saturday evening." Court St. Lawrence was the new Monmouthshire estate of Jane's father, Edward Berry.

wrote to Smirnove, offering Arthur (by his own desire), providing the sum granted for the purpose was adequate.[226]

As will be seen, the project came to fruition in the following year, and the Rev. Arthur Young spent most of the remainder of his life in Russia.

This account for 1804 can be concluded by several excerpts from the diary, throwing light upon Young's religious attitudes and problems, and showing that he was far from satisfied with his spiritual progress.

The sins of a journal are like those of life, much offense and a little repentance, minutes applied and months neglected. . . .

I have a conviction amounting to sensation in its truth, that everything but looking unto Jesus is weaker than water—vain and frivolous. This is the grand consideration, result, and object of religion in the soul, all beside is wide of the mark and without power and efficacy. Oh, my God, my God! write these truths in my soul, impress them in my heart, that by communion with Thee I may by Thy grace be purified, washed, and cleansed from every evil thought. Blessed be Thy Holy Name for keeping me from *acts* of sin. Oh! have· mercy on my mind and take away every *thought* of it.[227]

At the very end of the year he wrote:

I have been for months past, and am at present, in a dead, sinful state, remote from the only God of hope and consolation. . . . I am not in hell; but I find a horrible difficulty in coming to God, and a deadness of heart which hurts my prayers and plunges me more and more in sin and offence.[228]

[226] Young, 1898: p. 402. Novosiltsoff's mission was the Anglo-Russian alliance of 1805, and the very remarkable proposal to establish permanent peace through international organization. He had been educated in England, and perhaps had been another of the bright young Russians who had studied agriculture at Bradfield.

[227] *Ibid.*, pp. 391, 393.

[228] *Ibid.*, p. 403.

IX. Slowing Down, 1805-1808

On APRIL 25, 1805, Arthur Young listed in his diary the many blessings which he enjoyed through God's grace:

First, He gives me great health, at sixty-four, as good as at any time for twenty years past, and much greater than forty years ago. Secondly, He has been pleased to leave me two children. Oh, that He would call them to feel truly His faith, fear, and love! Thirdly, He has granted me an ample income very far beyond what I had, upon entering the world, the smallest reason to expect. . . . Fourthly, He has given me the power of being greatly useful to my country; it would be foolish not to reckon that which I know beyond the possibility of vanity deceiving me. Fifthly, He has given me a paternal estate and residence which I greatly love and never wish to change. . . .[1]

The above shows Young at the very height of his powers and influence, but the period in this chapter marks the beginning of a definite physical decline. In the summer of 1805 he was seriously ill. In 1807 his eyesight began to fail, and in 1808 he was forced to employ a reader. Although still active, his pace began to slow down. He had attended his last meetings of the Woburn and Holkham sheepshearings. In 1806 he resigned as secretary and treasurer of the Smithfield Society. The great days of the *Annals* were also over. In 1805 he reduced the number printed and the last regular issues appeared in 1808. Only at the Board did he continue active. He surveyed two more counties, Essex and Oxfordshire, and in 1807 began to read agricultural lectures before the Board.

The most important event in Young's life early in 1805 was the departure of his son and daughter-in-law for Russia in April. As noted previously, the project was broached late in 1804.[2] Count Novosiltzoff, whose chief purpose was to make an alliance with England, had also been directed to procure a trained agriculturist

---

[1] Young, 1898: pp. 410–411.
[2] *Cf. supra,* pp. 483–484.

to survey some Russian provinces on the model of the English county surveys. The offer was made to the Englishman probably best qualified for such a position, Arthur Young, who turned down the post. On his return to Bradfield for Christmas, however, he found that the Rev. Arthur Young was definitely interested. Consequently the father wrote to James Smirnove, chaplain at the Russian Embassy, whose reply was all that could be wished:

It is agreed, my dear Sir, that Mr. A. Y. your son is perfectly qualified to undertake the Plan, desired by the Russian Government, and no doubt, but that he will acquit himself with honor and to the Satisfaction of all the Parties concerned. His expenses to, in, and from Russia will be defrayed by the Government, but respecting the sum to be given clear of all Expenses, Mr. Novosiltzoff would wish you wou'd have the goodness to name it. . . .[3]

Smirnove also encouraged the hope that the surveyor might eventually be granted a Crimean estate and even a government loan. The Rev. Arthur Young's ultimate aim to make a fortune in the Crimea was thus in his mind from the very beginning.

Late in January, Young breakfasted with the Russian negotiators and presented his son's terms which were readily accepted:

. . . he and wife should go by land, getting out hence the beginning of March, in order to be at Moscow the beginning of May; that is, if out a year he should have a thousand pounds, and proportionably for a longer time, all his expenses paid to, at, and from Russia.[4]

After the meeting Young recapitulated the conversation and requested a written confirmation. When the Russian reply was slightly delayed, both Arthur Youngs feared that the whole thing was off, especially because another candidate had appeared who spoke French fluently. Since Arthur lacked such a capacity, Jane was going along as an interpreter with all her expenses paid. The Rev. Arthur fumed especially because he would have to dispose of his Bradfield farm if he went, and wrote a typically complaining letter to his father, blaming not only the Russian government but his father as well:

What thorough-faced insolence this Russian bear has shown to you

---

[3] Add. MSS. 35,129, f. 194.
[4] Young, 1898: p. 403.

thro' the whole of this transaction. & I am astonished that you can bear such studied delay, which after all will now most clearly end in a cold answer. Nor can I possibly imagine, what on earth could induce you to renew a negotiation which had ended in a bargain so remarkably to my advantage—which had been explicitly settled in the presence of witnesses (or one at least) This letter of yours has given the finest opening that could possibly have been required, to overthrow the whole business, by affording the man such grounds for his excuses! . . .

I hope my Mother has not been interfering underhandedly by writing any secret letters.

Jane's long postscript is just as typical of her:

Arthur is, as you see, angry about the Russian scheme being delayed, or what is most likely totally set aside—Novositzoff's behaviour is certainly most ungentleman like if not to say shameful, for an answer one way or the other is the least he could do. . . . tell him *Ar's* business is at a stand & on account of *preparations* & *setting off so soon* things must be finally arranged one way or the other. Those Russians are certainly very mean people & I am sure staying away is better than going, but as Arthur wishes to go, do everything in your power & let it above all be settled, for it worries & vexes him sadly. . . .[5]

All the worry had been unnecessary, for on February 1 Smirnove wrote that everything had been settled in favor of the Rev. Arthur.

On April 18 Arthur and Jane sailed from Harwich. Mrs. Young seems to have been ill in London, and Mary was probably with her, but Arthur Young almost certainly saw them off, for Harwich is only about thirty miles from Bradfield. Their departure left him in a profound state of melancholy. The risks of the journey, shortly before the renewal of fighting on the continent, were not slight, and Young had only the one son, cantankerous though he unquestionably was. Moreover he was deeply attached to Jane.

I have taken a long and melancholy farewell of them! Oh! may Almighty God give His blessing to the undertaking, and that we may all meet again in health and happiness. . . .

The undertaking, thus employed by a foreign sovereign to make a

---

[5] Add. MSS. 35,129, ff. 261–262. The manuscript guesses that this letter was written in November or December, 1805, but Young, 1898, shows that it must have been January, 1805.

report of one of his provinces, is the first thing of the kind that has occurred, and will either give Arthur a great reputation, or sink that which he has gained. It is a very difficult work, however, to produce a good book from a very ill-cultivated province, and in the large experience of all our own reports we see that very few are well executed. . . . Arthur goes with every possible advantage except language. I hope he will exert them.

Bradfield is very melancholy without them. Jane was always cheerful, always affectionate and kind to me, ever pleased with my presence, and never parted from me but with regret. The loss of such a friend with much conversation and an excellent understanding nothing human can make amends for.[6]

To dissipate his melancholy Young resorted to his many friends in the neighborhood. One day he walked to Bury, dined with the Oakes' and staid there over night. "It raised my low spirits a little—but badly—for such company is mere dissipation."[7] Another day John Symonds came to Bradfield for dinner and stayed over night. As usual he was full of ancedotes, but one of them was hardly calculated to raise Young's spirits:

He dined last year at Sir C. Bunbury's, where he met the rich Mr. Mills, the brandy merchant, who bought Mure's estate, and who said he found my "Annals" there, which were good for nothing. "What, nothing good in them?" said S. "No, nothing at all." "That is unfortunate with so many correspondents. But if the "Annals" are bad, have you read Mr. Y.'s travels?" "Yes, and very poor stuff they are. . . . I can see nothing in them."

The next day he wrote Symonds a letter with many apologies, as he understood that he had been talking to a gentleman who had contributed much to the "Annals."

So much for my rich neighbor.[8]

The *Autobiography* mentions two letters from Arthur and Jane after their departure. The first was from Berlin where they had spent about a week and had been kindly treated by the English and Russian ambassadors. The second came from St. Petersburg: "They give sad accounts of the treatment the Eng-

---

[6] Young, 1898: pp. 408–409. For further details on the Rev. Arthur's Russian adventure, *cf.* Gazley, 1956: pp. 372–405.

[7] Young, 1898: p. 409.

[8] *Ibid.*, pp. 412–413.

lish have received there if most specific agreements be not made before-hand for everything. I wish cordially they were well home again, and so do they, I believe."[9]

As usual Arthur Young spent most of the first half of the year in London, busy upon Board work. Apparently the Board accepted almost without question his recommendations about the award of premiums. Early in April he laid before the Board the summary of the replies to the circular letter about the increased costs of tilling arable land from 1790 to 1803, upon which a clerk had spent most of the summer of 1804. But all that the Board did with the compilation was to place it on file.[10] Young was probably much upset in May when Humphrey Davy was requested to draw up plans to convert a room at 32 Sackville Street into a chemical laboratory.[11] At the meeting on May 31 Young was commissioned to survey Essex that summer, and was also requested to ask his friend, neighbor, and protégé, Rev. William Gooch, surveyor of Cambridgeshire, "to re-examine the county & to render the work more satisfactory, as many subjects are but slightly mentioned, which ought to be more particularly explained. . . ."[12] At the last meeting of the session on June 5 the secretary was given his summer homework, to examine the various unpublished manuscripts of the Board and to recommend those he considered most suitable for publication."[13]

Young seems to have been quite worried in 1805 about a successor to Lord Sheffield as president: "this miserably constituted Board of Agriculture is ever in a dilemma when a new president is to be elected. Lord S. will keep it no longer, and he is in a difficulty to propose another."[14] Apparently Young feared the re-election of Lord Carrington. As it turned out, Sheffield was persuaded to stay on for one more year and was re-elected by a unanimous vote on March 26.[15]

During the Easter recess, Young heard disquieting rumors for 1806: "Nothing but bad news. Sir J. Banks writes me that Sir

[9] *Ibid.*, pp. 415, 418.
[10] Minute Books, **6**, f. 10.
[11] *Ibid.*, f. 30. Fortunately for Young on March 4, 1806, Davy wrote that he planned to continue his experiments in the Laboratory of the Royal Institution. *Cf. ibid.*, f. 57.
[12] *Ibid.*, ff. 34–36.
[13] *Ibid.*, f. 40.
[14] Young, 1898: p. 407.
[15] Minute Books, **6**, f. 3.

J. Sinclair is to resume the chair of the Board under promises of good behaviour."[16] Earlier on February 26 Young had written a very friendly letter to Sir John, two excerpts from which have survived:

I lament any thing you undertake out of Agriculture and Finance. The efforts of such a mind, diligent and penetrating, keen and indefatigable, would on one subject carry you to a great length. . . .

My son and his wife go to Moscow in a fortnight. The Emperor Alexander will have reports of all the governments of Russia like ours, and Arthur goes to do the government of Moscow for an example—thirteen times as large as Norfolk—they wanted me, but —the Board.[17]

The above would indicate very cordial relations, but Young's diary for May 23 reveals his deep anxiety should Sir John again become president:

I was awake at 2 A.M. and laid without sleep till 3 A.M. My thoughts were not edifying, so I jumped out of bed, and having prayed to the Father of mercies, I began with business. But the train of thought I had been in came again and interrupted me; it was upon the event of what would befall me as secretary to the Board. I have many reasons for thinking that several of the members do not like me, and should anything happen that gave them any handle, would be glad to get rid of me. This was not the case when I was one of themselves, but they know that I associate with religious people, go to the Lock (a very black mark), and read the Bible, and now and then words drop which I understand. Should Sir J. Sinclair become president or Lord Carrington, they might make it very unpleasant to me. Sir John is as poor as a church mouse, and would like well to have his lodging here. Should my family lessen, it would be quite unbearable, and if the idea was started, I must resist it, the question would probably be lost, and then I should resign; this would fix me in repose at Bradfield, and I should be to the full as happy as at present, but my family would not, and then—all this is very wild. I will have done with it for so much as I am persuaded that everything is in the hands of God. . . .[18]

---

[16] Young, 1898: p. 413. This proved an accurate prediction for 1806.

[17] Sinclair, 1837: 1: p. 186; 2: p. 73, note.

[18] Young, 1898: pp. 414–415. Sinclair's poverty at this time is borne out by a letter to Pitt on August 11, 1805, asking for a pension for Lady Sinclair because he had paid so much rent for a house in Whitehall while president of the Board, and because there were twelve children to educate. *Cf.* Public Record Office, Chatham Papers, **178**.

On March 4 and 5 Young attended Lord Somerville's cattle show in London, where many breeds of cattle, sheep, and hogs were exhibited. The show ended with a dinner at which nine cups were presented, and many toasts drunk, among them "The King, with grateful thanks for his patronage," "Mr. Coke, of Norfolk," "The plough, worked by good oxen, where the land is capable of carrying them," "The fleece, covering plenty of good flesh, and a proper quantity of fat," "Sir Joseph Banks, and thanks for his able Treatise on the Mildew in Wheat," and "Mr. Arthur Young, and may we profit by his admirable treatise on Manures." The Duke of Bedford proposed a toast to Lord Somerville, "which was drank with three times three, and great applause."[19]

In April Young wrote in the *Annals* in support of a fund to raise a monument to Luther, and perhaps slightly exaggerated the great reformer's influence:

. . . every one must be sensible how much (under God) we are indebted to the courage and piety of that GREAT REFORMER. There is not a field in Britain, which is not the better cultivated on account of his exertions,—not a manufacture which doth not flourish the more,—not an article of commerce which doth not, in some measure, spring from his labours,—not an effort of the human mind which is not more free, because LUTHER lived and wrote.[20]

On May 1 Young was elected to the Executive Committee of the British and Foreign Bible Society,[21] and henceforth he was always active in the Bible Society.

Young reached Bradfield on June 6, the day after the last meeting of the Board, "earlier than ever before or since the institution of the Board, which is a great blessing."[22] Such an early arrival was possible only by his refusal to attend the Woburn or Holkham sheep shearings. He was probably alone during most of the summer at Bradfield because of Mrs. Young's sickness, which was first mentioned in the diary on April 28: "Letters from London, and I am very sorry to find that my poor wife is much worse."[23] In the summer Mrs. Young and Mary were at

[19] *Annals* **43**: pp. 641–652.
[20] *Ibid.* **44**: p. 506.
[21] Add. MSS. 35,129, f. 221.
[22] Young, 1898: p. 415.
[23] *Ibid.*, p. 413.

Aldeburgh on the Suffolk coast, while Young himself was ill for six weeks in June and July, presumably with a bronchial infection.

Before he was taken ill, Young had been hard at work on his projected masterpiece, the *Elements of Agriculture:*

I have stuck close to my great work the "Elements," . . . What an immense labor has it been, and for how many years to collect and arrange materials. I could not have conceived how much it is necessary to do before I can fairly say, Now all is before me and in order, ready to compare and draw conclusions.

I mean it to contain everything good that has ever been printed. Till all that is collected and before me, how can I know what is already done, and what wants to be added?

But the labour, when continued year after year, is what I never dreamt of when I began. I have worked hard at the first division—*Soils,* and brought it into form; and it is a specimen of how much attention every division will demand. I have also begun the second, on Vegetation. I fear making the work too voluminous, and that by-and-by I must curtail greatly. Success is pleasant, and I should fear that if it exceeded two large quartos.[24]

For recreation during the early weeks of his summer vacation Young was reading a volume of Cowper's letters, which Jane had left at Bradfield:

Cowper is invaluable to a country gentleman that would enjoy his residence without the world's assistance. Reading his letters has made me more attentive to every beauty of this place. . . . there is something very amiable in the manner in which he converts every flower, tree, and twig to enjoyment, and I walk out better prepared for this pleasure from the persual of that most agreeable writer. There is but one danger from which, poor man, his poverty secured him, and that is the mind insensibly running into speculation of improvement. I have made many here, and the taste is very insatiable; this may without a guard lead to much expense. But I endeavour to correct the wanderings of imagination, and to dwell on the beauties of every single tree, shrub, and spot, and to be content with them all as they are. The laburnum in the back lawn is more beautiful that I ever saw it, so entirely covered with rich clusters of its golden flowers, that I can admire it for an hour

[24] *Ibid.,* pp. 415–416.

together. This enjoyment, however, is very poor and fading if we do not, with Cowper, turn our minds habitually to the great and beneficent Author of all these beauties. . . . A reading habit is a great blessing. I am sure I find it so, for though I have risen at 3 A. M. since I have been here, and not once in bed at four, still I am not tired at night. A walk is a refreshment to be had in the country in a moment, but at London half a mile of street thronged and noisy, and then only a crowded park, with sights to wound or tempt.[25]

Young realized, however, that he could not just enjoy Bradfield all summer. The survey of Essex had to be made!

I could write the Report from materials before me, and from a long knowledge of the county, and produce a valuable work, but that would not be honest. I shall take their money, and will therefore travel as much in it, and give as much attention to it, as if I had no materials at all to work upon.[26]

On August 5 he was planning to start soon. On October 20 he was still on tour, and probably did not return to Bradfield until some time in November. Shortly before his departure, he wrote to Thomas Ruggles:

I am sorry in one respect (that in as far as your body is concerned) that you also have been ill—but I am not at all sorry on account of your soul. Illness is very beneficial to the mind, and turns it more to God. It is well for me that I have [been] afflicted, six weeks of sick solitude has done me no harm at all; nauseous to bodily feeling as the dose has been; it is useful to the soul. My cough is yet heavy; but I am rather stronger. I hope to be able to start the end of this week but whether to Essex or to Aldboro I know not till I have a Lr. from my daughter who is there with her sick mother, & so badly yt. I think it may be proper for me to go to her. You shall hear from me the day I leave home for Essex, not yt. it is necessary for all I ask is a bed that has been slept in. . . .

I hope soon yt. we shall talk over the matter—for whom a sickman your Lr. has not enough of Christianity in it: it is one that a philosopher might have written. If the tongue & the pen do not express strongly, the heart feels insufficiently. I have long doubted your state, & it has made me low spirited more than once. Is yr.

---

[25] *Ibid.*, pp. 416–417.
[26] *Ibid.*, pp. 417–418.

hand drawn to books almost insensibly—that dissect, & probe & cut & slash the deceits of ye human heart till ye reader cries out how stands ys. with me! God bless you my dear fd. & by increasing his Grace increase yr. comfort & indifference (not philosophical) to life.[27]

The two thick volumes of the finished work attest that Young did cover Essex thoroughly. Although his journal of the trip has not survived, the printed work makes it apparent that he visited most of the county's agricultural notables. He probably visited Ruggles at his beautiful red brick Elizabethan manor house, Spains Hall at Finchingfield. Almost certainly he stopped at Castle Hedingham with an old contributor to the *Annals*, Lewis Majendie. At Great Chesterford he visited his old friend, the Rev. William Macklin, whom he portrayed as a farming parson on a fairly large scale. He certainly went to that superb mansion, Audley End, where Baron Braybrooke was busy adding to his properties and making agricultural innovations. In central Essex he stopped at Felix Hall near Kelveton to see Charles Western, M. P. for Malden, a member of the Board of Agriculture, and for many years to come the outstanding parliamentary champion of the landed interests. Along the Crouch estuary at Burnham he visited Edward Wakefield, then in his early thirties and aspiring to have his own sheepshearing like those at Woburn and Holkham. Young spent three nights with the famous Earl St. Vincent at Rochetts in southern Essex.[28] The former first lord of the Admiralty received him with complete cordiality and turned over Lady St. Vincent's dressing room to him as a study with a fire. "He showed me every acre, cow, ox, pig, and talked sensibly enough on farming, as far as he knew of it." One night Young's very old friend, Montagu Burgoyne of Marks Hall near Harlow, was a guest and much politics was talked, "for Burgoyne is a desperate politician."

Since Young did not return to London in December to attend the meetings of the Smithfield Society, he remained at Bradfield well into the new year.[29] In December, John Symonds became so ill that his life was despaired of. For three successive days

[27] Letter VIII in the Ruggles Collection.
[28] Young, 1898: pp. 418–419.
[29] The account of these meetings for 1805, *Annals* 44: pp. 360–362, is very brief and impersonal and shows that Young was not present.

Young rode the seven miles to St. Edmunds' Hill, the beautiful Adam-type home of his friend, where twenty years earlier they had enjoyed the company of François de la Rochefoucauld and especially of Lazowski. Young was now a very different person whose main objective was to convince his friend of the immanence of God and the need for repentance. On December 8 Symonds talked of Naples and Bonaparte. When Young made four or five attempts "to turn his mind to Christ," the sick man "was silent, not one religious or serious word came out of his mouth. . . ." "Awful is such insensibility." On December 9 Young came again, "through such a day of rain that I hope I shall not take cold." Symonds was still in the same mood, and "talked of the water gods, smiled, and joked." In vain Young attempted to turn his attention to religious things, and urged prayer read in his room. "It made no impression, and soon after closing his eyes as if for sleep, said, 'Your servant, I only keep you,' so I left him." When Young arrived on December 10 an excuse was made to prevent him from seeing Symonds: ". . . but he was getting up to have his bed made, and White, the physician, thought I had better not go up."[30] Apparently Young did not suspect that Symonds was unwilling to undergo another nagging attempt at a death-bed Evangelical conversion. But he must have wondered when in February, after a partial recovery, Symonds wrote frankly but without bitterness:

Though I have shewn no one your Letter, the advise [sic] you gave me *ora tenus,* or by word of mouth, during my illness, got abroad, and people blamed you for it. To one who spoke rather freely to me about it, I replied, that I always commended the conduct of Louis XIV, who when he was desired by his courtiers to disgrace Bourdaloue, who in his sermon had condemned some favourite vices of the King, made the memorable answer: "The preacher has done his duty, let us do ours equally well."[31]

.By 1805 it was clear that the change of printers for the *Annals* in 1803 had not worked:

. . . the "Annals" with Phillips are certainly at an end. They do not answer with him, and he had demurred at settling the account,

[30] Young, 1898: pp. 420–421.
[31] Add. MSS. 35,129, ff. 305–306.

with 100*l.* or more due to me. This will be a loss of 180*l.* a year. My plan is to print four numbers a year on my own account, for the sake of selling old stock; by this I shall lose 40*l.* more.[32]

Young's own contributions to the *Annals* in 1805 lack either distinction or importance, based chiefly again as in 1804, upon notes taken on enclosures during his trip in 1800.[33] There was another article minuting still more experiments on the application of various substances as manures.[34] It was probably during August when Young visited his sick wife at Aldborough that he also spent a day at Mr. Vertue's fine farm at Noddishall nearby, and found that, at least agriculturally, Mr. Vertue's name was appropriate. He had banished fallows, cultivated both potatoes and carrots on a large scale, and drilled many of his crops. He was a strong advocate of the use of long, fresh dung instead of the compost heap, and indeed regarded manure as the heart of his success. Young commented: "Good muck is his forte."[35] There is also a very brief article describing, with the help of a plate, the Ass Car "which I have found extremely useful on many small occasions,"[36] a shallow cart about four feet square, with the front end entirely open and the rear end appearing like a kind of rick.

The year 1806 was not particularly important for Arthur Young. There was no great crisis in his public or private life. No outstanding work appeared, although his last important political essay, "The Example of Europe a Warning to Britain," was published in the *Annals*. At the Board the most important development was the return of Sir John Sinclair to the presidency. The year is not very well documented, except for numerous letters. The diary contains no entry after July 14, and the *Annals* give little clue to his activities. Young spent only about four months of 1806 in London, two months before and after the Easter vacation of the Board. During the remainder of the year he was at Bradfield. His unwillingness to return to London in December led him to resign the secretaryship and treasurership

---

[32] Young, 1898: p. 405.
[33] *Annals* **44:** pp. 39–62, 174–201, 288–307.
[34] *Ibid.,* pp. 344–359.
[35] *Ibid.,* p. 265.
[36] *Ibid.,* p. 366.

of the Smithfield Society in which he had been so active since 1798.[37]

During the entire year the Rev. Arthur Young and Jane Young remained in Russia. That the reverend land surveyor and his wife were having difficulties is all too clear. The Russian interpreter proved to be "an ignorant puppy of a nobleman who is too lazy to do anything." In the spring Arthur had a week's illness with a fever "caused by want of sleep, owing to bugs, lice, fleas, &c., fatigue and vile food." After receiving one letter Young confided to his diary: "Of all the Governments I have heard of, it seems to be the most stupid, the most ignorant, and the most profligate. . . ." After a second he wrote that the Russians ". . . are horrid savages, and five centuries behind us in all but vice, wickedness, and extravagance."[38]

One of Jane's letters contained a story not calculated to enhance her father-in-law's happiness or to increase her popularity with her mother-in-law. Had Jane had any tact, she should have suppressed the story. Young's account in his diary follows:

The governor took Jane into a window and told her that he was informed she disapproved of all her husband's farming ideas as much as anyone could do, and ridiculed all his schemes. Upon explanation it came from Marshall Romanzoff, who had it from a German baron that had been at Bradfield, who, admiring the number of experiments, Mrs. Y. told him that she detested them all, and that I had ruined myself by them. A true report I will answer for, for this was her conduct through life. Lamentable it was that no enemy ever did me the mischief that I received from the wife of my bosom by the grossest falsehoods and the blackest malignity; of just such anecdotes of her conversation I have had instances from every part of the world.[39]

The letters also contained the alarming news that the Russian nobility were almost uniformly hostile to Arthur's mission, mistakenly believing that the ultimate aim was emancipation of the serfs and maintaining that all the inquiries might lead to serf revolts. Arthur claimed that he had been threatened with murder.

---

[37] Add. MSS. 35,129, f. 363, a formal letter of regret from the Smithfield Society, dated December 19, "on finding they are to lose the benefit of your assistance as secretary. . . ."

[38] Young, 1898: pp. 428, 432.

[39] *Ibid.,* p. 429.

The unpopularity of his survey is borne out in the journal of Martha Wilmot, the young English protégé of the famous Princess Dashkov:

. . . Mr. Young, the son of Mr. Young the great agriculturalist, who came over to make observations on the productions, culture and capabilitys of the Russian soil on the State of the Peasants &c. &c. call'd here this Eveg & was very badly received by the P. who is insensed at the *nature* of his employment which she thinks tends to overturn the Government & excite discontent in the people. In this idea she is join'd by almost all the noblesse, so that Mr. Young finds himself involved in the unpopularity of his profession to a degree that is often highly embarassing & disagreeable. I cannot say I am very much pleased with his manners as there is neither dignity nor elegance in them, but he stay'd a very short time. . . .[40]

Unfortunately Martha Wilmot's description of the Rev. Arthur only reinforces the impression that he lacked refinement and grace.

Early in 1806 the Board of Agriculture seemed almost moribund. The president, Lord Sheffield, did not attend at all, and meetings were repeatedly adjourned for lack of a quorum. On March 25 an unusually large meeting voted Sir John Sinclair back into the presidency by a vote of twenty for him to ten for Sheffield.[41] In spite of Young's fears, everything indicates that Young-Sinclair relations during the latter's second term as president were at least outwardly correct and probably much more cordial than earlier. On April 22 Sinclair outlined his new program of thirteen points. The county reports must be completed and a general report should be made on them. He cited the secretary's Suffolk survey to show how useful the surveys could be, and instanced Young's advocacy of the abandonment of unnecessary spring plowing. The agricultural lectures should be continued and opened to the public. The Board should start its own cattle show near London and support the formation of a Joint Stock Farming Society with a capital of one million pounds, a sort of land bank to aid agricultural progress. It should also establish experiment stations and eventually an agricultural col-

[40] Wilmot, 1935: p. 264. Dated January 4, 1806. In a letter to her sister, *ibid.*, p. 215, dated February 18, Catherine Wilmot mentioned Alexander I's unpopularity among the Russian nobles because of Young's survey.

[41] Minute Books, **6**, f. 74.

lege. Sinclair announced his intention of attempting to get necessary additional funds.[42] Surely here was a forward-looking program, anticipating many key developments of the twentieth century.

After the death of the under-secretary, Sir John Talbot Dillon, William Cragg was elevated from chief clerk to under-secretary, and Alfred Vigne from second clerk to chief clerk.[43] At this time the Board also employed George Quinton to draw engravings of the machines in the Board's collection and paid him £91 from March, 1806, to March, 1807.[44] On June 6 the General Committee requested Young to "correct and improve" his son's Sussex Report and £50 were appropriated for this purpose. He was also requested to give two lectures for the following spring on Tillage and on Farm Yard Buildings.[45]

On February 22 during a walk Young encountered his old agricultural rival, William Marshall:

. . . I never see and converse with him, but I think I see the haughty, proud, ill-tempered, snarling, disgusted character which he manifested in his connection with Sir John Sinclair. A thousand pities that so extremely able a man, for of his talents there can be no question, should not have more amenity and mildness. Government, however, should have promoted him without any doubt; and it is a blot in their scutcheon that they have not done it.[46]

Within a week the General Committee acted favorably upon Marshall's request to make a "Selection from the Reports":

Resolved that a Letter be written to Mr. Marshall, to inform him that the Board are extremely desirous to listen to an application from an author, who has so much distinguished himself by his agricultural publications. . . . they have no objection that a person so competent to the Task should examine the Reports which have been published, or are now printing, & should publish the result of that examination, and that the Secretary is instructed to furnish Mr. Marshall with any information or assistance, which may be necessary for the purpose.[47]

[42] *Ibid.*, ff. 81–99. Sinclair's program is also summarized in Clarke, 1898: pp. 33–34.

[43] Minute Books, 6, f. 74.

[44] *Ibid.*, f. 65. On March 14 nearly £50 were voted to Quinton, but the Cash Book cites the larger sum for the entire year.

[45] *Ibid.*, ff. 136–137.

[46] Young, 1898: p. 427.

[47] Minute Books, 6, f. 54.

One wishes that the Board's records had minuted the discussion, but in light of his diary comment Young probably supported the request. Eventually Marshall published five volumes in the period, 1808–1817, under the title, *A Review of the Reports of the Board of Agriculture.*

The *Autobiography* for 1806 contains many examples of Young's typical Evangelical ideas. The very first insertion for the year is a confession of sin for reading a French novel lent him by his neighbor and friend, Peggy Metcalfe of Hawstead: "Oh! the number of miserables that novels have sent to perdition!"[48] A few days later he had a call from the Rev. William Gooch, who had just become curate of Whatfield under the Rev. Robert Plampin, brother of Betsy Plampin Oakes. Gooch complained that Plampin had evaded the curate's law and cited many examples of rectors cheating their curates while insisting on the last turnip in tithes. Young was properly shocked, "Most lamentable is this for those who should be the ministers of Christ's gospel. But in no country that I have heard of is there such a set of clergy as in this neighborhood. Dreadful." Shortly before returning to London he went up to Bobbin's room, wiped the mold from her books, and recalled that she would now have been twenty-two years old, and commented, "But what a world is this for a girl of that age."[49]

On February 15 he dined in London with the Duke of Grafton, "against my will, for I think it wrong to go to the house of a Unitarian, or have anything to do with them. . . ."[50] He made one of the few references to his personal habits when he commented on his observation of a general fast day: ". . . I determined to take no snuff, of which I every day take much, and am almost uncomfortable if at any time I forget my box. This was more a fast with me than abstaining from food. . . ."[51]

In March, Young experienced a sense of religious peace which was the aim of all his striving:

March 17.—Yesterday morning, at 4 A. M., I came down to pray according to custom, and it pleased God that I should pray with more than usual fervency. I then meditated a little and fell asleep.

[48] Young, 1898: p. 421.
[49] *Ibid.,* p. 423.
[50] *Ibid.,* p. 426.
[51] *Ibid.,* p. 429.

I awaked with a certain sweetness of frame that I noticed at the time, and a transitory idea crossed my mind that God had heard my prayers, and that what I felt might possibly be His grace, or an effusion in some small degree of the Holy Spirit in my soul. . . .

23rd. Thanks to the ever blessed God, I think that I spent last week in a more satisfactory frame of heart and mind than any for an age past—more upon the watch against sin—more in contemplation of the greatness and goodness of God; vile thoughts have intruded, but I dismissed them by struggling, and my prayers have been more fervent.

But the old Adam was persistent, as seen in two June entries:

June 3rd.—I was up again this morning, at 3 A. M., and by it escaped falling into evil imaginations. This is an evil I thus fight against and struggle to avoid. I think the Lord will hear my prayers, and free me from this buffeting of Satan. . . .

4th—I found the devil at work with me this morning, and jumped out of bed at twenty minutes before 3 A. M., dressed, came down to prayers.[52]

Young's justification for his refusal to attend the Holkham sheepshearing is interesting:

. . . there is not one feature which could carry a Christian there for pleasure, but a thousand to repel him. . . . The Norfolk farmers are rich and profligate; of course oaths and profanations salute the ear at every turn; and gentlemen and the great, when without ladies, are too apt to be as bad as the mob, and many of them much worse. I am never in such company, but the repugnance of my soul to it is so great, that much as I love agriculture I can renounce it with more pleasure than I can partake of it thus contaminated.[53]

Young's diary for June 13 made its first reference to his school for poor children at Bradfield:

Every Sunday I hear sixteen or eighteen children read the Scriptures and say their Catechism, and I pay for the schooling of all that will come. They are sadly careless and inattentive, but still they come on. If it pleases God to turn it to account by their reading the Scriptures it will be well.[54]

---

[52] *Ibid.*, pp. 430–431.
[53] *Ibid.*, p. 432.
[54] *Ibid.*, p. 433.

Young saved a number of letters dating from 1806, some from old friends, some primarily agricultural, some asking favors or even begging. In the summer there was a very cordial invitation from Thomas Ruggles to spend a few days at Spains Hall:

Reading a Book naturally puts one in mind of the author, not that *Travels in France,* are necessary to put me in mind of my friend the Traveller, yet I follow my feeling, when I tell you that I wish to see you, or if that cant be at present, to hear from you, we both of us get old, but dont let us be forgetful of each other, the forgetfulness may be a symptom of age. . . . if you will spend a few days with me you shall live on fish as I have reserved a Pond to draw down the day after you come, if this is no temptation I can give you no other but the assurance of a friendly welcome.

I have heard but not seen by the Papers that we have lost our old school-fellow Jonathan Carter, a better man I believe is not easily to be found, you & I are now I believe the only school-fellows of Lavenham School contemporary & now alive.[55]

Another friendly invitation came from the Bristol family late in September:

Mr. Young is hereby summoned to Ickworth on Thursday next at ½ past five, & his bed will be ready for him. He will be under the controul of Miss Upton the whole of that day & Friday; & not be allowed to stir beyond the Park Paling till Saturday morning at soonest.[56]

If Young accepted he would have certainly have regarded it as yielding to temptation, for Miss Upton was young and attractive, and Italian airs would have been among the amusements.

Early in May a rather important letter came from Richard Phillips, the publisher of the *Annals,* in reply to one from Young about a possible contract for *The Elements of Agriculture.* Phillips' letter was at best non-committal:

I admit the loftiness of your pretensions relative to the Elements of Agriculture, & the price of 2000 £ is a fair commission for a Bookseller. Besides yourself & *Charles Fox,* no author living could command such a bidding.

---

[55] Add. MSS. 35,129, f. 348. Ruggles had been misinformed, for Carter lived until 1817.
[56] *Ibid.,* f. 352.

REALLY, however you must deliver to me a complete work both as to plates & Ms.—and all I can concede to you is to *pay* any artist money on account which will be charged to you.[57]

Several letters came to Young directly or indirectly as secretary to the Board. One was from Ireland from Humphrey Davy, relative to his experiment on some chemical fertilizer which was publicized by the Board. One paragraph of Davy's letter must have pleased Young: "You have been of great & durable service to Ireland. I have met with a number of persons who have been enlightened by your labours & who now follow an enlightened system of agriculture."[58] Not every one was successful in using Davy's mixture, however, for the Earl of Egremont wrote a complaining letter:

I was fool enough to send . . . & get six gallons of Mr. Davies mixture to accelerate vegetation & I steeped all my Turnip seed in it & the consequence is that not one seed has vegetated & I have the trouble of sowing a hundred acres over again. . . .[59]

At the end of the year James Mease, secretary of the Agricultural Society of the Philadelphia area, informed Young that he was sending him a list of their premiums. Mease poured on the flattery pretty heavily: "You have rendered such important services to agriculture, that any attention is due you from its friends." He also contrasted a new English book, which had been very critical of American agriculture, with Young's *Travels in France*: "Why do not Travellers copy the excellent model you have given in your tour throughout France, a work which has afforded me more amusement, than almost any book of the kind I ever read!"[60]

Most of Young's contributions to the *Annals* in 1806 were not very significant. There is a fairly long and extremely technical article on tithing. Young was trying to determine the basis upon which tithes should be assessed for poor rates when taken in kind. In general the rector should be rated upon his net income after deducting only the cost of collecting the tithes. The clergy could hardly have objected to the article's tone, for Young showed no animosity to tithes in general, although he had reservations about lay improprietors:

[57] *Ibid.*, f. 329.
[58] *Ibid.*, f. 347.
[59] *Ibid.*, f. 350.
[60] *Ibid.*, ff. 364–365.

The Rector's right to the full value of his tithe, in kind or compounded, is so clear that it ought to be paid fairly and honourably, liberally and cheerfully: it must be fairly paid to the improprietor, but it cannot be there discharged with the same cheerfulness, for reasons too many to enumerate.[61]

There is a very brief article by a Bury shoemaker on the average weekly costs for a family of six in that town. At the end of the estimate Young made an appeal for help for the poor cobbler:

It is a case of real distress that falls on a worthy man, and a sincere Christian. Should these papers be read by any person who would assist in relieving him, either by donation, or finding some employment in writing, at the very cheap rate of one penny for eighty words, it would be thankfully received: the poor fellow's shoemaking ruins his health, and copying is somewhat more profitable.[62]

It should be noted that one claim of the shoemaker for help was that he was a "sincere Christian." Certainly this is a different Arthur Young from that of the 1760's and 1770's, and one can only admire the changed attitude. It must also be confessed, however, that this was also a different *Annals of Agriculture,* and here the change was not so desirable.

Without much doubt the chief achievement of Charles James Fox's short ministry in 1806 was the abolition of the slave trade. Young celebrated the event by a short article in the *Annals,* hardly more than an editorial:

The most glorious event in the Annals of Britain! and more worthy of a day of thanksgiving than any victories that fleets or armies could achieve.

The Abolition removed a load from my conscience: as an individual I shared the general guilt, in not abhorring this detestable commerce so deeply and perpetually as I ought to have done; . . . and though I employed a weak pen in the same cause, how poor the effort, when measured by the call to every latent energy in the soul. . . .

But the trade is abolished—the load is removed. . . .

And are we engaged in a fearful and tremendous contest with a

[61] *Annals* **45**: p. 200.
[62] *Ibid.,* p. 214.

dreadful enemy, who overturns the thrones of sovereigns as if but cobwebs before the hurricane? Are we the peculiar objects of his vengeance? . . . Then let us rejoice and be thankful for an event which is worth a dozen Trafalgars.[63]

The above reveals Young's serious concern with the international situation in 1806. His nationalism in face of the French threat is well documented in several brief excerpts in the diary, his long article in the *Annals*, "The Example of Europe a Warning to Britain," and a long manuscript draft letter to J. C. Worthington. He was greatly shocked by the death of William Pitt in January:

This has struck a damp into my soul that I cannot shake off. . . . The providence of the Almighty has taken the two men perhaps the most necessary to our worldly prosperity in order to show us that it is on Him only we should depend, and to convince us that vain is the arm of the flesh. May He protect us! Providence is better to depend on than a hundred Nelsons and Pitts if we consider them in any light except that of means in the hand of Him who governs the fate of nations.

His deep love of country was also clearly expressed in a diary passage comparing England with Russia: "England! England! thou art the first of countries! Oh! that thou wert grateful to Heaven for the multitude of thy blessings."[64]

"The Example of Europe a Warning to Britain" is Arthur Young's last really important nationalistic work. He wrote it in January, 1806, and it appeared in the April issue of the *Annals*.[65] It does not seem to have been printed separately, nor to have had much influence. The similarity of its title to his most influential pamphlet of 1793, *The Example of France a Warning to Britain,* is of course striking. In originality, vigor, and clarity of expression the article of 1806 seems about equal to the pamphlet of 1793. Yet the pamphlet went through four English editions within a year and was translated into French, German, and Italian, while the article fell completely flat. Why the difference? The pamphlet came at just the right time and was completely in line with the ideas which the government was anxious to

---

[63] *Ibid.*, pp. 211–212.
[64] Young, 1898: pp. 424, 426.
[65] *Annals* **44:** pp. 385–410.

spread at that moment. And there is one basic weakness in the article. It was written in the mood of profound discouragement which the defeat of Austerlitz naturally induced, and it aimed to arouse the British people to prevent any possible French invasion from being successful. But there is absolutely no reference to Trafalgar and no sign of any realization that Trafalgar had made the danger of invasion much less acute. Had the article appeared late in 1804 or early in 1805 it might well have created much more of a stir.

Although his diary at this time emphasized God's Providence, the article is predominantly secular in tone. The following comment on the recent Austrian defeats is clearly activist:

. . . there can be no doubt but it has been the will of the Almighty that they should come to pass. He certainly rides in the whirlwind, and directs the storm; but this does not in the smallest degree lessen the duty of every power resisting, to the uttermost, the attacks that are made upon their liberty and independence.

England must arouse herself to avert the shocking collapse of the continental nations: ". . . what a spectacle is it to see so many countries conquered, or crouching, with Spanish imbecility, under the foot of a tyrant: and the people of the West, except one, the beasts of burthen to the French!!!" In this "most fearful moment Europe has seen for many ages," England remained "the last refuge of liberty, property, and religion," and yet the "peculiar vengeance" of France was "whetted" against Britain. The burden of a French conquest would fall especially on the landed interest:

. . . a conquest would transfer the soil of the kingdom to French landlords: Buonaparte would portion it out gradually with more than Norman rapacity; and the farmers would be the slaves, the *villains,* of the new possessors. . . . the evil of final defeat would be such as this country never yet experienced.[66]

The most tragic feature of the continent's collapse was that the regular army's defeat in one major battle had resulted in the conquest of a whole country:

. . . the fact remains great and glaring: Europe has trusted her defence to troops of the line, and Europe is conquered. Forty mil-

[66] *Ibid.*, pp. 386–388.

lions of men, ten millions of whom are able to bear arms, are now trampled on, as if they were sheep and pigs, by two hundred thousand Frenchmen!!. . . .

Whatever the evil might have been, the whole amount was the loss of an army; a loss great enough, without doubt: but the defence of a country rests on a foundation of straw, if the loss of an army is the loss of a kingdom.[67]

The conclusion seemed obvious: the whole adult male population must be trained. Two and a half million men should be put under arms and organized into twenty-five armies of one hundred thousand each. The day of volunteers had passed and now conscription must be used.

In the present situation of the kingdom, its defence is the first *business* of every man that can carry arms, and the necessity of exertion is such, that every man should be forced to bear his share of the burthen; and those whose years exceed or fall short of the limited age, should pay a personal tax, that the burthen may fall universally.[68]

No exemptions but "the most absolutely necessary ones"[69] should be given, and no substitutes permitted.

Such a conscription should be made in the cheapest manner that would be effective. Although nice, uniforms were unnecessary. Since muskets and bayonets were expensive, the conscripts should be armed with pikes alone. The important thing was to create such great masses of trained men that any invading army could be completely surrounded and that army after army of Englishmen could be thrown at the invaders and make their victory impossible. During the first month the recruits should be trained for an hour every evening, and afterwards one day a month. Officers and men should receive the same pay of one shilling a day, "a circumstance which would render the measure more popular than making the common distinction."[70] Close order drill at the company level was less important than exercise of larger bodies on open ground. Young also advocated the construction of a 100-mile-long entrenchment to cover the coasts

[67] *Ibid.*, p. 390.
[68] *Ibid.*, p. 409.
[69] *Ibid.*, p. 393.
[70] *Ibid.*, p. 395.

of Sussex, Kent, Essex, and Suffolk. Twenty-four-pound cannon should be placed six yards apart for the whole distance, and behind the entrenchments there should be a parallel broad road for quick troop movements. Of course Young made elaborate estimates of the costs of such a program which he believed might very easily be borne.

Young admitted that his whole plan might sound wild, but he claimed that wildness might be necessary:

How many men, when they read a proposition of this sort, will be sure to cry out out, *all this is very wild*. If wildness be an entire departure from that system which has hitherto been depended on for the defence of Europe, I hope it is *exceedingly* wild: it cannot be too wild in that respect. Troops of the line have lost Europe: in the name of common sense, let us not trust to them alone.

If it was within the verge of possibility to bring into the field five or ten armies of troops of the line, it might be very well to rely on them; but we have not an hundred thousand such, if the debates in Parliament are to be relied on: that is, we have a number sufficient for one battle. Lose it—and the kingdom is gone.

But the great principle for which I contend, does not depend on the arms, or on the description of the troops, to be raised: LET EVERY MAN BE ARMED AND EXERCISED; if with musquets well; if not, with pikes. Permit not the nation to be in a state of Austrian imbecility; a regular army defeated, and the foot of the conqueror on the neck of the nation.[71]

Young expressed many of the same ideas in a letter, dated March 6, 1806, to J. C. Worthington of Southampton, who had sent Young his pamphlet, *An Address to the Right Hon. W. Windham . . .*, dealing with military problems. One paragraph of Young's letter is so forcibly expressed as to deserve extensive quotation:

. . . You state, and most ably state, the causes . . . of the superiority of the french army— now inherent in it, and from the nature of our government, impossible or nearly impossible to be introduced in our own; & thus having proven the vast superiority which must remain with a french over an english army—what is your conclusion? Why to trust our all to the defence of 100,000 regular troops, because France cannot probably bring more into an english field.

[71] *Ibid.*, pp. 403–404.

Here seems to be a direct contradiction between your premises and your conclusions. You state the superiority of the french army as terrifying; . . . and then you go on. We must be out general'd, officered, and manoevered, and therefore do not aim at numbers . . . but trust the Kingdom to ye event of a battle . . . which full half your book is calculated to prove must inevitably be lost, Well—what then have we to trust to?—why nothing. Plainly, litterally and positively nothing. . . . Now Sir this seems to be neither more or less than absolute insanity. It is . . . this plan that has lost Europe. Her governments only have been conquered—not her people—nor can a people be conquered that are armed. . . . The baseness of governments has betrayed the people; you describe that almost inherent baseness—and then you bid us trust to 100,000 troops so officered & so led!! [72]

The documentation for Young's life in 1807 is quite complete, with considerable material in the *Autobiography,* a large number of letters, and full records of the Board of Agriculture. He came to London late in January. February and March were busy months, as usual, at the Board. The Easter vacation was spent at Bradfield, but most of April in London. There seem to have been no Board meetings in May or June, so Young had the unusual experience of being at Bradfield during those months. There were two Board meetings early in July which he presumably attended, but most of July and August were spent at Bradfield. The two following months he was touring Oxfordshire for the Board. By early November he was back at Bradfield for the rest of 1807.

There is little material about his family for 1807. Arthur and Jane remained in Russia, but none of their letters home have survived, nor are there any references to them in Young's diary. There does exist, however, what is probably a copy of a letter by James Smirnove, dated January 13, in response to Arthur's complaining letter of September 12, 1806. Smirnove took great pains to conciliate the irritated clergyman who was encountering obstacles to his proposed surveys. He urged patience and pointed out that when the Russians were convinced that the surveys had

---

[72] Add. MSS. 35,129, ff. 309–310. This is a first draft, corrected and in some places illegible. Windham made certain proposals in 1805 which produced a space of pamphlet replies in 1806. *Cf. ibid.,* f. 333, Worthington's letter to Young on May 5, stating that he was sending Young the second edition of his pamphlet and inviting Young's comments.

no ulterior motive, the opposition would gradually disappear.[73] Young's diary contains one sad and rather typical entry about his wife. On August 29 when expressing his regret that he had to make the Oxfordshire survey instead of remaining quietly at Bradfield, he added: "Mrs. Y. going to the sea for seven weeks, and therefore seven weeks' peace if I stayed."[74]

The diary and letters are full of Young's friendships. In February came the news to Young in London of the sudden death of his very old and dear friend, John Symonds, after a paralytic stroke. Symonds's will contained the following clause: "To my old & worthy friend, Arthur Young, Esqr. I bequeath my gold snuff box, & the Portrait of Soame Jenyns."[75]

Young spent three days in November at Euston as the guest of another old friend, the Duke of Grafton. On Grafton's request Young read Priestley's autobiography while at Euston, but complained that Priestley seemed self-satisfied and complacent, with no sense of sin and perfectly willing to be judged on his own merits:

I had no conception of any man having it in his power to review . . . his life . . . without much self-condemnation, and beseeching God to try him in any way rather than by an appeal to his life or his own merit, but (if he be really a Christian) by the merits only of the blood of his Redeemer.

Naturally Grafton the Unitarian agreed more with Priestley the Unitarian than with Young the Evangelical. "We were as far as the poles asunder." Interestingly enough, Young criticized the Duke for living too economically, in a way below his "rank and fortune." As always he felt it necessary to apologize for consorting with a Unitarian:

I do not feel comfortable, though he always receives me well, and desires me to come again. It is long since I was there before, and will be long before I go again; such visits, however, have very little of dissipation in them, and so much the better.
The less we like them the safer they are.[76]

There were other invitations too, several from prominent ag-

[73] *Ibid.*, ff. 366–367.
[74] Young, 1898: p. 438.
[75] Add. MSS. 35,129, ff. 371, 375, 377.
[76] Young, 1898: pp. 439–440.

riculturists. Charles Western invited Young to spend several days
late in May "with some agricultural friends."[77] The invitation
from Montagu Burgoyne, an old friend with whom he had much
in common and who sympathized with his plans for allotments
to the poor, was almost supplicating in its tone:

I am sorry to be so urgent but I must beseech you to come to me
on the 18. You have so often disappointed me, that I shall be quite
hurt if you do not come now. A very little exertion is necessary.
I have really a good deal to show you more than I ever shall again.
You will meet many of your friends & the Essex farmers will idolize
you for your arguments in favor of leases. Let me entreat you to
grant my request. [78]

Everything indicates that people enjoyed Young for his own
sake, but it must be remembered that any squire would gain
prestige among his neighbors if he could produce for a farming
party a lion like the celebrated secretary of the Board of Agriculture.

A few weeks after Burgoyne's farming invitation came an
Evangelical invitation from the Rev. Thomas Fry, late chaplain
at the Lock Hospital, now rector of Emberton near Newport
Pagnell, and Young's guest at Bradfield in 1804. Fry's appeal was
quite different from Burgoyne's:

Pray let me know when you can come as I long to see you here.
There are some things here which will please you & many more . . .
which would give you pain. . . .

Oh my dear Sir it is matter of astonishment to me that I believe
as I do, & yet the wonderful realities of another world make no
stronger impression on my heart! A Preacher & yet of all I feel
a need of exhortation! Do come & give me a shaking.[79]

A considerable portion of Young's correspondence in 1807 was
with three members of the Worgan family, all previously unknown to him. On February 16 George B. Worgan wrote a hard-
luck letter. At one time a prosperous and progressive farmer in
Cornwall, he had been turned out by a grasping landlord and was

---

[77] Add. MSS. 35,129, f. 420. Young had visited Western during his Essex survey in 1805.
[78] *Ibid.*, f. 429.
[79] *Ibid.*, f. 433. There is no indication that Young accepted the invitations
from Western, Burgoyne, or Fry.

temporarily a surgeon's assistant at Hampton, but separated from his dear wife and four children:

Walking one evening solitary and dejected, by the side of the Thames at Hampton a sudden internal impulse suggested—Write, immediately to Mr. A. Young, I found I could not resist the Impulse urging, among his respectable connections it is possible he may know a Gentleman possessing a neat cottage with a few acres attached, and which being in sight of his mansion, he may wish it to be neatly cultivated.

Worgan also added he had dared to write Young because his writings had revealed him as "a thorough Christian."[80] A few days later came the first of many letters from Worgan's sister, Lady Charlotte Parsons,[81] who became the intermediary between Young and Worgan. She was certainly devoted to her brother, but she also recognized his lack of practical business ability.

Young spent much time and effort to help this completely unknown man, and eventually secured for him a commission from the Board of Agriculture to survey Cornwall. This survey had apparently been kicked around early in 1807 among several candidates, including Richard Parkinson, 1748–1815, a former employé of George Washington at Mount Vernon, and Humphrey Davy who refused it.[82] Finally on July 3 the General Committee of the Board voted to pay G. B. Worgan £100 to survey Cornwall.[83] Certainly the initiative must have been Young's. Again he had given a survey to a protégé of questionable competence. Of course Lady Parsons was gratified: "You dear Sir have been the cause of all the innocent joy herein depicted—and Heaven will I am sure Bless you and yours for it."[84]

The appointment also prompted a letter of gratitude from a third member of the family, a brother Richard Worgan: "Such disinterested services though strictly the duty of every Christian, are not common." Richard Worgan seems to have been a rather remarkable man and more congenial to Young than his brother

[80] *Ibid.,* ff. 369–370.
[81] *Ibid.,* ff. 373–374, 384–385, 386–387, 389–390, 396–397, 403, 419, 455. Letter 389–390, dated March 20, is probably earlier than the two undated letters, 384–385, 386–387.
[82] *Ibid.,* ff. 398, 400.
[83] Minute Books, **6,** f. 205.
[84] Add. MSS. 35,129, f. 435.

or sister. In his letter he invited Young to visit him at his modest home, "Laurel Cottage" on Lake Windermere, and told Young something of himself:

Perhaps you will think me but a stupid clod, when I tell you that altho' I love a country life, yet I do not fish, hunt or shoot though only a middle aged man, but for these sixteen years have devoted my hours to the study of Divinity, Physic & farming and by way of recreation music. . . . With the first I am endeavoring to save my own & the souls of others—with the second I am enabled to attend all the poor sick people around who amidst our mountains are dreadfully off for medical advice. . . . Farming I have given up least like my brother I should burn my fingers, so that now altho' my income is very small, yet I live independent not liking to follow any profession for profit.[85]

Of course such a man interested Young, who had replied asking for more details about his beliefs and way of life, and extending a very cordial invitation to visit Bradfield. In his second letter Worgan revealed himself as a typical Evangelical in many ways. The books which had most influenced his thinking were all Evangelical in tone—Milner, Doddridge, Scott, Newton, and Law.

I am rather against the methodistical errors of Predestination, Experiences, & a building too much on faith without works . . . but our own church is quite enough for me if your clergy would only preach Xtianity instead of cold morality— . . . from your letter as I rather expect to find in you the exemplary Christian I want no other inducement to visit you that we may see how we like one another, therefore next spring or summer I will with much pleasure do myself that honour. . . .

In the long postscript Worgan explained that he was having the letter franked by the Bishop of Llandaff, the famous pluralist Dr. Richard Watson, who had for many years lived at Rydal in the Lake Country. "I have just been passing a day & night in his house." Of course Llandaff wished to be remembered to Young for they were old acquaintances with a very close mutual friend, John Symonds. Worgan described Watson "a fine hearty looking hospitable cheerful old man," but added, "I do not quite agree with him on subjects of Divinity."[86] Young must

[85] *Ibid.*, ff. 440–441.
[86] *Ibid.*, ff. 482–484.

have been convinced that here was no ordinary man, for he lived in a cottage, had no means, was a typical Evangelical, and still was apparently a welcome guest in the home of Bishop Watson.

Four letters in 1807 throw considerable light upon Young's relations with Betsy Plampin Oakes of Nowton Court. One letter was from Betsy herself and three were from her dear friend, Margaret Metcalfe of Hawstead. The Metcalfe family had resided at Hawstead for some time. Margaret, or Peggy as she was generally known, was nearly fifty years old. In 1807 Young began to keep her very numerous, very long, and quite interesting, but rather stilted and sentimental letters. Many were unsigned, or merely with her initials M. M., and since many give only the day of the week, the attribution of month and year is often no more than a guess. There is no hint of anything like a romance between Arthur Young and Peggy Metcalfe, who was devoted to Betsy Oakes whom she regarded as her best friend.

Peggy's first letter was probably written just before Young's return to London after his Easter vacation at Bradfield. It well illustrates her epistolary style:

My *poor, dear,* DEPARTING Friend, I wish I could speak a few words of comfort to you, but I cannot, for I am in too much grief, & affliction myself to have the power of administering consolation to others: I consign you once more to your living Tomb, with heart felt regret, and at the *same time abuse you for a Beast,* for not *once* giving me the opportunity of saying what I had to say about *our Friends,* as I wished to have done. . . .

Alas! why must you go dear Spirit of our own Spirits? . . . for in good truth your society renovatés. . . . *gentle* gales waft you from us, return, return, on Zephyrs wings, & find us what we *are,* all *Innocence & truth.*[87]

Peggy's second letter, dated June 29, written while Young was in London, is interesting because of its description of the Oakes family:

Yesterday the O. O.'s arrived en famille to go to our Church with us, and a very pious Party we made—O. O. drove his chariot, his Lady sat beside Him, and the Olive Branches were within it, not excepting sweet little Tig Wig, in his first suit a l'homme.

[87] *Ibid.,* f. 402.

. . . they all appeared so cheerful and happy. O. O. was more *himself* than I have lately seen Him. I do flatter myself that the thick cloud of folly, & idle imagination is dispersing.[88]

Peggy's third letter was written in very late fall from London while Young was at Bradfield. She was taking care of her rather famous uncle, Sir Philip Metcalfe. Her account of his bad eye inflammation and her reference to Young's failing sight can hardly have been consoling:

God preserve your's! it is a sad and affecting . . . sight to behold a man heretofore so extraordinarily independent, now feeling his way about, feeding himself with difficulty, confined as it were in a Dungeon & at the mercy of those who are around him!

Peggy continued with her lament for having to remain in London and ended with an imaginary Bradfield dinner party:

. . . it is just now 3 oclock—& as dark as Erebus; I can hardly see—you, on the contrary are light & glad . . . you are in the country—and therefore living—I am in the Town, and barely existing. . . . O! why was the love, the real, pure, unadulterated love of the country so strongly implanted in my nature?. . . .

from 4 o'clock, on Tuesday last, I was in Spirit, at Bradfield, that essence escaped the load of flesh embodying it, & in defiance of snow, and distance, reached the Hall, fluttered by the roaring fire then so crackling & joyous, entered the drawing room, expanded, & embraced the dilettanti party assembled; took its seat at the festive well furnished Board, sipped the cup,—joined in the mirth, fell in extasy—and died in despair. . . .[89]

Betsy's letter, the first from her to Young to have survived, attempted to console him for the death of his "Christian" servant, a man named Caufield. Another letter of sympathy reveals the close relation between Young the master and Caufield the servant: "Whenever I visited him, during his long confinement, I always heard him speak, and sometimes with tears of joy, of your pecuniary bounty, of your friendly Letters, and of your condescending visits to him."[90] Betsy's letter was probably written late in July or early in August:

---

[88] *Ibid.*, f. 431.

[89] *Ibid.*, ff. 496–497. This letter was only dated "Saturday," but the postmark looks like December 14.

[90] *Ibid.*, ff. 443–444. This letter from James Dove is dated August 12.

Tremendous indeed!! I am petrified!! to say I am at a loss for expression is not sufficient, for no words could convey to you how much I feel & am struck upon this most awful event—from the inmost recesses of my heart do I lament & commiserate the very great loss you have sustained in that most excellent & invaluable servant, it is indeed a severe blow to you, but the mercy & goodness of the Almighty *to him* in having made *him* your servant is to me very striking! . . .

Orbell too is sincerely affected, we lament I assure you deeply for your loss but the mercy & goodness of the Almighty in this event to the happy deceased is the consolation of all who feel for him.

Surely you had better not let Mrs. Y. come at present—if Cobbins can be of use to you till you can get a servant, we will spare him— so pray let me know & don't scruple taking him—it would be unkind.

Adieu my dear friend for my sake be comforted for you cannot conceive how much I feel for you—I dreamt the other night that you were dead! I went into mourning, & could neither see, hear, or speak to any Being!! , , ,

God bless you
Yours ever[91]

The beginning of the letter seems so stilted and exaggerated as to cast doubts on Betsy's sincerity, but as it continues her real sympathy is apparent enough. Betsy's emphasis upon the goodness of God's providence would certainly have appealed to Young.

The *Autobiography* also contains a rather touching statement about Young's relations to Betsy in the summer of 1807:

I associate only with Mrs. O. Oakes. Having written till I am tired, I go once or twice a week to relax with the mild green of her soul, because I can be free and do as I like, and I try hard to make her a Christian, but hitherto in vain. If evil ideas at any time plague me, then I keep away, and, thanks to God, I have of late had an unusual command over my imagination, which for years plagued me terribly. Prayer is my refuge.[92]

That Betsy Oakes was at this time, and had indeed been ever since Bobbin's death ten years earlier, the dearest person in

---

[91] *Ibid.*, ff. 438–439.
[92] Young, 1898: p. 438.

the world to Arthur Young there can be no doubt. That the beauty of her person, as well as the beauty of her soul, still appealed to him enough to be regarded as a temptation, is brought out in the above quotation.

Young's private life in 1807 can be concluded with a few more details about his religious views and activities. When Wilberforce faced a very expensive parliamentary election campaign, Young contributed fifteen guineas to the fund raised by his Evangelical friends.[93] Young saved a printed manifesto supporting the famous religious fanatic, Joanna Southcott, and urging him to join her organization.[94] Since the *Autobiography* shows that he frequently speculated about the Apolcalyptic prophesies in *Revelations*,[95] Joanna Southcott's millenarian prophesies may not have been entirely repugnant.

The Evangelicals frequently attempted to convert convicted criminals just before their execution. In 1801 John Symonds had ridiculed Simeon's efforts on behalf of some Cambridge thieves and had written Young, "Should it ever be my lot to be condemned for execution, I will immediately apply to you for consolation."[96] Young's only recorded effort for such a case was in 1807 in behalf of a convicted forger:

I had a good deal of conversation with him, and prayed with him. He was very ignorant, and had no feeling of religion till he was condemned. . . . I gave him the best instruction I could, and urged little more than (which is his only possible hope) faith in the Lord Jesus. He prayed aloud for some minutes, . . . thanked me much for coming, and as he begged me to come again I . . . accordingly this morning went on purpose. It comforted him much. I prayed . . . fervently for him, and his amens and ejaculations seemed to come from his heart. I left him tranquil, and . . . I think he may without presumption hope strongly for pardon at the throne of mercy.[97]

And yet the man who prayed so earnestly with the condemned forger in the Bury jail was probably the most famous

---

[93] Add. MSS. 35,129, f. 410. This is a printed list of the subscribers. John Scott acknowledged Young's gift on May 28, *ibid.*, f. 424.

[94] *Ibid.*, ff. 476–477.

[95] Young, 1898: pp. 349 (March 22, 1801), 425 (February 2, 1806), 445 (June 3, 1809).

[96] *Ibid.*, p. 369.

[97] *Ibid.*, p. 436.

English agriculturist alive, a man with an international reputation. When Lord Hardwicke wished to help his bailiff find a position in Russia, he wrote to Arthur Young for letters of introduction.[98] Count Dohna wrote Young about conditions in Prussia in 1807.[99] A Swede applied to Young to help find English gardeners and artisans to make agricultural machinery in Sweden.[100] David Humphreys, 1752–1818, famous American poet, military man, and agriculturist, wrote Young on March 13: "Mr. Humphreys flatters himself his long & intimate connection with the illustrious farmer of Mount Vernon in America will be received as an apology for his being ambitious of an acquaintance with Mr. A. Young."[101] Richard Phillips wrote to introduce a Sicilian nobleman, the Marquis de Saloo,[102] who reported how his teacher, the famous Abbé Paulo Balsamo of Palermo, regarded Young as his master in all things agricultural. Young's diary commented on the emptiness of such fame:

If there be glory in this sort of fame, oh! my Father, let me have done with glorying save in the Cross of the Lord Jesus. I have been glorying in foreign and domestic fame for forty years in true fleshly vanity.[103]

Still he was unable, however, to conquer such temptations, for in the very same diary entry as the above he could not help boasting that he was the first Englishman to cultivate cock's foot as a pasture grass.

Young was very busy in 1807 with Board affairs. On February 13 he was given three commissions, to make abstracts suitable to print from Count Reventlow's letter on Spanish sheep, to recommend the best method of sowing a new sort of wheat from Virginia, and most important, to make suggestions for advertising the Board's publications.[104] Two weeks later, after talking with several booksellers, he proposed a definite ratio of 60s. of advertising for every shilling of sales price for each publication,

---

[98] Add. MSS. 35,129, f. 487.
[99] *Ibid.*, f. 422. Dohna asked to be remembered to Sir John Sinclair and to Young's family, almost as if he had been Young's pupil.
[100] *Ibid.*. ff. 494–495.
[101] *Ibid.*, f. 382.
[102] *Ibid.*, f. 379.
[103] Young, 1898: p. 434.
[104] Minute Books, **6,** ff. 149–150.

and that advertisements should be divided between country and London papers in the proportion of 8 to 12.[105] Later that spring Richard Phillips purchased the dead stock (16,000 volumes of Board publications) for £3,246 to be paid in gradual installments and agreed to print future Board publications at his own cost, to give the Board 100 copies of each, and to be free to sell the other copies himself.[106]

Early in March, Young was worried by a possible food shortage and high grain prices resulting from the closing of the Baltic by the Continental System. He suggested that the government should grant £2,000 to the Board for premiums for potato production as cattle food. After he found that Nicholas Vansittart inclined to favor the proposal, the General Committee requested him to draw up more detailed plans.[107]

Young was to have delivered his lectures on tillage and farm buildings in May, but actually they were postponed until 1808 because there were no Board meetings in May or June, 1807. The reason for this unusual recess is not certain, but a guess may be hazarded. On April 29 Young received the following note from Sir John Sinclair, the president:

A most extraordinary circumstance has happened to me. A rascally saddler, whom I employed at Edinburgh to furnish accoutrements to my regiment of Invincibles, brought in so exorbitant an amount, that I refused to pay it; and instead of bringing an action against me at Edinburgh, he has arrested me for 750*l.* So many of my friends are out of town, that I must trouble you to give bail for my appearance.

Young and another friend signed a bail bond and Sir John was released from prison, but Young was disgusted:

Sir John's regiment has been disbanded nine or ten years, and consequently this rascally saddler has been at least so long kept out of his money. What can these people think of themselves! To live quietly while thus depriving tradesmen of their right for such a number of years![108]

[105] *Ibid.,* ff. 160–161.
[106] *Ibid.,* ff. 182–183, 191–194. This problem was discussed at meetings on April 17, 21, 24, and 27.
[107] *Ibid.,* f. 170; Young, 1898: p. 435; Add. MSS. 35,129, f. 388.
[108] Young, 1898: p. 437.

May it not have been thought undesirable to hold Board meetings when its president was in such a predicament?

On March 6 the General Committee voted: "Took into consideration the M. S. of the General Report on Enclosures, drawn up and laid before the Committee by the Secretary. Resolved that this M. S. be printed with broad margins for additional remarks and observations." Two weeks later the General Committee voted Young £100 for his "Chapter on Enclosure."[109] Sinclair's letter of August 18 to Young indicates that Young had not yet submitted the finished manuscript. After stating that the Ministry was interested in a general enclosure bill, Sinclair continued:

It will be desirable therefore to complete the General Report on Inclosure, with as little delay as possible, that it may be circulated among the Ministers and those who may have influence with them, as soon as possible. The state of Europe, and the dispute with America gives us some chance of success, and we must avail ourselves of circumstances as they arise.[110]

At the Board's last meeting on July 7 it was:

Resolved that, as it would be material to have an able agricultural survey of the County of Oxford, which would be a great inducement to those who attend the University of Oxford, to pay more attention to Agriculture than otherwise they might be inclined to do,— Mr. Arthur Young be requested to draw up the said Survey, with an allowance of £ 200. . . .[111]

Whatever may have been his attitude at the time of the appointment, he certainly faced the trip with reluctance: "As the time approaches to go this Oxford journey, I dislike it more and more, and wish I had firmly rejected it."[112] He probably left Bradfield early in September, and went through Essex, stopping to see Ruggles. He made his headquarters at Oxford itself, where several letters were directed to him at the post office. It was during this trip that Young first noticed the eye trouble which led eventually to total blindness, as Mary Young's note in the *Autobiography* attests:

[109] Minute Books, **6,** ff. 164, 175.
[110] Add. MSS. 35,129, ff. 450–451.
[111] Minute Books, **6,** f. 206.
[112] Young, 1898: p. 437.

It was upon this journey that Mr. Y. first perceived the approach of that dimness of sight which afterwards terminated in its total eclipse. His first suspicion arose from looking at the planet Jupiter, and perceiving what appeared to him to be two very small stars near him, at which he was much surprised, as he knew that the satellites were invisible to the naked eye, and nobody saw these stars but himself.[113]

The *Annals* for 1807 contain only three articles by Young. One was an account of three more experiments with manures, at the end of which he concluded: "Charcoal and indigo, in all my trials, are sure to prove beneficial. Burnt straw is of decisive benefit. Fresh leaves of vegetables have a considerable effect."[114]

The second was an account of the agriculture of his neighbor, Bernard Howard, Esq., of Fornham, an article filled with detailed accounts of the costs, expenses, and profits of Howard's sheep. These accounts were so seriously questioned by many readers that the Rev. William Gooch agreed to be their mouthpiece and wrote two long letters which Young printed.[115] Gooch's criticism was restrained, but apparently the men whom he represented had violently attacked Young's whole system of accounting. The controversy was extremely technical, but it revealed the degree to which these farmers thought in terms of profit as well as the complete chaos in accounting methods. One example will illustrate just how technical the problem became. Should the cost of a crop of turnips be charged entirely to the sheep which consumed them, or partly to the succeeding arable crops which benefited from them? If the turnips were charged only to the sheep, the profit of the sheep would be reduced. On the other hand, if the turnips were charged to the succeeding crops, the profits of arable land would be lessened. These arguments reflect the famous controversy between farmers whether sheep or arable were more profitable.

Young's most interesting article was "On Hemp," and was an attempted reply to some queries addressed to Young by the Office of Naval Revision.[116] The problem arose through the French control of the Baltic and the probable interference with

---

[113] *Ibid.*, p. 440.
[114] *Annals* **45**: p. 339.
[115] *Ibid.*, pp. 298–316, 430–435, 475–487.
[116] Add. MSS. 35,129, f. 442.

the normal purchase of Russian hemp averaging £600,000 per year. Young made it clear that it was much better to import cheap Russian hemp than to grow hemp in England at great cost, but, if Russian hemp were cut off, then England must grow her own hemp, at any cost. "Hemp we certainly *must* have; and if we cannot import it, we *must* raise it." Grasslands could be broken up and would produce excellent hemp, but Young did not believe such heroic measures necessary, for bog lands, if cleared and drained, would produce just as good hemp without the sacrifice of wool and meat which breaking up grasslands would entail. He admitted that to break up grasslands would cost less immediately than to drain bogs, but the ultimate benefit of clearing bogs would be much greater. Young ended the article by pleading eloquently for a general enclosure bill and the breaking up of the wastelands:

The events to which I have alluded, cannot fail of turning the reader's mind to the enormous wastes which are suffered at such a momentous period to disgrace the policy of the kingdom. Is it, can it be possible, that a general enclosure bill will be longer refused to the earnest desires of men, who crave of you leave to cultivate these wastes; ready to undertake and effect it; but crave in vain to make your wilds, moors, heaths, bogs, and commons, produce wheat for the support of your people, and, if you offer price enough, hemp for your shipping?[117]

The year 1808 was not especially important or eventful in Arthur Young's life, but the documentation is fairly complete. Although the *Autobiography* is thin, there are the Board records and numerous letters. For all 'practical purposes the *Annals of Agriculture* ended in 1808.[118] There were great triumphs, notably the lectures read before the Board, resulting in the award of the Board's Gold Medal. There was his important Report on En- closures, and the long, worrisome distillery business, involving a controversy with William Cobbett in the *Political Register* and testimony before a parliamentary committee. Young stayed at Bradfield until late in January, was back there late in April for a short visit, and returned in June for the remainder of the year.

Unquestionably his greatest worry during 1808 was the pro-

---

[117] *Annals* **45:** pp. 328–330.
[118] The final volume, **46,** began in 1809 and was completed in 1815.

gressively worsening condition of his eyes. The first entry in the *Autobiography* for the year reads: "In February I was obliged to take a reader as my sight was failing fast." By means of a clerk's "great hand and black ink" Young was able to read his own lectures before the Board in early April. On July 25 he wrote: "I can see to write a little, but can read scarcely anything." By the end of the year even his signature was almost illegible. The man chosen as reader was William St. Croix who seems to have been a pleasant young man of good family who was accepted as a member of Young's family and circle of friends. It seems almost as certain that his intellectual ability was limited. The job wasn't exactly a sinecure for Young stated that, though he continued to rise at 4 A.M., he did "not call St. Croix before five."[119]

Many friends commented on his affliction and offered advice of varying kinds. Perhaps Peggy Metcalfe's was as good as any, but certainly Young did not follow it:

I hear with the most sensible & lively concern what you say of your eyesight—preserve it, if you possibly can, & pray to God to . . . punish you in some way less dreadful than this; spare your Eyes, . . . and instead of tearing them open at a time of the morning when the creation is not ready for you, suffer those lids, which nature has so kindly placed to guard, and nourish them; if you cannot sleep, reflect, meditate, commune with yourself in your own chamber, and be still.[120]

The Rev. Thomas Fry sent Young a very cordial invitation to visit and referred to Young's eyesight:

You shall be your own master, and if your eyes are bad I will be your guide & my wife who is very ready at her pen, offers her time as your amanuensis. I know your habits & they shall be mine. Indeed I have now an interested motive in renewing my invitation. Your agricultural reports has [*sic*] induced me to make various

[119] Young, 1898: pp. 441–443. Various friends of Young refer to St. Croix with respect in their letters. His own letters or comments lack distinction. The Bradfield church contains a tablet in his memory, indicating that he probably lived there after Young's death.

[120] Add. MSS. 35,130, f. 4, dated January 16, from London to Young at Bradfield. Peggy also implored Young to influence Betsy to go to Cheltenham for her health: ". . . exert . . . every influence you have over Her to prevail upon Her to sojourn yet a little in this Vale of Tears."

experiments to better the worldly conditions of my poor neighbours, & I have various questions to ask in regard to them.[121]

During 1808 Young received four letters from Jane Young's sister, Louisa Jones, who had apparently visited the Youngs for a fairly long period and felt much indebted. In June she married the Rev. John Jones, a young unbeneficed clergyman. Unquestionably Louisa was the most completely Evangelical obscurantist in Young's whole acquaintance. Which is saying a great deal! Her letters were very long, very difficult to decipher, and very stupid. Her obscurantism is revealed in her attitudes both towards Young's increasing blindness and towards her sister's continued residence in Russia, as seen in two excerpts from her letter of August 10:

I am very sorry that your sight is so very bad, & I shd be more so did I not know that the Lord has a good intention toward the afflicted one. . . . the fulfilling of your prayer may be, by shutting out a great deal of the carnal sight, by striking the offending members; but be not careful much, as to the event sh'd it turn out according to yr fears, the mortifying the whole man must be necessary before it is suffered to enter into glory. . . .

I am uneasy ab't Jane every day I think & fear they lose the opportunity of coming to England when they might; where is it that sin will not drive us to! I trust that God will have mercy on them & bring them safe to our longing arms![122]

Again in another letter Louisa wrote:

I think if Jane & Arthur remain in Russia out of obstinacy, they will assuredly repent of it, what a strange thing that they cannot come home. I am sure if they prayed earnestly for it, God w'd not deny their request. . . .[123]

Two other letters throw a little light upon the Rev. Arthur in 1808. The first was one of Martha Young's secret letters, this time to Lord Hardwicke, apparently in the hope that he might in some way intervene to speed up her son's return to England:

My Lord
    I have the honour to plead for my excuse the following reason!
My only son Arthur Young has been three years in Mosco, at the

[121] *Ibid.*, f. 108.
[122] *Ibid.*, ff. 96–97.
[123] *Ibid.*, ff. 100–101, not dated.

desire of Monr Novoziltzoff . . . to make an agricultural Report
of that government (Mosco) altho' it has long since been finished
amidst the most untoward & miserable difficulties, he is still unable
from the well-known tardiness of that Nation to get it presented
to his Imperial Majesty! his situation of course is rendered in-
discribably anxious in addition to which he has suffered four or
five times by a violent fever & severe sore throat &c from a climate
hostile I think to every stranger, which at this time prevents all
comfort in society or prospect of return.

I heard your Lordship had honoured Mr. Young this day with
a call & I am surprized he forgot to mention the above circumstances . . . as Mr. Young's various & daily pursuits occupy him
so constantly I really fear he will not remember it, which is the
sole cause of my doing it. I had very nearly lost my life when
Arthur first left us, as Dr. Reynolds can testify. I am now so unwell
&c. &c. that I little doubt every thing relating to this world will
pass from me before his return.

. . . Mr. Young would I know attribute my miserable state to
weakness only & therefore as I shall spend a few days with Lord
Coventry I take the further liberty of begging a line in answer
directed to me Mrs. Young under cover to Earl Coventry Piccadilly. . . . I take the freedom to believe this letter will not be
mentioned.[124]

The second letter was from James Smirnove to Young:

. . . I am very much pleased with your Son's success in the Business,
it does him & you great honor; it will be of great Benefit to my
country. . . . I think he does Right to make an Excursion into
Crimea and I am certain his doing so, on the present occasion
placing in a manner a great confidence in the Emperor by wishing
to continue longer in Russia, whilst the two countries pretend to
be at variance, will be very agreeable to his Imperial Majesty. For
I am still of opinion that not many of our Ploughshares will be
made into spears against England—much more Ink, than Blood
I hope will be spilt on the occasion.[125]

At the end of July, Richard Worgan paid his long awaited
visit to Bradfield. His letter of July 22 was typical of his forth-
rightness:

[124] Add. MSS. 35,647, f. 158 (Hardwicke Papers). The post office mark is
clearly 1808.
[125] Add. MSS. 35,130, f. 110, dated September 13 from London. Two days
later Young replied, inviting Smirnove to visit Bradfield, and on Sept. 12 Smir-
nova accepted for about a fortnight later. *Cf. ibid.*, f. 112.

My first motive for wishing to know anyone, is there [*sic*] being *real* and not nominal Xtians. My second is that [they] be pleasant & communicative companions. . . . In the country only can rational society be enjoyed—and truly happy shall I be to pass a few days at Bradfield Hall, when if we like each other the pleasure will be reciprocal, if not no harm can befal us—and we need not meet again.

Apparently Young's reply was cordial, for Worgan's second letter announced his arrival on July 30: "You certainly have the brightest gem in the Christian crown—viz. *Humility*. The *virtue* of all others I am constantly striving for. . . ."[126] There is evidence that Worgan's "few days at Bradfield" were extended to five months.

Three letters from Thomas Ruggles testify to continuing cordial relations between the old friends. In September, Young and his guest Worgan spent a day or more at Spains Hall. On September 8 Ruggles wrote:

Sorry indeed am I my good friend at the account I read respecting your eyesight in the first paragraph of your letter, yet willing to hope from the continuation that it may not be so bad as you apprehend. . . .

I shall be truly happy to see you & your friend Mr. Worgan on the day you mention. he, coming from the Elysian fields of this happy Island, will find I fear nothing worth seeing in this tame country, but he will find a family happy to see him for his friends sake & I doubt not when we do . . . see him, for his own. All my Daughters are in some degree musical my two eldest much so, and our Piano is as I may say *untimely grown old* by the practice of the younger part of the family. I cannot get its power renovated but will try before you come to get it tuned. . . .[127]

Very late in the year Young invited the entire Ruggles family to be his guests for a Bury ball. In accepting Ruggles wrote:

My good friend does not know the pleasures that I find in this life of vegetation wch he supposes me with my family to pass at Finchingfield, if I find Books in the running Brooks, sermons in stones & good in everything; what can, or rather, what ought man to wish for more? besides I am rather better off; I have a wife a son and Daughters who administer to my Comforts and therefore

[126] *Ibid.*, ff. 92–95.
[127] *Ibid.*, f. 104.

am even in a better situation than the melancholy Jacques; . . . My
wife and Daughters desire me to say they think a little of Mr.
Worgans voc instrumental musick would be of great service to
them in their supposed state of vegetation. . . .[128]

Although there are no letters from Betsy Oakes for 1808, there
are several from Peggy Metcalfe and from another neighbor,
Philippa Affleck. Both of Mrs. Affleck's letters reflect Young's
theological ideas. The first dates from July 12:

. . . I fear we shall never agree upon the subject of Baptism—
you appear to think it of so little effect that I begin to suspect
you are a Quaker—We are quite impatient to wine & see you—
and only wait for neighbour Betsy to fix the day—which I know
will be a delightful one to us—Adieu.

The second was apparently written after the party:

Your poor Infidel friend returns your catechism with effusions of
gratitude for the anxiety you appear to have for her conversion
. . . . I must ever most obstinately adhere to my favorite opinion
that our original sin is blotted out the moment we become Christians & that it is the world makes us wicked, not our own nature
. . . . I hope we shall meet soon again—but we will not talk upon
religion, as it is the only subject upon which my opinion would
not bend to yours. . . .[129]

His curiosity aroused by his article on hemp in the *Annals* in
1807, Young asked Joseph Scott, an engineer from Chatteris who
had made the original Cambridgeshire report, to travel about the
fenlands making inquiries about the possibility of raising hemp
there. Scott spent nearly three months in these inquiries and
wrote seven letters to Young about the business,[130] always starting
with the formal salutation, "Honoured Sir." From the beginning
Scott's financial reward for his work was left all too vague. In his
first letter, he wrote: "And as to my expense . . . I will leave it
entirely to your Honour to allow me just what you think reasonable. . . ."[131] After Young had requested that the journeys be
ended, Scott wrote again:

---

[128] *Ibid.*, f. 147.
[129] *Ibid.*, ff. 86, 88.
[130] *Ibid.*, ff. 33–34, 35–36, 43–44, 61–62, 98–99, 106–107, 121. The dates run
from March 18 to Nov. 11. Young apparently wrote six letters to Scott.
[131] *Ibid.*, f. 34.

And your Honour desired a note of what your Honour was in my debt. But I would rather leave it to your Honour to send me what your Honour pleases, after I have just informed your Honour of my Journies in this business.[132]

Nevertheless he was greatly disappointed with Young's offer of £5 for his exertions:

And as I always act with integrity in every case and wish to oblige you as far as I am able I will take the £5 for my Journies about the Hemp although that small sum will not half pay Horse Hire and my other expences about the business. But I hope that you will maturely consider the pains I have taken and the use my journies have been . . . and either send me a larger sum or procure me some other employment as soon as possible.[133]

His last letter makes it clear that Scott only got the £5.[134] He may have made some profit from the sale to the Navy Office of some hemp purchased during the trips, and he published an article in 1809 perhaps based upon the information gained in them. But £5 for two and a half months' work and travel does seem pretty niggardly.

The year 1808 marked a new headache for Young which continued for several years. He had tried repeatedly to improve the situation of his protégé, the Rev. William Gooch. In May, 1808, Lord Templetown of Castle Upton in Ireland and of Pakenham, Suffolk, was searching for a new agent for his Irish estates. The job was a good one, and Young suggested Gooch who received the position. From the beginning both Templetown and Gooch regarded Young as an intermediary. The arrangement proved satisfactory to none of the three and ended in Young's almost complete estrangement from Gooch.

On December 1 Templetown addressed the following memorandum to Young:

Lord Templetown begs to present his best respects to Mr. Young— The enclosed is the amount of the remittance made by Mr. Gooch since his arrival in Ireland viz. *Two thousand three hundred and twenty-seven Pounds, nineteen shillings*—and will if not steadily added to, be a very serious inconvenience to Ld. *T*.[135]

[132] *Ibid.*, f. 98.
[133] *Ibid.*, f. 107.
[134] *Ibid.*, f. 121.
[135] *Ibid.*, f. 134.

Immediately Young dictated a very long letter to Gooch:

. . . as it appears to me . . . that you have not clearly understood
his intentions, I am induced through mere friendship for you, to
trouble you with a few observations, which strike me as being the
more necessary, because I perceive a reluctance in his Lordship, to
be so explicit with you, as he would be, if he was not corresponding with a gentleman: this is a delicacy very honourable to
him, and advantageous to you, provided it does not lead you into
any erroneous conceptions of his meaning. . . .

Lord Templetown's first expectation as a sine qua non, is a half
yearly regular remittance of £ 4,500 . . . and you will observe that
all improvements, repairs, payment of bills, things purchased, salaries,
in a word, all payments whatever, & however authorized by himself
in conversation or by letter, must all be secondary to this first
& great standing order in your operations; keep this well in mind;
to slate a stable, to stock a farm, to build cottages, to make dona-
tions, to pay bills, &c. &c. may or may not be proper *after* the
regular remittance, but they are utterly inadmissable, *before* it . . .
I was very sorry to find, that in the end of November he is above
£ 2,000 short in the receipt; . . . Now to come to a few particulars. . . . You mention furnishing offices, and buying things for
that purpose; his Lordship does not know what this means, and
that is sufficient proof to me that it ought not to be done: & as
to all improvement by buildings, it is much to [*sic*] early to think of
t, such things if ever done at all, must come out of your improvements. . . .

. . Lord Templetown remarks that you have had a great deal of
trouble with the late agents business, & that you managed the
matter of the arbitration with much industry & address; & I should
also observe, that it gives me much pleasure to find, that he has a
high opinion of you & has no doubt of your being able to do his
business entirely to his satisfaction. It is my seeing this, that makes
me the more solicitous that there should be a perfect understanding between you, which can only arise from an explicit mention of
what is expected.

All bills against Lord Templetown should be sent directly to
him, for him to endorse . . . then you are no longer answerable for
them, either to him or to the tradesman. . . .

. . you must send many about their business, & gradually establish a better tenantry, but this must not be done at Lord Temple-
own's expence; . . . Nor should you forget, that a non-resident

Landlord, & a resident agent have in one respect a contrary interest; the former wants as much money as he can get; the latter to live agreeably amongst people that are under obligations. . . .

. . . I take for granted that his linen is all locked up & never touched; to say the same of wine would be needless. . . . this applies still more to certain beds, particularly a state bed, but writing as I do ignorant of the state of the Castle I can no more than hint these things. . . . all your remittance to this date, amount only to £ 2,377, 19. this is instead of £ 4,500 &c. is such a defalcation, that you are roasting him before a slow fire, my expression not his, but to use the mildest, he really is in hot water. . . .

His Lordship laid open to me his affairs in England in such a manner as brought conviction with it, that a regular remittance of the whole net income is essential to his existence, or all idea of comfort must fly from him. . . .[136]

Young's letter shows why even enlightened English estate agents could do little to improve Irish conditions. Here was a young man enthusiastic for the new agriculture, being advised by the very high priest of that new agriculture, to forget all about it in Ireland until the absentee landlord's rent had been met. One wonders whether Gooch knew that Young himself had hardly been a success as an estate agent, and for precisely the same reason that he was going to fail. At any rate Young's letter resulted temporarily in better relations all around.

On January 26 the Board voted that Young's report on Oxfordshire should be referred to the General Committee which in turn voted to send the manuscript to John Fane, Esq., for possible corrections and additions, and to pay Young £100 on account for the work already done.[137] Fane was a member of the Board and a very prominent Oxfordshire agriculturist to whom Young constantly referred in the printed report. Whether the General Committee's decision indicated any dissatisfaction with the manuscript is impossible to determine. Certainly the final document bears no evidence of having been materially revised by Fane.

The Board was not especially active in 1808 and several meetings in February had to be adjourned for lack of a quorum. On March 22 Sir John Sinclair was unanimously reelected as presi

[136] *Ibid.*, ff. 132–136.
[137] Minute Books, **6,** ff. 208, 210.

dent.[138] That he was still indefatigable, and to some purpose, is shown by the following note to Young:

I have the pleasure to inform you that I have just moved an address to the crown, which passed *unanimously,* for granting *an additional grant* of £ 1,500 to the Board of Agriculture for enabling it to complete the County Reports *this year.*

Is not that a great event?[139]

April 5 and April 12 were proud days for Young. The Board minutes for those dates read:

April 5—The Secretary read his Lecture on Tillage and explained the principles thereof in a very satisfactory manner, to a numerous and respectable meeting, consisting of the Board and a number of Visitors, several of them from Foreign Countries.

April 12—The Secretary delivered a Lecture on Farm Yards, and exhibited to the Board several models explanatory of the principles of that important department in Husbandry:—Resolved that the thanks of the Board be given to Arthur Young Esqr. for the two Lectures . . . and that he be requested to prepare these lectures for immediate publication, accompanied by the necessary Engravings for explaining the principles thereof:—the Board being of opinion that by the publication of these Lectures, great advantage may be derived by the farming interest of the Kingdom.[140]

In spite of the vote, there is no indication that the lectures were published. They do not appear in the *Communications,* nor have they survived elsewhere. The *Elements of Agriculture* mentions one point covered in the second lecture: "I proposed a small canal to pass from the farm yard to as great a distance through the fields of a farm, as the extent of it would admit for the conveyance of dung in narrow barges. . . ."[141] The *Autobiography* commented briefly on the first lecture: "The room was well filled, and several of much ability and more of rank; but the day was a bad one, which kept others away. I found that they were well satisfied."[142] Apparently the above is an understatement. After the second lecture Sinclair wrote: "I enclose the Resolutions regarding your lectures, which gave very

[138] *Ibid.,* f. 234.
[139] Add. MSS. 35,130, f. 82.
[140] Minute Books, **6,** ff. 242, 245.
[141] Add. MSS. 34,861, f. 50. *Elements of Agriculture* **7.**
[142] Young, 1898: p. 442.

general satisfaction. Mr. Pinckney and others said they were never more gratified."[143]

Three weeks after the second lecture, Sir John made the following announcement:

The President . . . gave notice that at the next meeting of the Board, . . . a Ballot would take place for the Gold Medal to be given to Arthur Young Esqr. for the ability with which he had explained in his Lectures to this Board the advantages of Tillage and the best mode of conducting that most essential operation, and for the new and valuable light he has thrown on the proper construction of Farm Yards,—two most important branches of Husbandry, which have never hitherto been so ably illustrated: and also for his long and faithful services in the cause of agriculture.[144]

Sir John presided over the meeting which was well attended. At noon they "proceeded to ballot for the Gold Medal being presented to the Secretary." Following the balloting Humphrey Davy read a lecture. "At 3 o'clock opened the Balloting Box, when it appeared that the Gold Medal was unanimously voted to Arthur Young, Esq."[145] Unfortunately the minutes do not state whether he was then called in and given the medal on the spot, whether Sinclair made a congratulatory speech or Young himself an acknowledgment. Regardless of the merit of the lectures, the recognition was surely due to one who had served the Board faithfully for fourteen years.

On May 20 the General Committee met. The minute book started out to record the meeting, "Took into," after which two words the account breaks off.[146] And there is no manuscript record of the Board''s work from that date until 1817. Unfortunately there is no clue for the break. Perhaps it was connected with Young's eyesight, but the under-secretary could have taken the notes. Perhaps a new book was started which has been lost. The mystery remains unsolved.

Volume 6 of the *Communications* contains a very short article by Young entitled "Oeconomical Dwellings for small Proprietors

---

[143] Add. MSS. 35,130, f. 45. William Pinckney, American Minister to London, was elected an Honorary Member of the Board on May 3. *Cf.* Minute Books, 6, f. 250.

[144] Minute Books, 6, ff. 248–249.

[145] *Ibid.,* ff. 253–255.

[146] *Ibid.,* f. 258.

of Land." The text, less than a page long, was really only an explanation of the two plates accompanying it:

Such a house as this, will not be burthensome to a gentleman with £ 800., (clear income), and a prudent man, with from £ 2000 to £ 2500 per annum, will not regret the want of a better. In a country where taxes are much on the increase, the *dwelling* ideas of gentlemen should be reduced, or distress must be the consequence.[147]

The plates show a front elevation of the house, severely plain except for a handsome front door with a fine fanlight decoration. It was a two-story dwelling with two large chimneys. The plates also show the ground plan for both stories and the arrangement of the farm offices and barnyard. The ground floor contained a dining room, drawing room, study, sitting room, dressing room, store room, butler's pantry, and water closet. The second floor contained seven bedrooms, two with adjoining dressing rooms. The kitchen was outside the house, but connected with it by a covered passageway. Among the offices and farmyard buildings were a scullery, poultry house, woodhouse, dairy, coal cellar, brewhouse, bakehouse, granary, cow-house, coach-house, stable, servants' hall, and laundry. One can agree with Young that only a great noble or a man seeking ostentation could have wanted more.[148]

The *Report on Enclosures*, which had been laid before the Board and accepted in March, 1807, was published in 1808. Young had received £100 for it. It had been planned that the report should serve as the basis for further public contributions on the subject, and the Board had offered prizes for the best ones.[149] Sir John's brief introduction made clear that the report

[147] *Communications,* **6**: p. 261.

[148] The dining room and drawing room were 24 x 18 ft., the study and sitting room 18 x 18 ft., and the two large bed rooms 18 x 16 ft. It is not certain that the construction of the offices and barnyard on the scale pictured could also be borne easily by a gentleman of £ 800 annual income, but the presumption is that it could.

[149] *Cf. supra* p. 520. P. vii of of the *Report on Enclosures* lists the Board's premiums. The first prize was a piece of plate worth £50. The Young MSS. contain two criticisms of the Report. Montagu Burgoyne (Add. MSS. 35,130, f. 21) warned Young to take precautions in behalf of the poor, "for if they are injured & the Poor rates increased, the benefits of enclosures will fall to the ground." A critical note (*ibid.,* f. 14) from William Pitt—not the statesman who had died in 1806, but one of the county reporters—to Sinclair, pointed out specific errors in the Report: ". . . I have already noticed some very material errors, of the press or otherwise; it is I should suppose of great importance that the numerical tables particularly should be free from errors."

could not have been made without the digested material in the county reports.

It has been prepared "by a very intelligent Agriculturist, who seems to have done ample justice to the plan above suggested. . . . Such a paper, formed with so much care and attention, ought to be considered as a species of Code, or Standard, regarding all points connected with Enclosures; and indeed must set almost every question regarding it at rest."[150]

Later similar reports would be made on implements, management of grasslands, cattle and sheep.

Naturally the Report is primarily a compilation. Only a very small portion of the 392 pages was newly written by Young, whose fee of £100 seems quite generous for the amount of work involved. About two-thirds of the volume consists of appendices of various kinds, tables, statistics, texts of acts. Large excerpts were taken from the county reports. Young also quoted copiously from other books and reports, and from his own articles in the *Annals*.

The first chapter was devoted to the lands which should be enclosed, both waste and arable. Under each heading he discussed the extent of such unenclosed lands, and the prospective results of enclosures on farmers, landlords, the poor, the church, and the public in general. The second chapter treated in general terms the methods by which land was enclosed, especially of course the act of Parliament. The third chapter analyzed in detail the enclosure process under such headings as the survey, the valuation, the award, fences, roads, drainage, and expenses. The short last two chapters emphasized the insufficiency of England's agricultural production and the necessity for a general enclosure act, making separate acts unnecessary and hence greatly reducing the costs.

Arthur Young could write little new about enclosures in 1808. He had been advocating them for forty years! The arguments differed somewhat for wasteland and arable. He summarized the benefits from the enclosure of wastelands in the following panegyric:

[150] *Report on Enclosures*, pp. iv-v. Nowhere in the volume is there any direct reference to Young as the author. The Board's records make it clear, however, that it was his work. Amery ascribes it to him. Internal evidence supports the view of his authorship.

. . . to convert tracts really or nearly waste, into profitable farms; to change ling for turnips; gorse for barley; and overstocked and rotting grass lands to wheat . . . is a real acquisition of territory, pregnant with every advantage attending the husbandry of the kingdom; food, population, wealth and strength.[151]

The chief benefit from the enclosure of arable open fields came from the abandonment of a routine dictated by the least progressive cultivators:

What system of barbarism can be greater, than that of obliging every farmer of a parish possessing soils perhaps totally different, all to cultivate in the same rotation! What a gross absurdity, to bind down in the fetters of custom ten intelligent men willing to adopt the improvements adapted to enclosures, because one stupid fellow is obstinate for the practice of his grandfather! To give ignorance the power to limit knowledge, to render stupidity the measure of talents, to chain down industry to the non-exertion of indolence, and fix an insuperable bar, a perpetual exclusion, to all that energy of improvement which has carried husbandry to perfection by means of enclosure! Yet is all this done by the common-field system.[152]

Young attempted to meet all the commonly expressed objections to enclosures. Obviously they almost always benefited the landlords, and he rejected out of hand all complaints by the clergy: "That the interests of the church have been well attended to upon enclosure, is very generally known. . . . whoever may have lost by enclosures, the church has been sure to gain."[153] He admitted that many farmers had a real grievance, owing to the length of time required and the expenses incidental to the process of separate parliamentary acts. These evils would be largely remedied by a general act. More serious was the complaint of the small farmer who sometimes lost his holding in the process. Wherever possible such small farmers should be helped to survive, but they must not stand in the way of progress. They were bound to lose out in the long run anyway, since they were unable to meet the competition of men with greater capital.

That it is a great hardship, suddenly to turn several, perhaps many of these poor men, out of their business, and reduce them to be

[151] *Ibid.,* p. 22.
[152] *Ibid.,* p. 219.
[153] *Ibid.,* p. 36.

day-labourers, would be idle to deny; it is an evil to them, which is to be regretted: but it is doing no more than the rise of the price of labour, tithe, rates, and taxes, would infallibly do, though more gradually, without any enclosure. These little arable occupiers must give way to the progressive improvement of the kingdom.[154]

That Young should have been so callous towards the small farmer is somewhat shocking, considering his very real solicitude for the laboring poor, but he had always been an enthusiast for capitalistic farming and had little sympathy with subsistence farming. He repeated his contention that the laboring poor, including cotters, should be left enough land for a garden patch and a cow. Humanitarian considerations were of course important, but even more so was the fact that such a scheme would reduce poor rates. Here was speaking, not Arthur Young the Evangelical humanitarian, but the secretary of the Board of Agriculture, representative of the landed interests.

The humanity of such a measure is a very strong recommendation; but in a general system, such a motive must be sunk, as too weak to expect any general operation. The grand object is the reduction of poor-rates; a burden which has of late years proceeded with so rapid an increase, as to threaten very heavy evils to the landed interest.[155]

To the argument that the enclosure of wastelands had not resulted in increased production and that such lands often remained wastes, Young answered that all too often lands had been ruined by attempting to raise corn on lands only suitable for grass. He frankly admitted that the enclosure of arable open fields had frequently resulted in converting arable to grasslands, but contended that enclosures so greatly increased the yield per acre that any decrease in total grain production was doubtful. He continued:

If the lands in question are much better adapted to grass than to corn, that is, will yield a greater neat profit, they certainly ought not to be under corn. The reason why they yield such profit, is clearly the price of the product of grass-land being raised by the

---

[154] *Ibid.*, p. 33.
[155] *Ibid.*, pp. 14–15. He also wrote, p. 130: "But one word more—take care of the interests of the poor; they will pay for wastes (in saving rates) treble the rent of all your other improvements put together."

demand; and where is the writer who will contend that this demand should not be satisfied?[156]

Since separate parliamentary enactments for each enclosure
were cumbersome and expensive, the obvious remedy was a
general enclosure bill. To the argument that Parliament should
not delegate such important authority, he answered that it had
done so in the case of the Poor Laws.

Are precedents wanting for such delegation of power over private
property? What think you of delegating an unlimited power of
taxation to every vestry in the kingdom, filled with as low and
ignorant people as are to be found in it?. . . . You give a power to
an ignorant, and even to an interested set, to tax with much failure in the object of the intention; and yet you will not delegate a
like power into better hands, in order to attain an end in which
it is impossible you should fail![157]

He concluded "a General Enclosure Act is absolutely and
essentially necessary to the prosperity, peace, and safety of the
kingdom."[158]

Fully as important as his *Report on Enclosures* were Young's
letters and articles on the "Grand Distillery Question." In the
summer of 1808 Parliament passed a measure prohibiting the
distillery of barley into spirits. Behind the bill stood the West
India lobby which hoped to substitute sugar for barley in the
distilleries. It was also argued in support of the measure that it
would increase the amount of grain available for food in case
a poor crop should combine with Napoleon's Continental System
to bring about famine conditions. The agrarian interests fought
the proposal vigorously but in vain, primarily because it would
lower the price of barley.

The whole controversy was very technical and complicated.
Naturally there was a great deal of repetition in the arguments
used. Corn Laws and enclosures became involved in the discussion. Young was consistently hostile to the proposal. He wrote
several letters to William Cobbett's *Political Register*,[159] which
in turn Cobbett attempted to refute. These letters were not
confined to the distillery question, but ranged over the whole

---

[156] *Ibid.,* p. 38.
[157] *Ibid.,* p. 129.
[158] *Ibid.,* p. 111.
[159] *Political Register* **13:** pp. 288–303, 375–378, 568–574, 768–776.

gamut of national economic policy, especially over Corn Laws and enclosures. In March he testified before a parliamentary committee, and reprinted his testimony in the *Annals*.[160] He sent out a circular letter on the subject[161] and inserted twenty-five answers in the *Annals*. A short article in the *Annals* summarized his main objections to the measure.[162] Young's role in the agitation was heightened by his arguments with Cobbett, for both men were able controversialists and masters of the English language.

Young's first letter in the *Political Register* had nothing to do with the "Grand Distillery Question," but rather attempted to refute Cobbett's attacks on commerce in 1807 in a series of articles entitled "Perish Commerce," and similar views advocated in William Spence's pamphlet, *Britain Independent of Commerce*. Both Spence and Cobbett had maintained that foreign trade was unnecessary, for national wealth depended almost solely upon agriculture. Young likened these ideas to those of the Physiocrats which he had opposed more than thirty years earlier in his *Political Arithmetic*. He agreed that agriculture was more basic than commerce, "for a nation may exist without commerce or manufacture, but not for a moment without agriculture." He continued:

The habits of my life for forty years, have given me a decided preference for agriculture; but, Sir, it is for an agriculture animated by a great demand; and, when I hear the sister employments depreciated which constitute that demand, I must readily confess that I am alarmed, lest the first and greatest basis of our national prosperity should suffer as much from its friends, as ever it did from its enemies.[163]

Young had no patience with the views that most imports were luxuries which could well be replaced by domestic products:

. . . that those who drink tea might substitute sage or balm; that those who drink wine might drink water; that tobacco being a vile weed men might leave off smoking; in a word, that those who consume commodities because they want and desire them, might become much wiser and consume other commodities they neither

---

[160] *Annals* **45**: pp. 573–604.
[161] *Ibid.*, p. 514.
[162] *Ibid.*, pp. 605–608.
[163] *Political Register*, **13**: pp. 289, 291.

want, nor desire. Such speculation may shew . . . what great talents can effect in confounding . . . the plainest dictates of common sense.[164]

In his second letter Young apologized for a statistical error in the first letter about wheat imports in 1806, an error excused by his eye trouble and dependence on an amanuensis. The letter was very alarmist about the outlook should a serious shortage develop. Where could England make up the shortage? ". . . to me it seems just as probable to procure it from the moon as from Prussia or Poland." On the other hand,

To expect our bread from America, would be to look for it from a country whence it never came . . . Who can contemplate the consequences of a short crop, a mildew, or a wet harvest, without terror? Manufactures and commerce inactive; and ill disposed minds gathering discontents into petitions for peace![165]

Young saw the best hope in the encouragement of the cultivation of the wastelands through the passage of a general enclosure bill.

In his reply to these letters, Cobbett praised Young highly: "Great praise is due Mr. Young for his researches, for his accuracy in detail, and for the ability with which he discusses all the subjects of which he treats. . . ." Nevertheless he vigorously disagreed with Young about the necessity for a general enclosure bill, which he dubbed "a wondrous monument of national folly."[166]

In his third letter Young attempted to refute Cobbett's objections to a general enclosure bill, but he exchanged compliments with his adversary:

. . . if you would give the necessary attention to such a question, nobody would discuss it more powerfully, and I must esteem it a misfortune to the public, that the editor of a paper, the circulation of which is so considerable, should have declared himself explicitly against a measure, which I must esteem as essential to the public welfare. [167]

Young denied that a general enclosure bill would lead to endless litigation. To Cobbett's contention that enclosures might

[164] *Ibid.*, pp. 299–300.
[165] *Ibid.*, pp. 376–377.
[166] *Ibid.*, pp. 366–367. These pages are in same issue with Young's second letter.
[167] *Ibid.*, p. 568.

increase corn production and hence lower corn prices—already too low—Young answered that prices were reduced by corn imports and that enclosures tended to lessen imports. He concluded by completely denying Cobbett's charge that a general enclosure act would be an act of folly:

It remains, Sir, for your readers to judge whether the measure of a general inclosure act would indeed be a monument of folly; or the basis of wealth of power and prosperity. I think it might be so framed as to prove a decided means of public security;—an irrefragible proof of attention to great and important interests. That it would improve the morals and animate the industry of the people! encrease the revenue of the public and prove in the event a MONUMENT OF NATIONAL WISDOM.[168]

The last regular number of the *Annals* was devoted almost entirely to the "Grand Distillery Question" which Young stated as follows: "Will stopping the Malt Distillery be injurious to the Agriculture of the Kingdom?"[169] A transcript of his evidence before the parliamentary committee was included. Young was very thoroughly examined by the Committee. His main point was that the proposed bill, by announcing that the malt distillery would be stopped at a certain date, would reduce barley prices by decreasing the demand for that grain. Young maintained that it would be fair to stop the distillery of barley only after a general grain shortage had developed. He was also suspicious that the prohibition, once passed, would become permanent. He admitted that the West Indian sugar interests needed relief and suggested a reduction in the import duty so that sugar might be used to fatten cattle. He denied that oats or spring wheat could be substituted as a crop for barley. On certain lands, "peculiarly proper for barley"[170] in Suffolk and Norfolk, no substitute was acceptable for barley, which also constituted a vital step in the crop rotations.

Young had his answers ready to several questions about the best policy to meet a real food shortage. To build up a grain reserve would only depress prices. Rather, cultivate the wastelands and raise potatoes. He repeatedly attacked the Corn Laws

---

[168] *Ibid.*, p. 574.
[169] *Annals* **45:** p. 513.
[170] *Ibid.*, p. 581.

which had made England dependent upon imports by failing to give adequate incentives to domestic agriculture. England could still produce all the food she needed if only the incentives were given. When asked whether such an end could be obtained without breaking up pastures which would in turn depress the grazing interests and thus raise meat prices, he answered:

It would be a very great benefit to break up pasture. A great deal of bad pasture remains so, for want of encouragement to plough it. . . . The grazing of the country does not depend upon bad pastures, but upon good; and landlords would certainly take care that good grass should be supplied. . . .[171]

On April 5 Young sent out a circular letter on the distillery question to a large body of correspondents. The two vital questions were: (1) Could oats, hemp, or spring wheat be substituted for barley without injury to the farmer? (2) Would the prohibition of the distillery reduce barley prices? Most answers to the first question agreed with Young that no substitution for barley could be made, but a few felt that oats or spring wheat might be substituted on some kinds of land. The answer was unanimous to the second question that the proposed measure would seriously reduce the price.

Young's position was typical of the agrarian interests. Two letters from Sir John Sinclair show very clearly that the landed people were making a concerted effort to defeat the measure and that Young was playing his part. On April 15 Sinclair wrote: "I hope you are getting Suffolk put in motion."[172] When Young replied that Suffolk was asleep, Sinclair commented: "I hope you will be able to rouse it, before it is too late—Nothing but unanimity & exertion will give the landed interest the least chance of success."[173]

Young's short article for the *Annals,* entitled "Plain Facts," may have been in response to Sinclair's appeal. There was nothing new in the brief numbered theses or assertions. The substitute of any other crop for barley would hurt succeeding wheat crops in a rotation. The enclosure and cultivation of waste would increase the output of wheat and oats, not that of barley, and

---

[171] *Ibid.*, p. 590.
[172] Add. MSS. 35,130, f. 45.
[173] *Ibid.*, f. 48. Letter is undated, but postmarked April 26. It was sent to Bradfield where Young was spending the Easter vacation.

hence enlarge the available food supply. The farmers never complained of low prices caused by plentiful crops, "but when low prices are caused by the import of foreign corn, or by mischievous prohibitions, like that now in contemplation, they have the greatest reason to complain." The measure would raise sugar prices and hence affect the poor adversely. His final proposition was a call to action: "To the Freeholders of this Kingdom the proper address is:—Gentlemen! this measure demands your attention; for its direct object is to *raise* the price of your sugar, and its direct effect is to *sink* the price of your barley."[174]

In several issues of the *Political Register,* Cobbett strongly supported the proposed measure, and repeatedly hammered away at the basic inconsistency of the landed interest, who on the one hand expressed alarm at the dangers of a scarcity and on the other hand opposed a measure which might ameliorate such a scarcity. In his issue of May 7, Cobbett attacked Young's testimony in detail, quoting liberally from it. He dubbed Young "the grand war-horse of the corn men" and wrote: ". . . it is painful to see a gentleman of great and acknowledged talents, and of experience . . . surpassing that of, perhaps, any other man living, thus hampered by the influence of a more than sectarian bigotry to one particular pursuit."[175] At another point he burst out, "Respect for Mr. Young's talents and zeal restrains me; but, really, this is almost too much to bear."[176] His concluding blast was equally bitter:

. . . as Mr. Young is evidently the oracle of the country gentlemen, and of all the patrons of high prices of corn, it seemed to me necessary to show, that, either he is a gentleman of very unsettled opinions, or is carried away by a misguided zeal for the interest of that particular class of the community amongst whom he has had the greatest intercourse, and with whom he has long been an object of admiration and respect.[177]

Even before these blasts could have come to his attention, Young attempted to refute Cobbett's arguments in favor of the proposed legislation. He was pretty devastating, for he could turn a phrase almost as well as Cobbett:

[174] *Annals* **45:** pp. 607–608.
[175] *Political Register* **13:** pp. 718, 720.
[176] *Ibid.,* p. 723.
[177] *Ibid.,* p. 728.

Political economists and common sense tell us, that if a large portion of a demand be withdrawn from a market, price must fall; that a fall of price discourages production, and that eventual scarcity will be the consequence. . . . Now, Sir, a considerable portion of your paper is employed in stating a train of consequences that militate with these first principles of political economy: they are erroneous, or you are wrong. For, Sir, what is the grand object of your reasoning? But to prove that an immense demand for barley may be withdrawn from the market, and yet the farmer not suffer; this is the position, turn and twist it as you please.[178]

Young also accused Cobbett of "confounding the terms *corn* and *barley*."[179] to explain his charge of inconsistency against the landed interest. Again Young expressed his fear that the measure would not be a temporary one, but would lead to a prohibition against barley made into beer as well as into spirits. His fears were intensified by his knowledge of the Corn Laws during the past generation.

To tell us, therefore, that this measure is an experiment, and that government *may* give relief, is to feed us with a very thin diet indeed; whipt syllabub has cream and sugar in it, and water is an wholesome beverage, but the police of corn in England is framed (to use a farming expression), to starve a lark.[180]

Young ended this letter with an extravagant panegyric upon the blessings of the English government and upon God's providence in protecting England:

Give me leave, Sir, to add, that I could not contemplate the transactions of the committee-room without delight. When and where did the sun ever shine upon a country that exhibited such a spectacle? Planters and merchants, agents and revenue officers, landlords and their plain tenants, nay even dabblers in political economy, all listened to with patience and candour, as if but one motive animated every bosom,—a wish to ascertain the truth. What a spectacle! and whence has it arisen, but from the beneficent providence of a Deity that has poured out on this happy country the unexhausted blessings of matured freedom. Who that lives in such a kingdom but must draw in gratitude to heaven with the very air he breathes? . . . Let Britons be true to their God, their king,

[178] *Ibid.*, pp. 769–770.
[179] *Ibid.*, p. 773.
[180] *Ibid.*, p. 775.

their country, and themselves, and, that unseen, but mighty Hand, which has rendered us the envy of the world, will, with infinite wisdom, protect what infinite goodness bestowed.[181]

Never had Arthur Young made a more complete identification of his patriotism and his religion.

[181] *Ibid.*, pp. 775–776.

X. More Tragedy,
1809-1811

Two BITTER BLOWS came to Arthur Young during these three years. His eyesight became steadily worse. In 1811 he was operated on for cataract, but the operation was unsuccessful and total blindness followed. In the same year came the loss of his dearest friend, Betsy Oakes. She had not been strong for several years. After hemorrhages in 1809 it became clear that she had tuberculosis. Although she rallied in 1810 the end came in 1811 and Young was indeed alone.

The documentation also is much less complete for these years. Since there are no manuscript records for the Board of Agriculture, Young's activities there become shadowy. The *Annals* had also come to an end. There are only ten pages in the *Autobiography* for the entire three years and the gaps are frequent. Fortunately the letters are quite numerous, perhaps because he now had a secretary who kept them more systematically.

Young spent the greater part of 1809 at Bradfield. He went to London late in January, returned to Bradfield for the Easter vacation, and was back in the country by mid-June. During the summer and fall he was alone with St. Croix, for Mrs. Young and Mary stayed at London until September and then spent the fall at Aldborough on the Suffolk coast. His eyesight became steadily worse. In May his lectures at the Board had to be read for him by the assistant secretary, William Cragg.[1] By July he was apparently considering a cataract operation,[2] and noted in the *Autobiography* his trouble in viewing the scenery near Bradfield and in making out faces. Early in August he started a journal letter to Jane Young in his own handwriting, but abandoned the attempt after a few lines.[3]

[1] Young, 1898: p. 443.
[2] Add. MSS. 35,130, ff. 277–278. Letter from Wilberforce, July 18. ". . . & though I cannot welcome a Cataract or admire its Beauties & Perfections with a professional eye, yet I shall be glad to hear that your Hope of Relief, tho' through ye operation of couching, continues unimpaired."
[3] Young, 1898: p. 446.

Young still possessed enough fire to compose the following remarkable self-portrait for Peggy Metcalfe:

My own Portrait

I came into the World a fine looking, thriving Boy, and thus gave an early promise of becoming in due time what is called a proper looking handsome man; such at least my glass told me, and I did not *quarrel* with it as a Flatterer, my height was above the middling stature, being about 5 ft. 10 the *then* standard of perfection. . . . there was however something in my air, and figure *imposant,* and attractive enough to secure as much of the notice, and attention of the part of the creation I most adored as was sufficient to satisfy my amour propre—My eyes were of the Hawk-kind,—*Piercers!* such as could read the heart & soul of Man,—and Woman! a Nose somewhat long perhaps, but of an aquiline form, and, in good harmony with its companions, the mouth and chin:—the general outline, or contour of the face,—*not* bad! My countenance, "the index of the mind," of course full of fire, of animation, —of energy—of enthusiasm,—and, if my smiles were pleasing, my frowns were *horrible*. So much for my Exterior, and now in a few words, to give,

My Character

Bold, ardent, impetuous, enthusiastic, it was difficult to dissect, or analyze any part of it; the Lion, and the Lamb had so whimsically blended themselves . . . that it seemed impossible to dissolve the union. . . . All that my mind seized upon, it *grasped*—my energies were not to be confined,— . . . what I *chose* to be,—I *was*—I quitted my Study, only to take the Field—it was my passion, my pursuit, but, far from straitening [*sic*] my "scythe into a sword," I bent my sword into a Sickle! I invoked Ceres—she was my Goddess—my genius—she smiled,—I succeeded; my Ambition mounted aloft,— I was the *Ploughshare* of Great Britain! and if, *as such* Harrowed up by Envy, I was in the end, rolled smooth by Fame!—the exuberance of my knowledge burst forth in Annals, inscribed to my Goddess— my discoveries gained me applause—my Experiments were never ceasing—I dibbled wheat! I scalded Hogs! I yoked Asses! and, I measured Counties, as men measure Ribbons! my temper, was "sweetly mutable," now grave, now gay,—not an Iota of sans souci about me—*Apathy* was my abhorrence, she dared not approach me. *Enthusiasm,* my Idol! . . . Whether I turned my Plough, or trailed my Harrow, whether I toiled for church, or State, whether I was at the feet of my Mistress, or *scraping my own* at *Her* door, all, all was, with Enthusiasm. My prejudices were strong, & my *judgement* at *times,* so warped by my fancy, that it often betrayed me into great errors—then it was the Lamb subdued the Lion—I was mild, & gentle under correction, open to conviction: *enthusiastic* in repentance—and—in short, I was an original.[4]

No wonder that Peggy was enthusiastic in her reply:

I *cannot thank* you *sufficiently* for the *exquisite treat* you have given me by sending your *Portrait* & *Character!* Incomparable!!! the hand of the Master is indeed visible throughout, though it is impossible not to lament that any of the glowing tints should be in the least degree softened down by His innate modesty:—no mortal can mistake the Portrait, or doubt the Character![5]

Such self-portraits were common in the age of sensibility, but Young's is something of a masterpiece, accurate and corresponding closely to everything we know about him. That he could have penned such a sketch at the age of sixty-seven shows that he could still on occasion be as amusing and gay as he had appeared to Fanny Burney forty years earlier.

Surely he had little reason for gaiety in 1809. Arthur and Jane remained in Russia. The excerpt from his letter book, dated December 1, ended pathetically: ". . . and the idea of your coming in autumn is now all past by, and I am precluded from the possibility of seeing you till next summer, by which time I shall have no eyes to see you."[6] Poor Martha Young also yearned for her long absent son. Much of her very doleful letter to her husband of July 14 is unintelligible, but apparently Mary had just received a letter from Jane stating that they would not be returning soon. Martha Young's laments reflect much of the tragedy of the family history:

Not a line is there from Arthur which is unlucky, but as it is the kind, benevolent hand of God, that he was well when *she* heard from him *I must not* be so wicked to go near breaking my heart because he chuses to stay where he is, I suppose; altho its afflictive to know one's child lives in such a strange country as she tells Mary he is in. I can say little about it, for it is daring to blame their going to Russia when it was thot for the best & who could

---

[4] Add. MSS. 35,130, ff. 163–164. This self-portrait has no date. The only basis for ascribing it to 1809 is its arrangement in the Young MSS.

[5] *Ibid.,* f. 161. Addressed to Young and dated only Friday night, 10 o'clock.

[6] Young, 1898: pp. 448–449.

dream for such a length of time. . . . his education was so enormous-
ly expensive (& *that, I know to my heart* was with the best intentions) that it formed perhaps too strong notions of independence . . .
& I see plainly a splendid life caught him with the compliments
he has been paid on yr account & in some lesser degree I think
it may be on his unremitted attentions as Jane did say on his own
. . . . Still an excessive damp is over me not merely because I do
not expect to live to see him again, but at the thought of what
he may incur both in body & soul. I pray the Lord in his great
mercy to protect, pity & forgive him. . . . I fear Jane is a little
displeased with me, but I hope I have not given cause. . . .[7]

This letter lifts at least a corner of the curtain which covers
so much of the Rev. Arthur Young's life and character, revealing
probably differences over his education as well as his vaulting
ambition. Young's disappointment that his son had not returned
must have been intensified when Lord Egremont brought to his
attention late in 1809 the possibility of exchanging Arthur's
Irish living for one in Dorsetshire worth from two to three
hundred pounds a year.[8] Unfortunately action had to be taken
promptly and the Rev. Arthur was in Odessa.

Four of Martha Young's letters to her husband in 1809 have
survived. They reveal her as a physical and mental wreck. Pity
goes out to her but even more to poor Mary, the daughter who
had to care for her. In the worst of the series, Martha requested
that she and Mary might go to Aldborough for a couple of
months and expressed an almost psychopathic reaction against
Bradfield:

The Peculiarity of this season for several weeks has pressed most
strongly on my nerves, causing an indescribable depression, so that
I dread Bradfield at *present* the length of time to come, cold of
every room terrifies me, high winds, amongst those fine trees affected
me, with such fear last time with every door thro the house shaking my bed &c. &c. . . . nor have I been in the church there these
six years which is most shocking, resolution I have none. . . . I
wish to try Aldboro' for about two months, such a portion of sea
air may brace me as it did two years ago. . . . you have no poultry
either to enliven nor to eat. . . . a joint is what does not agree

[7] Add. MSS. 35,130, ff. 275–276.

[8] *Ibid.*, f. 340. This letter is undated, but envelope is marked December, 1809.
The postmark seems to be 1810, so it was probably written at the very end of
1809.

with so relaxed a stomach. The only walk of any length I can encounter is on the sea beach—& the expence if too great I shall relinquish if you disapprove it without complaining. Thank God amongst all my old symptoms I have no palpitation of heart! The damp weather & the interdiction of Russian letters throws a ponderous load all over me enough to produce what I really suffer. . . . I pray God for dry weather. I am so cowardly that without a servant before him I dread even Jack (old as he is). . . . If we were to go to Aldboro I would board you & Mr. St. C both for two guineas together per week & coals every day . . . might not bathing be of service to your eyes, what think?

I do not find the ability to see Bradfield yet & my allowance ends on ye 12th Wednesday such a season only rain, rain, is very sad & we are dull indeed. . . .[9]

A comment in Martha's letter of July 14 quoted above about her son, shows that her husband had consented, probably very willingly, to the Aldborough scheme for her and Mary. Yet on September 6 they were still in London as shown in Mary's note to her father, referring to the recent death of Lord Coventry: "Mr. Coventry (in answer to a letter of condolence my Mother wrote him . . .) says, his Father has left Mrs. A. Young an annuity, but does not mention particulars—previous to his death he gave my mother Janes Portrait. . . ."[10] The connection between Mrs. Young and the family of Lord Coventry, evidenced in these letters, remains a mystery. Why should Mrs. Young and Mary be welcome guests both at Coventry House in London and at Croome? Why should Lord Coventry leave an annuity to Martha or have Jane's portrait in his possession? Martha Young's last letter to her husband in 1809, dated November 4 from Aldborough, thanked him for sending them four "beautiful fowls" and two "fine partridges." In return she sent her husband two wild ducks and two teals and included long and confusing directions on just how they should be cooked. She had gone out in a "Dickey cart" to procure them, but unfortunately had caught cold and had a sore throat.[11]

Without much doubt Arthur Young's greatest concern in 1809 was Betsy Oakes's very serious illness. She had been sick

[9] *Ibid.*, ff. 273–274. Dated July 10. There was an earlier letter from Martha, dated June 28, *cf. ibid.*, ff. 269–270.

[10] *Ibid.*, f. 295.

[11] *Ibid.*, f. 325.

at Bath in the winter of 1808–1809. The return trip to Bury late in April via Bristol and London proved too much for her. Peggy Metcalfe blamed Betsy's London indiscretions, her "dreadful exertions" and "the madness of her going to shops in a N. E. wind."[12] Apparently the last seventy miles of the journey was made without stop in a chaise. Two days after her return Betsy wrote Young:

We arrived very safely at Home on Sunday eveng but it was very late & I think I never was so tir'd with a journey in my life, I kept in Bed yesterday till near 4 Ock & was late again today—the rest has been of use to my aching legs, but I still feel the affects of the cold & journey. . . . My stomach is very indifferent & my appetite is not good. . . .[13]

After seeing Betsy, Peggy Metcalfe wrote Young her impressions: "I do not find her thinner than she was, when she left Nowton, nor, unless they skin skeletons in the Places she has been to, do I conceive how she was to be so. . . ."[14] The very day Peggy wrote the above, Sunday, May 7, Betsy suffered a severe lung hemorrhage. During the remainder of May she was seriously ill. Nearly every day Peggy went to Bury to get firsthand reports for Young, but she was not permitted to see Betsy for more than three weeks and became very impatient:

I seldom see O. O.: and it is evident that He has no comfort or relief in talking to me, . . . I *could not* have *believed* the possibility of my being in the way, in her very Home. and she very ill, & being excluded from seeing Her, when Mr. O. goes in, & out of her Room 10 times in the day, & *she likes* to see Him—well then, she does not like to see me—is that the inference I am to draw? *Certainly not,* for I *must* believe just the contrary—but. . . . they make no distinctions, which, as I have ever been considered . . . as her *most particular Friend,* I am extremely hurt at. . . .[15]

Her writing, however, makes it easy to imagine Peggy as a tiring guest.

[12] *Ibid.,* f. 229. The date on this letter is not at all clear, but is sometime in May.

[13] *Ibid.,* f. 214. This letter was probably written on May 3.

[14] *Ibid.,* f. 212. Dated Sunday, May 7. I believe that the order between these two letters in the manuscripts should be reversed.

[15] *Ibid.,* f. 233. This letter is dated only Friday. Probably it is May 19. O. O. is of course Betsy's husband, Orbell; Mr. O. is his father, James Oakes the banker.

Young was naturally greatly depressed by the news as shown in his diary entries:

May 16. . . . Mrs. Oakes arrived at Bury last Sunday fortnight; and on Sunday se'nnight she broke a blood-vessel and brought up two spoonfuls, and on Saturday evening last had another smaller attack. My own fear and opinion is that it will end fatally. Her fatigue coming from Bath and Bristol, and at London, contrary to advice, has caused it. I have prayed most earnestly for her. Oh! may the Lord of all mercy hear and grant my petitions. My thoughts are all employed on the state of her soul. . . .

May 18. I am very, very unhappy, and cannot think of her without wretchedness. In every worldly respect what a loss will she be to me! A placid, sweet temper, with a good understanding; that ever recd. me with kindness, and attention, and preference, with whom I was at my ease, and where I could be at any time; a resource in blindness fast coming on that would have been great. The hope has fled and a sad and dreary vacancy, which freezes me, is in its place.

May 19. . . . Poor thing! she has been bled twice more, and the blood as highly inflamed as ever; bleeding gives relief to her lungs.[16]

Betsy wrote Young within a week of the attack:

I do believe you will not be disappointed to hear that I have been really frightened, tho' at the same time I flatter myself it will be no little satisfaction to you to know that I am better. . . . The spasms I have had since I came home have been more violent than in London. . . . I was certain it was impossible to have all that pain without inflammation—it was the cold coming from Reading that has caus'd all this. . . . in religion I trust it will accelerate the progress I am truly anxious to make, but without the assistance of God's holy spirit I can do nothing. . . . Considering that I had leeches yesterday, blisters the day before, & bleeded today I have done much to write this, but I am very tir'd—I hope it will give you pleasure to hear from yrs ever,

E. F. O.

On the reverse side of the sheet a note from Orbell shows that Young had been lecturing Betsy severely:

Betsey [*sic*] desires me to say she admits the truth of every word of your last letter, but she wishes you would treat these subjects in a

[16] Young, 1898: pp. 443–444.

gentle manner, what a pity it is you will not, as she is really desirous of attending to every word you say. You will readily believe me, when I tell you that I have been miserable about my dearest Betsey, but I hope now, through the blessing of God, she is in a fair way of doing well.[17]

Young's letters to Betsy at this time were sent to Peggy who took them to Orbell who allowed Betsy to see them if he thought they would not upset her. Betsy acknowledged one of his letters as follows:

Too well do I know the fervency of your religious zeal on my behalf & the sincerity of yr affectionate friendship to doubt a moment of what your feelings must have been on my account. . . . Pray that my Faith may be strong & above all, my repentance *perfect.* . . . A thousand thanks for your letter, I see it is a cautious one, you are right and continue that caution—I am afraid of myself & I am afraid of everything. . . . I earnestly wish Orbell had a prayer, that the affliction, for it has been a great one to him, may be so sanctified to him, that it may bring him nearer to God. I earnestly wish you would send him a proper prayer and me one too.[18]

A later letter from Betsy shows that he sent the prayers:

I received & read all your letters & prayers with much pleasure & satisfaction, but they fill my mind with so much that I am anxious to say to you in return, that not being able to do it fatigues & vexes me—but do not let this prevent yr writing. . . . My illness may continue long, I suffer pain & painful remedies . . . but be assured I do not repine, but am thankful to God for his goodness—I hope he will spare my Life, but if suffering longer will make me better I am content to do it. . . . Thanks for yr Prayers. I am to have another large Blister again tonight on account of Cough & pain— my strength is wonderful with all I have gone through—My arm is very sore with the 4 times bleeding & the last Blister not quite well yet—I sat up for 3 hours in the afternoon very well—Poor dear O. wants a good garden to ransack for me because vegetables are recommended, but I care for nothing of the kind . . . Adieu, Dr. frd. Yours ever,

E. F. O.

if writing does not give you pain, why not send a few lines in yr own hand.[19]

[17] Add. MSS. 35,130, ff. 221–222.
[18] *Ibid.,* ff. 243–244. The date is not clear, except that it was after May 20.
[19] *Ibid.,* ff. 245–246. The date may be May 26.

Young's diary for June 6 refers to another letter from Betsy:

Yesterday's letter bad. Poor thing, she has been bled eight ounces, and much inflamed; had been out twice, and I suppose took cold. She wrote herself, and there are some comfortable expressions relative to the state of her mind with regard to religion. I have little hope of her recovery.

After his return to Bradfield the letters of course stopped for he could see her, but several notes in the diary for July refer to her:

July 4. Betsy continues just the same, whether better or worse, the cough does not go, and therefore I conclude the case bad. But, thanks to God, her mind, I hope, goes on; she has the Testament read to her by all the four children.

July 8 . . . Mrs. O. the same, and the weather unfavourable to her; I drank tea there on Thursday.

July 19. Drank tea with her last night. She was bled in the morning, and going to have a blister. All her symptoms worse, I suppose she has caught cold.[20]

In line with the doctors' advice, Betsy went in October to Exmouth in Devonshire for the winter. Her oldest son, Ray Orbell, was in school, but the other three children accompanied her. Her husband took her down but returned about December 1. Several of Betsy's letters to Young from Devonshire reveal the depth of her affection for him. On October 23 she described a long water trip along the shore to Teignmouth:

I dispatched a very long letter to you only 2 or 3 days ago but I am so truly sorry to hear of your illness that I cannot forbear writing to tell you so & that I most sincerely hope you are quite recovered again.

We talked of you frequently the other day. . . . We wished very much for you—You would have enjoyed it & I really believe it does me good. . . .

With love believe me yours,

Very affecly,

E. F. Oakes

if you make a wound you know how to heal it. [21]

[20] Young, 1898: pp. 445–446.
[21] Add. MSS. 35,130, ff. 314–315.

On December 8 she wrote a very long letter inserting a list
of daily temperatures at Exmouth since November 1:

Your entertaining and very agreeable Letter came most acceptably
—it reach'd me the morng after Orbell had left me. His being
obliged to go is a sad business but however I have reconciled myself
to it much better than I had conceiv'd it possible. . . .

You cannot imagine how much I enjoy'd your Letter. . . . My
O. O. I hope reach'd home in safety this day! & he was to see my
poor Ray Orbell just from school & you perhaps he will see tomorrow!! how I envy him! when shall I be able to see my friends
again at my own House? and shall I ever? I feel this banishment
but I will not complain. . . . I wish there was not one obstacle to
your coming here—what comfort & pleasure it would be to me.

Several additions were made in the following days. She was
trying to go out each day, but while mild, it usually rained, and
then she caught cold.

Monday—Thermtr 44 too cold for me to go out more particularly
as I have more pain in my chest if it is not better tomorrow I shall
put on a Blister—if you see Orbell tell him I desired you to say I
was the same as when I wrote to him—I begin to think I must
give up going out. . . . Let me have a long letter—if you and O. O.
write ye same day you might enclose as one letter. . . .

I wish you wou'd come & bring me Ray Orbell with you—tell my
O. O. what I say—I hate to send you so stupid a Letter, & I know
why you will think it stupid. . . .[22]

Such a letter is utterly inconceivable from a mistress to her lover
and is more like that of a daughter to her father.

In 1809 Young had invitations from both his old schoolfellows
to visit them. On March 28 the Rev. Jonathan Carter invited
him to spend a night with him during the Board's Easter recess,[23]
and, late in September, Thomas Ruggles invited Young to
Spains Hall in very cordial terms:

. . . it would much gratify me and my family for we all have a
regard for you as my old friend; as to myself, I am in health,
thank God, very well, better so, than in spirits, wch flag very much;
if you and your friend Mr. Worgan would come & spend a few

[22] *Ibid.*, ff. 331–332.
[23] *Ibid.*, f. 197.

days here I do think your company would operate better than any cordial or syrup from any Quarter of the world. . . .[24]

Young's letter of acceptance follows:

I am very sorry to find by your letter that you continue low spirited, but this must be the case with every person arrived at our time of life, who has not a joyful dependence on the scripture promises of God, and who derive their faith from their good works; and not their good works from their faith; those who will put the cart before the horse, must make a hobbling journey of life.

Respecting your invitation, you are the master of it; if you and your wife and daughters will come to me to give a look at Bury Fair, & the great Ball in October, in that case, I will come to you next week; and in expectation of your agreeing to this, I am ready to accept your offer of sending your carriage to meet me at Clare, which will enable me to send back my own Whiskey; I will be at the half moon at Clare on Tuesday next, at 12 o'clock. And I shall order the Whiskey to meet me at same hour on Thursday, which will give me a whole day with you.

Mr. Worgan is not with me, he has been for some time at Wilberforces; but Mr. Cragg, under Secretary to the Board of Agriculture is here, & will accompany me on Tuesday.[25]

Back in May, Young had offered a cottage at Bradfield to the Wilberforces for the summer, an invitation which they had declined.[26] Instead they went to Pemberton with Thomas Fry. Since Worgan had been with Fry when the Wilberforces arrived he also met them. In September, Worgan wrote to Young: "I am now in the truly enviable society of Mr. Fry & Mr. & Mrs. Wilberforce." In speaking of Fry he continued: "Never did I meet a man so humble, so active, so pleasant & impressive in all he says. Mr. & Mrs. Wilberforce & Mr. Fry all desire to give their kind love to you, they are planning to get you over to Emberton to visit them." Wilberforce added a brief note to the letter which he franked:

. . . our friend Worgan has just left us—I never saw a man who

---

[24] *Ibid.,* f. 300.

[25] Letter IX in the Ruggles correspondence. Only the signature is in Young's hand writing. This letter is the only indication that Cragg visited Bradfield that autumn.

[26] Add. MSS. 35,130, f. 251.

appear'd more restive to devote himself to ye service of his fellow creatures. He made us a little idle by his musical powers. In general it is only at & after meal times that I indulge in social intercourse & in recreations, but as there [are] 3 meals daily, ye opportunities occur pretty frequently.

The real purpose of Worgan's letter had been, however, to prod Young about the remaining £100 due his brother George for the Cornwall survey:

And now let me ask you, can you do anything to forward the payment of the other hundred pound promised my brother George; for he is now in the greatest distress. Might not a letter from you to Sir John Sinclair representing the hard case of my brother being led by the Survey into debts far beyond the hundred pounds he has received, induce him to give an order for the money—or at least some reason why the money is not paid?[27]

Whether Young wrote in behalf of George Worgan is unknown, but actually he was paid £80 in the fiscal year 1810–1811, and the remaining £20 in 1811–1812.[28]

Relations between Rev. William Gooch and Lord Templetown were apparently improved this year, and Templetown expressed himself on the whole well pleased with Gooch's performance and agreed that Gooch could use the linen and purchase some new furniture. Still at one point in his long letter in March, Young reprimanded his protégé very sharply:

There is one passage in your letter, which does not run quite so smoothly, and that is that you should & & *if the usual poundage was allowed you;* that passage is quite unworthy of you, and as a sufficient answer to it, could be only a rap on the knuckles, I will give no answer at all to it.[29]

Far more of a nuisance in 1809 was James Powell who had caused so much trouble in 1799–1800 over his Ryeland sheep. Early in 1809 he was Lord Sheffield's steward at Sheffield Place in Sussex, a post very likely procured for him by Young. Between February 10 and April 13 Powell wrote five letters to Young complaining of bad treatment.[30] The house was unsatisfactory,

---

[27] *Ibid.*, ff. 297–299.

[28] Board of Agriculture, Cash Book (folios not numbered).

[29] Add. MSS. 35,130, f. 188. This is of course a copy of Young's letter. St. Croix frequently made and kept copies of Young's more important letters.

[30] *Ibid.*, ff. 168–170, 180–181, 190, 199–200, 205–207.

he was not allowed to keep poultry or pigs, he had insufficient firewood, the other stewards all plundered Lord Sheffield and ganged up against him because he was honest, and the bailiff's wife complained that his "old, harmless, beautiful Newfoundland Dog"[31] ate her eggs. He could say nothing bad enough against Lord Sheffield. He constantly appealed to Young not to desert him. In his first letter he wrote:

I have no Intention of giving you a *moments Trouble*—not even of answering this letter. . . . but if unfortunately I should be shipwreck'd—I trust in God, you will yet stretch forth your Friendly arms, to enable me to reach the shore.[32]

On April 6 he attacked Sheffield very bitterly:

I am very sorry to tell you that I intend leaving this place immediately. Had Lord Sheffield been a man of Honour and feeling, when out of his vagaries, it might have been borne with. But, it is melancholy to reflect that he keeps his Word with no one—To describe his system and proceedings would make one shudder. His workmen whom he persecutes were formerly many of them his Tenants, who were ruined under him, and he will now hardly give them bread.

If I lose your Friendship & support, 'twill be hard. Indeed—*I do not deserve it should be so.* I would expend the last drop of my Heart's blood in the service of any Gentleman who would treat me like a man—and I know that I am capable—but to be spurned & spit upon is not to be borne. . . .[33]

There is no indication that Young answered these letters or that he did anything to help when Powell left Sheffield's service. Sheffield expressed his opinion of Powell very succinctly to Young on July 7: "A more worthless fellow than your protégé Powell never was employed by any man."[34] One wishes that he had been more explicit.

About June 1 Young was having the *Edinburgh Review* read to him and confided to the *Autobiography* two of his reactions.[35] He was indignant at Sydney Smith's somewhat flippant review of Hannah More's *Coelebs in Search of a Wife:* "Wretched stuff;

[31] *Ibid.,* f. 180.
[32] *Ibid.,* f. 170.
[33] *Ibid.,* ff. 199–200.
[34] *Ibid.,* f. 272.
[35] Young, 1898: pp. 444–445.

false and frivolous, reasoning on cards, assemblies, plays, &c." He communicated his irritation to Peggy Metcalfe who answered a little later:

I denounce, & renounce, & pronounce S. S. a *Puppy*—farther I cannot say until I have read his criticism, *although I am determined to execrate* the injustice of it—I suppose it attacks the fame of Holy H—whom I am resolved to believe spotless, notwithstanding being every day told, "that she proves what she is, by her efforts to prove what she is not;" do you understand this?[36]

Was Peggy perhaps being a little sarcastic? The other comment shows all too clearly how obscurantist Young had become:

I was much struck with a reference . . . saying that 7,000 pedigrees of the nobility . . . were destroyed in the Revolution. It occurred at once to me that this was the exact number of *names* of *men* slain in Revelations. It is in the eleventh chapter: "The tenth part of the city fell in a great earthquake, and 7,000, &c." The commentators all seem to have reckoned France as the tenth part of the city; but of the names of men they knew not what to make, but 7,000 pedigrees answer to a wonderful degree, and the coincidence of numbers is truly amazing.

It seems almost unbelievable that this was the same man who wrote the *Travels in France*. So strongly did this coincidence affect him that he wrote about it to Thomas Scott and to one of the apocalyptic prophets, G. J. Faber.[37]

Young described in some detail his life at Bradfield at this time in his letter of August 4 to Jane:

I rise from four to five in the morning, pray to God for half an hour, more or less, according as *He* affords me the spirit to do it. At half-past five I call Mr. St. Croix, who comes to me at six, and reads a chapter in Scott's Bible with notes. I then dictate such letters as want to be written, after which we sit down to my "Elements of Agriculture," which have been more than thirty years in hand, and at which I have worked for two years past with much assiduity, wishing to finish it before my sight is quite gone. At half-past eight the servant brings me the water to shave; from nine to ten we breakfast, and sit down again to work for two or three hours, as it may happen, as I take the opportunity of sunshine for a brisk walk of an hour, very often backwards and forwards on the gravel

---

[36] Add. MSS. 35,130, f. 254.
[37] For Scott's and Faber's answers *cf.* Add. MSS. 35,130, ff. 259, 285.

between yours and the round garden. I wish much to have my thoughts during that hour employed upon death and the other world, but my weakness and want of resolution are lamentable, so that I sometimes think on every subject except that which I intend should occupy me. We then sit down to work again, till the boy and his dicky arrive with the letters and newspapers. When they are read we work again, but usually catch half an hour for another walk before dinner. When alone we dine at four, . . . but if any person be in the house, as it prevents an evening work, five is the dinner hour. What is read afterward is usually some book not immediately connected with work. At eight we drink tea and go to bed at ten, but the Sunday is an exception; you know there is service but once a day. At the church hour, whether morning or afternoon (when no service) about thirty children from Bradfield, Stanningfield, and Cuckfield come to read in the Testament, and repeat their Catechism, and undergo some examination from Mrs. Trimmer's "Teacher's Assistant." Whatever is well done receives a mark against the name; the girl or boy that has fewest marks receives nothing, the next a halfpenny, next a penny, and so on, all which does not amount to more than two or three shillings. I cannot boast much of their progress, though I pay for most of them as constant scholars. In the evening, between six and seven o'clock, forms are set in the hall to receive all that please to come to hear a sermon read, and the numbers who attend amount from twenty to sixty or seventy, according to weather and other circumstances. Such, my dear Jane, is the tenor of my life both in summer and winter while I am in the country.[38]

Young's first publication in 1809 was his survey of Oxfordshire, the trip for which had been made in 1807. Since Young did not know Oxfordshire very well, his survey is probably less valuable than those for East Anglia. It is certainly less interesting than his Lincolnshire. Some of his earlier surveys had been severely criticized for merely presenting large quantities of undigested facts.[39] Although noting the criticisms, Young justified his method of piling up such masses of facts:

Without minutiae of this sort, the reader, from the first page to the last of a book, has always the writer between him and the farmer, and not as a transcriber, but as a calculator of effects or

[38] Young, 1898: pp. 446–448.
[39] Thus the reviewer of his Norfolk survey (*Farmer's Magazine* 7: p. 370) had written: "Information is accumulated in wagon loads; but the carriages are upset, and their contents thrown on the ground without order or regularity."

averages. I know not how other readers may feel, but to myself, in examining the description of a district, I wish to see authority named for all that is possible; as it is not only a proof that the writer has had such conversations on the spot, but that there exist sufficient judges resident in the country, who know whether the farmers named were, or were not, proper men to apply to.[40]

Consequently the Oxfordshire report contains page after page of the crop rotations or the stock of cattle or sheep of many farmers with whom he talked. Moreover, Oxfordshire contained no great agricultural innovators and few of Young's old friends. Hence the report lacks color.

Young was especially interested in the Bishop of Durham's cottages for the poor at Mongewell and was also impressed by a village shop where the poor could buy their produce at reduced prices.[41] Hence these cottagers were thrifty and respectable and never needed any poor relief. He took issue with the Malthusians by declaring that a larger population in rural areas meant the production of more food relative to total population.

By and large, Young found more to condemn than to praise in Oxfordshire husbandry. He praised the widespread use of oxen for plowing in the Burford district, especially since this practice had declined in England as a whole. He was greatly impressed with their haycocks and stacks of straw:

They perform so perfect a contrast to the ragged heaps called stacks, by the courtesy of Suffolk and Norfolk, that I have returned to my own county and farm with no little disgust; and must walk into five-and-twenty acres of well-dibbled wheat, in order to refresh an eye wearied with the deformities of the stacks.[42]

Oxfordshire had also made great progress with two crops. Sainfoin was more widely and perhaps better cultivated than in any county he had visited. Oxford farmers were also almost pioneers in the extensive cultivation of the kind of turnip known as rutabaga or Swedes.

On the other hand, Young found much to criticize. There seemed to be a blind, unreasoning opposition against long-term

---

[40] Young, 1809: p. vii. The volume has a large number of plates at the end, chiefly of agricultural implements.

[41] *Ibid.*, pp. 24–27.

[42] *Ibid.*, p. 19.

leases. Paring and burning was seldom practiced in southern Oxfordshire. Farmyards were inconveniently arranged and not planned around the threshing mill. Many farmers followed poor rotations and practiced fallowing. He bitterly attacked their prejudice against long, fresh dung, and their preference for *mining* the manure:

I beseech you, gentlemen, to go on carting, turning, mixing, *mining* and rotting; give your manure to the sun and to the winds; continue to expend no trifling sums in the reduction of four to one, in order, by studied operations, to render your one less valuable than the fourth of the original four: it is a wise conduct, therefore stick close to it, and argue strenuously for it over the next bottle you drink. *Mine* away; see that the heap lies light; keep the carts off; take care that the air pervades it, and let the sun shine and the winds blow—Who's afraid?[43]

In the appendix to the Oxfordshire survey Young discussed a plan to make the Ratcliff Library a great center of agricultural enquiry. He commented sarcastically on the lack of a chair of agriculture at either Oxford or Cambridge: ". . . but that Oxford and Cambridge should be the only two Universities in the enlightened part of Europe without Professors for teaching this most useful of all the Arts, is a circumstance that must excite some degree of surprise." Young proposed that a large, unexpended fund in the control of the Ratcliff Library should be spent in the fields of natural history, chemistry, mineralogy, agriculture, and rural mechanics. He envisioned the completely empty vestibule of the library as filled with agricultural implements. Young almost certainly talked these matters over with George Williams, professor of botany, who became the Ratcliff librarian in 1810 but apparently spent the funds, not as Young had hoped, but rather on medicine and physiology.[44]

Young's second publication was *On the Advantages Which have Resulted from the Establishment of the Board of Agriculture*, based on his lecture which was read by the under-secretary, William Cragg, on May 26. Only a slim pamphlet, it still gives the most complete contemporary summary of the Board's work

---

[43] *Ibid.*, p. 255.

[44] *Ibid.*, pp. 343–346. Williams' dates are 1762–1834. In the preface, p. v, Young stated that it had been hoped that Williams would make the Oxfordshire survey.

to that date by the man who knew more about the subject than any other with the possible exception of Sir John Sinclair. Its chief defect is its almost completely laudatory tone. According to Young all the criticism against the Board was unjustified. The Board was not to blame if individual farmers had not adopted its proposals.

When originally established it had probably been expected that the Board would be a rather passive body upon which Parliament or the ministry could call for information or advice. The smallness of the grant for its support implied that no more positive exertions had been expected. He then enumerated specific steps taken by the Board in line with such a conception, the many substitute breads suggested in the crisis of 1794, its repeated proposals for a general enclosure act, its initiative in the establishment of standard weights and measures, its success in removing the tariff upon oilcake importation and the excise on draining tiles. Young pointed proudly to the report advocating that lands be attached to cottages for the benefit of the poor, the publicizing of Elkington's system of draining, and the 350 memoirs on breaking up grasslands and laying down again without destroying the value of the lands. It was also in accord with the original purpose that the Board should become "an office of intelligence" for the public:

It is certainly the duty of the Secretary to give at all times, and to all persons, every species of information in his power; to make whatever inquiries may be necessary, with that view; and to introduce such persons to each other, as can best supply their mutual wants; and it is no exaggeration to assert, that this has been done to the amount of some thousands of cases.[45]

But the Board's greatest achievement, the publication of the County Reports, had not been originally envisaged and could not have been completed with the funds originally appropriated. Young claimed that the first hastily prepared reports had been unfairly criticized, since they had only aimed to elicit contributions from practical farmers which in turn would be used for the revised and more definitive reports. He pointed out that other governments had copied the plan, and quoted George Washington on the reports' utility. He showed how the re-

[45] Young, 1809: p. 12.

ports had converted certain local practices into more general ones—for example, warping, the sowing of winter tares upon bad grasslands, the planting of spring wheat without any spring plowing, and the use of long fresh dung.

Young paid graceful compliments to some of the Board's leaders by inventing a suppositious speech which Pitt might have addressed to some of the original members of the Board:

. . . you, Sir John Sinclair, have been active in whatever could promote the prosperity of Scotland; you, Mr. Coke, are happy in efforts for the improvement of Norfolk; you, Lord Egremont, act with energy for Sussex; you, Duke of Bedford, manifest decided talents, to bid a happier cultivation adorn the district that surrounds you; you, Lord Winchilsea, are alive to whatever is beneficial to Rutland; . . . Somersetshire looks to you, Mr. Somerville, for your skill in live stock; you, my Lord Bishop of Llandaff, whether in the paths of science, or the fields of cultivation, never applied to any pursuit that did not flourish under your masterly hand; . . . and you, Sir Joseph Banks, who have, so beneficially to science, presided at the head of another Institution, you, Sir, have contributed, by enlightened efforts, to render philosophy subservient to the promotion of domestic industry; . . . Such men as you are, cannot meet and deliberate without effect; the result must be beneficial to the Agriculture of the kingdom.

He went on to show how absurd were the attacks upon a Board which included such talents:

. . . nor could any person contrast the applause and reputation, which have been given individually to these several persons for their agricultural exertions, with the jealousy, suspicions, and misrepresentations, under which the Board has laboured, without amazement. It should seem as if those talents which blazed at Woburn, became extinguished in Sackville-street; as if the genius which illuminated Petworth, became a common mortal by association in this room; as if a Coke lost all knowledge of turnips, and a Somerville all his skill in cattle, by entering these doors.[46]

Young's peroration was almost too smug:

That you well know how to draw wise conclusions from the premises you have created, you have given repeated and convincing proofs; no advice offered by you has been acted upon, without decided

[46] *Ibid.*, pp. 7–10.

success; none has been rejected, without the mischief coming in full relief to the eye of the Politician; you would have remedied one former Scarcity, and you would have absolutely prevented another. . . . I know not an interest in the community, to which this Board has not been a benefactor; and he must be a poor reasoner, who will attribute such effects to the parsimonious Grant of 3000 *l.* a-year. No, Gentlemen! money would not effect this; it is the union of great talents, of unquestioned patriotism, that assembles harmoniously in this room—men of every party and description, anxious for no other emolument than the pleasure of doing good: —this is the spirit which has animated your exertions—this has merited success, and it is this spirit that will command it.[47]

Surely the Board members must have felt a warm glow of self-satisfaction that day, whatever the Cobbetts might say. Surely Young's main point was correct, that the Board had been and still was a very useful institution. The admission of a few shortcomings, however, might have helped to disarm those still critical.

Following his usual routine, Young could not have been in London for more than four months in 1810. The year's documentation comes almost entirely from his correspondence. The only entry in the *Autobiography* is a letter. The only record in the Board of Agriculture papers consists of two brief letters in the *Letter Book*.

The letters make no mention of the condition of his eyes, but certainly no improvement had taken place. During much of the year he was in severe pain from a sciatic complaint, as the following reply to an invitation from Thomas Ruggles attests:

It would give me much pleasure to accept your kind invitation and pass two or three days with you at Spanes Hall, but since I saw you a severe change has taken place in my health. I have a bad Sciatica from the ankle to the hip which has continued from last Novr and bids defiance to all medicine internal & external, electricity, bathing, &c—it make me almost a cripple and I am nearly always in constant pain, in bed only excepted: I look upon it, & hope I shall receive it, as a blessing from the hand of God, to break the remaining chains which attach me to the world, and to turn all my views to a better. . . . I employ my time a good deal in putting the last hand to a work, which has much occupied me for more than 30 years, and if I am able to finish it by next Christmas,

[47] *Ibid.*, pp. 69–70.

I shall have wiped my hands of worldly pursuits; and it is high time, as I may look for a failing memory, as well as you. . . .[48]

There is evidence that Arthur Young was still working unsuccessfully to increase the return from his son's Irish tithes. In May there was a discouraging letter from the Rev. William Gooch: "The V—l—ny of persons calling themselves Gentlemen, in the county of Clare, respecting Tythes, is shocking. . . ."[49] Almost the only legible sentence in a letter from Egremont confirmed Gooch's point: "There is very little hardship about Tythes in that part of Ireland except to the owner of them. . . ."[50] How much was received from Ireland is not clear, but certainly the Rev. Arthur frequently requested advances from his father. On February 25 he wrote:

Theodosia, February 25, 1810

My dear Sir!
The present note is sent merely to inform you, that I have drawn upon you for the sum of £ 363. 11. 10 sterling, which you will be so kind as to pay to the bearer 30 days after sight, and place the same to the account of,

Dear Sir,
Ever yours affectionately,

Arthur Young, Junior.[51]

It does seem as though he might have written a little more, even if he was counting upon Jane to relay the news. When a similar request for £200 came in December, there was only a notification from a Moscow agent.[52]

By October at the latest Jane Young had returned to England alone, for her husband remained in Russia. He was interested in settling permanently on an estate in the Crimea, while Jane was probably sick of Russia. Late in January Betsy Oakes wrote: "I am truly concerned that Arthur shou'd leave poor Mrs. Arthur in such a country! & she wishing to come home! & to part with

---

[48] Letter X in the Ruggles correspondence. For Ruggles's invitation, *cf.* Add. MSS. 35,130, ff. 428–429. Many of Young's friends interpreted his illness as ordinary rheumatism. On November 17 Rev. Roger Kedington wrote Young a note and humorously addressed it to "Arthur Young, Esq., Europe," *ibid.,* f. 492.

[49] Add. MSS. 35,130, ff. 397–398. Dated May 14.

[50] *Ibid.,* f. 525. This letter is undated, is not even clearly of 1810.

[51] *Ibid.,* f. 360.

[52] *Ibid.,* f. 506.

such a woman extraordinary."[53] Late in February, Louisa Jones reported a recent letter from Jane: "I received a letter from Jane lately, & she mentions in it of her intention to return this spring. . . . poor thing she talks of ill health, I think it a great pity that she has stayed there so long. . . ."[54]

On August 1 Sir John Sinclair commented on the rumor that Arthur was receiving an estate in Russia:

If the Emperor of Russia gives your son a good Estate, it will be some inducement for me to send to that country a young Highlander; but if he pays no attention to the Family of the Secretary, I am afraid that the President's family would be as little attended to.[55]

On October 17 Peggy Metcalfe wrote from Brighton:

Let me . . . be the first, or among the very first to give you joy, *great* and *hearty* joy, . . . upon the actual arrival of Mrs. A. Young . . . I do not know when I have heard of anything that has given me so much pleasure, because I am certain that it will give you new life, & spirits. . . . Have you led Mrs. Arthur to Her Seat? & is she not transported with joy at finding herself in it? how did she fight her way to *dear blessed England?* . . . Adieu, my incomparable Friend—hate me if you can—I have *read your life*—you know it, although, from *sheepishness,* I suppose *Merino-sheepishness,* as being somewhat new, you do not chuse to notice *my flattering* criticisms.[56]

There was also a congratulatory letter from Betsy:

Most sincerely do I congratulate you my Dear Friend upon Mrs. Arthur's arrival, the news of such joy to you does me good—give my kind love to her & tell her how much I long to see her. . . . I sincerely congratulate Arthur upon his great acquisition if the Crimea and 7500 acres are more to his taste than Bradfield and the society of those most dear to him.

I shall be much vexed if you dont come on Tuesday—but I think you wont disappoint me in this—I do not see why a friend's friend

---

[53] *Ibid.,* f. 354. Date of this letter is "24," but the month is not clear. From internal evidence it was probably January, for it refers to his trip to London, which would certainly have taken place before February 24.

[54] *Ibid.,* ff. 362–363. Dated February 27.

[55] *Ibid.,* f. 445.

[56] *Ibid.,* ff. 480–481.

sh'd keep you from an older friend than all—& not less sincerely affecy.

I am prétty well but cough if a bright day troublesome.[57]

The above was the last of Betsy's eleven letters to him in 1810 which have survived. There are also two letters from Arthur to Betsy for this year. Their correspondence, by far the most interesting and important material for the year, deserves extended quotation. Her first letter was written from Exmouth on January 5:

. . . You are angry that I do not fill my letters with religion & I could wish that you knew every thought & feeling that I have on that as well as every other subject as I never yet did conceal anything from you & it is not likely that I should begin a plan of reserve at this time of day—so far from it, I am hurt & mortified that you do not know exactly the state of my mind, but you have that discontented discouraging way of speaking & writing upon religious subjects that I have no comfort in writing or speaking indeed I fear always to name it. . . .

. . . it is very nearly 2 years that I have had entirely an invalid life . . . with your damp sheets & shirts how can you expect to escape the rheumatism—it is impossible but you would never listen to me on that score. I cannot bear that you should suffer pain or have an uncomfortable feeling. . . . I cannot express how deeply I regret the loss of your society surely I may hope for a happy day or two at Bradfield in the summer.[58]

She wrote again late in January just as Young was departing for London:

. . . You will say I am very perverse—I wrote a sheet full of religion to Mr. Oakes & wont utter a word to you, but so it is & so it will be. I cannot help it—but this I must & will say it is not *my* fault. . . .

. . . What a wild day you have for travelling. I am glad you are going to London for you were shut up so completely at Bradfield that it made you melancholy & I hope yr rheumatism will be better & that you will let me hear from you oftener, tho I find you will write only letter for letter which is hard upon me—I

[57] *Ibid.*, f. 487.
[58] *Ibid.*, ff. 342–343.

have no amanuensis—writing hurts me & I hate it & you dislike my letters so much that I am quite discouraged in taking up the pen.

I am comforting myself with the hopes of soon seeing Orbell. . . . the conversation he has heard from you has made an impression upon him beyond what you can conceive and I never in my life heard anything of his age express such thorough esteem, respect & *real affection* for any person as he does for you—he says, Mamma, what should we do if Mr. Young was to die & I am sure I should go into mourning for him tho he is not a relative. . . .[59]

Early in April, Young described the Burdett riots in London in a letter to Betsy, interesting both for its detail and its rigidly conservative attitude:

. . . . I have been witness to such a scene as I hope, through the blessing of God, will not occur again. On Friday night, the mob was extremely agitated in Piccadilly, especially near Sir Francis Burdett's, and they took the unaccountable whim of forcing everyone to illuminate. I lighted up as other people did, and when I went to bed left orders with the servant who sat up to be sure to keep the candles burning till daylight, instead of which, when others put out their candles, ours were extinguished also. At two o'clock the mob returned and broke many windows, and ours among the rest. The servant ran into my room and waked me out of my sleep to tell me the windows were smashing. We hurried the candles out again, and upon examination, found the alarm exceeded the damage, for only three panes were broken. All Saturday passed in a very quiet manner, and in the evening the illumination was more general, but troops pouring into London from all quarters, we hoped to be secure without violence. The mob, however, were so determined, that by twelve o'clock we heard platoons firing in Piccadilly, and a few in other directions more remote, the Riot Act having been read. . . . In five minutes Piccadilly was cleared. We afterwards heard a little more distant firing, and a party of horse scoured up Sackville Street, firing in a scattered manner at the flying mob; but, from the reports of the pieces, I believe with powder only. . . . During Sunday the agitation of the streets threatened a bad night, but Government had brought in so many troops, that had we known it we need not have been alarmed. . . . It really is a tremendous moment, for if they do not carry it with a high hand, as a means of prevention, we shall have an organized mob and

[59] Ibid., ff. 353–354. Orbell was the second son in the family and was probably about ten years old at this time.

great mischief will follow. It is expected that the gallery of the House will to-day be cleared by acclamation and a Bill brought in to suspend the Habeas Corpus Act, and the rascally authors, printers, and publishers of those inflammatory papers which have done so much mischief seized and imprisoned. . . . I am much inclined to expect that good will result by drawing close to the Ministry all the honest men in both Houses, with all others that might be wavering; for it is a question now whether we are to be governed by Parliament or the mob. Many circumstances, however, are unfortunate, and not the least, that though the public revenue amounts to 62,000,000*l.* yet the expenses of the year will rise to above 80,000,000*l.* and must be made good by means that will occasion the necessity of having additional taxes. This will cause a yell for peace—and such a peace as must be ruinous if made.[60]

Betsy reacted as follows:

You really have made me tremble, the alarm & scene altogether that you have experienced must have been tremendous & frightful—it is indeed an awful moment!. . . . I feel full of apprehensions & the more so knowing what we all deserve—We are told here that this spirit for riot is spreading every where—Bristol, &c.—& even in the villages around this place they all seem ready for it. Orbell being in the local militia I am frightened for fear of a letter coming to call him away—if I must not go home, I would not be left here for the world. . . .

. . . do you think the rising of the Heart to God is as acceptable as prayers when kneeling? My mind is for ever lifted up but I don't kneel half as often as I ought—in short I can do nothing that I ought to do—as my desires become stronger, I feel myself weaker—What an admixture I am—sometimes to have much religious feelings & thoughts, & at others to take up with nonsensical employments. . . .

if Peggy was frumpy you are right not to go near her again, for I am sure she can have no reason to be angry with you but I know no Lady so soon displeased or so ready to show it. so far she is very honest but it does not make her more agreeable. don't fail to let me know yr movements—in short you must tell me a great deal or I shall be as grumpy as Peggy. . . .[61]

[60] Young, 1898: pp. 449–451. Young stated that this account of the riots was sent to Jane on April 9 and was a copy of what he had written Betsy.

[61] Add. MSS. 35,130, ff. 379–380.

Betsy's next letter of April 21 came from Bath:

. . . . we this day dine here, No. 41 Milsom St. excellent Qutrs. for this place is so hot, if I had not any large rooms I should be suffocated— . . . to have had to leave Exmouth without Orbell to settle & pay wd. have been dreadful & the journey here odious, because the Inn at Taunton so vile—I regret the quietness of poor little humble Exmouth & the beautiful country—There I enjoyd the beauties of nature, here I detest those of Art. . . .

I have not been to Church this twelve month tomorrow I hope to have that comfort, for I am close to the Octagon, which is a very warm chapel. it will be delightful to get to church again & of course the Sacrament being Good Friday. . . .

I hope to come Home well enough to surprise you & if you are not pleased I shall be angry. . . .[62]

On May 15 Betsy wrote that she hoped to leave Bath soon for London:

Nothing but the grace of God would have enabled me to have listened with such devout & fervent attention—with such lively interest, such accute [*sic*] feeling, that I was almost choaked with a sort of agitation but then when over, the world presses in upon you & apparently as if every thing & person was endeavouring to drive all good impressions away—& thus it ever is & will be with me—in reading this, people in general wd say, why she is quite religious, and I don't think I have any religious state, for any nonsense of the world, dress, company, & so forth all drive it away in a moment. . . .

. . . . be assured I have *no one* symptom of diseased lungs—therefore I am not in a consumption. . . . this is the first word you have said of longing to see me—I catch at it with much pleasure— are you sure you do? but would be ungrateful if you did not. do you think there is chance of your being able to get rooms for me— I can put [?] as long as they are not stinking & filthy. . . . Weather too hot will be as bad as too cold, so there is no knowing what to do with such an animal.

Adieu Ever yrs—very affectionately

E. F. Oakes[63]

---

[62] *Ibid.*, ff. 387–388.
[63] *Ibid.*, ff. 399–400.

On May 25 she was still at Bath but very anxious to get away:

. . . I had rather you had get me rooms, but James Oakes offering I thought perhaps it would be expected I should accept. I have therefore written to him. . . . pray inquire where I am to be as I shall depend upon your coming to me. I hope to be in London on Monday & shall stay only one night if possible. do *not fail* coming *instantaneously* to me . . . I shall be distracted to see you. . . . let me see you as soon as possible on Monday. . . . Adieu & believe me ever yours affectionately. . . .[64]

On Sunday May 28 she dispatched a brief note en route from Reading:

. . . I shall be sadly disappointed if I dont see you the very first moment possible on Monday—& I shall want you for 100 things— so dont disappoint me. . . . Adieu till I see you.[65]

Young almost certainly saw Betsy in London on May 29, but it was a very busy day for him, since he delivered a lecture on manures that morning before the Board. James Oakes had attended the lecture as his guest and wrote a nice note of thanks: "It was written, according to my poor judgment, in your best style; and I hope to see it when it appears in Print. . . ."[66]

A few days later Betsy wrote a very moving account of her homecoming to Nowton:

I arrived at Nowton by 8 or ½ past—the two Orbells met me beyond Bradfield—every soul young, old, lame & blind were all out to greet my return, the men with oak branches in their hats, the Bells ringing & poor old Grumpy making as much music as he could with his Flute—all absolutely shouting—it was altogether quite overcoming . . . we had the whole Parish in the mile before our house—every man woman & child had a pint of beer each—& the same of milk from the Cow—after dancing in *their* way, & showing every mark of real pleasure they all returned with their full pitchers highly delighted, but I think I may say, not so much so as myself for it was the most gratifying sight I ever had. . . .[67]

Shortly after his return to Bradfield, Young of course went to

[64] *Ibid.*, ff. 403–404.
[65] *Ibid.*, f. 405. She also requested Young to have Dr. Chilvers call while she was in London.
[66] *Ibid.*, f. 407.
[67] *Ibid.*, f. 410.

see Betsy and then wrote the only complete letter to her which
has survived:

My dearest Friend,

You are always better according to your own account, and yet the bad
symptoms [*sic*] if they abate one day, seem to return the next. I
was with you only one hour, yet you coughed twice or thrice, and
your hand was often at your chest; I must say, that my anxiety
is great, and nothing can give me peace, but to see, either your
complaint giving way, or your Soul in the full vigour of faith and
repentence: to give a helping hand to the latter, I shall make
a few observations, collected from writers of great abilities, but
chiefly from scripture itself.

There is no repentance without a heart felt sorrow for past sins,
trying to recollect all, and self abasement on account of them:
to hate, loath, detest, abhor, and abominate yourself, for the ingratitude of them against a Saviour who loved you so much as to
die for you. I loath myself for not feeling this contrition, so deeply
as I ought to feel it: it is easily excited, for great sins but not
easily for a life of carelessness and inattention; which wants it if
possible still more: if a person, does not feel, a good measure
of this contrition, there is great reason to doubt his repentance:
and without much of this feeling our faith in Christs atonement,
must be weak and inefficient: for we go to him to be cured for only
half our malady, when the whole of it must be completely done
away. A penitant person, who thinks himself in danger, has not
one thought of anything, but of, sin as the evil, Christ as the cure,
and the other world as the end: they speak of nothing else; every
other topic is insipid, but the load and burden of sin, and the
glories of redeeming love. It is very terrible when such meditations
which ought to be deeply felt, pass lightly through the mind; but
skin deep for a little present comfort. I hope and trust that this
is not your case but that every fibre of your heart is tremulously
responsible to the words you lately repeated, *the burthen of them
is intolerable*. Why should not repentance be thus deep and universal seeing that we are told, that though our sins be as crimson
and scarlet, yet shall they be as wool. I think all day long of your
case, I read, study, and hunt about it and meditate upon repentance
and saving faith, as if I were going to my own death bed, and
many are the prayers I offer up for you, and I am sure, that you
ought to be praying all day long for yourself, and find the day
too short to offer your ejaculations.

I have drawn from my manuscript collection, the coppying [*sic*] of which, I believe, has cost an £ 100—one or two articles pray keep them carefully in order to return them, but read them with attention.

God bless and preserve you; and grant you every day an increase of grace.

Ever

A. Young

June 20th, 1810.[68]

No wonder that Betsy had no comfort in writing to him about religion, or that she complained of his nagging.

Two other brief undated notes complete Betsy's letters for 1810, both apparently written during the summer or early fall. Both show a warm, affectionate relationship. The first follows:

The boys are so crasy [*sic*] for a row in your Boat that I have let them ride over to you which I hope you wont think too trouble-some—*One* is big enough one would suppose to take care of himself—but I hope Mr. St. Croix will have the goodness to go with them—send me word how you are today. I am tolerable but sadly tired of con-finement. if Thursday is a very fine day I hope we shall be able to come to you—but all is such an uncertainty that I must entreat you will make no preparation more than is *positively* necessary for such hungry hounds.[69]

The second letter is equally interesting:

Orbell wishes you would have the goodness to be here by *eleven* ock *tomorrow morning* to assist him in settling some business with Lord Bristol—I desired Orbell to write to you himself, but he seemed to think you would come so much more readily if *I* made the request that he insisted upon my writing—therefore *see to it* that you do not let me have the credit of influence with you *for nought*. . . . I am better this evening thank God, & with care I yet hope I may do well & I assure you I have felt too ill, to suffer me to be careless again—my cough is better which is a great matter.[70]

[68] *Ibid.*, 35,131, ff. 99–100. This letter is out of place in the collection, being arranged under 1811. The date is clearly 1810 and by June 20, 1811, Betsy was no more. It seems logical to assume that this letter has survived, alone of the very numerous ones he must have written, because it was returned with the articles which had accompanied it, after her death. It is typical of St. Croix' carelessness in spelling and punctuation. The letters in his handwriting are full of mistakes which Young would never have made.

[69] *Ibid.*, 35,130, f. 450.

[70] *Ibid.*, f. 484.

Arthur Young's public and business affairs in 1810 are neither very important nor interesting. He was elected an honorary member of a newly established Literary and Philosophical Society at Preston and of the Cork Institution.[71] There was a very flattering letter from his publisher, Sir Richard Phillips, who wrote about the Board of Agriculture:

I fear however it will not survive its founders—but their Glory will be the greater, & the age of YOUNG and SINCLAIR will be to agriculture what that of Pythagoras & Plato was to Philosophy. I feel much disposed to spread this sentiment in a dedication to you at the head of my little Grammar & I want your permission.[72]

Phillips also wrote again that he would pay £200 for Young's lectures.[73] Why this project was never completed is uncertain. The Board had expected to have them published. Surely the preparation of the manuscripts for publication would not have been very onerous.

During the summer Young had frequent letters from Sinclair. He wrote from the Holkham sheepshearing, enclosing some manuscript from his General Report on Scotland for Young to examine. Since grass might be in short supply, he also requested Young to suggest substitutes for the Board to recommend.[74] On July 15 he wrote from Edinburgh, enclosing a lecture for Young to criticize, and adding some queries about converting pease and bean straw into dung, and using cabbage for sheep and horse food.[75] On August 1 he sent still more queries, some about the exact meaning of the term "soiling."[76] On August 30 he sent more proof of the work on Scotland and an account of his being made a privy councillor:

Many thanks to you for your friendly congratulations on an event, which took place yesterday in a manner highly gratifying to me. I have drawn up a short account of it, a copy of which I inclose for yours and Mrs. Young's perusal, but not to be shown to any one else. . . . It certainly is an event that is creditable to the Board, and, I hope will be attended with useful consequences both to it and to

[71] *Ibid.*, ff. 372, 377.
[72] *Ibid.*, ff. 370–371.
[73] *Ibid.*, f. 413.
[74] *Ibid.*, ff. 424–425.
[75] *Ibid.*, ff. 435–436.
[76] *Ibid.*, ff. 444–445.

the agricultural Interests of the Kingdom, the promotion of which has cost both you and me so many weary hours.[77]

In reply Young wrote on September 4 that he awaited "with great impatience, to see a work complete which will form a new era in agricultural knowledge."[78]

In 1810 Francis Horner's famous Bullion Committee recommended the resumption of specie payments and the return to a metallic currency. Sir John Sinclair published two pamphlets opposing the report, upon which Young warmly congratulated him: "I have read with great pleasure your most able and satisfactory refutation of the Committee. It is in my opinion unanswerable, and does great honour to the rapid application of your talents, ever ready to answer the spur on any occasion that demands the exertion."[79]

Except for his lecture on manures, the only knowledge of Young's activity at the Board of Agriculture in 1810 consists of two letters to A. Bell, whose Hampshire survey was rejected by the Board. Young's first letter was dated May 31:

I am very sorry to find by your Letters that you seem to forget the transaction in which your Hampshire Ms. originated. It was left absolutely to the Board to accept or decline . . . and I cannot but hope that you will on recollection be persuaded that no injury can have been sustained by you in this business: but that on the contrary the inquiries you made might have the effect of introducing you to persons who according to your expectation might be serviceable to you in your profession.

Since this letter failed to mollify Mr. Bell, Young wrote a second on June 18, stating his opinion even more forcibly:

In answer to your Letter I can no more than repeat in the most decisive manner that I made no agreement with you on behalf of the Board that did not depend conditionally on their approving your Report; but on the contrary told you, that if your Report was not approved you would be paid nothing: this is the fact and if you remain under any error I am sorry for it—any further application must be by Letter to the President Sir John Sinclair Bt. who will I

[77] *Ibid.*, ff. 451–452.
[78] Sinclair, 1837: 2: p. 76.
[79] *Ibid.* 1: p. 266. The reviewer in the *Quarterly Review* 4: pp. 518–536; 5: pp. 120–138, took quite a different view towards Sinclair's pamphlets.

suppose take the opinion of the Board when they resume their meetings.[80]

In July, Young received news that a banking firm with which he had an account had closed its doors, but a later letter congratulated him that his balance had been less than he had figured. The explanation for the difference does not shed a very favorable light, however, on Young's care in business matters:

We are glad to find that you had no more money at Messrs. Devagne & Co. & still more glad to find that you are not even so badly off as you thought, for on applying to them to know the Balance of your Account they say it is not £ 39.11.4 as you think but only £ 15.9.7 which they suppose arises from your having drawn some small checques to the amount of £ 23.10.5 & not having set them down in your Book. . . .[81]

The manuscript letters for 1810 contain a proposal from John B. Edwards about renting part of the Bradfield estate:

Mr. Edward's proposal to Mr. Young.

Jermain Field—8 acres—Rent £ 2. 10s. per acre—£ 20.0.0
Free of Tythe, Parochial Rates & Tax's & allowance of any kind.
Covenants of croping & quiting it, as the last hire.

Barn Field. 6 acres—4 £ per acre £ 24.0.0
with Liberty to plant where the chickory now grows with seeds, or Tares.

Church Field—Mr. E. declines offering Mr. Young any Terms, not having stabling for six Horses.

Mr. E. will provide Mr. Young with 2 cart loads of Barley Straw yearly for £ 1.10s.

If Mr. Young is determined to let Church Field Mr. E. has no doubt of geting [sic] Mr. Young an excellent Tenant on fair Terms. . . .[82]

Except for 1797 when Bobbin died, 1811 was the most tragic year in Arthur Young's life. On March 10 he was unsuccessfully operated on for cataract and permanent blindness resulted. Two months later on May 8 his dearest friend, Betsy Oakes, died after

---

[80] Board of Agriculture, Letter Book 2: ff. 20, 33. These are copies of original letters. From Sir Ernest Clarke's table, it seems that Bell's report was never accepted.

[81] Add. MSS. 35,130, ff. 440, 448.

[82] *Ibid.*, f. 465. The Bradfield Church contains a tablet to "John Bidwell Edwards, Esq. died 7th August, 1824, aged 66 years." Apparently relations between the Youngs and Edwards' were cordial, although not those of complete equals.

a prolonged and very painful illness. In one sense her death must have been a relief to her friends for now her sufferings were at last over. On the other hand, Young had great hopes that the eye operation would be successful. Hence the disappointment was the greater. Moreover the convalescence was not only prolonged but also quite painful. Both tragedies occurred in the spring. In the summer, however, a new figure fortunately entered Young's life, the amusing and lively Marianne Francis, a girl of twenty-one, who in some measure filled the void left by Betsy's death and in a sense compensated at long last for Bobbin's loss.

Young's own account of the operation for cataract is in a letter to Thomas Ruggles:

You know the Cataract which for three years past has threatened me with blindness; being sufficiently advanced for couching, that operation was performed by Phipps on the 10th of March, under circumstances so apparently favourable, that he was full of expectation that I might be walking about the Streets in three weeks; in bed only one day, and two more in my chamber. I was in as perfect health as at any time for 30 years past, but we did not sufficiently consult the will of the Lord, my prayers had been in a measure fervent and regular, but I did not enquire of him whether such an operation should be performed, taking it for granted it should;—a violent inflammation came on, and I was in my bed a month very ill indeed: I am still a poor creature, but blessed be God part recovering my common health, but my eyes do not promise to be healed for a month to come, but the worst of all I have yet to announce, I am blind, in all probability irrevocably so, the heaviest calamity next to great pain that can befal a man of an active mind, but I thank God for his mercy. I receive his will with submission, and I fully believe that it is for the benefit of my Soul, at the age of 70 I attended too much to worldly pursuits, and I did not use my sight to the glory of the giver. . . .[83]

Perhaps it was just as well that the old man believed that the operation had failed because it was God's will and because he was still so sinful that his soul could be saved only through more suffering. Mary Young's account in a letter to her reverend brother, still in Russia, throws a more mundane light upon the operation and its outcome:

It seems that the poor patient was very intractable, and that the operator said, "Indeed, sir, if you are not more patient I must leave

---

[83] Letter XI in the Ruggles correspondence.

you." . . . Mr. Wilberforce, with the best wishes imaginable, called after the couching, and was shown up to his bedroom; and the very first words he said were, "So we have lost the poor Duke of Grafton!" then began and continued in his mild, soft manner a most pathetic dissertation on the duke's pious resignation, &c. &c., till your father *burst into tears,* which was Phipps (the occulist) vowed, the worst thing possible, and which anyone knew in his lamentable state of inflammation was *destruction.* It flung him back, only a week after the operation. Oh, Ar., as I greatly believe he will be entirely blind, do try to come to him.[84]

Although Mary's original letter has disappeared and has been cut, it seems as though Phipps was at fault in not warning the family of such a possibility. It is uncertain whether Wilberforce ever knew what his unfortunate call had done to his friend, but he was a man of the world and should have known better. The following passage from his letter to Young of July 21 grates uncomfortably upon the modern mind:

O, my dear Sir, Let not a single doubt remain in yr Bosom, as to the gracious design of your bodily sufferings. . . . We, poor erring mortals, however kindly we may design, may take wrong methods of accomplishing our purpose. But we may confide no less firmly in ye wisdom than ye goodness of the Lord Almighty. Why my dear Sir would He by his Providential Disposition abt ye year 1797 have called you out of a State of utter forgetfulness of him & have disposed your Heart to seek him, but from a gracious Design to promote your Happiness. . . . But his ways are not our ways, nor his thoughts as our thoughts. . . .[85]

Yet one cannot doubt Wilberforce's genuine friendship for Young. Probably shortly after arriving at Bradfield for the summer, Young requested Wilberforce's advice about possibly publishing some of the excerpts on religion which he had been collecting for some time. Wilberforce's reply made his customary apology for being so tardy, expressed his willingness to examine and criticize Young's plan, and ended with the following note: "My dear friend—It is a matter of standing regret with me that I see so little of you & I should be glad of any opportunity of spending a few quiet days in your society."[86] Incautiously Wilber-

[84] Young, 1898: p. 454, footnote 1.
[85] Add. MSS. 35,131, ff. 123–126.
[86] *Ibid.,* ff. 114–115. Dated July 11, from London.

force also asked Young whether he knew of any retreat where he and Mrs. Wilberforce could spend six months or so in the country. Young responded by sending him the topical outline of his proposed book and suggesting a cottage at Bradfield. But Young could never entice the great Evangelical to spend any length of time at Bradfield. His reply was a courteous refusal:

The house near you . . . strongly tempts . . . I love you, I love even yr Rooks & the voice of both each in its season would be music in my Ear—but there are other vocal performers who would altogether make up such an assemblage as to preclude all Hopes of Quiet. . . .[87]

Several weeks later Wilberforce returned Young's outline with several suggestions for improvement but with a cautious encouragement to go ahead. Again he added: "I wish I were just now with you hearing in ye pauses of our conversation the cawing of yr Rooks. . . ."[88]

In September, Young wrote to Wilberforce requesting his support of the application of Mr. B. McMillan, the printer, for one of the posts in the projected government printing office. Young backed his request by stating that McMillan had "printed much both for the Board of Agriculture and myself, and has executed every thing with so much ease, assiduity, and correctness, that I think I may recommend him as a person capable of any exertions in his profession." Young continued that McMillan "has earnestly requested me to make application to you in his favour as he thinks that any recommendation of him from you to Mr. Percival would be most effective."[89] A later letter from McMillan to Young[90] indicates that Wilberforce absolutely refused to support the application in any way. Nevertheless in

---

[87] *Ibid.*, f. 124. Whether Wilberforce was referring to Mrs. Young's tongue or to the fact that Young's home was always full of guests is not clear.

[88] *Ibid.*, ff. 156–157. This letter is dated August 14. The manuscript proposal for the book on religion, corrected by Wilberforce, is in *ibid.*, ff. 116–119.

[89] *Ibid.*, ff. 178–179. This was written after receiving McMillan's request, *ibid.*, ff. 176–177, dated September 8.

[90] *Ibid.*, f. 206, dated October 11. McMillan's letter also reveals the financial difficulties of Sir Richard Phillips, Young's last publisher of the *Annals*, on whom Young had apparently made a legal claim. McMillan wrote: ". . . I have been to Bridge Street, & found the following written in pencil under your account by Sir R. P. 'R. P. does not, as he thinks, owe Mr. Young ONE FARTHING, and the Balance is rather the other way. The stock should be taken, & returned . . . and Mr. Y. debited for payments & on account of the above. . . .' This I do not comprehend. . . . I trust you have preserved all Documents respecting these transactions." The correspondence contains no other references to this business.

early November he requested Young's support for a friend of *his* for the vacant chaplaincy at the Lock Hospital.[91]

The Young correspondence for 1811 includes letters from his two old schoolmates, Thomas Ruggles and the Rev. Jonathan Carter, who were in almost as bad a physical shape as Young himself. The four letters from Ruggles to Young each contained a most cordial invitation to visit Spains Hall. On March 24 he wrote, ". . . there is more satisfaction in taking one old friend by the hand than twenty young ones, we can walk together arm in arm, and can talk together thank God as well as ever. . . ."[92] After receiving Young's tragic letter announcing his blindness, quoted above, Ruggles renewed his invitation on May 5. He referred to the gout in his feet, "wch. if it increases much will occassion [*sic*] a supporter to be as necessary to me as a guide to you & nevertheless we can find places of walking I lead you while you support me in the meantime our Tongues may do their office. . . ."[93] In August the Rev. Jonathan Carter also invited Young to visit him at Flempton and promised to return the visit: ". . . then I will use every contrivance to spend a night or two with my oldest friend; we may mutually comfort each other. . . ."[94] Other letters late in November and early in December refer to Carter's impending visit to Bradfield, but since he had to be carried upstairs by his servants, it is not certain that he ever arrived.[95]

In the summer of 1810 Betsy Oakes had seemed considerably better, able to be at her beloved Nowton and even to visit with old friends. With the advent of winter, however, she grew worse and finally died on May 8. When Young left Bradfield late in January, he was certainly contemplating the eye operation, but hoped for a quick recovery so that he could return to Bradfield for the Board's Easter recess as in previous years. The long illness after the operation, however, prevented his return, and hence he never saw his beloved Betsy again.

Young saved many letters on Betsy's last agonies, three from her, four from Peggy, and several each from the Rev. Henry

[91] *Ibid.*, f. 233. Dated November 9.
[92] *Ibid.*, ff. 52–53.
[93] *Ibid.*, ff. 77–78.
[94] *Ibid.*, f. 152. Dated August 14.
[95] *Ibid.*, ff. 245, 251, and Berg Collection, Folders A. Y. to M. F.

Hasted and her Bury physician, John Smith. Unfortunately very few are definitely dated and the exact chronology is therefore just guesswork. On February 13 Hasted wrote that although there was no hope of recovery she might linger for several months.[96] Early in March she had convulsions and several days of delirium, but again she rallied, and late in April was so much better that Peggy saw her at least twice, the last time only nine days before her death.[97] The end apparently came quite suddenly on the morning of May 8 with very little suffering.[98]

Betsy's three letters are not dated, but presumably were written early in February shortly after Young's return to London. Excerpts follow:

May the Almighty restore your eyesight! & that you may live to be a comfort to my Dearest Orbell & my children & to pray for the soul of your friend—do not regret my death, I leave a world of trouble and sorrow tho God Almighty in his goodness has bestowed upon me more blessings happiness prosperity & comfort than I could in any way deserve. . . .[99]

. . . I rejoice to find yr eyes are in a state to admit of the operation— you are sure of my fervent prayer—wou'd that I were a better Xtian, that they might avail, but I trust to the goodness of Providence for your success. . . . Myself I confess I CERTAINLY shou'd have the eye in the *most fit state* for the operation *done first—by all means* because in case of failure you wd have nothing to regret in comparison. . . . dont have it done on a friday & by *no means* let me know the time, for I cou'd not endure the suspense & agitation. My stomach continues so bad I am distracted to get to London. . . . to think of yr fondness for an old Coach always make me smile   x

God bless you my Dear friend
Ever yours most affectly

E. F. Oakes[100]

x which I have not done lately

I am well aware of all your kind anxiety & that it claims my utmost exertions. you wish for a line from me & you shall have it. I think

---

[96] Add. MSS. 35,131, ff. 19–20. Hasted also urged Young to write a few lines to Betsy and added that Orbell thought it would please her.

[97] *Ibid.*, ff. 67, 71–72.

[98] *Ibid.*, f. 73. Letter from Dr. Smith to Mary Young, undated.

[99] *Ibid.*, f. 13.

[100] *Ibid.* ff. 15–16.

there is a weakness of stomach never to be overcome & all I pray for is that perfect submission which I know God requires—his kindness to me is beyond all expression. . . . tho I suffer past description. . . . The good Mr. Hasted comes to me every day unless I am too ill—you will be glad to hear this. . . . sun would cheer me a little my kindest regards & affecte love to Peggy. . . . I trust these lines will be some comfort to you. kind love to Mrs. Arthur Adieu, heaven bless you

ever yours with the same affectionate friendship

E.F.O.[101]

After her visit with Betsy on April 29 Peggy wrote Young:

I was with Her yesterday till between 9 & 10, & we had a great deal of conversation between whiles; she asked much about you, of whom I gave as favorably [*sic*] a report as I could consistently with truth, & as she did not ask positively whether the operation had succeeded, of course I said nothing particular about it but she desired me to remember her most kindly to you when I wrote, & to bid you take care of yourself. . . .[102]

When the end finally came Dr. Smith wrote to Mary Young to ask her to break the sad news to her father.[103] On May 8 Young wrote in his diary:

Twelve o'clock at noon my dear friend Mrs. Oakes breathed her last, after a long severe illness, and many and great sufferings. Thanks to God she was attentive throughout this sad period, as I am well informed, to the state of her soul with God. Thus is terminated in this world a very intimate friendship of twenty-six years, with a temper so mild and cheerful, with manners so gentle and persuasive, that had it pleased the Almighty to have spared her, she would have been the source of great comfort to me in my melancholy state.[104]

A week after her death Betsy's father-in-law, James Oakes, wrote to Young, presumably in reply to a letter of condolence:

I am well aware of how much you esteemed, & what a sincere regard you had for the afflicted sufferer. . . . you who had known her from her Infancy knew how to appreciate her most excellent accom-

---

[101] *Ibid.,* f. 17.
[102] *Ibid.,* f. 71.
[103] *Ibid.,* f. 73.
[104] Young, 1898: p. 453.

plishments & have no doubt felt keenly, indeed all who had any
knowledge unite in lamenting her loss.

The last solemn Rites were performed yesterday at her favourite
Nowton; Mr. Plampin himself read the service, it was indeed with
much difficulty that he got thro, but it was at the request of my son
in the absence of Mr. Hasted.

My son desires to be most kindly remembered to you & earnestly
wishes to hear your Health may not have suffered from the late calamity & consequent long confinement.

Poor Orbell's spirits are terribly reduced, the voice of Grief must be
heard. . . . The unwearied & unremitting attention so costanly
[*sic*] paid the dear deceased by every word of tenderness & affection
in his Power to soothe the Bed of Sickness, will be its own reward.[105]

It is utterly inconceivable to believe that such a letter at such
a time could have been written to Young by James Oakes if the
relation between Arthur and Betsy had been a guilty one. It
is almost as inconceivable that their relationship could exist in
the twentieth century without scandal.

In the attractive little church at Nowton there is a fairly pretentious monument. It is in gray stone, with a woman weeping
over the tomb, a cross, and a book entitled, "Thy Will be Done."
The inscription is as follows:

To the Memory of
Elizabeth Frances, wife of
Orbell Ray Oakes, Esqre. of Bury,
    This Tablet is erected
by her affectionate, afflicted Husband!
In the works of Faith, Hope, and Charity,
and in her Conduct through the several
    Relations of Life,
She approached as near Perfection
as the State of human nature can permit.
    She died May 8th 1811
in the 42nd year of her Age.

Unfortunately Arthur Young was deprived of the affectionate
attention of his daughter-in-law in the spring of 1811. Late in
April she left Sackville Street to visit her sister and father in
Wales. From there she accompanied her sister, Louisa Jones, to

---

[105] Add. MSS. 35,131, ff. 81–82.

Bath where she spent most of the summer waiting for Louisa's confinement and did not return to Bradfield until early autumn. Jane wrote eight letters to her father-in-law, running from May 4 to August 30, which throw much light on her and her family. Louisa Jones was probably a very poor housekeeper and her husband a shy, kindly man. Jane approached Young about a possible curacy for Mr. Jones in Suffolk and he actually came to Bradfield in the summer hoping to gain the Whatfield curacy from the Rev. John Plampin, but the plan fell through. Edward Berry, Jane's father, was a wealthy man engaged in building a pretentious house near Usk in Monmouthshire. His illegitimate grandson was part of the entourage but was recognized as only a poor relation. Jane traveled in some style with her own maid and a pet squirrel in a cage. She was trying to pay off a debt to Young as quickly as possible. She never heard from her husband in the Crimea. She was very friendly with a Mr. Boward, who had aided her while in Russia, and whom she entertained while visiting her father. Worst of all, she was on very bad terms with her mother-in-law, who apparently had accused her of indiscretion with Young's secretary, St. Croix. All in all, Jane's letters must have caused Young much worry and vexation. She seems to have been very sensitive and quite affected, but there can be no doubt of her deep affection towards her father-in-law and sister-in-law. A few excerpts from the letters follow:

May 4. Re her pet squirrel:

. . . To hear of your daily convalescence is indeed the most welcome and interesting news I can be informed of. . . . My Betty & that sweet little Charley arrived both safe and in good health—he is happy enough and has plenty of nuts, wheat, bread & milk but he is deprived of his liberty and therefore not so well off as in Sackville Street where he had besides a companion to amuse and play with him and a whole room to run about in. I keep him in my room chained to his House & he sometimes visits downstairs to the great delight of my little Jane. . . . Poor little fellow, I am very fond of him on *many accounts.* . . .[106]

May 16. After describing her father's home and estate:

. . . . remember my Dear Friend that I do not chuse Mrs. Y. should know any of these particulars. . . . *Mrs. Y. is to know nothing at all.* Pray tell me in answer if you burn my letters, and if you have kept

[106] *Ibid.,* ff. 75–76.

all the intelligence from her. . . . When I think of you I wish I had wings to fly to you, but when the remembrance of Mrs. Y. presents itself before I shrink with horror at the idea of being again within the sound and reach of her tongue. I feel a most *decided aversion* against her, and cannot overcome it, I have suffered too much from her behaviour to me, ever to forget it. . . .[107]

May 23. More on her mother-in-law:

. . . What I said about Mrs. Y. I am sorry for, as it made you uneasy, but having suffered so much from her, can you wonder at my horror of her. . . . be assured that my dislike of her shall not be greater, than my love for you, and that I will not stay away from you on *her* account. Her behaviour to poor St. Croix, was even more painful to my feelings than all her abuse of me, it was vile & mean to a degree, and his conduct was always that of a gentleman and I will ever stand up as his friend and advocate, and maintain that he is one of the best behaved, best tempered, and most obliging young man [*sic*] I ever met with. . . .[108]

June 15. Re Mr. Boward's visit to her:

Boward arrived here on Wednesday evening; my Father is much pleased him. . . . I feel much pleased in this opportunity of introducing Boward to my Father & Sister, and of his becoming acquainted with every part of my Family. he has done so much for me that I feel it a real obligation to try all in my power to convince him of the *wish* at least that I have, of being grateful. . . .[109]

Jane's last letter on August 30 expressed her fear that the Bradfield people were becoming indifferent towards her:

I must trouble you with another letter which will make the third one unanswered! . . . By a letter which I receive from Boward a few days ago, he informs me that after much trouble and expence he has at last managed to clear the things from the Customs House & has deposited my Trunk & the Box of Seeds in Sackville St. . . . Arth's Crimea Journal is in my Trunk & when we meet we can look it over. . . . Have you heard from Arth?. . . . Surely Arthur might have written by Odessa but it always was with him, "out of sight out of mind"—So Mrs. Y. is at Bradfield. . . . I hope Mrs. Y. is in good health & spirits and has recovered her indisposition. Pray give my comps. to her. . . . If you do not answer this letter my Dear Friend, I shall be seriously offended; as I never neglect you, I think myself

[107] *Ibid.*, ff. 85–87.
[108] *Ibid.*, f. 88–89.
[109] *Ibid.*, f. 96.

entitled to the same attention from you & Mary. . . . I have every reason to suppose that my absence will be more agreeable to every one—should that be the case, a return to Russia is not *impossible* & never would I remain with any one who is not desirous of my society. . . . Is Mrs. Y. in good spirits? pray give me a particular account of her—God bless you! . . . I fear that I have written what you wont like, I am sorry for it, but indeed indifference & neglect are not what I am accustomed to—make allowance for offended feeling. . . .[110]

There is one letter from Martha Young to her husband which must be quoted in full:

Dear Mr. Young,

Finding myself so much worse, I have determined to set out for Bradfield after breakfast, & should it please God expect to be there on Saturday by the coach. Will you have the goodness to order two or three lean chickens to make some strong broth; two calves feet to be made into Jelly. I must also have 3 pounds of shin beef to make into very strong soup No addition to the beef of vegetables, salt, or any else. I expect to be down by Saturday, but if I feel too ill I must be longer.

I am
Dear Mr. Young
Yours affectionately
M. Young

32 Sackville Street,
15 August, 1811
My whole frame is so terribly alarmed & with an irritation exceeding all expression that I am forced to take 2 stages, i.e. To Ingatestone friday night, London most dreadfully hot & quite a different impression on me to Hayes & that is a very low & terribly moist situation.

Adieu. Pray for me & be very good in soothing my disastrous case.

The lodging at Hayes had no sort of accommodation whatever. I *can* scribble but to dictate & write too *I could not do at all.* If the cats either of them fly at Mino it will throw me into agitation incurable. No whiskey unless it rains! I beseech not to have Mrs. A. Y's room it's so *damp* & so dismal.[111]

What a revelation of utter misery on the part of the writer and of impending irritation and friction for her husband!

Before Jane returned to Bradfield, Young had entertained two

---

[110] *Ibid.*, ff. 169–170.
[111] *Ibid.*, ff. 160–161. This letter is in two distinct handwritings. The first part was probably dictated for the handwriting is fairly good, the latter part, probably written by Martha, is just a scrawl.

visitors, Mrs. Charlotte Broome and her daughter Miss Marianne Francis. Young had known Mrs. Broome since childhood as Charlotte Burney, Dr. Burney's youngest daughter by his first marriage. She had had a somewhat tragic life. In 1786 she had married Clement Francis, by whom she had three children before his death in 1792—Charlotte, Marianne, and Clement. Six years later she had married Ralph Broome, who had gone insane and died in 1805, and by whom she had one son, Ralph or Dolph, as he was known in the family. At one time well to do, by 1811 her circumstances were very modest indeed. As she wrote to Young late in 1811: "Mine is a stage coach income, & I always endeavor to lower my wants to my fortune . . . & be as happy with one servant as when I had seven . . ."[112] Charlotte Broome's oldest daughter Charlotte, a beautiful and clever girl who later edited her famous aunt's diaries, had married in 1808 a man nearly twenty years older, Henry Barrett. As will be seen, the Barretts spent several summers at Bradfield. Clement, younger than Marianne, became a student at Caius College, Cambridge, in the fall of 1811.

Born in 1790, Marianne was just twenty-one when she first met Arthur Young. In good Burney tradition, she began a diary when only thirteen. At sixteen she began to correspond with the famous Mrs. Piozzi whom she adored and to whom she had signed a letter in 1807: "I am ever my dear Mrs. Piozzi's madly attached Marianne Francis."[113] At seventeen she had mastered Latin, Spanish, and French, and was working on Greek. In 1808 she started Dutch and by 1811 was busy on Hebrew. No wonder that her grandfather, Dr. Burney, described Marianne to her aunt Fanny as "a monster."[114] She was also devoted to the piano

---

[112] *Ibid.*, f. 264. Details about Charlotte Broome's family are to be found in the Burney Scrapbook in the Berg Collection, ff. 3, 64, 66, 68.

[113] There are sixteen holograph diaries of Marianne Francis in the Berg Collection, running from 1803 to 1809. For the letters from Marianne to Mrs. Piozzi, *cf.* English MSS., John Rylands Library at Manchester. For this letter, *cf. ibid.*, 582, f. 5.

[114] Johnson, 1926: p. 358. In one letter to Mrs. Piozzi Marianne wrote, Rylands MSS. 582, f. 53: "I believe no man ever yet sincerely tolerated a dead language in a live female. . . ." In another, *ibid.*, f. 24, she complained of the very superficial education commonly given to girls. If women had the same education as men, "*do* you think the difference between the two sexes would continue to be so great as hardly to allow the wisest woman to equal the silliest man? That there certainly is a *difference* of powers & intellect, I believe. But that there would be the great *superiority* if women were allowed fair play, I can hardly believe. I am rather inclined to think they would be nearly on a par."

and by 1811 was a zealous Evangelical. She was short in stature and probably plain in appearance. Unfortunately no portraits have been found. She relied on her sister to help her with her clothes and admitted that she lacked taste in that field. She was a prodigious walker[115] and most methodical in her habits. Her correspondence hints at several unimportant love affairs. Such was the girl who became Young's chief correspondent late in 1811, who spent much time at Bradfield from that time, and who gave Young a new interest in life. At times she was almost a second secretary, at other times more like a granddaughter.

For many years Young's former close intimacy with the Burneys had lapsed, but in 1810 Charlotte Broome requested his support of a candidate for clerk to the Royal Society, and in the same letter expressed her wish to renew the friendship.[116] As a result Young invited both Charlotte and her daughter to Bradfield the next summer. While there Marianne wrote three letters to Mrs. Piozzi, giving her impressions of Young and of Bradfield. On September 8 she wrote:

This dear Arthur Young is a delightful old man—so good & kind & patient under his sad *poetic* complaint of blindness. . . . I scout about in the grounds, which are delightful, with magnificent trees, & beautifully laid out by Mr. Young himself—the shade & the air—& the good Library—& a kind welcome—make amends for want of a Pianoforte. . . . Arthur Young believes in *steam engines*—& says it is the maxim of some famous farmer that the science will never be brought to perfection till every farm is a circle & the steam engine in the centre to do all the work.[117]

Her second letter was on September 24:

You *would* like Arthur Young, & he would adore you—has such true enjoyment for excellence of any sort—admires you & Dr. Johnson as much as the Plough & the Steam-Engine, which is saying a great deal for *him,* devoted as his active life has been to agricultural pursuits—& famous, *very* famous I believe he *has* been in his time—Medals without end, for his services; the king of England sending him rams & the Empress of Russia snuffboxes, for his labours. . . . But a Pond & a Boat & an island to row to, are delightful to me. . . .[118]

---

[115] On March 16, 1809, she wrote of walking from London to Richmond with her brother-in-law, Henry Barrett, one day and returning the next. *Cf. ibid.,* f. 34.
[116] Add. MSS. 35,130, f. 489.
[117] Rylands MSS. 583, f. 82. For Mary Young's picture of Bradfield, see fig. 7.
[118] *Ibid.,* f. 83.

The third letter was on October 18, shortly after Marianne's return to London:

. . . & till the leave taking & packing up moments arrived, all I had were spent in reading, & writing for Mr. Young, whose constant amanuensis was unfortunately cut down with a fever—of course I felt but too happy in devoting my poor services to our kind & hospitable friend. . . .

He is very religious—happily for him—for no other support but that of divine grace, could have enabled him to submit so beautifully & completely to the deprivation of sight—a calamity which he so bitterly felt—& often so feelingly describes—talks of the constant *blackness* that is before him; asks *if* the sun shines—*if* candles are on the Table. . . . I saw some beautiful letters from Wilberforce to him—there is a great Friendship between them—He has Wilberforce's picture in his drawing room—his book in his Library—his name upon his lips—and, better than all, his *spirit* in his heart. . . .

30 or 40 poor neighbouring children are cloathed & educated by Mr. Young, & on Sunday they always come to be instructed in religion. I am very fond of teaching poor children, though they *are* such fools: & he used to make me hear their catechism—the answers they made were sometimes quite irresistible—only the disappointment that after so many years tuition they should know no more, changed the smile into sorrow before it could *merge* in a laugh. . . . every Sunday evening, from 80 to 100 poor Tenants & Villagers come to the Hall—forms placed for their reception; St. Croix reads a sermon; Mr. Young comments upon it afterwards & talks to them admirably; then makes a beautiful prayer & they go home. He is so innocent & humble about all excellent devices; says he has no other ambition but to be a *rival to the Alehouse;* that it is better they all come to him than go there. . . .[119]

There followed the long correspondence between the clever and very pious girl in her twenties and the old, blind, and still more pious man in his seventies. Much of our knowledge of his last years is derived from these letters, very numerous and many of them very long.[120] She wrote much more frequently than he. Usually there was a light, bantering tone in her letters which Young really loved, even when he condemned it. Several

[119] *Ibid.,* f. 84.
[120] Young's letters to Marianne are in the Berg Collection in folders entitled "From Arthur Young to Marianne Francis."

of her letters in November contain interesting tid-bits. Very early in the month she wrote:

If you are not like Cassius "aweary of the *world*"—you certainly are weary of Marianne—this day week & longer did I write to you,—no answer even *yet.* Why do you *know* I am extremely affronted with you, & only write *now,* for the pleasure of saying so?

You may think this mere matter of "moonshine" to use your own favourite metaphor. . . . Come! I will write no more, I am in such a rage with you—so farewell, which is no more than you deserve from your forgotten Friend & ci-devant correspondent.[121]

Still just a few days later she wrote again:

Take my advice, & leave your pond alone—it is *quite* deep enough for beauty & for fun—nobody rows there but me—& I am quite content with it—you will only waste all your substance in vain efforts & all your patience in lamenting this failure.[122]

On November 16 she wrote: "SCOLD indeed! Pray what encouragement have I to scold, if you are so incorrigible as to pretend you *like* it? . . . As to scolding *prettily,* that's impossible; the *effect* cannot be so different from the design. . . ."[123] A week later again:

What! angry in *right-earnest,* my dearest Sir? Oh no—not unless you wish to *break* my foolish heart. . . .

Charlotte said yesterday, "as for Arthur, he is *quite* an *Angel!*" As for *me* and my letter I suppose you always fall fast asleep, the minute Mr. St. Croix begins, & dont know, at this hour, whether you have heard from me or not. . . . I must beg you to keep awake just while my *questions* are read, because I always like to have them answered. . . .

In my early prayers & my solitary walks—when I am gay & when I am grave—when my fingers are intent upon the keys, or even my eyes upon my books—the recollection of your kindness comes across my memory like a refreshing breeze, but does not desert me so soon. The remembrance has a double use—I think of your opinion, &

---

[121] Add. MSS. 35,131, ff. 227–228. The date on the outside looks like November 4.
[122] *Ibid.,* ff. 229–230. Dated November 5.
[123] *Ibid.,* ff. 238–239.

feel humbled by my deficiencies; I think of your conduct & am stimulated by your example.[124]

Three of Young's letters to Marianne, written in December, 1811, have survived. The first on December 2 has only a few interesting passages:

I have long had a desire to write to you in a serious strain but when your letters come they are so full of vivacity that I am always in my replies hurried away from my design, and foolishly attempt to follow you as a capering cow does a Greyhound. . . .

I am delighted to hear that you are pleased with Baxter. I have the book and read it with great satisfaction: he is a capital writer, I mean in the way that pleases me most. . . . So good morning to you. . . . may you instruct your sister better in her knowledge of what Angels are, and not be ready to assign the title to the vilest of sinners, could she look into my heart, she would be far enough from making such mistakes.[125]

His letter of December 6 deserves more extensive quotation:

I intended to have filled a sheet by answering your last excellent letter, but the frost is so nipping, that it forces my attention to itself, the thermometer is now 7 in the morning at 27½ and I am setting [sic] before the fire in the first library with a great coat over my robe de chambre, two pairs of stockings on, shoes lined with flannel, gloves and two night-caps on; all which shows that I ought to have a sort of parental authority over you, as you are young enough to be the youngest of a very numerous family by a very late marriage: therefore my dear child I desire that you will attend to me in whatever I have to say to you relative to that part of you over which I claim an influence; to wit your Soul. . . .

My day is thus past. I rise between 5 and 6 pray to God the best I can walk in the Hall &c 25 yards, till 7, then Mr. St. Croix reads a prayer and a chapter in the Bible with Scott's practical observations; then the religious papers for arrangement, till half past nine; dress and breakfast at ten; then wool gathering till two; walk in the rope walk with Jane till three, then newspapers and letters, then wool till five; after dinner Jane reads a Chapter in Bennett's intermediate state, and too often a nap till tea at 8 but not always; after tea, that is from nine till ten Sully's memoirs, having finished Thibauts recollec-

---

[124] *Ibid.*, ff. 247–248.
[125] Berg Collection, A. Y. to M. F. folder.

tions of Berlin: at ten a chapter in the Testament and prayer; bed by half past ten: such is the day, how do you like it? As to preaching I muse a good deal on what I ought to say, and having as you know a wretched memory talk to them for half an hour on something perhaps very different from what I proposed: The Parson of a Parish not far distant will not permit parochial assistance to be given to any that come here on a Sunday night, but I have not heard of a like prohibition against an evening of drunkenness and swearing at an ale house. . . . I continue thank God free from the rheumatism which is a great blessing for which I am not sufficiently thankful. . . .[126]

By the time Young wrote the above, Marianne was on the way to Ireland with her friend, Anna Maria, Viscountess Kirkwall who had been called home by the serious illness of her father, John, Baron de Blaquiere who had had a fairly notable and quite profitable career in Ireland. On December 17 Marianne wrote two letters from Ireland, the first describing the difficulties of the trip to Dublin by water, and the second giving details of how she spent her day when at home.[127] Young's letter on December 26 completes the correspondence for 1811:

Two letters from you are on the table are truly corn, wine and oil: they have quite appeased my wrath, for I was agitated at reading of nothing but storms and shipwrecks in the papers, and hearing the wind howling among the trees, without one line from you to calm my ruffled soul, but you are safe and so all is well. . . . I do not like the plot that is to last till May. . . . your not being in London all the winter, will make a fine mess for me. . . . as to your returning by yourself, I should advise your Mama immediately to put you in Bedlam; the world would say that Greek had made you mad. . . . Now for your second letter; work after dinner, at London, Dublin and Bradfield, right, very right, because it will give you infalably [*sic*] a red nose, of which I have heard a hint before, and there is something so extremely mischievous in every thing that tends to beauty that I heartily commend the practice. . . . I highly approve of your method of passing your time; it is truly worthy of a reasonable creature, all except the Piano, which is an utter loss of that which must be accounted for to God. . . . half an hour digging

---

[126] *Ibid.* I do not believe the reference to "wool gathering" means that he was idle; more likely he was working, perhaps on the *Elements*, perhaps on work for the Board. *Cf. supra*, pp. 558–559, for his daily routine in 1809.

[127] Add. MSS. 35,131, ff. 271–274. For Blaquiere *cf. Dictionary of National Biography*, and *Gentleman's Magazine* **82** (1812): Part 2, p. 298, for his obituary which also identifies Lady Kirkwall's name and title.

is far better for exercise: I will have a very nice spade ready for you, and you shall dig, sow and rake; and if a pig dares to put his nose in the crop he shall be hanged immediately. When Lord de Blaquiere was minister his was the most famous table ever kept at Dublin. I hope you have nothing of the kind now, or mutton and turneps at Bradfield will not go down. . . . I shall be much provoked if you do not receive gratification from your two new acquaintances always under your eye, the sea and the Wicklow mountains. . . . Yesterday I dined 40 children on plumb pudding roast beef and mutton, and after dinner distributed the rewards according to the number of cards from six shillings to sixpence; it was a scene that would have pleased you.[128]

Again there is relatively little material on Young's business and public life. There is considerable evidence of his great interest in the Bible Society and of his quite generous financial support for a man of his means.[129]

There is not much precise information about Young's activities at the Board. On September 15 Sir John Sinclair requested Young to examine his manuscript on Scottish agriculture for corrections or improvements in style or organization, and also brought to Young's attention a complaint by an employee of the Board against Under-Secretary Cragg.[130] Without much doubt the most gratifying event in Young's public life was the Board's vote on June 6 after his lecture on the husbandry of three great farmers—Bakewell, Arbuthnot, and Ducket:

Resolved unanimously that the thanks of the Board be given to ARTHUR YOUNG Esqr. for the valuable information contained in his very excellent Lecture read this day and that a piece of PLATE of the value of TWENTY GUINEAS be presented to him as a small mark of their esteem. and of the sense the BOARD entertains of the assistance given by Mr. YOUNG to their Exertions for the Improvement of the COUNTRY.

The piece of plate was a very handsome square silver vegetable dish, on the cover of which were engraved the arms of the Board of Agriculture, a reclining sheep in high relief, and the Board's resolution.[131]

[128] Berg Collection.
[129] Add. MSS. 35,131, ff. 9–10, 235.
[130] *Ibid.*, f. 182.
[131] In 1931 this dish was owned by Mrs. Rose Willson, niece of the last Mrs. Arthur Young, who kindly furnished me with a photograph of it. For illustration, *cf.* fig. 9.

His lecture was published later in the year. It was hardly more than a pamphlet, slightly over fifty pages long, and embellished with several pictures of agricultural implements. Young had known all three men personally, Arbuthnot and Bakewell very well. Indeed for nearly a decade Arbuthnot had been his closest friend. He had described the agricultural practices of Bakewell and Arbuthnot in his Eastern Tour, those of Arbuthnot exhaustively.[132] Young had also written two major articles on Bakewell and three short ones on Ducket in the *Annals of Agriculture*.[133] The *Annals* had also published the two brief articles on Ducket by George III under the pseudonym "Ralph Robinson,"[134] to which Young gracefully referred in his lecture, "said to have been written by a Great Personage, whom we all revere and love, and whose present situation excites the deepest feelings of our hearts."[135] There was thus no need for fresh research, for he had only to boil down what he had already written. Nevertheless he wrote to Bakewell's disciple, George Culley, for possible additional material.[136]

Of course Young praised highly the work of all three men. He analyzed briefly their chief contributions to agricultural science, Bakewell's breeding methods, Arbuthnot's and Ducket's modes of cultivation and improvements in agricultural machinery. He pointed out the folly of not accepting the work of such great experimenters and improvers, and of still arguing over points which these men had conclusively proven. Young flattered his audience by calling agriculture "the first and most important of all arts,"[137] and also specifically praised the Board's work which "crowned all preceding efforts and secured to the nation the invaluable means of concentrating, in the best possible manner, the result of past and future exertions." The great improvements in agriculture dated from George III's accession in 1760 and the succeeding period was "the epoch most remarkable in the history

---

[132] Young, 1771: **1**: pp. 110–134 for Bakewell; **2**: pp. 250–500 for Arbuthnot.

[133] For Bakewell, *cf. Annals* **6**: pp. 452–502; **16**: pp. 480–607; for Ducket, *cf. Annals* **10**: pp. 186–198; **17**: pp. 129–178; **38**: pp. 625–630.

[134] *Ibid.* **7**: pp. 65–71, 332–336.

[135] Young, 1811: p. 39.

[136] Add. MSS. 35,131, ff. 21–22, 46, letters from Culley on February 18 and March 9.

[137] Young, 1811: p. 1.

of British Agriculture."[138] That he still knew how to appeal to the agricultural interests is shown by the following remark on Bakewell's work: "And some statues are to be found in Westminister Abbey, directed to the memory of men, who merited far less of the public, than did this humble cultivator of the earth."[139]

The publisher, B. McMillan, wrote Young on October 11, "The Lecture has sold tolerably well. . . . I am sorry it was so cheap, for one advertisement swallowed all sold."[140] It was reviewed briefly but favorably in the *Monthly Review* for January, 1812:

The conciseness of this pamphlet will prevent it from frightening the farmer; who, if he has any good sense, will perceive the importance of the hints which it contains, and, if he has any regard to his own interest, will be guided by them.[141]

As soon as he reached Bradfield in June, Young began to collect materials for another lecture to be given to the Board in 1812 on the relation between prices and paper money, and between prices and agricultural prosperity. Once he got into the subject, however, he found the materials so voluminous that he decided to print a pamphlet rather than give a lecture. Apparently he spent most of his spare time during the latter half of 1811 on this subject. In order to have his material up to date he sent out a questionnaire in August which requested answers on the prices of various food commodities, wool, coal, timber, and the wages of labor. Some inquiries were for 1810, some for 1811, and some for both years. Apparently the Bishop of Durham protested to Young because he had used the official residence of the Board at the head of the questionnaire and had signed it "Arthur Young, Secretary." In his reply Young admitted the questionnaire was "merely a private one," but claimed that he had hoped through it to secure for the Board "many useful tables on the progress of Prices."[142]

[138] *Ibid.,* pp. 29–31.
[139] *Ibid.,* p. 17.
[140] Add. MSS. 35,131, f. 206.
[141] *Monthly Review Enlarged* **5** (**67**): p. 106.
[142] Add. MSS. 35,131, ff. 173, 194.

XI. St. Martin's Summer, 1812-1816

The year 1811 had indeed been tragic for Arthur Young, but the following five years were comparatively happy. His general health was remarkably good. He remained busy and carried on as secretary to the Board of Agriculture. In 1812 and 1815 he published important pamphlets in the related fields of money and prices. *Baxteriana,* a collection of excerpts from the writings of Richard Baxter appeared in 1815. The Rev. Arthur Young was home from Russia from April, 1814, until October, 1815. While he was home Martha Young died in April, but her death could hardly have been regarded as a tragedy by any of the family and certainly not by herself. Throughout these years the happiest season was the summer at beloved Bradfield. Marianne Francis was usually there part of the time, and in 1815 and 1816 Henry and Charlotte Barrett spent part of the summer in a cottage at Bradfield. Young continued his schools and his Sunday evening services at the Hall, and also became increasingly active in the Bible Society. All this kept him busy and he remained, in spite of his blindness and other ailments, remarkably cheerful.

The year 1812 was not very important in Young's life, nor very well documented. The *Autobiography* contains but a single long entry of less than four pages. The only documentation for the work of the Board of Agriculture consists of numerous letters from Sir John Sinclair. Fortunately a considerable number of letters have survived, especially from Marianne Francis and her immediate family.

Young was in London from about the middle of January until June 11 when he left for Bradfield. For some unknown reason he returned to London much earlier than usual, about December 1.

The one entry in the *Autobiography* related Young's last attempt in May to restore his sight. When a new occulist, a Dr. Adams, appeared in London, Young's friends urged a consultation. The tragic account of the interview follows:

From the reluctance I showed to name the day even after resolving to go, the dread of hearing my doom, and the natural desire to enjoy a little longer the precious glimmering of hope, may be inferred. At length the long-wished-for dreaded morning came. . . .

I was shown into a room, where I waited a few minutes (they were painful ones), and Mr. Adams appeared. "I wish, sir, to be informed, what is the state of my eyes," looking very attentively at him. "You have not, sir, undergone an extraction for cataract?" "That you must decide." "Why, yes, and I fear unsuccessfully." "Is there any hope of recovery?" Mr. Adams started, and looked down with evident marks of bitter disappointment the first instant he saw me. "I grieve, sir, to say that the eye itself is destroyed, the cornea gone, and there has been such an excessive discharge of the vitreous humour, that the coats are collapsed." "No chance, then, of course?" "I fear, sir, *none*"; then, after a pause, "I believe I am addressing Mr. A. Young?" I bowed, "I have heard your case differently reported; it was the subject of much conversation, and excited unenviable interest last spring when it happened, and I had hoped that it would have been possible to relieve you, but I now see the contrary."
. . . . On the way home, I was for a few moments depressed. "How happy," I cried, "are those beings who can see, no one can tell the misery of blindness, . . . If it were not for religion, I should wish to be the poor man who is to be hanged next Monday; but, thank God, I can consider the whole affair as . . . intended not for a curse but a blessing, and can reconcile my mind to it completely as His will. "You will see," I added after a pause smiling, "I shall be as cheerful and happy as ever," and so I was.[1]

At the beginning of 1812 Marianne was still near Dublin with Lady Kirkwall visiting the latter's father, Lord de Blaquiere. Her first letter in 1812 to Young announced that she would return soon to England and complained that Young had advised her mother how to handle her:

Mind you send no advice to Mama about me, because you dont know her constitution so well as I do; but write it all to me, yourself, as much as you choose. . . .

As to the music I am of H. More's mind & yours that it is a waste—but —I could sooner for all that, tear out my heart with my own hands & present it to some young fool of an Irishman, as you propose,

---

[1] Young, 1898: pp. 453–456. Either Young was not stone blind at this time, or else others must have told him that Dr. Adams looked down.

than give up my Piano. . . . If anybody tells you I have a red nose, as I find somebody has, don't believe it. . . .[2]

Young's very long reply deserves fairly extensive quotation:

When you give the reins to your fancy, you can dilate into such temerity of imagination, that I know not what to compare it to, unless it be the streaming lights of northern electricity; they wave and flow in so many directions that the eye strives in vain to follow them. . . .

I am very sorry to find that you are so obstinate in respect to the Piano. I want all the activity of your mind to be kept in full exertion, and none in your fingers, but when a pen is in them; sewing with your nose to a gown and rattling pieces of Ivory, are exactly adapted to thousands, but not to minds of your calibre. Mrs. Young is rather worse than better, and full of apprehension of the effort she is about to make to go to London by short stages . . . Lingley is here every day, Sundays excepted, to enable me to finish the wool gathering which has given incessant labour every day from breakfast during six months, and is not yet finished, and I must have him at London for that purpose, wanting also to consult thirty or 40 folios at the british museum: there are two Latin household books also to consult . . . but I can borrow them of Sir Joseph Banks or the Royal Society. . . . I have just finished the first reading of my religious papers, and ready to begin the sub-arrangement at London; this is always the employment from the time of rising till breakfast. On your return, I beg you to be very attentive to all you see and hear on your Journey. . . . You should never be without a journal, especially to mark the course of your reading. . . .[3]

When Marianne acknowledged the above, she was with Mrs. Piozzi in North Wales, and on her way home:

Brynbella, 19 Jan. 1812

Your kind letter, my dearest Sir, reached me just in time, for I left Ireland 3 hours after it arrived—had a quick passage, 9 hours, & arrived at St. Asaph's by post chaises, early on Sunday morning . . . an old nurse &c of Lady K's for my chaperon. . . . But here we are—dear Mrs. Piozzi all that is kind & charming. . . . Do you think you can prevail on yourself to get so far as our abode any time on Saturday, after one oclock? I am almost ashamed to ask it—but still I *will* venture, in case it may be either in your power or your inclina-

[2] Add. MSS. 35,131, ff. 284–285.
[3] Burney Papers, Berg Collection. Dated January 11.

tion, or both—it would give me great pleasure to see you . . . & I have so many things to say, to ask, to thank, to reproach you for, that I think it high time, considering the shortness of life to begin—*Do* come, if you can—& I will write again if *I* can. Dont you think I am a gallant little Traveller, to be going all over the world, with nothing but an old lady—forced to manage & contrive everything myself—& never losing either baggage or my purse? I have set my heart on going to church with you next Sunday—I wonder will you let me? . . . Adieu—I *am* tired—but there is *one* thing I shall never be tired of saying that I am dearest Sir—ever yours most gratefully & affectionately

Marianne Francis[4]

Almost as soon as Marianne arrived in London, probably on Saturday, January 24, she dispatched a note, presumably by hand, urging Young to call that day if possible:

Safe arrived in London my dearest Sir, at last. . . . I forgot to say that in Ireland I read your travels in France. . . . I tried hard for your Irish Tour but cd not get it—nor cd I help smiling at the instability of the human mind, when I was following you through France & reading in every 5th page sentiments opposite to those you at present indulge I was much amused on the whole by your book. Some parts of it I thought foolish, but *more* wise, & all entertaining—some a little untidy. . . .

Mind, I depend on the pleasure of seeing you today (Saturday) any hour or hours that you can command—my best respects to Mrs. Young, & compts to Miss Young & Mr. St. Croix[5]

A strange girl, this Marianne, only twenty-two years old, and yet urging a man of seventy-one, old enough to be her grandfather, to come to see her as quickly as possible after her return to London.

Marianne and her mother had lodgings in the spring of 1812 in Charles Street in Bloomsbury and there was considerable visiting back and forth between the two families. That indefatigable pedestrian Marianne would not have minded a jaunt down to Sackville Street. In March she wrote to Mrs. Piozzi that she had little time for either reading or writing: "Arthur Young makes me do *both*, in plenty, for *him;* he has a good right to my poor services, kind as he is to me, & I am always most happy

---

[4] Add. MSS. 35,131, ff. 289–290.
[5] *Ibid.*, f. 293.

in rendering them."[6] Two months later Marianne regaled Mrs. Piozzi with Young's predictions on the crop prospects: "Arthur Young seems not so distressed with the thought of a scarcity as many others. Fears he says are exaggerated; & the way to *make* a famine is to raise *apprehensions* of it."[7]

Marianne and her mother accompanied Young to Bradfield in June. Charlotte Broome only decided to go at the last moment and only stayed a short time, but Marianne remained at Bradfield until late November. Early in her visit Marianne wrote: ". . . How agreeable it is to find people whom one can love disinterestedly: I mean for their own sake, not for their appendages; as I trust I should Mrs. Piozzi & Arthur Young, if they lived in a garret in hog Lane. . . ."[8] After Charlotte Broome returned Young wrote her:

. . . You hear from Marianne that she is very well, though she has not dasht about the country in the manner which before made the rustics stare; she is more quiet, and pursues her studies with the utmost regularity and is most industrious in attending the schools. Thank God, I have been very well, marvellously so for my great age.[9]

While at Bradfield Marianne wrote fairly frequently to Mrs. Piozzi. On August 10:

[Mr. Young] . . . desired me, with his best respects, to tell Mrs. Piozzi, that he . . . finds just *seven* grand causes of human misery, which he denies any ingenuity, (even her's, to multiply.
Will you hear what they are, & try?

1. Disease of body
2. Disease of mind. (madness & Idiocy)
3. Oppression
4. Ignorance
5. Death— (of those we love—
6. Poverty
7. Last & greatest—*Sin*

. . . . How is it, that *good men*, from the days of Job, to those of Arthur Young, are plagued by their wives. . . .

Three weeks later, on September 1:

[6] Rylands MSS. 583, f. 94.
[7] *Ibid.*, f. 96.
[8] *Ibid.*, f. 98.
[9] Berg Collection. Dated July 17.

Curiosity is my predominant feeling, for to pretend affection for a person one never saw, is ridiculous; & I have not the most remote recollection of my celebated aunt. . . . A young idle Cantab is lately come here, & at 5 oclock this morning, crack went his gun for a partridge! He is [not?] the clergyman of the place, but took a cur-acy . . . for sporting, which greatly enraged Mr. Young, who is Lord of the Manor, & expects to have all his game shot by this *poaching Parson* (a new description of character). . . . Direct your next here, tho' I must go to Town as soon as Mr. Y. will let me.

Late in September:

. . . but poor Mr. Young has been so unwell. . . . It is melancholy at his age to see him bending with suffering. . . . But he is better now. . . . Mr. Young is in despair at the report that Buonaparte has forced the Russians to a peace now he says there will be nothing for him left, but pouring into England. . . . We are annoyed here with vipers & hornets.

On October 23: ". . . poor Mr. Young is so unwell, Mama says she thinks I ought not to leave him yet—but I want to go home, & must soon & see my celebrated Aunt for the *first* time." On October 31: ". . . one only motive prolonged my visit so greatly—poor Mr. Young's extreme indisposition. But he is now much recovered, & grown cheerful again, well & strong. . . ." On November 28, just after her return to London:

In Suffolk I was much employed in Mr. Young's schools for poor chil-dren, superintending the mistresses & endeavouring to introduce the new mode of monitors: making the children teach each other. I had many more than 100 scholars, & felt some regret at leaving the poor little things.[10]

If Marianne was curious about her famous aunt, that lady was also curious about the reports of her niece, which she passed on to her husband in Paris:

. . . sweet Charlotte Barret [*sic*] is just what she was, i.e. charming. Marian is on a visit at Mr. Arthur Young's, & I have not yet seen her, but all agree she is a *prodige,* though some with praise, some with censure, & all with wonder.[11]

---

[10] Rylands MSS. 583, ff. 99–104. The references to her "celebrated aunt" are of course to Mme d'Arblay.

[11] Berg Collection. Her spelling of her niece's name at this time is interesting.

Marianne's last letter of 1812 to Mrs. Piozzi was written from Kensington Gore, William Wilberforce's magnificient home in South Kensington:

The Wilberforces are acquaintances only of a year's standing; but they kindly invited me to take up my abode with them during Mama's absence from town. . . . My friend Arthur Young . . . is 73—strong again now, as a castle. The Wilberforces are *his* friends, & he introduced me to them. Mr. Wilberforce is one of the most charming characters I ever knew. So very cheerful & animated—& a kindness in his manner as superior to common activity, as gold to tinsel. Very amusing, instructive, & always ready for a kind action or agreeable conversation. . . . The W.'s have a little boy Sam—I teach him Latin—& a school over the stables of 45 poor children. They have six themselves, very intelligent, & to see Wilberforce divesting himself of the statesman & orator, & playing like a child with his own young family, is very interesting & beautiful.[12]

Such was the beginning of a close relationship which meant that Marianne spent much time with the Wilberforces in the next decade, often accompanying them for a whole summer to the seashore. Consequently Marianne saw much less of Young than he would have liked, but his biographer benefits, for separation meant more letters.

On Christmas eve Henry Barrett, Marianne's brother-in-law, requested Young to become a sponsor for his new baby:

. . . you will oblige me greatly, if you will be so good as to become one of the sponsors for my youncker; Mme d'Arblay and Clement are the two others; when baptism will be performed . . . I shall be quite pleased to put on my stock the graft of your sweet name Arthur. . . . perhaps you will do me the favour to give me another card of admission for the Lectures of the ensuing year at the Board of Agriculture. I began last night to thumb over Sir John Sinclair's Book on Scotch agriculture—what an indefatigable man. . . .[13]

The previous chapter noted that Young had spent much of the latter half of 1811 collecting materials for a pamphlet which

[12] Rylands MSS. 583, f. 105. The "little boy Sam" became "Soapy Sam," the great Bishop Wilberforce of later fame. There is no reference to Marianne in the three-volume life of Bishop Wilberforce.

[13] Add. MSS. 35,131, ff. 422–424. The baby was baptized on January 20, 1813, as Richard Arthur Francis Barrett, but he probably did not live long. *Cf.* Burney Scrapbook, f. 79.

was duly published in 1812. The original aim had only been a lecture before the Board, but as his materials accumulated he decided a pamphlet would be better. He had opposed the famous Report of the Bullion Committee in 1811 recommending the return to hard money and his pamphlet was essentially a defense of paper money. Much of it consisted of statistical tables on prices and wages since 1700, based on his own tours, on various surveys made by the Board, and on the returns from his questionnaire distributed in 1811.[14]

Young's basic point was that increasing prices were largely responsible for England's great prosperity. These prices in turn were not caused primarily by any excess of paper money in circulation but rather by increased demand in relation to supply. On the other hand, and somewhat inconsistently, he predicted that, if the amount of paper money in circulation were drastically reduced, prices would fall and stagnation would result. At one place he declared, ". . . low prices are absolutely inconsistent with prosperity," and continued:

. . . there is a stagnant sluggishness, a torpor and imbecility in cheap times, that are inconsistent with the spirit of exertion; whereas, when products are at a high price, the dearest labour cheerfully employed; all is animation; and the landlord, the farmer, the labourer, and the public, thrive. . . .[15]

Young ended the pamphlet with several pages of eloquent peroration tying together his belief in paper money, his sincere conviction that economically England was basically sound, his deep patriotism, and his Evangelical religion. A considerable portion is worth quoting:

One of the greatest questions in political economy is this; should we measure the prosperity of a country by its liberty, population, agriculture, commerce, shipping, and commodities: by the ease of its internal communications, its roads, canals, and harbours, . . . by the inclosure and improvement of its land, and by the circulation of a currency, adequate to the demands of its people and its industry? . . . or must we, before we can answer that question, demand whether the currency be silver or paper, gold, or shells? . . .

If paper bids cultivation spread over our wastes, and the looms of

[14] Young, 1812: pp. 74–75.
[15] *Ibid.*, pp. 89–90.

Britain are busy, while Buonaparte is burning their fabrics; we may
surely say, let Bankers coin and Frenchmen burn . . . we still remain
unmoved, resting on the secure basis of an animated industry; agriculture productive, manufactures prosperous, and commerce flourishing; the edifice of our greatness is founded on a rock, around which
domestic alarms and foreign threats will play impassive; let the people
of these happy Isles be true to themselves, and it will be the immediate Providence of the Almighty alone, that can impede the uninterrupted progress of British greatness.

It is not paper but profligacy, that forms the predominant evil of
the present period. The spirit of insubordination; ripening into discontent; and that into hostility against all law, government, and
order, and accompanied by acts of such atrocity, that parallel crimes
are not to be found in our annals. . . . Infidelity has spread from
the palace to the workshop; the Sabbath neglected, or outraged; the
evangelical doctrine of the National Church ridiculed as Fanaticism,
or Methodism; the duties of morality relaxed, and those of Christianity vilified. . . .

. . . . yet is the prosperity of this great country founded . . . on a
basis far more substantial, than any measures of value—not paper
but liberty—not silver, but industry—not gold but national energy,
are the foundations of the greatness of Britain; and the permanence
of that felicity which has been the envy of the world, may be expected
. . . from the perfect conviction, which ought to animate every
bosom, that whatever difference of political opinion may separate
public men in this country; relative to domestic arrangements, all our
parties revere the public welfare; and would unite heart and hand, to
bleed in defense of the national interests; upon this true patriotism,
this animated feeling, this best hope of posterity, we under God,
repose; . . . here is the foundation of British greatness. . . .[16]

The pamphlet did not make much of a stir. The leading reviews did not notice it. Young sent a copy to Nicholas Vansittart
who politely acknowledged its receipt and expressed his gratification that Young shared his views on currency questions.[17]
Young also received a lengthy epistle from John Prince Smith
who praised Young's diligence and his life long devotion to

---

[16] *Ibid.*, pp. 131–134.
[17] Add. MSS. 35,131, f. 364. Vansittart had become Chancellor on May 20.
His letter was dated May 29.

the promotion of agriculture, but as an advocate of sound and hard money he politely disagreed with Young's conclusion.[18]

Young's only other published writing in 1812 was an essay or letter urging a greater cultivation of potatoes. The letter was summarized in the *Bury Post* as follows:

Arthur Young . . . has published a letter recommending an extended cultivation of potatoes. Half an acre in every hundred, added to the present space under this crop, would (he says) produce human food sufficient to answer the purposes of all the foreign corn imported into this country, . . . one acre of potatoes being deemed equal to two of wheat.[19]

The Board of Agriculture held no regular meetings in 1812 until nearly the middle of March because Sir John Sinclair had not arrived in London. He was in Edinburgh throughout February, working hard at his *Husbandry of Scotland* and the *General Report on Scotland,* and deluging Young with eight letters.[20] Several enclosed manuscripts inviting Young's comments and criticisms. One inquired about the cultivation, use, and cost of green beans as food for horses, and another about the average English produce of the leading crops.[21] Sinclair also asked Young to arrange with Humphrey Davy a series of lectures for the spring, and urged Young to deliver three or four himself, including the one on farm yards, already given in earlier years.[22]

Seven more letters from the indefatigable baronet, plus two from his son, reached Young in the summer and fall. Late in July, Sir Francis Burdett attacked the Board of Agriculture

[18] *Ibid.,* ff. 365–366. This was John Prince Smith the elder (1774–1822), less famous as an econom'st than his son. He urged Young either to publish the raw materials for his statistics or to deposit them in the British Museum where they would be available to scholars. Young had offered in his pamphlet to let anyone examine his papers.

[19] *Bury Post,* April 15, 1812. A note from the *Farmer's Journal* office of April 22 states that they are publishing an essay by Young on the value of potatoes for cattle food. *Cf.* Add. MSS. 35,131, f. 347.

[20] Add. MSS. 35,131, ff. 300–301, 302, 307, 309, 311, 313–314, 321–322, 323, 326–327.

[21] *Ibid.,* ff. 309, 313–314.

[22] *Ibid.,* ff. 321–322.

and its large appropriations as a job.[23] To avert further attacks Sir John hoped that by the time Parliament met in the fall the remaining county reports could be completed as well as the General Report on Scotland: "The last in particular must be *very complete*. I rely on your examining *most carefully,* all of the Chapters. . . . By that Report the Board of Agriculture must rise or fall. It ought to be a most capital work . . ."[24]

On July 13 Sinclair urged Young to undertake a fairly important work for the Board:

I see there is a great wish to have a complete paper drawn up comparing the produce of grass and arable Land. It is a most important subject, and I know no man so able to do it justice, as you are. . . . You might draw it up by way of Lectures to the Board, and I hope you will have no objection to undertake it. The Plan herewith sent, may furnish you with some hints regarding the arrangement, but you will please order it to your own mind.

In order that the paper may do you and the Board credit it would be necessary to circulate some printed queries among your friends and correspondents in England, and I will procure you any answers you may think necessary from Scotland. . . .

I hope that you will exert *your utmost ability* to make this work a creditable one both to yourself and the Board.[25]

Unfortunately Young's reply to this request has not survived, but apparently the job was never completed, perhaps due to his illness of the late summer and fall. Sinclair seems to have been less effected by sympathy with his secretary's sickness than by fear that the illness would postpone the completion of certain Board projects:

[23] *Parliamentary Debates* 23: p. 1264. The pertinent statement in this speech follows: "The next head included sums which might be much more fitly included under the title of jobs; among these were 80,000*l.* to the Board of Agriculture. . . ." This seems a great exaggeration. Through Sinclair's friendship with Spencer Perceval, the normal appropriation of £3000 had been increased by special appropriations in the years 1809–1812, to a total of £8500. George Sinclair, Sir John's son, brought Young's attention to Burdett's speech in a letter of Aug. 4, Add. MSS. 35,131, f. 385: "I intended to have made some remarks upon that & other parts of his speech—but as his motion was not seconded, I had no opportunity. Indeed the best answer to all his lucubrations [*sic*] was that no individual in the House was disposed to countenance and support them . . ." For the special appropriations, *cf.* Clarke, 1898: p. 34.

[24] Add. MSS. 35,131, ff. 383–384.

[25] *Ibid.,* ff. 374–375.

I regret much to hear that you have been unwell. I have no doubt that we shall get through the General Report of Scotland in less than six months; but what can we do with the General Report of England without your aid. You must therefore rouse yourself, and get well, that you may have the satisfaction of taking the principal part in that great undertaking. . . .[26]

Young's easy and friendly relations with his under-secretary, William Cragg, are revealed in Cragg's letter of January 1:

. . . We feel ourselves greatly indebted to you for the happy, tranquil month we passed under your hospitable and friendly roof which we shall ever consider as one of the happiest of our Lives. Our acknowledgements are also due to you for your friendly recollection of us at this festive time of the year, as well as for your kindness to us in Septmr: the Birds then and the Turkey now proved excellent. . . .

. . . . I know nothing of the President's movements; the moment I shall, I will acquaint you. He has sent the translation of the Italian M.S. to Bulmer. Will the enclosed preface be proper? if not, be so good, as to substitute one. . . .

P.S. I hope you have received a Barrel of oysters & that the [*sic*] proved good.[27]

The Young correspondence contains a considerable number of other letters which help to complete the picture for 1812. He was still serving as arbitrator between Lord Templetown and his Irish agent, William Gooch, and with some success, as Gooch's letter of February 15 indicates:

I have received your letter of the 6th, and observe your resolution, to which I bow most cheerfully. It is a great self-congratulation to me . . . that *your* Idea of the subject is *more unfavourable to Lord Templetown* than my own, and *equally favourable to me.* . . .[28]

Several letters show that neither Young nor Mrs. Young were very well during the spring months. One in March from Lewis Majendie said: "I was grieved to see you so unwell when I last met you with Miss Young in Piccadilly. . . ."[29] In April one from his son-in-law, the Rev. Samuel Hoole, added: "I hope poor Mrs. Young does not suffer as she did."[30] A few days

[26] *Ibid.*, f. 399.
[27] *Ibid.*, f. 280–281.
[28] *Ibid.*, ff. 317–318. *Cf.* also *ibid.*, ff. 295–298, 303.
[29] *Ibid.*, f. 328.
[30] *Ibid.*, ff. 336–337.

later Jane Metcalfe of Hawstead expressed her regret that she had failed to find a suitable attendant for Mrs. Young.[31] In September there was a request from Ray Orbell Oakes to furnish him with "a few good *public Toasts*" for his alderman's banquet in Bury St. Edmunds.[32] In August a note came from the Earl of Galloway, written from 32 Sackville Street where he had called to ask Young's aid in procuring some Suffolk milch cows. The postscript added: "I cannot say much for the Pens & Ink of the Board during the vacation."[33]

One of Young's most prolific correspondents in 1812 was Edward Wakefield, who in that year published his great two-volume work, *Ireland, Statistical and Political,* perhaps the most noteworthy account of Ireland since Young's more than thirty years earlier. In his introduction Wakefield praised Young and his work in extravagant terms:

Properly to execute such a work requires greater talents and knowledge than is commonly to be found in the same individual. England, however, in Mr. Young, may boast of such a person; his labours will shed a lustre on her fame through future ages; but truth compels me to declare, although the assertion may reproach my country, that he has been ill requited for his exertions in her service, and that during the best days of his life, she seems to have been coldly insensible of his indefatigable and important labours.

A footnote on the same page elaborated on Young's important achievements and concluded:

. . . the friend and associate of the greatest men of the age, . . . generously imparting to all persons the result of his accumulated store of knowledge, Mr. Young had spent a long life in cultivating and promoting the arts of peace. Contemning all private emoluments, and serving the public without any view of adding to his private fortune, he has received, I believe, from his country, no other reward than that of being appointed to the office of Secretary to the Board of Agriculture, with the small salary of £ 400 per annum. Such, reader, is the extent of the boon conferred upon this benefactor of mankind! It is posterity now which must do him justice; and some future

---

[31] *Ibid.*, f. 341.
[32] *Ibid.*, ff. 392–393.
[33] *Ibid.*, f. 386. Later letters show that Young had acted as the Earl's agent in procuring the cattle and had even advanced the purchase price of £ 60. *Cf. ibid.*, ff. 388–389, 404.

biographer, in speaking of his services, may, perhaps, be inclined to remark, that his country behaved to him as Frederick boasted he had done towards Voltaire—"he treated him like a lemon; squeezed out the juice, and then flung away the rind."[34]

In spite of the success of this work, Wakefield, thirty-six years old and the father of a large family, was still completely unsettled in 1812. In one letter he wondered whether either Young or Sinclair could find him a place as an estate agent on a part-time basis.[35] Whether Young made any efforts on Wakefield's behalf in this direction is uncertain, but Young and Sinclair did invite him to lecture before the Board of Agriculture, an assignment eventually accepted for the following year.[36]

There is no entry for 1813 in the *Autobiography*. Young published nothing. As in 1812 there is almost no information about the Board of Agriculture except in Sinclair's letters. Again the chief documentation is in Young's correspondence and even that is relatively scanty. Nor was it a year of great interest or importance in Young's life. There were no triumphs and no catastrophes. The blind old agriculturist continued his tranquil existence in grooves already well worn. His health was comparatively good and there is no evidence of any serious illness. He spent most of the year in London, with only July, August, September, and October at Bradfield.

There is very little information about Young's family during 1813. Arthur remained in Russia and there are no letters from him. There is absolutely no mention of Jane in the correspondence, nor any letters from her. None of his friends asked to be remembered to her as they did to Mary or even to Marianne Francis. What is difficult to understand is why there were no letters from her, if she was not with him. The only reference to Mrs. Young came in Dr. Richard Valpy's letter of February 19: "How is Mrs. Young? I hope you assist in preparing her for the awful change, which she is likely soon to undergo. . . . You have an amiable and sensible daughter to soothe your declining years. Give my kind regards to her. . . ."[37] Vague as it is, the above is

---

[34] Wakefield, 1812: 1: pp. vii-viii. Wakefield's mother was the famous authoress, Priscilla Wakefield, and his son the even more famous—and infamous—Edward Gibbon Wakefield, the colonial statesman.

[35] Add. MSS. 35,131, ff. 377–378. Dated June 29.

[36] *Ibid.*, ff. 371, 377–378, 410–411.

[37] *Ibid.*, ff. 456–457.

one of the few evaluations of Mary Young. Usually the reference reads: "Please remember me to Miss Young." In 1813 she was forty-seven years old. For many years she had combined the functions of her father's hostess and her mother's nurse. She was the true "Martha" of the Young household, holding it together through self-effacement.

As in 1812 Marianne Francis furnishes most of our knowledge about Arthur Young, as she also probably contributed more than any other individual to his happiness. Since she was at Richmond most of the year and at Bradfield in the summer, only four of the letters of the Evangelical blue-stocking of twenty-three to the Evangelical agriculturist of seventy-two have survived, and only two of his to her. Some additional information is to be found in Marianne's twelve letters to Mrs. Piozzi. Nevertheless all these letters contain relatively little that is important for Young's biography.

Late in January, Marianne went to Richmond for the christening of her nephew for whom Arthur Young had agreed to be a godfather. In her letter describing the event, she remarked that her cousin, Alex d' Arblay "told me, spontaneously, that he learnt more from your example than from all the books he ever read." She also requested information about the steps necessary to get his home licensed for preaching. There was a postscript about her little niece: "Little Julia is at my elbow. She desires her best love & a kiss to Arthur."[38] Young's reply to Marianne's letter had a few points of interest:

My dear little girl,
    I am very glad to hear such good accounts from Richmond; first of your sister and Mr. Barrett and their little ones, and I desire you to give half a score kisses to little Julia for me. . . . respecting the License there is no trouble or difficulty in it: the act requires the house to be registered at the Quarter Sessions, the Bishops Court, or the Archdeacons Court, and the last as being at Bury I registered my house in that; a certificate is given of the register, for which a fee of half-a-crown is paid. . . . Our friend Alex. has much too good an opinion of me, it is only for want of knowing me better. . . . What will you say to this dull epistle, my papa is become as stupid as he is blind, why then it is high time to bid you good night, saying if

---

[38] *Ibid.*, ff. 442–443. Little Julia was Charlotte Barrett's oldest child.

I had my eyes I should notch a stick till you return. God bless you my dear little friend.[39]

Marianne's letter to Mrs. Piozzi in June referred to her approaching visit to Bradfield in terms which imply that she might have preferred going elsewhere:

I write from Mrs. Wilberforce's, with whom I am spending a fortnight, previous to our all leaving London for the summer. The Wilberforces for Sandgate, whither I should accompany them, were it not for my kind friend, Mr. Arthur Young, who insists on my going with him & his family into Suffolk; which I suppose I shall.[40]

On June 24 Marianne pleaded with Young to postpone the trip to Suffolk for a few days so that she could meet Hannah More at the Wilberforces:

I look forward with great pleasure to being at Bradfield, myself—But if you find a half distracted, almost out of my mind & moping, mournful, & melancholy you must not wonder. In my last note I wrote you word that Hannah More was expected to be at K. Gore, for a day & night, next Tuesday or Wednesday—& if I find we are setting off the very day she comes there, that by staying till the Thursday coach I would have seen *her*, whom I have no other chance to see & was invited for a day to meet—after wishing it so long & so avidly—I cannot answer for the consequences on my mind & spirit. . . .[41]

Of course Marianne had her way and saw the Evangelical saint whom she described to Mrs. Piozzi:

I was delighted with her. . . . Much as I expected from Hannah More, I was *not at all* disappointed. . . . I had a long tete à tete with her, which Mrs. Wilberforce contrived for me, charging me to lock the door & keep her to myself till dinner. . . . I followed her about like a little dog, too happy to be in the same room with her—& not knowing any other ostensible way of shewing my joy, covered her with roses; for she is very fond of flowers. . . .

In another letter of August 23 to Mrs. Piozzi Marianne summed up her Bradfield life as follows: "But between reading to Mr.

[39] Berg Collection. Dated January 26. *Cf.* Add. MSS. 35,131, f. 394, for letter from E. Sparks, September 1812, referring to Young's application for a license in the Archdeacon's court.
[40] Rylands MSS. 583, f. 114.
[41] Add. MSS. 35,131, ff. 488–489.

Young, minding the schools, exercise, & the necessary study for
the recruit of my own spirits, I seldom find a vacant ½ hour,
& go to bed quite sleepy at ½ past 9." On October 4 she described
to Mrs. Piozzi Young's recent feast for his school:

Our humble theatre was a barn; the scenery branches of oak & ash with
flowers, wh concealed the walls, & made it look like Rosamund's
bower; the *actors,* Mr. Young's school, 130 children. The parts they
had to perform, were all alike, & had more to do with the teeth than
the tongue, being the consumption of beef & plumb-pudding. . . .
We waited on the children ourselves, who, tho' they did not *look,*
like Despair as if they *"never dined,"* certainly *eat* as if they never
had before. After this, they sung a hymn; & when they had enough
eating & playing to their heart's content, we all went to church. Our
Rector preached an excellent, simple sermon on the benevolent words
of our Saviour: "Suffer little children to come unto me": . . . It
was melancholy to see dear Mr. Young in the midst of so much innocent gratification, all of his own creating, unable to behold it: yet
it was *delightful* too, at his great age, 73, to see him so anxious to
contribute to the happiness of others, & taking so much pleasure in
hearing that they were pleased.[42]

The correspondence also indicates that the very pretty Suffolk
village of Ampton became an Evangelical center that summer.
Many Cambridge students, probably influenced by Charles
Simeon, studied and tutored there under the Rev. Joseph Cotterill
and his wife. At least some of them seem to have discovered
Bradfield. One such was John Babington who wrote to Young
on October 9:

Notwithstanding the doubts I expressed when at Bradfield, whether
I should be able to spend another Sunday there, I propose, if all is
well, to accept your kind invitation for next Sunday. I cannot reconcile it either to my conscience, or to my feelings to leave Suffolk without seeing you once more. . . .[43]

Another expressed his regret at being unable to visit Bradfield
again before returning to Cambridge:

I need not say how much I am grieved at this, for your kindness to
me, and many interesting circumstances which I met with at Bradfield have worked upon my affection in no ordinary manner.

[42] Rylands MSS. 583: ff. 115–117.
[43] Add. MSS. 35,131, ff. 519–520.

Permit me, dear Sir, before I take my leave of you, to request that when your views are strongest, on the awful responsibility of the Christian minister, and the value of eighteen thousand souls, which I suppose it will one day be mine to take charge of, you will not fail to pray for the blessing of God on

> My dear Sir
> Yours gratefully & affectionately
> Roger Carus Wilson[44]

Young and Marianne must have left Bradfield very early in November, for she wrote him from Richmond on November 8:

A thousand thanks for your kindness in getting my books so nicely done up. They arrived on Saturday, & are all in their places.

. . . . I was at the school yesterday, & am going with Charlotte to the workhouse today. But if an idea of our being *Methodists* gets about we shall be ruined. . . . Charlotte & I are in daily expectation of being *blacklisted*. . . . Poor little Dolph has not lost his cough. He is to perform 2 hours every day, Latin, Greek, French, with me; which will . . . please Mama, wh I am very anxious to do in every way. . . .[45]

On November 10 Young replied at length:

I am much obliged to you for so early an acknowledgment of the arrival of your books. . . . it gives me much pleasure to find that all your relations are well at Richmond; and I suppose that Ralph is better, or you would not think of cramming him two hours a day with languages, which I hold to be very imprudent, and have seen the bad effects of it in several instances; you are all except Mama so young and ignorant of these things that you think it sufficient to intend well, but in all consumptive cases, all calls upon attention are mischievous, and learning must absolutely sleep, until health and strength are restored. . . . You ask good advice, the best I can give is to [?] at Richmond as you did at Bradfield, and I hope to hear that you have the honor . . . of exclusion for Methodism, as I hold those to be in a questionable state who are not stamped with that character in the world;—and the gospel is so preached at Richmond that I should tremble if they did not hold you in the utmost contempt. . . . My spirits have not been good, and no wonder on the loss of the most perennial chearfulness and vivacity that ever was experienced. I am

---

[44] *Ibid.*, f. 547. Why the 18,000 souls, unless he was preparing to be a missionary?
[45] *Ibid.*, ff. 551–552.

reading Wilberforce to my great delight for the third time. . . . Pray write again as soon as possible, for you know that of all your friends none are in such want of your letters as your poor old blind papa. May the mercy and grace of the Lord Jesus be ever in your heart, and never fail of directing you to the way in which you should go. . . .[46]

When Marianne wrote Mrs. Piozzi on December 21, she was visiting at Kensington Gore. In reply to some point Mrs. Piozzi had made Marianne commented:

As to what you say of the national wealth increasing by war & taxes —Mr. Arthur Young says it ought to be changed into *notwithstanding* war & taxes—also that Population encreases *notwithstanding* the death of thousands. That an accession of commerce & agriculture will account for the great influx of wealth; & that additional riches will always be followed by additional population.[47]

The manuscripts of the Board of Agriculture contain only one item for 1813, Young's letter to Sir John Borlase Warren, commander-in-chief on the North American Station, asking him to procure if possible some rams and ewes from Smith's Island at of the mouth of the Chesapeake, which had just fallen into British hands. These famous "Arlington" sheep, a cross between a Persian ram and a Bakewell ewe developed by Washington, produced very fine and valuable wool. Sir John was informed that the Board of Agriculture would pay all expences. A later entry shows that Sir John complied and that the sheep actually arrived in England.[48] The whole episode throws a new light upon the War of 1812.

All other information about the Board depends upon Young's correspondence. Early in April he was asked to testify about wages before Sir Henry Parnell's Parliamentary Committee on the Corn Trade.[49] The secretary of the Board was also waited upon at 32 Sackville Street, probably by Sir Henry, to obtain his estimate of the number of acres of wasteland possibly convertible to tillage and also the number of acres actually under the plow.[50]

[46] Berg Collection.
[47] Rylands MSS. 583: f. 119.
[48] Board of Agriculture, Letter Book 2: ff. 173, 203.
[49] Add. MSS. 35,131, ff. 467, 468, 470. There are three separate documents here, dated April 1 and 2. Young probably testified on April 7.
[50] *Ibid.*, f. 482. The Committee's Report is given in *Parliamentary Debates* 25: appendix, p. xcviii. Mention is made of certain information received from the Board of Agriculture, but Young's testimony is not identified.

Late in November Young received a complaint from Edmund Cartwright which probably had some justification:

I last year sent in to the Board of Agriculture an account of experiments on the use of contaminated sugar in fattening hogs. It was sent in claim of a premium. I have reason, however, to believe that none of the members gave themselves the trouble to examine it. . . . I wish to have the paper returned as I did not keep a copy of it. I mentioned some of the facts which it contained to Sir H. Davy, who was of opinion that they were too important to be suppressed. . . . I attended the Board twice or thrice when in town last spring and was sorry to observe that its spirit and energy seemed much abated. We will trust that both will revive on the return of peace. . . .[51]

The Young correspondence also contains fifteen letters from Sir John Sinclair, who continued to send Young everything he wrote for possible correction or improvement. The Board probably began its meetings late in February, when Sir John finally reached London. His last letter before arriving, unfortunately undated, requested the secretary to call a meeting for the following Tuesday: "It is absolutely necessary to have a Board on Tuesday to get the petition signed, & presented the first day that the House meets. . . ."[52] A fairly heavy stream of letters from Sir John came from Edinburgh to Bradfield from July to December. On July 12 he requested Young to go through his *Husbandry of Scotland* and prepare a series of detailed questions still needing a final answer.[53] On September 1 he asked Young to prepare queries about breeding and feeding livestock to circulate to prominent Scottish farmers, and on September 4 to abstract information from the Bedfordshire and Middlesex reports about the use of straw.[54] On October 31 he requested a report on the management of a large farm where all the straw was littered and the hay used for "soiling," and somewhat later wrote: "I have received your plan of converting all the straw into muck, which I hope you will explain more fully, & publish in a short paper to be circulated by the Board."[55] Several of Sir

---

[51] *Ibid.*, ff. 555–556.
[52] *Ibid.*, f. 454. I have been unable to find in the *Parliamentary Debates* any reference to such a petition. It may have concerned wool import duties.
[53] *Ibid.*, ff. 495–496.
[54] *Ibid.*, ff. 510, 511–512.
[55] *Ibid.*, ff. 545–546, 559–560. The latter letter is undated.

John's letters referred to plans for the coming year. On October 17 he told Young that it was unnecessary to return to London before Christmas, and on October 31 wrote even more strongly:

I think you might have saved yourself the trouble of a journey to London, as there is usually nothing to do at the Board; and it is very unusual for the Board to assemble before Christmas (indeed I do not recall an instance of it), that I scarcely think there will be a meeting.[56]

Later in the fall Sir John was hoping to hold Board meetings by the end of January, but on December 11 he had decided that he would not return until about March 1.[57] No wonder the Board was considered almost moribund.

Sir John's *Memoirs* furnish evidence that Young expostulated with him on his lack of religious interest:

On one occasion, his friend Arthur Young . . . ventured to remonstrate with him on his spiritual lukewarmness. "Your conduct," said Mr. Young, "suprises me beyond measure. You are a moral man. You do all the good in your power; you fulfill with great strictness all your relative duties; but you are not a Christian. You hardly ever attend the public ordinances of religion. You rarely, if ever, read the Bible, and you probably neglect private prayer. How can you, who know you ought to act differently, expect to prosper? Think of these things before it is too late."[58]

Marianne's letter of February 24 to Mrs. Piozzi contains an amusing account of what may well have been the same incident, but not necessarily so, since Sir John gave no date for his account:

There is a literary genius in London, who plumes himself on the versatility of his talents, one time writing on enclosures, another on longevity, next year publishing the corn reports, then writing an essay on the diversions of the rich—& then inquiring into the conditions of the poor. And pray said Arthur Young, amongst your various pursuits, have you ever thought *Religion* worthy an inquiry?—

---

[56] *Ibid.*, ff. 535–536, 545–546.

[57] *Ibid.*, ff. 559–560, 572. All the standard accounts mistakenly maintain that Sir John retired from the presidency in 1813, but the letters cited above show that he remained president during that year. Sinclair, 1896: p. 14, made this mistake about his father. Clarke, 1898: pp. 34–35, repeated the mistake and again in his article in the *Dictionary of National Biography*.

[58] Sinclair, 1837: 1: pp. 377–378.

"Why really, I have had no time to think of religion; but however some day, I assure you, I intend to *take up the subject,* & *write* a book upon it!" Need I say that this was Sir John Sinclair?[59]

Much of Young's correspondence in 1813 consists of letters from and about his three protégés, Edward Wakefield, the Rev. William Gooch, and James Powell. Young was tireless in his efforts to help Wakefield, and corresponded with Lords Egremont and Sheffield and with Wilberforce in his behalf. In the early part of the year Wakefield was striving to secure a position at the Naval Arsenal. Lord Sheffield suggested that he might become superintendent of one or more large estates, and in the latter part of the year Wakefield prepared an address to be sent out to great proprietors who might be interested. There was also much correspondence about securing Wakefield a cottage near Bury, and Wakefield bothered Young even about such details as the garden space, fences, painting, and papering. At least he had the grace to apologize for all the trouble he was causing.[60]

If Edward Wakefield was thus quite a nuisance to Young in 1813, the Rev. William Gooch was far more of an embarrassment. His affairs as estate agent for Lord Templetown at Castle Upton in Ireland reached a climax and resulted in serious financial loss to his employer and his own disgrace and death. When Lord Templetown discovered that Gooch had been charging fees for the renewal of leases, but without his knowledge or consent, he dismissed Gooch. After his dismissal Gooch was permitted to collect the August rents which he did not remit to his Lordship but absconded with them, something over £3000, to the Isle of Man. When Templetown, however, gave full authority to Young and a lawyer to make a settlement, Gooch indignantly refused to accept Young's mediation and bitterly reproached him for prejudice:

. . . . Your letter is so complete a proof of prejudging, is so full of unfounded crimination, unwarranted conclusion as to my motives, & conduct, shows such an *extreme concern* for the interest of Ld. T.— such absolute *indifference,* for mine. . . .[61]

[59] Rylands MSS. 583, f. 108.
[60] Add. MSS. 35,131, ff. 428, 429–432, 478, 497–498, 504, 515, 575–576, 579. *Cf.* also Garnett, 1898:pp. 5–6.
[61] Add. MSS. 35,131, ff. 553–554.

The dénouement of the tragedy was revealed in the following letter of December 13 from Gooch's daughter to Young:

As you were formerly the friend of my dear Father, and only withdrew that friendship upon so fragile a foundation, the injustice of which my Father was convinced sooner or later you would be sensible of; and as he died in peace with God & man, praying for the forgiveness of his persecutors; and as I trust you will not extend *your* animosity beyond the grave, so I hasten to inform *you* with the *rest of his Friends* of the melancholy event, which has deprived *us* of the best of Fathers, & the world of the best of men; my beloved Father was seized with cold on the third inst., and inflammation succeeded, which terminated his existence in this miserable world, on the twelfth.[62]

The whole mess could only have been torture to Young.

Just a few days before Young received the·above, there arrived a letter from James Powell, who had long been an intermittent thorn in his flesh. Powell told his usual hard-luck story, how he had always been misjudged and mistreated, what talents his fine sons possessed, and finally how he was now completely at the end of his resources.[63] The letter came at a bad moment when Young was not apt to forgive or forget. From Powell's second letter Young apparently offered to help his son, but absolutely refused to do anything further for him and probably recalled his failure in past jobs to which Young had recommended him. Powell's answer was typical:

. . . if you suppose me from anything that has passed, to be undeserving of your confidence, I am unfortunate indeed. . . . *God knows* I have never deserved ill at *your* hands. Then why forsake me—but even if I had, I should recall to your recollection, the words of our *blessed Redeemer,* when he says Thou shalt forgive thy brother not *seven* times—but *seventy Times seven.* . . .[64]

Were Young's biographer confined to the *Autobiography* for his information for 1814, his account would be meager indeed, a mere two lines: "1814—This year I paid much attention to the 'Elements.' My son came from Russia."[65] The records for the Board of Agriculture are completely missing. By far the most im-

[62] *Ibid.,* f. 573. For more details on Gooch case *cf.* also *ibid.,* ff. 446–447, 508–509, 525–534, 563–568.

[63] *Ibid.,* ff. 569–570.

[64] *Ibid.,* ff. 577–578. This letter is undated.

[65] Young, 1898: p. 456.

portant sources are the twenty-nine letters from Marianne Francis to him, and still more his twenty letters to her.

The first six months of 1814 were spent in London, the second at Bradfield. His health was unusually good, even in the very severe weather in the early part of the year, although he had a severe attack of rheumatism in February: ". . . as to myself I had a bad Rheumatism, and applied Mr. Wilberforces Remedy which almost flea'd [sic] me alive; and . . . so tormented me that the fatigue made me sleep for five nights like a top. . . ." In March, however, he wrote: "I have walked for near an hour almost every day. . . ." During the summer he answered Marianne's query: "My health . . . has continued much better than I have any reason to expect, but I have been strangely cold all this summer; so that except for a few hot days, I generally set [sic] in a great coat."[66]

Without much doubt, his son's return from Russia, after nine years' absence, was the outstanding event of 1814. The Rev. Arthur Young left his estate at Karagos in the summer of 1813. The trip home was difficult and costly. From St. Petersburg his journey had led through southern Finland and across the Gulf of Bothnia, partly in an open sledge across ice which was not too secure. He had drawn upon his father beyond what he had been authorized to do and also had to borrow from a fellow traveler. From Gothenburg in Sweden he had written a bitter, complaining letter, contrasting the Swedes very unfavorably with the Russians:

After having seen much of Russia, & always told to consider Sweden as far superior to her great rival, I never was so mortified and disgusted as I have been since the very first moment of my arrival in [torn] barbarous, savage, Gothic country the cleverness, activity, knowledge, & invention of a Russian is now well known; the stupidity, dullness, meanness, rascality, pride and ignorance of the Swedes surpasses all that I could have believed. Russia is a thousand years before Sweden in arts & inventions. I have been plundered & cheated & robbed by these brutes. . . .

To add to his bitterness the port of Gothenburg was icebound and he was compelled to stay six weeks in "this detested and

[66] Berg Collection. Letters of February 14, March 17, August 22, 1814.

inhospitable town perhaps as dear a place as can be found in Europe."[67]

He finally arrived in London early in April, for Marianne wrote on the eleventh: "I hope his coming will give you new life & spirits."[68] Whether her hopes were fulfilled seems rather doubtful. This father and son had never been close. The father's letters to Marianne evince little pleasure in his son's return. He tried to help his son prepare a manuscript for publication and even turned over one of his own amanuenses to give further aid, but he correctly judged that the book would never appear:

. . . I do not sleep well at night, and therefore am up generally at 4 o'clock, and call my son to give him some assistance for a couple of hours in the Journal of his travels, which however goes on so slowly that I know not if they will be ever published; he had very uncommon opportunities to have made a most entertaining book but sadly neglected them, when the notes ought to have been made.

The father had become ever more religious with the years, the son probably ever less so. It hurt that the Rev. Arthur Young was never asked to take a service at the Bradfield church although once a curate there: "My son has not yet done any duty, nor indeed has any application been made to him."[69]

Most disquieting of all, relations between the Rev. Arthur and his wife Jane seem to have been pretty strained. Although she had not seen her husband for three and a half years, there is no indication that she was in London awaiting his arrival. On the contrary she apparently permitted four months to elapse before joining him at Bradfield. Although Young's comments to Marianne were quite guarded, the following excerpt from a letter in July almost certainly refers to an estrangement:

. . . . and of the rest I yet know nothing, and I have little heart to make enquiries; the frame of my sons mind, after 9 years absence from every religious ordinance, is not calculated to make any amends; he is fixed here for the summer, and further about him and his wife I know nothing. . . .

The same impression is created in another letter early in August: "I have nothing to add to what I said before relative

[67] Add. MSS. 35,132, ff. 57–58. Dated March 6.
[68] *Ibid.*, ff. 83–84.
[69] Berg Collection. Letters of August 12, October 11, 1814.

to arrangements here: I make myself as easy as I can, but that is a string I do not desire to touch, there is no harmony in it. . . ." Once Jane had arrived at Bradfield the references are to her religious activities not to her relations with her husband. Her religious convictions had been strengthened by associating with a Moravian family at Bristol. She read the Bible and other religious works to her father-in-law and helped him to catalog the sermons in his library. By October she had already cataloged about 500 but the job was far from completion. He contrasted unfavorably her work with the schools and the poor with that of Marianne: "Jane has all the dispositions that can be wished, but she wants both the habit and the strength, and how can a person that weighs 15 stone supply the braced [not clear] activity of your light and agile frame."[70] If Jane indeed weighed over 200 pounds, she must have lost much of her former beauty.

One of Young's oldest and formerly most intimate friends died in 1814—Dr. Charles Burney, his brother-in-law and Marianne's grandfather. Both Young's and Marianne's chief emotion seems to have been regret that Dr. Burney had passed on without a true religious conversion. Marianne wrote:

I never knew that he was ill, or I w'd have persuaded you to go with me to see him, & pray with him, all which I fear was sadly neglected. Oh, my dearest Sir,—how happy would it have been, had he been in your state of mind![71]

To which Young responded:

At the great age of your grandfather, his death can surprise nobody. Yet in this fine weather it came so unexpectedly to me that I was struck at it, and the state of mind in which he probably was, makes the impression deeper; I have known him for 56 years: such cases should give us the clearest conviction of the immense importance of being in some measure prepared to meet our God.[72]

What memories might Dr. Burney's death have evoked in Arthur Young! Memories of King's Lynn, of the two sisters whom they had wooed and won, of the pleasant homes in St.

[70] *Ibid.* Letters of July 4, August 2, August 22, October 1, October 11, 1814.
[71] Add. MSS. 35,132, ff. 85–86. Dated April 13.
[72] Berg Collection. Letter of April 14, 1814. Add. MSS. 35,132, f. 88, contains a printed form, requesting Young to accompany the remains to the cemetery on April 20.

Martin's and Poland Streets where he had so often visited, of the musical evenings which as a young man he had so loved. But much of this now seemed at best frivolity, or even sin.

What a different circle it was in which Arthur Young now moved. He had always enjoyed feminine society and in his old age was attracted by and attractive to Evangelical ladies. There was Mrs. Elizabeth Wayland, probably about forty years old, wife of John Wayland, lawyer and writer on the Poor Laws. Mrs. Wayland wrote: "Believe me my dear Sir, there are few things which I enjoy so much as a quiet conversation with you upon those subjects we both love. . . ."[73] Far more interesting was Mrs. Julia Strachey, wife of Edward Strachey, who was only twenty-three when she penned the following:

I cannot resist the desire I feel to tell you how deeply I am impressed with love & gratitude for the Christian kindness you evinced to me during my *short* acquaintance with you—you have been continually in my thoughts & in my heart ever since I last saw you, & the little book you gave me has been my constant companion—I am sure you will rejoice to know that my intercourse with you has been most remarkably and providentially blessed to the good & comfort of my soul. . . .

I do not ask you to write to me, because I could not receive a letter from you dearest Sir without giving my husband offence.[74]

Whether Edward Strachey objected to his wife receiving letters from any gentleman or whether he was not entirely sympathetic to the Evangelical point of view is unknown, but the latter supposition seems the more likely.

Marianne was still of course Young's dearest friend. Since she did not go to Bradfield, the letters between them are especially numerous. Her letters somehow lack the sparkle of earlier years and seem more matters of duty than of pleasure. Perhaps her increasing religiosity led her to take less pains in her writing. Unquestionably the Wilberforces had become the chief focus of her existence, except for her immediate family. When in London she was often with them at Kensington Gore and in the summer she went with them to Sandgate.

[73] Add. MSS. 35,132, ff. 98–99.
[74] *Ibid.*, ff. 76–77. Dated April 2, 1814.

After July 1 Young's frequent letters to Marianne from Bradfield are full of regrets for her absence:

. . . the three last summers I had my little companion whose company
caused perpetual chearfulness [*sic*]: her various employments; her
benevolent attentions to the poor people, the steady regularity with
which she attended to the cultivation of her own mind; and the
constant spring of a vivacity which never failed, were so truly pleasing
to me, that the loss of all sits as a cloud as dark as blindness itself:
I have poaked [*sic*] about the Library and felt the little girls chair
and table, her desk and her drawer, and even the barley box from
which she fed the chickens,—all in their place, but unanimated by
her who rendered them agreeable.

He continued the day schools, Sunday Schools, and the Sunday night meetings in the hall, but the attendance fell off and
much of the spirit seemed to have disappeared without Marianne:
"the whole life, soul & spirit of the thing is fled with my little
girl. . . ."[75]
Shortly after Young reached Bradfield, Marianne wrote that
Thomas Clarkson had requested him to circulate a petition for
the international abolition of the slave trade.[76] Young immediately sent a servant "around the country three whole days and
procured 150 names."[77] In another letter Marianne expressed
Clarkson's gratitude:

He desires his best regards to you, & spoke in the highest terms of
yo kind prompt & masterly way of attending to his request abt the
Petition. He admired yours more than any other he saw. . . . I believe,
because it arrived in best time.[78]

Late in September Young had an invitation to a meeting of
the Bury Bible Society on September 30 and stating that they
wished to elect him vice president. On October 1 he wrote Marianne about it all:

Yesterday myself, Jane, Arthur & St. C[roix] went to the Anniversary
meeting at Bury of the Bible Society. Mr. Cotterell when he called
here . . . so much desired me to attend, that he made my de-

---

[75] Berg Collection. Letters of July 4, August 22, 1814.
[76] Add. MSS. 35,132, ff. 154–155. This petition aimed to abolish the slave
trade on an international basis.
[77] Berg Collection. Letter of July 21, 1814.
[78] Add. MSS. 35,132, ff. 206–207.

sign of absence to waver, and Jane finished it by her persuasion. I was conducted into the Committee room, where . . . they had come to all the resolutions to be moved, and among the rest one for electing me a Vice President . . . Cotterell led me onto the bench in the Shire hall, and seated me by himself: he made the first Speech in which he was pleased to speak of me in a manner which filled me with confusion; he had told me just before that I should of course thank the meeting when the resolution passed, and that as he found the speakers would be very few he hoped I would add something by way of a little supply of the deficiency. Was not this a very pretty business for a blind man of 75 who never spoke a word in public in his life? I had only to dash at it, taking care that my voice should be loud enough to be heard; and it occurred to me, through the Lords blessing that the only good I could do on such an occasion was to declare in the face of the world the comfort and consolation which I had recd in old age and blindness from the words of eternal life, this I did without reserve, declaring that "after 40 yrs of folly and sin, all depended with me on the truth of the bible, if that book be not true, I have not the ten thousandth part of a hope of pardon in God; but, Sir, Xtianity IS true, and I have joy and peace in believing." . . . . the Slave Trade had been mentioned . . . and this gave me an easy opportunity for a compliment to Clarkson. . . . I forgot full ten points in my confusion which I meant to touch upon . . . I was loud, and had some little energy; and therefore passed muster much better than I expected; and when I went down several of both sexes shook me by the hand. . . .[79]

In her reply Marianne was of course much thrilled:

. . . We were all quite delighted with the account of your speech & wish we had been there to hear it. Mr. W. was quite affected with the history of your proceeding . . . I . . . have not been so much struck with anything, for some time past, as with the accounts of your speech. I am sure you must reflect upon it with pleasure. . . .[80]

Every instinct of the old Young would have been pleasure and pride, but he knew that such feelings were sinful for a true Christian:

You should understand the deceitfulness of the human heart better than to praise me so lavishly for my poor efforts at the meeting: when I heard certain praises and a Quaker recommending to the

[79] Berg Collection. Letter of October 1, 1814.
[80] Add. MSS. 35,132, ff. 217–218.

audience to remember what I had said, . . . as it came from a man who had seen a great deal of the world, the devil was at work with me very early in the business, and afterwards when Jane told me that an old Lady a Quaker was crying while I spoke, my cursed vanity was at work and I felt so abominably that the moment I had an opportunity I prayed fervently against myself and the devil's suggestions: and is it not most lamentable that such detestable feelings should so mix with our duties as to turn them into direct and positive sins. I hope I have truly loathed and abhorred myself for this offence which is far from being a slight one; and you ought to give me a dressing instead of one word of praise.[81]

It is difficult to comment fairly upon such a statement. Was it just hypocrisy? Does he indeed protest too much? Would he not have been much disappointed if Marianne had given no praise? If he had not wanted and expected commendation could he not have barely alluded to the speech? But this is probably too hard upon him. Was it really so sinful to enjoy praise? The old Arthur Young who had gloried in being praised was not yet entirely dead, but the new Young realized that boasting was sinful and consequently had a feeling of guilt.

The letters between Young and Marianne reveal other aspects of the Evangelical outlook. Ever since his conversion Young had struggled to control his thoughts and in one letter to Marianne he announced a measure of temporary victory:

I am disposed to think that the Lord in mercy hears my prayers, as I hope without presumption I may say that I have for the last twelvemonth a far greater command of my thoughts than I was able before to attain: they were a great evil to me & I prayed night & day against them till I might almost be said to have wrestled as to have given God no rest. . . .[82]

As already noted, Marianne spent much of 1814 with the Wilberforces. When Young regretted that she would probably not be able to pursue her studies at Sandgate as at Bradfield the previous year, she answered:

I think the benefit of Mr. W's conversation amply compensates the want of time for extensive reading. . . . I think for opening the mind, enlarging the sphere of intellectual vision, gaining an acquaintance

---

[81] Berg Collection. Letter of October 11, 1814.
[82] *Ibid.* Letter of August 12, 1814.

with books & men & above all for inspiring a heavenly spirit & exhibiting a holy example, there is no society like that of your dear friend Mr. Wilberforce.[83]

In June Marianne accompanied the Wilberforces to Portman Barracks to see the Cossacks:

. . . if it had not been for his kind countenance & angelic manners, we should have been turned away; for they cannot stand to be looked at. We all thought it very fortunate to meet your son there, who served as a very able interpreter to dear Mr. W.[84]

Several weeks later she went with the Wilberforces to Woolwich to see a new warship:

Mr. W. took us all to Woolwich by water last Saturday to see a man of war, which was just finished. . . . The ship was very beautiful. . . . he remarked how great was the benevolence of God in making the avenues of innocent pleasure so numerous, & in forming us with a susceptibility for so many enjoyments, wch have so much in them of pleasure & nothing in them of sin. . . .[85]

A Quaker might have thought Wilberforce's example somewhat unfortunate!

At times the letters between Marianne and Young sparkle a little. Once she warned Young not to show her letters to Clement or any of his university friends at Ampton:

. . . . It will entirely take off the edge of my epistolary freedom, if I think what I write will have the misfortune to be inspected by all that learned college—an event not unlikely to take place if I got into any of their hands. Such things *have* happened. . . .[86]

At another time she recalled Bradfield's pleasures: "Does the screaming tree sing this year? And do the owls hoot, that I used to listen to with so much pleasure, & try to imitate? And how are the boat & the pond . . . ?"[87] Young's response was also playful:

The screaming tree sings to every gale as usual—I take it very ill that he does not reserve his music for you who are the only person that

[83] Add. MSS. 35,132, ff. 243–244.
[84] *Ibid.*, ff. 145–146.
[85] *Ibid.*, f. 185.
[86] *Ibid.*, ff. 162–163.
[87] *Ibid.*, f. 218.

every [*sic*] took notice of him.—I have not heard any owls this season, they pay you therefore a better compliment than the Acarian songster. The boat is quite neglected, so that for six weeks past I do not think there has been more than one party in it.[88]

Young had no objections whatever to working with Methodists in the common task of saving souls:

Every thing goes on at Bradfield in the old routine except a project now in execution of building a Methodist Chapel within one field of Stanningfield Green. . . . This Methodist Chapel will I hope be beneficial to our Parish; for none of the people belonging to it, 3 or 4 accepted [*sic*], ever come to the Hall, . . . they may however attend the Methodist. . . .[89]

Several weeks later he showed his toleration even more strongly:

We have had the Methodist preacher Mr. Gilpin, and his convert Mrs. Enraght to dine here, and at night he read the 22nd Ch. of the Revelations expounding it, and then prayed with great fluency and freedom, and a considerable expansion of mind; I hope the poor people of this village will attend him, for their want is terrible. . . .[90]

Young's letters contain relatively few references to the exciting international events of 1814. His comment to Marianne on February 14 revealed a kind of religious escapism all too common among Evangelicals:

The anxiety of the moment on the transactions in France is somewhat marked [*sic*] by every tongue that speaks and the general opinion is that Peace must soon be concluded: we live in critical times, but of what consequences to a real Christian are these and ten thousand other subjects, he has a battle to fight every day with his own corruption and a war to maintain that never ceases between his head and his heart. . . .[91]

---

[88] Berg Collection. Letter of October 11, 1814.
[89] *Ibid*. Letter of October 26, 1814.
[90] *Ibid*. Letter of November 14, 1814. Mrs. Enraght was the daughter of Young's substantial tenant farmer, John B. Edwards. Her letter in the summer, Add. MSS. 35,132, f. 149, shows the easy relations between landlord and tenant farmer: "I am commissioned by my Father & Mother to say that my Brother is just arrived & as his stay will be exceedingly short we shall consider it very kind if you, Mr. Arthur, Miss Young & Mr. de St. Croix will take a family dinner with us tomorrow."
[91] Berg Collection.

Young took a very dim view of the "frolics" to celebrate the end of the wars. At one near-by parish there were some "shabby fireworks," and in another it was rumored that the rector's wife had been dancing with the people until one in the morning.[92] Young summed it all up: "I am sorry to tell you that this country has been one universal scene of drunkenness by way of thanksgiving and dreadful it is [to ?] think, that a call to religious gratitude should be thus horribly perverted."[93] Needless to say, there were no frolics at Bradfield. From the point of view of international relations and public affairs, Young's letter of December 22 to Marianne is the most interesting. It has some of the fire of his earlier writing and recapitulates many of his deepest convictions:

Mr. Wilberforce's letter to Talleyrand and Birkbecks Tour in France came from the bookseller the same day; I knew Birkbeck to be a great and good farmer in Surrey, and it is so very uncommon for such men to take it into their heads to make a Tour to the Pyrenees and back, that my curiosity was on fire to read the account he gives of French agriculture and prices: there are some curious particulars in his book: it appears that the kingdom is almost frittered into small properties from one acre to 10 and as Buonaparte for the sake of breeding Soldiers procured laws for the division of all landed property amongst the children, the farmers of the kingdom will very soon become starving gardners. . . . but this man is a determined republican, or he would not have travelled with that fellow Flower the most notorious of the whole race of Jacobins, and therefore he must be read with some caution. . . .

What does Mr. Wilberforce say to the Congress at Vienna, I hope he considers it as a chapter of human depravity, and that the flames of war will speedily burst out from Madrid to Petersbourgh, and I suppose that instead of the Property tax being taken off it is more likely to be doubled.

Since the above was written I have . . . read the whole of Mr. W. I am delighted with it, it is a fine piece of eloquence, the conclusion logical, and the conviction it ought to convey, most decisive. . . . Who translated it into French I tremble to think of its not reading as well in French as in En. O! for the ghost of Gibbon from the shades below to have performed that office. If it got into the hands

<hr/>

[92] *Ibid.* Letters of August 2 and August 12, 1814.
[93] *Ibid.* Letter of July 11, 1814.

of a dull dog it was an abomination. Has he sent 500 of them to Nantz? As many to Bordeaux and the same to Marsailles Has he printed 5000 at Paris; of what consequence to be admired here? It should be read all over France; and instead of that it may be so managed as not to be seen by 100 persons in the whole kingdom.[94]

As mentioned above, there are no manuscript records for the Board of Agriculture for 1814. The only sources are in correspondence. The great event of the year was the retirement of Sir John Sinclair from the presidency. On March 9 he wrote Young: "I do not intend to interfere at all in the election, which might keep up animosities that ought to be buried in oblivion." He went on to express his esteem for Young:

I hope during our official connection together, that I never failed in showing that attention to your interests, & respect for your character, to which you are so justly entitled from your zeal in the cause of agriculture, and the advantages it has derived from your exertions.[95]

All during the spring Sinclair was sending Young chapters of his General Report on Scotland for criticism and suggestion. That he valued Young's criticisms highly is shown by his letter on March 11:

Upon looking over a letter from you in February I observe you entertain an apprehension that I was not pleased with the freedom of your criticisms on some of the Chapters of the General Report. Any idea of that sort is quite a mistake, for I think criticisms cannot be too free when works are in M.S. And I have never seen any remarks of yours that were not properly made. . . .[96]

Sinclair's reasons for resignation were very vague as given in a letter to "My Lord," a copy of which Young kept: "Various circumstances would prevent me from devoting to the concerns and interests of the Board, the time and attention which they

---

[94] *Ibid.* The author of the French tour was Morris Birkbeck. Young seems to have been mistaken about Flower, for the "Jacobin" was Benjamin Flower, while Birkbeck's companion was George Flower. Wilberforce's public letter to Talleyrand appealed to France to end the slave trade in the French colonies. Coupland, 1923: p. 400, declared that no French publisher would print Wilberforce's work, but that Wellington did much to further its circulation in France.

[95] Add. MSS. 35,132, ff. 45–46.

[96] *Ibid.*, ff. 51–52.

require. . . ."[97] Another letter expressed Sinclair's satisfaction with Lord Hardwicke's election in his place: "I am very glad to find, that Lord Hardwicke is elected President, as he is well calculated for that situation and I hope that every thing will now go on with unanimity and spirit."[98] Sinclair also thanked Young warmly for a public appreciation of Sinclair's contribution to the good cause:

I have received yours, with the printed paragraph in the Monthly Magazine, which is certainly abundantly flattering; and the value of the compliment is not a little enhanced, by the manner in which you, who are so well entitled to a similar eulogium, have expressed yourself upon the occasion. I think our names are likely to be handed down to posterity together, as zealous friends to the improvement of the country; whose efforts, on the whole, have not been unsuccessful, though we have not been able to effect all the good that we expected.[99]

Six letters from the secretary to the new president have survived. On May 14 Young sent Hardwicke a list of people who might furnish valuable information about labor and husbandry and included Edward Wakefield and Francis Place.[100] On June 15 he suggested that the Board subscribe to a fairly expensive publication on corn prices, exports, and imports, and five days later proposed that the Board study the rise of prices, both continental and English, and pointed out that his son could furnish much information on Russia.[101] On July 27 he informed Hardwicke that he had taken to Bradfield the replies to the Board's circular letter and hoped "to throw the whole into as clear a view as I am able for the use of the Board on its re-assembling."[102] Early in October, Young received a letter from Under-Secretary Cragg referring to him two matters from Hardwicke. To whom should some samples of wheat seed be sent for experimental purposes? A cask of sulphur muriatic ashes from Caen

[97] *Ibid.*, ff. 62–63. Sinclair, 1896: p. 14, states that his father resigned for financial reasons: "The great expense which he voluntarily undertook in connection with the office had considerably reduced his private fortune, and he felt obliged with great reluctance to resign."

[98] Add. MSS. 35,132, ff. 78–79. The biographical sketch of the 3rd Earl of Hardwicke in the *Dictionary of National Biography* does not mention that he was president of the Board for two terms of office.

[99] Add. MSS. 35,132, f. 106.

[100] Add. MSS. 35,700, ff. 183, 185. Hardwicke Papers, **352**.

[101] *Ibid.*, ff. 198, 204.

[102] *Ibid.*, f. 235.

should be studied to determine its efficacy "as a stimulant manure." Sir Humphrey Davy was abroad and Cragg doubted whether Mr. Brande would study them "after the disappointment he received in regard to delivering his lectures."[103]

During 1814 there was much ferment over the Corn Laws. Very early Young had a letter from Lord Sheffield who wished greater protection for the landed interest:

I am glad to observe the price of wheat is rising. It is absolutely necessary to raise the Importation price of grain very considerably, otherwise tillage will be ruined when our ports are open to America & the north of Germany &c. and so I tell the Ministers. . . . But the landed interest is a torpid race, & many of them not up to such subjects.[104]

On May 23 Young received a petition from Sir John Sinclair to support a highly protective measure. He immediately and very properly forwarded it to the new president of the Board:

I recd by this days Post from Sir John Sinclair the enclosed proposal, with a request that I would distribute copies of it among the members of the Board; as I do not think it proper to take any step in the business without orders from your Lordship, I content myself of course with communicating it.[105]

Lord Hardwicke's reply of the same day shows his prudence: "I am afraid as matters now stand, the circulation of the inclosed Petition would do more harm than good. . . . I agree to the reasoning of the Petition but think it imprudent to circulate it just at this moment."[106]

In response to a summons, on June 27 Young read a prepared statement before the Committee of the House of Lords which was enquiring into the Corn Laws. He addressed himself primarily to the question, "what price of white corn will pay" the farmers for producing it? In his calculations he assumed that the land rented for 40 shillings per acre, and that the farmer had invested £10 per acre, upon which he should receive a profit of ten per cent. He also assumed an eight-year rotation of crops, with two crops of wheat, and one each of turnips,

[103] Add. MSS. 35,132, f. 221.
[104] *Ibid.*, ff. 7–8.
[105] Add. MSS. 35,700, f. 192.
[106] Add. MSS. 35,132, ff. 121–122.

barley, clover, tares, oats, and beans. In answer to his own question, the price of wheat must be 87 shillings; nothing less would pay production costs. His testimony also included comparative costs of cultivation in 1790, 1803, and 1813, the two latter tables based on the returns from circular letters sent out by the Board of Agriculture. On every item of expence except manure, the amount in 1813 was nearly double that of 1790, and there was also a very appreciable rise between 1803 and 1813. The only question addressed to him was about his assumption on rents:

Do you not think that the rent you have supposed of 40s an acre is a very high one upon land, the average produce of which will be three quarters of wheat?—By no means; I let land higher myself that will not produce three quarters, and I have known many other persons do it also.[107]

One of Young's letters to Marianne shows something of the scope and scale of advanced agriculture in the early nineteenth century. She had been visiting Lord Barham's famous farm in Kent.

. . . but the enquiry whether £ 2000 be not a great produce for 260 acres takes my breath away, why it is nothing at all for that soil and country. £ 2600 would be but common management, and there were 40 acres of hops on that farm, which single article, *there,* has many times produced much more than the sum you mention, and I have I think some faint recollection of their yielding not far short of double that sum. . . .[108]

[107] For the Committee's invitation to testify, *cf. ibid.,* f. 147. The editor of the *Monthly Magazine* wrote Young on September 6 (*ibid.,* ff. 198–199): "I like your decisive evidence before the Lords' Committee & I shall transfer the whole of it into my next magazine." The testimony thus appeared in the *Monthly Magazine* 38: pp. 276–277. His testimony also appeared in the *Bury and Norwich Post,* August 31, 1814, but in a garbled form. The last wheat crop from the rotation was omitted, which threw the whole thing out. In the issue of September 7, Young's letter rebuked the editor in rather strong terms: "Now, Sir, I gave no such account to the Committee, and I hope could not have been so absurd as to propose a course for land of 40 s. an acre rent, yielding only one crop of wheat in seven years. . . . Such gross misrepresentations as I complain of are perfectly unaccountable, for I am very unwilling to suppose that editors and printers inhabiting towns petitioning Parliament against any alteration of the Corn Laws, should purposely misrepresent the statements of men whose opinions differed from their own." To which the editor replied in a note that he had copied his figures from a London paper and had no reason for misrepresenting the facts.

[108] Berg Collection. Letter of October 1, 1814.

All through 1814 Young was studying the writings of the seventeenth-century divine, Richard Baxter, a work which resulted in 1815 in his little volume of excerpts, *Baxteriana*. In March he wrote Marianne: "These folios of Baxter are a rich mine, but take a terrible portion of time for digging." And in October:

You ask me what I am reading? Jane reads Baxter to me for one hour every day, we have been for some time in his sermons, each of which is from three to four hours, and much, extremely tedious, but I mark *some* passages that are delightful. . . . I hope the whole of *Baxteriana* may still be comprised in a small duodessimo [*sic*]—I think it will be a very rich one.

He even used Baxter on his rector, but in vain.

Last Sunday sennight our clergyman dined here and stayed the evening & Jane had picked a passage from one of Baxters sermons which lasted fifty minutes & included a severe scrubing [*sic*] into the conduct of the Village clergy, but he *is not one* upon whom any good effect would be produced.[109]

In 1814 Young received four letters from Sir Richard Phillips, all concerned with a new edition, the tenth, of the *Farmer's Kalendar,* which actually appeared in 1815. In his first letter Phillips suggested, as desirable new features, tables of cost and produce per acre under various cultures, inventories of farming stock and implements, and several specimens of farm leases.[110] His second letter stated that Young had less new material than he had expected, and implied that he had bought up the copyright for £200.[111] His third letter urged Young to send him two or three months of the revision just as soon as possible.[112] His last letter told Young not to cut the material "because I value every line of yours as matter which never again can have the benefit of such experience & such maturity of judgment."[113] In the October issue of his *Monthly Magazine* Phillips gave advance notice of the new edition in glowing phrases:

A new edition is printing, with considerable enlargements, of Mr.

---

[109] *Ibid.* Letters of March 17, October 11, 1814.
[110] Add. MSS. 35,132, ff. 110–111.
[111] *Ibid.,* ff. 114–115.
[112] *Ibid.,* ff. 152–153.
[113] *Ibid.,* ff. 198–199. The tone of these letters indicates a very different relation between Phillips and Young than in 1811. *Cf. supra.,* p. 579, note 90.

Arthur Young's celebrated Farmer's Kalendar, the most useful and important volume which perhaps ever issued from the press. . . . Though the illustrious and veteran author is unhappily deprived of the enjoyment of his sight, yet his intellectual vigor continues unimpaired, and has been sedulously employed in the perfection of this favorite work, which, for its pre-eminent worth, might, with propriety, be called the Agricultural Bible.[114]

This [1815] was an eventful year, for my poor wife breathed her last after a long illness, and it gives me great comfort to be informed that she showed great marks of resignation and piety. My daughter was with her to the last.[115]

When her husband, son, and daughter-in-law left for London on January 31, "Mary remained at Bradfield with Mrs. Young, who was unable to move."[116] Marianne's letter of February 16 indicated that she expected Mary and her mother to come to London after a short delay: "My kind regards to yr family, for I conclude that by this time Mrs. Young is arrived."[117] But Young's reply on February 25 stated: "Mrs. Young and Mary remain at Bradfield, and I hear nothing of any intention of their coming to town." He seems to have been especially lonely at this time, for Jane had gone to Worcester during her sister's confinement. "Myself, Arthur & St. Croix poke on as well as we can, but I abhor a house without a woman."[118] The end came on April 8 and apparently the news reached Sackville Street the same day.[119]

Unfortunately no letter from Mary about her mother's last days or death has survived. Nowhere is there any indication of the nature of her illness. The letters of consolation seem strangely lacking in any warmth towards Mrs. Young and to be aimed primarily at convincing Young that his wife died in a state of grace. The Rev. Henry Hasted wrote two days after her death:

It may however be some satisfaction to you to know . . . that one of the last acts of poor Mrs. Young's life was that of prayer. I did not

---

[114] *Monthly Magazine* 38: p. 263.
[115] Young, 1898: p. 460.
[116] *Ibid.*, p. 457.
[117] Add. MSS. 35,132, f. 315.
[118] Berg Collection.
[119] The funeral monument in the Bradfield church gives the date of her death. She had been born January 31, 1740. Marianne's note of consolation to Young is dated April 8, the same day she died. It almost seems that a special courier brought the news to Young who then immediately let Marianne know.

administer the sacrament, because she hardly appeared to me sufficiently sensible, but in parts she did join & was aware of the office. . . .[120]

Wilberforce's letter was especially cold: "A thousand times, at least, very often, it has vexed me to reflect that I have suffered an Event in ye highest degree interesting to you to pass away without my taking occasion to assure you of my friendly sympathy."[121] Marianne's letter was very prompt, being written probably immediately upon hearing the news: "I hope poor Mrs. Young did not suffer much in her last moments & I hope most earnestly that she seemed to find consolation from religion. . . ."[122] Marianne's mother, Charlotte Broome, wrote along much the same lines:

It is a great gratification to me to hear from Marianne that the poor Soul died in so proper a state of mind, praying to her Saviour—and *you,* my dear friend, must feel comfort in the reflection, that you spared no pains to inspire her with Christian principles—and you certainly procured her every comfort & assistance you could think of towards alleviating her bodily sufferings.[123]

Arthur and Martha Young had been married fifty years. Certainly it could never have been called a successful or happy marriage. Her shrewish temper and his all too numerous flirtations were incompatible with true married bliss. Two of their four children had long since died. Their only son had surely been a disappointment. There were no grandchildren. There had been no separation and in the later years one senses a sort of mutual forbearance. Martha must have realized that her husband was a very distinguished man, while his increasing religious devotion fortunately made him more charitable towards her failings and more pitying towards her very real physical sufferings, whatever their nature may have been.

Nevertheless there has seldom been a less affectionate tribute by a husband to his wife than that on her funeral monument in the Bradfield church. The only thing mentioned was that she

[120] Add. MSS. 35,132, ff. 357–358. From 1808 to 1814 Hasted had been rector at Bradfield, whence he moved to become rector at St. Mary's at Bury St. Edmunds. He was a saintly man, and highly respected in the Bury area.

[121] *Ibid.,* f. 368.

[122] *Ibid.,* f. 353.

[123] *Ibid.,* ff. 362–363.

had been "the great-grand-daughter of John Allen, esq. of Lyng House in the county of Norfolk, the first person, according to the Comte de Boulainvilliers, who there used marl."[124] Thus spoke the great agriculturist, not a loving husband.

About two weeks after his wife's death Young had a letter from Arabella Enraght which exhibits the affection of the Bradfield villagers for him:

It is rumoured in the country that you intend henceforth entirely to reside in Town! I cannot myself believe you ever intended so much unkindness to us poor mortals at Bradfield—but, should there really be any ground for it, I propose, putting my great self, at the head of a petition, that shall cover Sackville Street—signed by all your friends & neighbours in the country—Should this fail we must come in a body & bring No 32 down to Bradfield in triumph—indeed we *cannot* give you up. . . .[125]

The other great event in the family was the return of the Rev. Arthur Young to Russia in October. He seems to have been very unsettled earlier in the year. There is an indication in May that he was thinking of engaging in Russian trade.[126] In June his father wrote the Earl of Egremont about having the Irish living exchanged for an English one. In this letter the stern Evangelical remarked about his son's Irish living: "In one respect Agassin is *peculiarly* advantageous that of being a sinecure. . . ."[127] Two letters from Egremont to Young in September show that the Rev. Arthur was considering selling his Russian estate and buying an English living with the proceeds.[128] Even in August, Marianne wrote Mrs. Piozzi that the Rev. Arthur was planning to sell his "Estate in the Crimea, of I think, *9000 acres,* with a fine stone house upon it, & a church," so "that he may live henceforth in his native land."[129] On the other hand, back in May the London *Times* printed the following note:

The Rev. Arthur Young, in an advertisement, published in the

---

[124] This monument is on a wall in the vestry. The same phrase is in Young, 1898: p. 32.

[125] Add. MSS. 35,132, ff. 369–370. Dated April 21. This rather clever letter is interesting as coming from a farmer's daughter and wife, and presumably not a "lady."

[126] *Ibid.*, ff. 408–409.

[127] *Ibid.*, ff. 429–430.

[128] *Ibid.*, 35,133, ff. 46–47, 52–53.

[129] Rylands MSS. 584, f. 135.

Suffolk newspapers, announces his intention of leaving England, and settling on an estate of 9,000 acres, in the Crimea, "the most beautiful province in the Russian Empire, where the proprietor (the Rev. Gentl. himself), during a residence of five years, *never saw the face of a tax-gatherer.*" He invites the farmers of England, whom he considers in danger of ruin, to accompany him, and is ready to receive proposals either for letting or selling parcels of the land.[130]

Probably it was this advertisement which had persuaded a certain Mr. and Mrs. Holderness and their family to return with him to Russia, although on what terms is not known. The preparations for the voyage seem to have been very hectic and to have taken more than a month. About the whole venture there is an aura of secrecy, smuggling, and illegality. Arthur took the title "Colonel" which led Marianne to write Mrs. Piozzi: "But the Rev. *Col.* Young, seems a strange anomaly."[131] Marianne wrote again just before he sailed: ". . . He had a bad quinsey, immediately before he started & was blistered & bled most fearfully. But he is used to quinsies & does not mind. . . ."[132]

The Rev. Arthur's own letter to his father on October 12 from Gravesend shows that he perhaps minded the quinzy more than Marianne realized, and also reveals his complete lack of scruple about smuggling:

In a state of convalescence, thank God, I started from Charing Cross at 1 this afternoon & reaching Gravesend at ½ past five, in one of the public stages, fearfully, most fearfully afraid of a relapse: & as fate would have it, we had contrived, as we thought very cleverly, to smuggle on board our ship a number of heavy packages, amounting in the whole to 34 very heavy cases & large trunks, besides lighter ones, off Blackwall; so we expected to meet with neither search or inquiry here; but we were wofully mistaken: In fact 12 of *my* cases underwent examination at the custom house in Thames Street; & about these therefore there was no question: but our other 34 were hauld over the coals. . . . I arrived, . . . having had nothing to eat since 8 in the morning, very weak, & wishing only for quiet, when the very moment after I got out of the coach, I was taken to the custom house here: a letter there was read to me of 34 packages sent on board . . . in my name & Mr. Holderness' which had not passed the custom

---

[130] *The Times,* May 8, 1815, p. 3.
[131] Rylands MSS. 584, f. 136.
[132] Berg Collection. Letter from Marianne to Charlotte Barrett, October 13, 1815.

house, all supposed to ·be illegal . . . we had, most fortunately, to
deal with a gentleman here, to my utter surprize, he requested me to
accompany him on board, all our baggage was hauld on deck, two
or three cases were examined, containing wearing apparel; books;
& sundries containing nothing contraband, or liable to duty: he took
the rest as granted for having the same things & the whole passed
safe thro the clutches of these sharks, & that most fortunately. . . .[133]

The sea voyage proved rather trying:

The Cabin from England, to this place has been in a state of the
most compleet confusion; squalling, crying, screaming; retching; has
sickened me most thoroughly of sea voyages. . . . The children's
shrieks and lamentations exceeded all that I had before met with; &
for a £ 1000 I would never encounter the like again. Mrs. Holderness has borne the voyage but moderately; her baby, & her squalling
brats have given me a distaste for all such travelling.[134]

His letter from **Riga** promised that the overland trip from
there to Karagoss could only be a difficult journey:

We arrived here Nov. 6 all well; since when; involved in immense
confusion—from 70 to 100 great trunks & other heavy packages of
the Holdernesses including only 12 of mine. We . . . have been trying to get out of town but our luggage is too frightful to expect that
we shall be less than 6 weeks from here to Karagoss. We go tomorrow
with 12 horses & 4 carriages. . . .[135]

Father and son were never to see each other again, for the Rev.
Arthur did not return to England until 1821. The letters just
cited show that the Rev. Arthur had a decided respect and
even love for his father. This trip to Russia also almost certainly
marks the final break between Arthur and Jane Young. They
had been together at Bradfield and London from late in 1814
until the late summer of 1815. But when the Rev. Arthur went
to London to prepare for his journey, it seems very doubtful that
Jane accompanied him. Marianne made no mention of seeing
her in London. Very likely Jane's severe illness in November,
1815, was partially induced by her realization that the final
break had come. On October 31 Young informed Marianne
that he could not return to London "for Jane had a violent

[133] Add. MSS. 35,133, ff. 66–67.
[134] *Ibid.*, ff. 85–86.
[135] *Ibid.*, f. 105,

rash out, and is under a course of physic of Smiths order, as she cannot be left here alone. . . ." Even as late as November 18 he wrote: "Jane is in a course of Medicine and does not stir from her room."[136] The whole thing sounds like nervous prostration resulting from the knowledge that her marriage had completely failed.

In Young family annals Mary Young is usually mentioned only casually. Her father never kept her letters and she apparently never kept his. In 1815 she stayed with her mother through the last painful month, but apparently went to London very shortly after her mother's death. For a moment Mary comes alive in a letter from Marianne to her sister: "Miss Young's Baby linnen is certainly a most useful piece of Charity & she has purchased many a comfortable feeling to herself into the bargain. . . ."[137]

Arthur Young seems to have kept reasonably well until late in 1815, but in October he developed pain and swelling in one hand and foot, which Charlotte Barrett thought evidence of gout and which continued to bother him well into November. Unusually cold weather at Bradfield late in November drove him back to London whence he wrote to Charlotte on November 25:

. . . and I might in addition plead a mind almost as much frozen as my crazy old body, though enveloped in additional coats, stockings, &c, yet, the Thermometer has not been lower than 24. I have felt enough to convince me that my probable departure from this wretched world will be by frost. . . .[138]

After his return to London he suffered at first from severe headaches and then from sleeplessness. Both Marianne and Charlotte urged him to wear a velvet cap in cold weather and the latter pointed out that Pope, Cowper, and Goldsmith had worn them. "Moreover the cap looks very well." Charlotte also suggested some wine for the stomach as St. Paul had advised and urged him to salt himself.[139] On December 20 Young described his sleepless nights to Marianne:

As to my health I have to bless God that I am entirely free from

[136] Berg Collection.
[137] *Ibid.*
[138] *Ibid.;* Add. MSS. 35,133, ff. 74–75.
[139] Add. MSS. 35,133, ff. 102, 134–136.

any bad headaches which weighed me down for many days: the chief
complaint that remained was the want of sleep in the night; for 3
in succession, I do not think I slept two hours, perhaps not more
than half an hour each night; yet I took hop pills, or laudenham
[*sic*]: but I have to be thankful for a most refreshing sleep last night
of 4 hours.

On December 26 he was much more optimistic: "The extreme
change of the weather has been unfavourable to my recovery,
but I am, thank God, much better, and gradually getting as well
as I have any reason to expect at so advanced an age."[140]

Young was in London from the end of January, 1815, to
the end of June. For some weeks after returning to London he attended as many religious services and meetings as possible. Marianne wrote him from Richmond: "It gives me great pleasure to
hear of your evening arrangements, because I know how much
more happiness it must give you to pass your time in a Christian
congregation than in a sleepy chair. . . ."[141] One of the very few
discordant notes in the idyll between Young and Marianne appeared in his letter of February 25:

My second reason for not writing was being somewhat displeased with
you; here have I been very near a month, and you within ten miles,
and you have never come near me; but the Ws being at Kensington
will turn ten miles into ten yards. I told you you might have a bed,
but every syllable tending to your coming here has been sunk [?] in
your letters in a manner which must have struck me unless I had been
a post; or was it necessary for me to tell you that I should be glad
to see your Mama as well as you; but I do not complain of anything
I meet with in this world. . . .[142]

Obviously Young was hurt and jealous of the Wilberforces.
Marianne replied:

I received your letter this evening, & you must forgive me for saying,
that I read it with sincere pain. You would not have reproached me
so bitterly, if you had known the circumstances that have prevented
my coming to see you. . . . But I have no friend to convey me to
Sackville St. my mother wd not approve of my travelling alone in
stage-coaches. . . . we both intend the happiness of dining with you

[140] Berg Collection.
[141] Add. MSS. 35,132, f. 314. Dated February 16.
[142] Berg Collection.

on Monday the 27th at your usual hour, 5 o'clock. I shall be with you, I hope, long before that time.[143]

Early in March she went to Kensington Gore for at least two months. She dined in Sackville Street about once a week, and Young frequently attended Wilberforce's Monday morning breakfast parties, where he met such Evangelical celebrities as General Zachary Macaulay, editor of the *Christian Observer*, and Legh Richmond the tract writer. On May 15 there was much conversation about Bonaparte and predictions that his great adventure would fail, and General Macaulay told Young of a Frenchman whose farming principles were those of "Monsieur Arthur Young."[144] Young attended these breakfast parties chiefly "for the pleasure of hearing his Exposition and Prayer; for the conversation at and after breakfast has been entirely desultory, and not once on any religious question."[145]

Young seems to have felt neglected in the spring of 1815: "It would be natural to suppose that a poor old blind man who, through the blessing of God, retains his health and strength might have received something more of friendly attention than this. . . ."[146] One high spot occurred on May 11 when he had a call from a young naval officer named Pakenham:

A few days ago, writing to Miss Francis, I used the expression, "If a Christian was to call on me it should be entered in a pocket-book with a mark of exclamation." Mr. Wilberforce saw this note, and yesterday morning Mr. Pakenham called on me, and introduced himself by saying that he came for some conversation with me, by desire of Mr. W. . . . I soon found that he was a firmly established Christian, ready to converse on the good subject, which he did with good sense and no inconsiderable energy. . . . Mentioning Miss Francis, he said he met her twice at Mr. Wilberforce's, and speaking in commendation of her, I told him that she was to dine with me at five o'clock, and that it would give me much pleasure if he would meet her; this he readily complied with, and came accordingly. I have not had so much religious conversation for an age past . . . I hope I shall hear more of this young man, whose determined avowal of his religious principles pleases me much.

[143] Add. MSS. 35,132, f. 323.
[144] Young, 1898: p. 462.
[145] *Ibid.*, p. 457.
[146] *Ibid.*

A few days later Young began another friendship with the Rev. Mr. Gurney, Rector of St. Clement Danes. After a sermon at West Street Chapel Gurney came into Young's pew and then returned to Sackville Street and told the old gentleman his life's history:

The detail was very interesting, from being not only well told, but, from the providence of God, clearly marked in many little circumstances, and attended by what to him were great events. . . . He is uncommonly lively and animated in conversation, and contrived to talk with little interruption, from drinking tea and smoking several pipes, till twelve o'clock at night. I much hope that we shall see him often.[147]

Early in June Young invited Marianne, her mother, and her brother, Dolph, to spend part of the summer at Bradfield. Marianne responded with suitable gratitude:

It would give me much pleasure to visit your peaceful & happy retreat once more, & to go about again, as I used to do, amongst the schools & poor people, & to rise & walk out at 4 & 5 in the morng. & I hope it may please God that this summer it may be accomplished.[148]

On June 20 Charlotte Barrett suggested that she also bring her family to Bradfield that summer to be near her sister and mother:

. . . and we, that is to say, Mr. Barrett & I, and all our little ones should much like to follow them and be near you; therefore provided we can find a cottage or cottages to hold us all, . . . if you encourage our scheme & our hopes, we should let this house & bring all the livestock of it into Suffolk—we should want about 5 beds in all. . . .

Apparently Young immediately offered one of his own cottages and Charlotte then told him more specifically: ". . . consider that we have four children, & 3 maid servants besides a wet nurse . . . ."[149] Shortly before arriving she gave more directions to the blind old man:

---

[147] *Ibid.,* pp. 460–463. "Mr." Pakenham was probably Captain John Pakenham, (1790–1876), later Admiral Pakenham, a son of Admiral Sir Thomas Pakenham, and a grandson of the 1st Lord Longford whom Young had known in Ireland in 1776. I have been unable to identify the Rev. Mr. Gurney. Young said that his father was a poor country laborer, so he did not belong to the great banking family.

[148] Add. MSS. 35,132, ff. 434–435.

[149] *Ibid.,* ff. 438–439, 440–441.

May I entreat that you will be so kind as to give directions for the furniture being put into the cottage . . . and then . . . will you let your housekeeper send some woman from the village to air the beds . . . to unpack and wash a hamper of crockery . . . this woman should also order in a few pounds of candles, & 6 sacks of coal, & some wood & some pounds of soap—& 3 quarts of milk . . . I am half ashamed to trouble you about such matters only I know how kind you are.[150]

None of the letters mentioned financial arrangements.

The whole tribe descended on Bradfield in August. Shortly after her arrival, Marianne described their life to Mrs. Piozzi:

For my Mother & Dolph are both on a visit with me at Bradfield Hall, & Charlotte & Barrett & their four little ones inhabit another house not far off. We ride, & row & swing, for Dolph's health; there are plenty of poor people who are happy to be visited, read to &c & a very good library of *readable* books within doors. So we are very happy here. . . .[151]

Young was greatly disappointed when Marianne's visit proved to be less than a month. Her other brother Clem got sick, alarming reports came from doctors and friends in London, so his mother and sister hurried back to be near him. Young's exasperation was increased when they took lodgings so near Sackville Street that Marianne could have seen him every day, had he been in London. He wrote her on October 16:

I cannot attribute Clems improvement to any of you being in Town, and it is utterly mortifying to me to hear you fixed in London without one shadow of reason, none whatever: it is most palpable that you might all have remained at Bradfield till I went to Town the middle of Novr: my ill stars will I suppose carry you thence as soon as I arrive there. . . . In my life I never knew anything of the kind more provoking, and I beg to remonstrate with your dear Mama for those fits of figgiting [*sic*] about Clement. . . .[152]

Nor was Marianne any happier in London. She was greatly worried over Clem and late in October had a bilious attack. She repeatedly urged Young to come to town earlier than he had expected and on November 6 her appeal was very moving:

---

[150] *Ibid.*, 35,133, ff. 16–17.
[151] Rylands MSS. 584, f. 135.
[152] Berg Collection.

It is so severe a mortification my dear Sir, to me especially in my present state, to be deprived of your Society in London, living, as I do, so near your own abode, & it would be vexatious to have you come up, perhaps the very day on which I go away, that I cannot yet quite exclude a hope, that you may possibly be here before the 14th of this month. . . . Do, try to oblige me if you can. You know not how great a consolation it would be to my mind: and I think I may safely say that under similar circumstances I should not refuse you.[153]

The almost desperate character of the above is explained in a very revealing note to her sister four days previously:

I think it right to tell you, my dearest, that all is entirely over between Capt. Pn. & me—yet I have not used him ill. But his going on distant, perilous destinations, & the constant anxiety attending this state of fluctuation & uncertainty wd. have destroyed my health & spirits, & made me perfectly useless. I will talk to you on the subject when we meet. Meantime, let me entreat you never to discuss or mention it to any human being, nor to let Mama know that I have startd it, even to you. . . .[154]

No wonder she was bilious! Particularly if the break had come, as her letter indicates, just because she could not bear to be married to a naval officer who might be called away frequently on dangerous missions. There are slight indications that Young may have known something about the situation, but there is nothing definite. After hearing of her bilious attack he wrote on October 31:

It is easier to preach against anxiety than to refrain from it, but when it generates bile, . . . it should be carefully guarded against, for which purpose the only effective means are an absolute trust in the goodness of God; if we are perfectly persuaded that nothing happens by chance we ought not to indulge in anxieties to the injury of the health. . . .[155]

Marianne's portrait becomes a little sharper from her letters during 1815. She borrowed many books from Young, both in the classics and in religion. One letter revealed that she had been reading Coxe's *Memoirs of Robert Walpole* and Adam Smith's *Wealth of Nations,* another asked Young for a history of the French Revolution.[156] Just before coming to Bradfield she wrote:

[153] Add. MSS. 35,133, ff. 93–96.
[154] Berg Collection.
[155] *Ibid.*
[156] Add. MSS. 35,133, ff. 101–102, 141–142.

Every Friday I go Incognito, to the little meeting here, where I hear sermons very different from those of our unhappy church on Sunday. I have no regrets in leaving Richmond, . . . & there are but 2 intelligent people in the place the Archdeacon & Dr. Gillies the Historian neither of whom are to my mind, except in matters of literature or curiosity. . . .[157]

That Marianne was not, however, just a disembodied mind and spirit is revealed in a half-humorous, half-pathetic little note in the letter to her sister announcing her renunciation of Capt. Pakenham. She asked Charlotte to purchase her a dress in the same color and style as her own. "And I mean always to adopt your devices, finding that I am never fit to be seen, except when I am a humble imitation of you. . . ."[158]

Hardly had the Barretts left Bradfield early in October than they began to plan for a long summer there in 1816. Henry Barrett's letter of November 3 demanding changes and improvements is pompous and utterly selfish:

I am happy to hear you have ordered already the necessary Improvements—permit me to say again, that it is almost indispensable that Anne's cottage should be cleared of its actual tenants; the compassion I feel in the ejection of them, is much diminished, when I reflect that we shall give them a good deal of profitable employment, when we come. . . . I hope it is still your intention to have the sections of the garden before the drawing room for a lawn in smooth grass, as we contemplate on such disposal much pleasure for the [?] and utility for the drying & bleaching plan. . . .[159]

Young seemed willing to do almost anything to get them and on November 25 he wrote to Charlotte: "My poor pocket has bled much more freely in order to fix you as an inhabitant at Bradfield than it would have done for any other person whatever."[160]

Young had other visitors at Bradfield in the autumn of 1815. In October the Rev. Mr. Gurney arrived for a week's stay. On the one Sunday he was there he preached to a tolerable congregation in the morning, to a full church of four hundred in the afternoon where he collected nearly four pounds for the Jews, and

---

[157] *Ibid.*, f. 35. Probably the reference to the archdeacon is to G. O. Cambridge, Prebendary of Ely, son of the more famous Richard Owen Cambridge. Dr. John Gillies (1747–1836) was an historian and classical scholar.

[158] Berg Collection. Letter of November 2, 1815.

[159] Add. MSS. 35,133, ff. 97–99.

[160] Berg Collection.

in the evening an hour and a half sermon in the Hall: "His voice and his surprising fluency struck all with amazement, and as he had no book they said he must have a most wonderful memory . . . he prays morning and night with great expansion, force and piety. . . ." Young's next letter bore witness to the pleasure which Gurney's visit had brought him:

. . . he is one of the most cheerful companions I have met with, with an immense budget of stories, which he tells admirably. I had much religious conversation with him, and especially upon his doctrine of Absolute Predestination. . . . He has no disbelief of supernatural appearances, or of occasional supernatural exertions, and had stories without end in support of his opinions.[161]

Early in November, Young had two visits from Wilberforce, but the first was more exasperating than pleasant:

A phenomenon has appeared in the Suffolk sky, Mr. Wilberforce. . . . he came accordingly, and having taken the walk of the paths, and exchanged a few words, he retired to himself in the dining room, and wrote 6 or 7 letters, then took a mouthful of cold Porridge and a glass of wine, and departed for Ampton. . . . I would not live in such a hurry, bustle, and worrying want of time . . . for all his fortune, though I would undergo anything to have half his piety. . . . if he ever comes to have a thousand things to do he will die of vapour, so much for our dear friend, who certainly is in every point an extraordinary being. For a year and half past I have not had one quarter of an hour of quiet conversation with him, and it is less and less to be expected as I advance in age and dulness.

Nevertheless, Wilberforce stayed over night at Bradfield a little later in November—a far more satisfactory visit, as Young reported to Marianne:

Mr. Wilberforce will have told you his last visit here, but he could tell you nothing of that pleasure which an evening of 4 hours with him conferred on me; he was quiet had not the least fidgeting and without any letters to write: think of that! it was indeed a treat. . . .[162]

Wilberforce's thank-you note shows that all Young's family put themselves out to please him:

[161] *Ibid.* Letters of October 16 and October 25, 1815.
[162] *Ibid.* Letters of November 4 and November 18, 1815.

With every good wish & kind Remces to all my Bradfield friends, for I cannot forget the kind attentions of all of them Mrs. A. Y's make me a partner of her Russian stories, Miss Y'g shew'g me yr out of door living & Mr. St. Croix's watchful services early in ye mg.[163]

Young's *Baxteriana,* published in July, 1815, was a small volume, consisting solely of excerpts from the writings of Richard Baxter, the seventeenth-century divine, except for a very short preface. The material was broken up into seven books, and arranged topically rather than by Baxter's writings. Young's preface stated that for many years he had been making a selection of impressive passages from his religious reading and that *Baxteriana* included all that he had copied from that author. The preface bears strong witness to Young's religious convictions:

Eighteen years ago he suffered a severe domestic calamity, which, for the first time in his life, led his mind into a train of thought very different from all that had formed, till then, his pursuits, pleasure, or occupation. For the first time, he began seriously to think of that which a Christian ought to think of every day of his existence—a future state.

. . . . I will not lay down the pen without most earnestly intreating those who are but entering on life, to be persuaded to pay a constant attention to the duties of religion, especially the four great means of grace, prayer, public worship, reading the scriptures of truth, and, as much as circumstances will permit, meditating on their contents. . . . In truth, there is but one principle that ought to govern mankind; to think, speak, and act in such a manner as will please God, and to avoid all that will offend him: not the *Supreme being,* the *great first cause* of modern philosophers, but the God of Revelation. O my young friends, let me with truth assure you, that though I have experienced some highly flattering, and partook of many brilliant scenes, yet would I not exchange the consolation and hope which Christianity gives me while blind, and quickly descending to the grave, for the most pleasing moments of my former life, with rejuvinescence to enjoy them.[164]

Marianne was of course enthusiastic. Early in November she wrote her sister: "I am very glad . . . that you are so fond of Baxteriana. I own I found it one of the most useful books I ever

[163] Add. MSS. 35,133, f. 133.
[164] Young, 1815: pp. iii-v.

read. . . ."[165] And on December 18 both Marianne and Wilberforce gave it praise which must have warmed Young's heart. She wrote: "It would do you good to see the delight Mr. W. takes in your Baxa. He keeps it with his Bible, & refreshes himself with it when he is weary. . . . I think I never saw him take more pleasure in any book whatever." And Wilberforce added: "Marianne bringing me a letter to direct to you unsealed, I avail myself of the opportunity of inserting a slip of paper, to assure you with how much pleasure I have been reading your Baxteriana. . . ."[166]

Young's letters to Marianne in the autumn of 1815 show that he was very much disturbed over the doctrine of Calvinistic predestination. The Rev. Mr. Gurney was a strong Calvinist who believed in "Absolute Predestination." "As I do not believe in this doctrine, I could not yield the point to him." Two letters in December discussed at considerable length the impact of Bishop Tomline's *Refutation of Calvinism* on him. On December 20 he wrote:

. . . and I must confess that the Bishop carried me with him in a considerable degree on Universal redemption: his extracts from the Fathers proves sufficiently that Calvanism [sic] was unknown in the first ages of the Church; and the more I reflect on the decrees of reprobation, the more horrible the doctrine appears to me. . . . I think that every thing which excites our heart to the love of God is valuable; but the system which supposes him to have decreed the damnation of millions before they are born has by no means any such tendency. . . .

Again on December 26:

I have had a large part of the Bishop of Lincoln's Refutation of Calvanism [sic] read to me. . . . I must confess that I go with him in his utter condemnation of the system of absolute decrees of reprobation before men are born, which always appears to me to be a horrid doctrine. . . .[167]

Young was probably closer to Methodism on predestination than to the dominant Evangelical theology. On two other points, however, the necessity for absolute trust in God, and a burning

---

[165] Berg Collection.
[166] Add. MSS. 35,133, ff. 131–133.
[167] Berg Collection. Tomline's famous book had appeared in 1811.

and fervent love of him, his letters were representative of the best of Evangelical thought. On October 25 he wrote:

. . . we are all equally called on to remember and admit that without holiness none can see the Lord, and that a principal part of holiness is putting an entire trust in God, which if well followed up will secure an entire resignation to his will, though manifested in misfortunes, or temporal evils of any kind under the persuasion that all things shall work together for good to those who love God. . . . we are then prepared in the best manner we can be for whatever events may happen to us; and we should take special care not to set our hearts upon any thing which God seems to deny, under the clearest conviction that he knows best what is good for us.

Could the above perhaps have been meant to help Marianne in the crisis of her relations with Captain Pakenham? On November 18 Young confessed how he still had to fight the old Adam:

It is dreadful to think how I am forced to kick, and spur and belabor my vile heart to bring it to any feelings attended by comfort and satisfaction . . . I want to have the love of God burning in every fibre of my heart; and instead of that, the whole of it is too often an icycle.[168]

Fairly early in 1815 the tenth edition of the *Farmer's Calendar* was published by Richard Phillips. In his advertisement, dated from Bradfield, October 20, 1814, Young wrote:

In various parts of the Correspondence published during twenty-four years in the Annals of Agriculture, I have been called upon for New Editions of this Calendar, and have as often resolved to give one; but the new improvements which have taken place, made so many and such great alterations necessary, that other and more pressing employments have prevented the undertaking. The tenth Edition is at last completed, and I hope the reader will find it, in the present form, worthy of his attention.[169]

The unwary reader might thus expect great changes in this tenth edition. It is somewhat of a shock to see an almost identical advertisement in the sixth edition of 1805. Only the number of years in the first sentence and the number of the edition in the last have been altered. Nor are the two editions very different in content. Two months have been examined with some care.

[168] *Ibid.*
[169] Young, 1815: p. iv.

In February, which covers nearly fifty pages, one paragraph on the culture of peas and one sentence on cabbages are omitted, and for additions there are one sentence on composts, one sentence on marling, and a section "Remedying the Defects of Old Enclosures." In July the changes are still less, the addition of a paragraph each on coleseed and on mowing grass. There are also several additions to the appendix, some pages devoted to "Sundry Circumstances Relative to Building," the draft of a Suffolk lease, and a recipe for making family wine. Altogether this was hardly the drastic revision that Phillips had originally envisaged. Of his suggested two items for the appendix only the Suffolk lease appeared.[170]

A more substantial change was in the October entry where sixteen pages were added under the heading "ascertainment of rent" which discussed also the costs of cultivation and the farmer's profit. Some of this material was based on his testimony before the House of Lords Committee in 1814. His most interesting proposals were that labor should be paid on a sliding scale depending on the price of wheat and that rents should also be calculated on grain prices. Admitting that his "corn rent" would be complicated, he thought it worthy of further study. He had experimented himself with such a sliding scale rent in 1814.[171]

A more important publication was *An Enquiry into the Rise of Prices in Europe,*[172] one of the many pamphlets arising from the Corn Law controversy in 1815. It was certainly not an outstanding contribution to the subject,[173] nor one of Young's great works. In general he defended the high English prices which were no higher than on the continent. He attacked those who urged that rents should be lowered. Since other prices were high, why should rents be reduced? Since the proposed Corn Law would keep agricultural prices and rents high, Young supported it strongly. He again urged that the poor agricultural laborers should be given a cottage and a cow: "I would give them a stake which should want no political explanation; positive and tangible. His

---

[170] *Cf. supra,* p. 633.

[171] Young, 1815: pp. 484–500.

[172] This pamphlet was issued as No. 271 of the *Annals of Agriculture* **46**, and also reprinted in the *Pamphleteer* **6**, No. 6. For the letter of July 10, from the editor of the *Pamphleteer,* asking permission to reprint it, *cf.* Add. MSS. 35,133, f. 5. Excerpts also appeared in the *Bury and Norwich Post,* October 11, 1815.

[173] For instance, Barnes, 1930, lists the pamphlet in his bibliography, but does not discuss it at all.

cottage, his land, and his cow would be a stake in the hedge, that
would make him value, respect and revere every twig in it."[174]
Young still showed himself a strong patriot. He frequently referred to "this happy country" and in one case to "the happiest
country the sun shines upon."[175] England had been preserved
"throughout the execrable period of the French Revolution, and
through that of the general devastation of Europe under the most
abominable of tyrants. . . ." The period had not brought horrors
to England, but ". . . successful industry and animated exertions; our fields smiling with cultivation; and the cottage of the
Peasant protected from every rough intruder as much as the
palace of the Prince." With such blessings Young felt "extreme
pain" at the spirit of discontent among the English people. He
was bitter against the immorality and drunkenness accompanying
the peace celebrations and also at the "infatuation" of the diplomats who had protected Napoleon "from the vengeance of an
enraged people, who were ready to save Europe from future
miseries. . . ." He also believed that patriotism and Evangelical
religion were closely connected. If England has been saved, she
has been saved by God:

> . . . we may with equal truth assert, that the best Christian is the
> best patriot; that the noble exertions of the British and Foreign Bible
> Society add a strength to our cause, not easily appreciated . . . that
> the great expansion of Evangelical Religion, which has taken place
> in this Kingdom, united with that spirit of educating the children of
> the poor . . . offer a fair reason for a similar reliance. . . .[176]

On January 17 William Pittman of Sandwich wrote Young
that he was addressing a public letter to him on the Corn Laws,
which eventually was published as *A Letter to Arthur Young, on
the Situation of the Growers of Corn in Great Britain*. In his
private letter to Young, Pittman's praise was fulsome:

> As a public character and as a gentleman who has done more in the
> support of and in the improvement of the Landed Interest of this
> country than any individual therein, I have taken the liberty of
> addressing a public letter to you on the present alarming state of
> the Landed Interest of the Kingdom. . . .[177]

[174] *Pamphleteer* **6**: p. 185.
[175] *Ibid.*, p. 167.
[176] *Ibid.*, pp. 183, 185.
[177] Add. MSS. 35,132, ff. 295–296.

The Corn Law problem continued to bother Young during the spring. On February 25 he wrote somewhat petulantly to Marianne: ". . . and I was half angry at a deluge of Corn Pamphlets which I must read. . . ."[178] Shortly after reaching London and about a week before the meeting of the Board of Agriculture, Young received a long letter from Lord Hardwicke, asking for information about the corn trade to lay before the first Board meeting. What was the proportion between imported corn and the total consumption in years when imports were the largest? In the past had the cessation of war brought about a rise or decline in corn prices? How much corn had actually been imported since January 1, 1814? He also suggested that the Board might discuss how to reduce enclosure expenses and how to relieve the farmers from the burden of poor rates.[179] Young's reply enclosed answers to Hardwicke's first two questions, but declared that he could not answer the third because the Board had stopped its subscription to certain periodicals in an ill-advised attempt to save money. The letter ended by some comments on enclosures and poor rates:

I am very sorry to hear that 140 applications have been made for new Enclosure Acts, as a bad use may be made of this fact in Parliament. The difficulty of restraining the expence after the Bills pass I fear will be very great. . . . To ascertain the present amount of Poor rates, can only be effected by legislative means, in a renewal of those requisitions which have been made more than once, and the returns printed: it deserves your Lordships consideration whether the Board could not take some step in recommending this measure: it is a very important point.[180]

The popular agitation over the Corn Law of 1815 reached its height in March. Petitions opposing the measure were presented in Parliament from many large towns and rioting occurred in London. The *Autobiography* contains nearly two pages on the agricultural distress of 1815 and the Corn Law Riots:

Monday, March 6, most execrable riots began in London, on account of the Corn Bill, . . . attended with circumstances proving decisively the abominable effects . . . of printing in all the news-

[178] Berg Collection.
[179] Add. MSS. 35,132, ff. 310–311.
[180] *Ibid.*, 35,700, f. 318.

papers those violent and mischievous speeches which are made as much to the Gallery as to the House, and can be intended for nothing else but to inflame the people, which they have done to a degree of desperation. Petitions from a multitude of cities and towns pour in to the Houses every day they meet. . . . 600,000 qrs. of French wheat of an excellent quality have been poured into our markets to meet a crop generally mildewed; this has reduced the price on an average of the kingdom to 59s. per quarter, and that average taken in so preposterous a way that the real price fairly ascertained would not amount to 50s.; 90s. per qr. would not pay the farmer in so bad a year. If importation was to be continued, at least half the farmers in England would be ruined, and wheat consequently must rise in a year or two to scarcity, and if importation should be prevented, by many probable events to famine. Country labourers throughout the kingdom are in the greatest distress, as I know from many correspondents. For want of employment they go to the parish, but these poor families never petition, even when starving, and a Legislature which attended not to their interest would deserve the abuse now vomited forth by towns. From thirty to forty houses have had their windows broken, many their doors forced, and everything in them destroyed; and after much mischief, with general anxiety and apprehension, the military were called forth; but it was the last day of the week before their numbers were sufficient to secure any tolerable tranquillity.

Monday, March 13.—I breakfasted with Mr. Wilberforce: a file of soldiers in his house, because his servants had been violently threatened that it should be speedily attacked.

The bawler bearing [*sic*] last week, in the House, read a denunciation in a petition from Carlisle against the Board of Agriculture, which made it necessary for me to hire a bedchamber elsewhere, as blindness would not permit an escape by the roof of the house.[181]

I wrote to Mr. Vansittart, transcribing a resolution of the Committee of 1774, proposing to lay the millers under an assize. . . .
In Mr. Vansittart's answer to me, he mentioned the difficulties in the way, but observed that as Mr. Franklin [*sic*] Lewis had taken up the business of bread and flour in the House, he would mention to him what I proposed. From Lord Sidmouth's speech it seems they intend

---

[181] Young, 1898: pp. 458–460. The editor pointed out that St. Croix must have misunderstood. "bearing" should be "Baring," and undoubtedly refers to Alexander Baring (1774–1848), member from Taunton who hated all restrictions on commerce. In time he became the 1st Baron Ashburnham. On the night of March 7 Young took refuge at Kensington Gore. *Cf.* Wilberforce, 1838: **4**: pp. 244–245.

to remove the assize of bread, which will leave in case of scarcity the bakers without protection in case of riots, and also leave the millers in full possession of their rascality.[182]

The riots were suppressed and the Corn Law passed, but the agricultural distress continued. Sir John Sinclair wrote Young on October 5: "I hope this will find you in good health, though you can hardly be in good spirits in times so dismal for the landed & farming interests of the country."[183] Further evidence of distress is found in letters from two of Young's tenants. On November 27 his very prosperous and respectable tenant, J. B. Edwards, wrote that he was willing to renew the old lease but that under present prices he could not possibly pay any increase in rent: "If we revert to the old lease . . . leting [sic] every Covenant be precisely the same for the Term of Seven or Ten years meets your approbation, I shall be happy to have the Honor of continuing your Tenant." At the end of the year the less prosperous Joseph Bird asked Young not to insist upon the full rent:

. . . I hope Sir you will take into your consideration the great pressure of the times, and that during the half year I could not obtain the price quoted, that I hope you will be satisfied with £ 75 for the half year Rent, you well know my family is Large, that considering all things I can barely provide for them. . . .[184]

Once in the late fall Young commented on affairs to Marianne:

The distress I left among all ranks in Suffolk had occupied my mind for sometime and I expected on my arrival in Town to hear some better account of the causes as well as the foundation of hope of better times to come in the latter of which I am sadly disappointed: Sir John has had conversation with Lord Liverpool who told him that the Bank must be brought back to their old mode of payment that is to say . . . fortunes are to be continued and increased; every evil we feel results from a want of circulation, and the remedy of our rulers is to diminish it. It is perfect infatuation. . . . but what do you care for these questions in political economy? however as they occupy my mind much more than they ought to do, you must submit to hear my groans.[185]

[182] For Vansittart's reply to Young *cf.* Add. MSS. 35,132, f. 346, which makes it clear that Frankland Lewis is meant.
[183] Add. MSS. 35,133, f. 58.
[184] *Ibid.*, ff. 112, 143.
[185] Berg Collection. Letter of December 5, 1815.

Except for the letters between Hardwicke and Young noted above, there is little information about Board activities in 1815, but it distributed a circular about the nature of the agricultural crisis, asking whether farms had been given up, whether distress was greater on arable or grassland farms, what was the state of the agricultural laborers, was the diminished circulation of paper money a cause for the distress, and what possible remedies might there be.[186] On June 12 Under-Secretary Cragg wrote to William Alcock, presumably in answer to an enquiry about Board finances:

The normal parliamentary customary Grant has been £ 3000 per ann but additional Grants have been made for specific purposes. . . .

The expenditure of the Board is regulated by a Committee summoned for the first friday in the month & all orders for payment of monies are now signed by five members. . . .

I beg leave to add that no addition has been made to the salaries of the Officers, since the first establishment of the Board. Gratuities have been occasionally given for extra work.[187]

The financial affairs of Sir John Sinclair seem to have been very critical in January, 1815. On January 16 he made a rather unusual request of Young:

I am carrying on some particular business at present, which obliges me to be in town. I have availed myself of your absence, to use the Back room, opposite to the Board Room: and as I hope to finish what at present occupies my attention about the latter end of next week, it would be obliging in you, if you could, without inconvenience, postpone yours & your family's journey till that time. The Board will not assemble till the 15 February.[188]

Twenty years earlier such a letter would have been regarded by Young as barefaced effrontery on Sir John's part, but his reaction in 1815 is unknown. Two days earlier, on January 14, a meeting was held in London to organize a public subscription to relieve Sinclair's financial embarrassments. A printed appeal was sent to members of Parliament, the Board of Agriculture, and various agricultural and commercial societies. The appeal mentioned Sinclair's parliamentary work and his numerous efforts

---

[186] Board of Agriculture, Letter Book 2: ff. 241–242.
[187] *Ibid.*, f. 239.
[188] Add. MSS. 35,132, f. 289.

to improve economic conditions in England and Scotland. Among the donors T. W. Coke gave £50, the famous farmers John Boys and John Ellman £10 and £20 respectively, and Arthur Young £31. 10s.[189]

Young's only substantial comment to Marianne on international affairs discussed a religious aspect of them:

As I suppose you are now in the habit of often seeing Mr. Wilberforce, I beg you will take an opportunity of enquiring his opinion relative to the circumstances of re-establishing the Jesuits and the Inquisition. I have read . . . a history of the Jesuits, with observations on the necessity of parliament awaking to the interests of the Protestant religion against the machinations of this execrable order who are pouring into Ireland: my blood runs cold. . . . I want particularly to know whether this new bull of the present infernal Pope has not occasioned a change in his opinion relative to the question of Catholic emancipation: I was for it till I read this detail, but I should not be very well pleased at the thought of Jesuits sitting in the British parliament; for they may appear in any situation of life. . . . I never was inclined to think so well of Buonaparts government as since I read this most horrible recital: he was by far the greatest enemy to Popery that has existed, a perfect contrast to the imbecilities and bigotry of the wretched Bourbons. . . .[190]

What a strange twist, for a devout Evangelical to preach anti-clericalism because of hatred of Catholicism.

One other letter of 1815 from his very good tenant, J. B. Edwards, deserves quotation:

As I understand the Bricklayers are coming to the Hall next week, I must request the favor of you to let them whitewash my House, the season is now proper for that work & it really looks so dirty. I am ashamed to see it. . . . The Females make a great cabal about the Oven, its very dangerous, the men might do that at the same time if it meets with your perfect approbation. You know of old I never ask for any repairs, but what are indispensably requisite. . . . I flatter myself the pleasure of hearing at the same time when we shall have the gratification of seeing & enjoying your enlivening conversation at the Hall. . . . I assure you Bradfield is as gay as a Bride in her wedding dress, the trees & flowers being in full bloom—& the meadows & corn fields look very promisingly.

---

[189] *Ibid.*, ff. 291–294.
[190] Berg Collection. Letter of October 31, 1815.

Mrs. Edwards & Mr. & Mrs. Enraght request to be remembered in the handsomest manner . . . not forgetting Mr. St. Croix; tho he has totally forgot his Bradfield Friends, they have not forgotten him.[191]

In the light of social history, this Edwards family is most interesting. They refused to pay a penny more rent, but they ran his schools and Sunday night meetings for him. They wrote more than barely literate letters. While respectful to the squire, they certainly were far from subservient, regarding themselves as friends, not social inferiors.

Arthur Young had come to London in November, 1815, and remained there until late June, 1816. Contrary to his practice in recent years, he stayed at Bradfield until early 1817. Apparently Mary and Jane were with him throughout the year. The Rev. Arthur Young was of course in Russia. Only one letter to his father, from Odessa on January 30, has survived. It started by stating that he had had to draw upon his father for £50. He had still not reached the Crimea and his letter described very graphically the hardships of the trip:

We have had a terrible journey, of 2 months & more from Riga: we left that place Nov. 17, N. S. & arrived here 3 days ago. The women & children have been overthrown, run away with, the carriages broken to pieces, all but drouned [*sic*] in passing the rivers; buried in snow, & dragged in the mud, & one of our unfortunate drivers hurld from his seat, & killd on the spot. We have been starved, frozen, baked—& almost roasted: . . . . Mrs. Holderness you know took 4. of her children, an infant of 2 months at her setting out has performed the journey with admiration. & the other children are in excellent health, as well as the rest of the party. but our pockets have woefully failed. We are 12 in number, & travel with four carriages & 14 to 16 horses. . . .[192]

Many letters which Young received in 1816 ended with a wish to be remembered to Miss Young and Mrs. Young, or to "the ladies." In December he described Jane's activities at Bradfield after the departure of Marianne and the Barretts:

I am glad to inform you that Mrs. Arthur continues her exertions unabatedly, goes every day at one, and is never home till dinner is on the

---

[191] Add. MSS. 35,132, f. 401.
[192] *Ibid.*, 35,133, ff. 336–337. For this trip *cf.* also Holderness, 1827: pp. 1–104.

table, though quite dark. I think I told you that Palfrey is ill . . . and she reads Scripture to him every day, and has twice or thrice a week a little congregation at Aldertons, and she sees Mrs. Enraght often, and they talk upon the good subject.

In another letter he told Marianne that Jane was giving "a Chaldron of coals" for the three winter months to all the poor families of Bradfield. As for Mary, the record is vague as usual. She must have been in his mind when he wrote to Marianne on June 13:

You are perfectly right in all you say on our absolute duty of attempting to convert every unbeliever. . . . but what do you say to the lamentable situation of having persons living with you and even nearly related and whose conversion must consequently be the earnest wish of our heart, yet without any impression being made: the only hope such a case admits is that God will in his mercy at some time or another turn things long since past to the benefit of such Souls. . . .[193]

He would hardly have referred to Jane thus, and Arthur was no longer living with him.

There is considerable proof of Young's good health in 1816. On January 11 he wrote Marianne: "I have thank God been much better and walk out every day." Five days later he wrote again: "My health is quite reestablished and as I have reading from 5 in the morning till 8 at night, naps only excepted, I am fully employed without time ever hanging heavy on my hands; what a blessing at my age and with some infirmities."[194] On November 28 Marianne wrote Mrs. Piozzi: "We left Mr. Arthur Young in admirable health & spirits blind as he is. . . ."[195]

Young's correspondence for 1816 throws much light on his somewhat extensive establishments at Bradfield and London. The Isaac Palfrey to whom Jane read the Scriptures in December was the old gardener at Bradfield, who wrote to his master on March 1, inquiring whether he should sow the Round Garden with oats or beans, on what terms he should sell a haystack, and how the land, which was to be used by the Barretts that summer,

---

[193] Berg Collection. Letters of June 13, November 4, December 14, 1816.
[194] *Ibid.*
[195] Rylands MSS. 584, f. 144.

should be improved.[196] Palfrey was also carpenter and painter and in previous years had prepared the cottage for the Barretts' use. A. Gardner was a sort of combination butler and caretaker who stayed at 32 Sackville Street when the family went to Bradfield. Early in 1816 their old cook Betty, who had served them twelve years in Sackville Street, left to become housekeeper for someone else. It was Betty who had somewhat acidly commented: ". . . the Suffolk girls come to London to pass the time in looking out of the window and are presently knocked up by the 80 steps from the top to the bottom of the house."[197]

A more important loss was that of Mrs. M. Heseltine who had served as a second reader to relieve St. Croix. She received a life annuity of £100 and hence in the summer was writing Young from Brighton:

I think very often of the beauties of Bradfield Hall and on Sunday was with you in Idia [*sic*] I could fancy I saw the smiles on the continences [*sic*] of all the dear school children I never think of them but with delight I also contemplated the eveng assembly rejoicing at your return. . . . Pray give my very best respects to Miss Young . . . my very kind respects to Mr. St. Croix. I hope I shall never forget with how much kindness as well as patience he instructed an adult. . . . My love to Mrs. Young. I shall write to her. . . .[198]

The difference in her salutation to Mary and Jane Young is interesting, as is her gratitude to St. Croix. Every reference to this young man is favorable, and there are no criticisms.

The Young correspondence contains two intriguing items from George Quinton, another of Young's "satellites"[199] who aspired to be an artist, painter and engraver. The first item read:

To Arthur Young, Esqr.
Sunday March 10th, 1816
                    Resolutions of G. Quinton
Henceforth
I will never be indebted to pawnbrokers—& I will settle my Rent & all other debts *every Saturday night*. And for the preservation of my

---

[196] Add. MSS. 35,133, f. 210. The Round Garden, in front of the house, was presumably a flower garden, and it is a mystery why it should be sowed with either oats or beans, except for the general badness of the grain crops.
[197] Berg Collection. Letter of December 5, 1815.
[198] Add. MSS. 35,133, ff. 342–343.
[199] The term was used by Charlotte Barrett, *ibid.*, f. 323.

eyesight, I will never work at my profession of either drawing painting or engraving later than *eight* oclock at night; and never attempt to do any engraving by candle light.

If I, G. Quinton ever break the above Rules, I forfeit my Word & the Honour of Mr. Young's patronage, Kindness or Friendship from the Date hereof.[200]

In October, 1815, Quinton had requested the loan of Young's miniature by Jaggers, to use for a portrait engraving for the publication *Gallery of Contemporary British Portraits*.[201] Apparently this was the miniature which Young had had painted for Betsy Oakes, for Orbell agreed to loan it provided "you will caution Quinton to take particular care that it is not damaged."[202] The loan was for two months, but more than a year later Quinton subscribed to another resolution:

I DO HEREBY SOLEMNLY RESOLVE, AS IN THE DIVINE PRESENCE, *without any excuse whatever,* IF MR. YOUNG WILL KINDLY EXTEND HIS LIBERALITY, *and allow me a* SIX WEEKS TRIAL *to finish the Portrait* COMPLETELY, and also to send finished proofs of it at the last day of that time: when, if I fail, I will acknowledge that I cannot do it—and will take Mr. Young's advice, and go down to my Father's cottage immediately.

Drawn up this day, Novbr 8, 1816

Geo. Quinton   A. Gardner
In presence of   A. Toovey[203]

One would like to know whether the promise was kept, but the correspondence is silent.

Again in 1816 Young's correspondence with Marianne and her family furnishes a most important source for his activities and thoughts. For January when she was with the Wilberforces at Brighton, three of Young's letters to Marianne and two of hers to him have survived. Surprisingly enough, both Young and Marianne approved of Wilberforce's acceptance of the Prince Regent's invitations to parties at the Brighton Pavilion. On Janu-

---

[200] *Ibid.,* f. 220.

[201] *Ibid.,* ff. 72–73. There are several miniatures of Young, one of which was used as the frontispiece to the *Autobiography,* and another to the original edition of Miss Betham-Edwards's *Travels in France.* The latter is reproduced as fig. 4.

[202] *Ibid.,* f. 87.

[203] *Ibid.,* f. 399.

ary 1 she wrote: "I hope, as I always do, that it may please God, some good may result from his producing himself amongst the poor prince & his gay companions. . . ."[204] Young was also pleased:

I have been delighted to read in the Newspapers, that Mr. W. is so often with the Prince his name seems to be in all the lists: to think that such parties will permit his presence is surely a very good sign; for I have no fear that they can corrupt him; and to discover that a very religious man can be agreeable in such company, may do more for religion than we can dream of.

In the same letter he continued:

When you return to Kensington I must draw on your German talent to read me the Deploma [*sic*] by which I am elected an Honorary member of a Mecklingburg [*sic*] Society. The time was when my vanity was much fed by many of these elections, but I have at present I hope much better things to think of: a far more important election should be near my heart.[205]

Young's protestation was probably sincere enough, but he did not need to tell her about the diploma. On January 9 Marianne not only commented on the narrow views of her brother-in-law, but also mentioned her joy in the friendship between Charlotte and Young:

I was indeed disappointed at dear Charlotte's plan of coming hither being so strangely frustrated. . . . Barrett . . . is very particular about Ladies not travelling alone. I believe he thinks none but actresses should till they are 100 years old. . . . I cannot say how much I rejoice at the affectionate friendship between you & my dear Sister. I am sure it is perfectly mutual: & that she highly values, & greatly enjoys the opportunities of intercourse with you. . . .[206]

What better example could be cited of the changed usage of a word between the nineteenth and twentieth centuries? On January 16 Young reported that he had nearly completed his *Oweniana:*

. . . . I have drawn out from my mass of extracts all that from Owen, which would make an *Oweniana,* almost as bulky as Baxteriana: if

[204] *Ibid.,* ff. 163–164.
[205] Berg Collection. Letter of January 11.
[206] Add. MSS. 35,133, ff. 171–172.

nothing comes of these labours, they afford me an amusement and no less instruction in the reading: but when I attempt to compose any thing of my own I feel the decline of faculty and facility.

He also commented on Marianne's statement that eloquence in the pulpit meant nothing to her: "I like your notions of preachers and preaching; but still I am foolish enough to prefer eloquence in the pulpit, and have not the measure of your patience for humdrum prosers."[207]

From late in January when Marianne returned with the Wilberforces to Kensington Gore in London until late in May when she joined her family at Richmond, there are no letters between Young and Marianne, for presumably they were seeing each other fairly frequently. While still at Brighton she had written: "I hope you will always let me dine with you once a week, my dearest Sir, at least. I look forward to our meeting with great pleasure."[208] On May 1 she wrote Mrs. Piozzi about the excitements of the coming months: "This is a glorious week. I wish you were in Town to go to any of the Bible Meetings, Missionary Meetings, &c with us. Not a day, almost, in the month of May, without some large, well supported Society, for benevolent purposes."[209] Here was the young Evangelical blue stocking's compensation for the loss of balls and parties. Also in May her aunt Mme d' Arblay wrote to Marianne's mother: "Is Marianne of the same vigorous inflexibility to all suitors?" She continued that there was no harm in "being difficult, provided it is not from an ambition of obtaining perfection in everything; and then they only wait to wish vainly the return of lost occasions."[210]

From late May to late July while Marianne was at Richmond three of her letters to Young and two of his to her have survived. On May 29 she asked him to visit her uncle, Captain James Burney, and talk religion to him.[211] On June 4 Young replied that the Captain, his wife, and daughter had taken tea with him the night before:

. . . and I was glad to find that he gave me a patient hearing upon the subject of the Xtian religion, while I urged the importance of it

[207] Berg Collection.
[208] Add. MSS. 35,133, f. 172.
[209] Rylands MSS. 584, f. 140.
[210] Berg Collection. Letter of May 15, 1816.
[211] Add. MSS. 35,133, ff. 320–321.

in consolation for the evils of life and for comfort in the decline of it, but I am sorry to say that I cannot flatter myself with having made that sort of impression which my heart earnestly wished for. . . .

A little later he answered Marianne's request about a religious book for one of her nephews:

As for a good book for young Phillips, my shattered memory is become such a sieve that I know not what to recommend; nor am I quite clear that any such book exists exactly adaptable to such a case; and I am in doubt an entertaining narrative is not more likely to do good, than any formal treatise; such for instance as Newtons narrative, or the Diary of Blackadder, or Doddridge's Colonel Gardner. You speak of this young mans father being my friend, but he was barely an acquaintance: his Mother Mrs Phillips when Susan Burney was a most intimate friend and quite an object of my admiration. . . . And when I very often visited in St. Martin's St. she was so kind to me as to dispose in the pocket of my great coat lying in the passage small bundles of her private Journal, which . . . was most highly interesting, as every word came of course from the heart.

This last passage must have astonished Marianne, if it did not shock her. This letter also provided a list of some of the leading Evangelical churches and chapels in London:

In order to avoid sleeping from dinner till tea I have accompanied Jane . . . to evening service every day when it is performed, especially Mr. Saunders of Broadway Chapel, whose preaching I am glad to say makes much impression on Jane: also Mr. Hyat at Tottenham Court Road and at the Adelphi; and have also been at Stevens Baptist Chapel York St. St. James' square: Dr. Nicols and a Wesley Chapel [not clear] St. Manchester Square. . . . What a blank will Bradfield be after all this. . . .[212]

Young probably left London for Bradfield during the last week in June. Charlotte Barrett and her mother came at the end of the month, her husband and children about ten days later, and Marianne arrived on July 25. There had been a question about the danger of riots in Suffolk, but on June 1 Henry Barrett wrote Young:

[212] Berg Collection. John Newton's autobiography was one of the great personal documents of the Evangelical movement. The only church I can identify was the famous London Tabernacle in Tottenham Court Road, started by George Whitefield, and at this time under John Hyatt.

Mrs. B. is not apt to be disturbed with imaginary fears and we consider ourselves in perfect safety at Bradfield. . . . if you had expressed an opinion that there was real danger, and your own journey had been suspended by the alarm I would likewise have acted in a similar manner. . . .[213]

Young was still spending money on the Barretts' cottage, and instructed the stone mason "to put up a chimney peice [sic] and Bath stove in the Cottage bed chamber, and also bells as desired by Mr. Barrett."[214] Charlotte had also requested Young to procure coals and a nine gallon cask of ale before their arrival.[215] A letter from Marianne to Mrs. Piozzi shows that the Barretts did pay rent for their cottage.[216] Charlotte also wrote quite frankly to her famous aunt d'Arblay about spending the summer at Bradfield:

Mr. Arthur Young is so good in wishing us to be there, and has incurred so much trouble & expence even, in preparing his cottage for us that I cannot but wish to undertake our interment in it, notwithstanding all chances of the Suffolk rioters interrupting our slumbers. Else—I must confess that a very few months of such complete retirement would satiate me & Mr. Barrett who adopted the plan from hopes of economizing finds those views frustrated by the expence of travelling & furnishing, & the uncertainty of our letting this house during our absence.[217]

Marianne and her mother presumably remained at Bradfield until late November and the Barretts until December 5. In August, Marianne wrote Mrs. Piozzi: "I am now endeavouring to contrive an adult school, for the 3 neighbouring Villages, to teach all the old People to read the Bible. . . ." In another letter Marianne said her brother-in-law constantly threatened to go abroad and added quite acidly: "I wish he wd threaten to get some employment in England, to maintain his family."[218] On August 27 Mme d'Arblay wrote to Charlotte Broome about the recent birth of another child to Charlotte Barrett:

What is my new Nephew named? Arthur, I hope, for I like the name,

[213] Add. MSS. 35,133, ff. 327–328.
[214] Berg Collection. Letter of June 4.
[215] Add. MSS. 35,133, ff. 334–335.
[216] Rylands MSS. 584, f. 141.
[217] Berg Collection. Letter of June 15, 1816.
[218] Rylands MSS. 584, ff. 142–143.

& the two principal men of the age who adorn it; both Field Marshals; though one to wield the sword, & the other to guide the Plough,—Duke of Wellington & Arthur Young.

I beg you to remember me affectionately to my old friend Mr. Young & tell him I hope he preserves the kindness for me I never lose for him.[219]

On the other hand, there is a letter (October 11) from the Rev. Charles Burney to Mme d'Arblay in quite a different tone about Young and his influence:

Is Arthur Young out of his senses? There has appeared such a philippick [*sic*] against the Established Church and Churchmen, from the Agricultural Society, signed by him as Secretary, as no enemies to our good old Church ever yet dared promulgate. He is cracked for certain. Oh! my poor dear nieces![220]

After Marianne and her relatives returned to Richmond there was considerable mail between that town and Bradfield for the remainder of 1816. In her very first letter Marianne mentioned the rumor that Richmond might have a new rector of Evangelical outlook.[221] This rumor, which proved to be correct, formed the chief subject for both parties in the correspondence. Conservative circles at Richmond were shocked at the thought of a Methodist or Simeonite; Evangelical circles were jubilant. The candidate proved to be the Rev. Samuel Gandy who remained rector for some time. In one letter Marianne quoted Wilberforce on Mr. Gandy's ". . . superior piety, most amiable temper & manners, & . . . high eminence as a scholar even at Eton."[222] Two of Young's letters to Marianne in December are worthy of short quotations. In the first, Young was commenting on Charlotte Broome's high opinion of him as quoted by Marianne:

Your dear Mama's opinion of me proves that she is able to form most incorrect ideas of the characters of mankind, and is a mortifying

---

[219] Berg Collection. *Cf.* f. 82 of the Burney Scrapbook for a reference to Arthur Charles Barrett, born August 12, 1816, at Bradfield.

[220] *Ibid.* Burney was probably referring to the statement on tithes in the Report of the Board. *Cf. infra*, p. 673. That statement hardly seems violent enough to justify such an outburst, but I have found nothing else to which he could have been referring.

[221] Add. MSS. 35,133, f. 406.

[222] *Ibid.*, ff. 419–420. The Berg Collection contains Marianne's will made in 1830 in which this same Rev. Samuel Gandy was executor and chief beneficiary.

proof that the measure which friends take of each other is liable to
desperate errors. What does God think of me is the question we should
all be ready to ask ourselves, and the result of that omniscient view
will depend entirely on how much of Christ is to be seen in us; and
when we strictly examine our own hearts, *humility* must be the
result. . . . Mr. Gandy your Rector! From Mr. Ws expression this
must be very good news indeed, and seems to give all that was wanting in Richmond, it will truly be food for your Soul, but if he
keeps you from Bradfield I shall growl in no very melodious ac-
cents. . . .

The second letter furnishes an excellent example of the earnest-
ness of his Christian belief and yet his essential moderation in
comparison with the real fanatics:

I trust that your next letter will give an account of your new preacher;
I really think it providential that the mercy of God should send a
real Xtian to preach at such a place as Richmond. . . but I think
the new Rector should move gradually, and lead his congregation
as it were step by step from mere morals to vital Xtianity; if he bursts
upon them at once in the full splendour of gospel truth they will
be disgusted and class him at once with the enthusiastic seceders. . . .[223]

There are also some other sources for Young's religious beliefs and activities in 1816. There was another change of rectors
at Bradfield in 1816, but whether for better or worse is hard
to determine. When the non-resident incumbent, the Rev. John
Morley, exchanged Bradfield for a living elsewhere, the patron,
the Rev. Henry Hasted, appointed in his stead the Rev. Robert
Kedington, whom he described as "a most respectable man &
both willing & able to do all he ought as to Parochial
Duties. . . ."[224] Young's tenant, J. B. Edwards, was much disturbed by the news, for it meant the loss of Morley's curate
named Fenton who was apparently doing an excellent job and of
whom Edwards was very fond:

. . . now our church is so well fill'd it grieves me to know that
Fenton must give up this duty on Trinity Sunday. . . . I heartily
wish as does every Inhabitant of this village Fenton could be con-

---

[223] Berg. Collection. Letters of December 14 and 27.
[224] Add. MSS. 35,133, ff. 273–274. The Kedingtons were an old Suffolk family
and lived quite near Bradfield.

tinued, as he gives such general satisfaction. he has treated us with excellent sermons in your absence.[225]

Although Edwards and his family were devout Christians, he was as much opposed to tithes as any atheist and was worried by rumors about Mr. Kedington in this respect:

Our Rector makes his first appearance in propria persona next Sunday & I hear intends perambulating in a few days I am not informed at present what is the day fixed & also propose having his Tythes valued, as he wishes to do ample justice, he has fixed . . . on his Brother-in-law this will be doing justice with a vengeance. I hear he does not mean to be covetous, but will have all he can get,— these last two years have been highly distressing to the Farmers & my good Sir if you do not stand our Friend we shall be all undone. . . .[226]

The reading of Edward Gibbon's *Miscellaneous Works,* edited by Lord Sheffield, led Young to some pretty gloomy ruminations:

. . . but, alas! the whole volume has not one word of Christianity in it, though many which mark the infidelity of the whole gang. Lord Sheffield never had a grain of religion, and his intimate connections with Gibbon would alone account for it. . . . Nineteen in twenty of the persons mentioned are gone to their eternal state, and of what account is it at present whether they were celebrated authors, splendid orators, great ministers, or successful generals or admirals? . . . . how little are they to be compared at present to the case of a poor Christian whose employment was sweeping the streets! Without doubt the propriety of such observations depends entirely on Christianity being true; but what a dreadful situation is that man in whose safety is attached solely to the falsehood of that religion. The reflection makes my blood run cold. . . . .[227]

In the succession of Young's Evangelical lady friends, 1816 added a rigidly stern Calvinist, Miss Caroline H. Neave, in comparison with whom Marianne Francis appeared liberal. The very long, very illegible, very dull, and very gloomy letters of Miss Neave are almost unbelievably obscurantist. Parts of her letter of May 13, written from the Circus at Bath, are quite typical:

. . . at all times communion with the Children of God is a blissful privilege, but when the soul is suffering drought in a dry & thirsty

[225] *Ibid.,* f. 269.
[226] *Ibid.,* ff. 313–314.
[227] Young, 1898: pp. 469–470.

Land which the River of God does not water nor make glad, it is peculiarly refreshing—Oh, that the Salvation of Israel were come out of Sion!. . .

. . . Miss Johnston . . . is indeed a very interesting monument of divine grace & her acquirements have been so great, her assent so perfect in so short a time, that I am inclined to think with you that the Lord designs to call her home at an early period, or else to give her some active work to do that is awaiting her immediate agency. . . .

I do not desire that your existence sd. be prolonged here below, how glorious will be the moment when you will look back to your present affliction & all else that has befallen you as less than nothing, when praise & praise only will fill your heart & tune your voice. . . .[228]

Young's speech at the Suffolk Bible Society at Bury on October 4 was reported at some length in the *Bury and Norwich Gazette:*

He said he ought to apologize . . . for consenting to take any part in the business of the day at his advanced age, and labouring under the infirmity of blindness; but he felt the value of the Bible so deeply in his own heart, that he thought it would have been wrong to negative the proposal. . . . he rose for the sole purpose of declaring that the Bible was the only comfort of his life, and therefore he was anxious to impart it to all. He thought this Institution the noblest that was ever formed on the earth. . . . In evincing his decided attachment to the Scriptures, he showed that he agreed with the great writer who described the Bible as "having truth without mixture of error for its matter, God for its author, and salvation for its end."[229]

The positive, active side of the Evangelical conscience is shown in Young's decision to stay at Bradfield until February, 1817, instead of returning to London in the late autumn of 1816:

. . . nor shall I be in London before the 2nd Feby, for it certainly is my duty to remain where I can do more good, but at London I can do none. Our spinning wheels at Bradfield are all most busily at work as I have engaged to double all the earnings for the month of Decr, and if it comes to no more than I expect I will do the same through the other dead months of Jany & Feby but in March their field work begins. Mrs. Arthur has also promised to give a Chaldron

[228] Add. MSS. 35,133, ff. 304–307. For other letters from Miss Neave, *cf. ibid.*, ff. 308–309, 352–353. The Miss Johnston mentioned was Helen Johnston, whose father gave Young's *Travels in France* to Napoleon. *Cf. infra,* p. 676.

[229] *Bury and Norwich Gazette,* October 16, 1816.

of coals ye month for 3 months so that upon the whole our poor neighbours will be well assisted through the worst period of the year. . . .[230]

As noted above, agrarian distress became acute late in 1815 and remained so through much of 1816. Late in the year Sir John Sinclair wrote Young: "You never expected to see the English farmer reduced to such a state. . . ."[231] And in November, Lord Sheffield wrote: "There never was such a miserable wheat season—surely we must have a famine next year."[232] Several letters from Young's tenants in the first half of the year depicted conditions in Suffolk and at Bradfield. On January 11 J. B. Edwards wrote:

It would certainly have been more agreeable to me to have known for a certainty whether I am or not to continue longer at the Lodge than next Michaelmas, having many offers made me of situations, much more easy in point of Rent. . . . I will frankly acknowledge I would not give the Rent I have offered but that my good old wife & myself are very partial to you & yours & the situation a pleasant one. The Village small—I farm the Land precisely as if I was proprietor instead of Tenant. . . .

I assure you I have not received a single Guinea of Rent this Christmas, nor do I know any acquaintance of mine who has rec'd their Rents, you are the only exception, thanks to the little Village of Bradfield. I believe without Jokeing [*sic*] it will hold as long as most, but if these times continue that must go likewise.[233]

Somewhat later the less literate William Greene wrote:

. . . there is now in Bury Gaol upwards of ninety Persons chiefly— Farmers and Trades-people for Debt and Labourers are Runing [*sic*] about for work . . . the Farmers have no money to pay them & in some Parishes they give their Labourers only nine pence a Day. . . .[234]

On May 13 Edwards described the riots in Suffolk and his efforts to prevent their spread:

<hr />

[230] Berg Collection. Letter of November 4, 1816.
[231] Add. MSS. 35,133, f. 407.
[232] *Ibid.*, f. 402. There are fairly complete accounts of the agrarian crisis of 1816 in Ernle, 1927: pp. 322–324, and Smart, 1910: 1: pp. 489–490, 512–538.
[233] Add. MSS. 35,133, ff. 173–174. It should be noted that although Edwards was Young's tenant, he was a proprietor to others, or else sublet part of his land.
[234] *Ibid.*, f. 213.

I am this moment returned from calling upon Mr. Collville the result of our Conversation is merely this he recommends having two of the special constables who were sworn last week to watch by Turns nightly, I want words to express my feelings of a view of ruins of a Fire which took place at Lawshall yesterday afternoon. . . . I intend calling upon Mr. Oakes. . . . I think I shall sleep the better for knowing we now have watches we can depend upon. . . .

He wrote again a week later:

Methinks I hear you say how are my Premises guarded, Why Palfry goes to bed very late, & his wife told me yesterday he generally gets up 2 or 3 times the little while he is a Bed, & I am up soon after 3 of clock so between the one & the other, we may be said to watch the whole time, now the nights are so short. . . .[235]

Young's entry in the *Autobiography for* February 17 shows that he spurred on the Board of Agriculture to action:

The Board met for the first time last Tuesday, but had no business whatever before them. I suggested the propriety of sending a circular letter throughout the kingdom, in order to ascertain by facts the real state of the farming world. They approved the proposal, observing that not a moment should be lost, and I retired in order to draw out a letter with *Queries*. This they examined and altered to their mind; it was immediately despatched to the printer, and all the rest of the week has been employed in drawing up lists of persons . . . to whom these letters have been addressed. . . .

The replies have just begun to come in . . . the probability is that much important information will be gained, and a basis laid for a very interesting publication, and I suspect that it will disclose so lamentable a state of distress, that it may prove dangerous, or, at least, questionable to make it public. What are we to think of the infatuation of Government in laying on a property tax at such a moment, rather than borrow a few millions to avoid the necessity, one of the great evils resulting from our Government being in all money matters little better than a Committee of the Bank?

Answers to the circular letter of the Board . . . flowed in rapidly till about April 10, and they describe such a state of agricultural misery and ruin as to be almost inconceivable to those who do not connect such a defect with the utter want of circulating medium; the ruin of the country banks, and the great want of confidence in

[235] *Ibid.*, ff. 302–303, 314.

those that remain, with an issue of Bank of England notes utterly insufficient to fill up the vacuity thus occasioned, has made the want of money so great as to cripple every species of demand.

It is difficult to pronounce what the consequence of the present ruined state of agriculture will prove, but I must confess that I dread a scarcity, which must have dreadful effects, coming at a period when such multitudes are almost starving for want of employment, even with such cheap bread. What must be their situation should it be dear? To my astonishment, Government seems utterly insensible of the danger, and has not taken one single step to prevent it, or to meet it should it come.[236]

Among the Board's queries were, how many farms were unoccupied, how many had been re let at lower rents, how many tenants had quit, just how bad was the farmers' distress and was it worse on arable or grass farms, to what extent was the diminished circulation of paper money the cause, what was the state of the laboring poor, how had poor rates been affected, and what remedies were proposed. In all 326 letters were received from correspondents, a very substantial basis upon which to draw conclusions.[237] As soon as the returns had come in, Young drew up a memorandum, presumably for some member of the government:

I should be greatly wanting in executing the duty of the office I hold if I did not take the earliest possible moment in communicating to you the general result of the information rec'd by the Board, in reply to a Circular Letter. . . .

It is evident from the letters recd. that many farms in various districts are abandoned; some run to waste, and many thrown on the hands of the landlords, who are very ill able to stock and cultivate them. . . . It is simply impossible to cultivate any indifferent soil at present, except by tenants who have very ample capitals; nor is it to be expected that any poor or indifferent land will be sown with wheat next Michs., except by men whose means are extremely ample; the consequence must be a scarcity in the harvest of 1817; and if matters go on as at present this scarcity must be truly alarming.

The state of the labouring poor is generally described under [not

[236] Young, 1898: pp. 465–466.
[237] *Annual Register, 1816*, pp. 459–469. The Board's letter was dated February 13, and was signed by J. Fane, vice president, presumably in Lord Hardwicke's absence.

clear] circumstances of deficient employment, so that numbers of healthy young men are either idle, or employed by overseers on roads &c to keep them from the parish; but this is a burthen on the rates at a time when many farmers are unable to pay them. . . .[238]

Eventually the report was published as a pamphlet under the title, *Agricultural State of the Kingdom*. The first edition of 5,000 copies, selling at 15 shillings, was nearly exhausted by the end of September and a second edition was in preparation, while a pirated edition was being advertised at 9 shillings. In the meantime much of the material had also appeared in the press. On September 30 Under-Secretary Cragg wrote Young:

Under all circumstances I am not sorry that the work has made its appearance *now,* as considerate people will feel that the Board were not inattentive to the State of the Country, and I have no doubt that the Institution will become a favourite with the opposition.[239]

Somewhat later Sir John Sinclair wrote: "What a noise has been made about the answers to the Board's Queries. I think however, that it will do the Board more good than harm though everybody almost abuses us at present."[240]

The report summarized the replies to each query. The vigor of some of the language and the views expressed sound very much like Young:

The distress of the present period will scarcely permit of a doubt . . . that the mere occupation of farms, free of all rent, is considered as a benefit. . . .

Bankruptcies, seizures, executions, imprisonments, and farmers become parish paupers, are particularly mentioned by many of the correspondents. . . .

Much mischief is noted from the failure of country banks; many of the correspondents are of opinion, that agriculture suffers much from want of a larger and safer circulation; and not a few complain heavily of the deficiency of paper being so extreme. . . .[241]

The report contained long quotations from the Earl of Winchilsea, describing his famous system of giving cows with a cot-

---

[238] Add. MSS. 35,133, f. 239.
[239] *Ibid.,* ff. 384–385. This letter included information about the various editions.
[240] *Ibid.,* f. 407.
[241] *Annual Register, 1816,* pp. 460–461.

tage. Other excerpts showed that where this system prevailed the poor were much better off and hence the poor rates much less. The burden of poor rates was brought out impressively:

. . . as in a variety of instances, the farmers who lately paid to these rates, have been obliged to give up their farms, and are actually become paupers themselves . . . many apprehensions are expressed of the system being permitted to continue, and increase till it will absorb, in union with tithes, the whole rental of the kingdom, leaving nothing more to the landlords of it, than that of acting as trustees and managers for the benefit of others.

He was not very precise on tithes, but said enough to show how severe a burden they were:

. . . the general complaint against the weight of tithe, would open too wide a field to permit more than a solitary remark: it appears from the correspondence, that 10s. in the pound rent is taken as a commutation in Dorsetshire; and 9s. an acre for grass-land is paid in Berkshire.[242]

The whole business shows Young active and influential in Board affairs. He had suggested the circular when the president was absent and no one had provided an agenda. He drew up the queries apparently within an hour or so. His office had acted with great dispatch. The Board meeting was held on February 12, the letter was dated February 13, it was printed immediately so that many were mailed by February 16 and the remainder on February 18. The number sent out had been large and the returns were impressive. As noted above, Young almost certainly wrote the report which was out in time to do some good, had the government taken prompt action.

Young's strong support for the Earl of Winchilsea's experiments in granting a cow with each cottage has been mentioned above. The *Bury and Norwich Gazette* contained an excerpt from Young's letter of September 2 on the same subject:

In the counties of Rutland and Lincoln, the practice is to attach land to cottages, sufficient to support that number of cows which the cottager is able to purchase, they are tenants to the chief landlords, and not sub-tenants to farmers, yet these latter are very generally steady friends to the system; well they may be so, for the poor rates

[242] *Ibid.*, pp. 467–468.

are next to nothing, when compared with such as are found in parishes wherein this admirable system is not established.[243]

Many of Young's letters to Marianne in 1816 commented on financial conditions and policies. In December he wrote about the need for increased paper money in almost Radical tones:

I think it more likely that bread will be dearer than cheaper, and the season for putting in the new crop has been so extremely bad that I am not without fears for the next year also: Government is fast asleep: read a new pamphlet called *"the Remedy"* which well explains the bearing of circulation upon prices, and shows how beneficial to the poor high prices are when proceeding from plenty of circulation, that is of paper, and not from a deficiency in the supply; bank notes should be doubled in order to raise prices and enable the farmers to set the poor to work; the good or bad state of the poor, is a mere question of employment, and not at all the price of bread: fix this well in your mind, for it is the grand question of Political economy, as applicable to the present period: we want paper, which would infallibly [*sic*] give employment, but our beastly ministers plainly know nothing of the matter by adjourning Parliament to so late a day, as to postpone most dangerously the application of any medicine to the present evils.[244]

As usual Young had several letters in 1816 from and about Sir John Sinclair. In September he was using the Board's facilities as his office while working on a new book, a code of agriculture. In October he sent excerpts from this new manuscript to Young: "Pray tell me candidly what you think of the Plan, and the execution, & point out, without reserve, what mistakes I have fallen into. . . ." Later in the year he wrote Young again:

I have sketched out some letters for the newspapers, explaining the causes of our distresses, and their remedies.—But what can I do with them? The landed interest will give me no support,—The ministers will consider it as an act of hostility to them—and I shall be abused by Cobbett, and all the hireling newspapers.

I wish much however, that you would consider them carefully, and

---

[243] *Bury and Norwich Gazette,* October 9, 1816.
[244] Berg Collection. Letter of December 14. On May 26 T. R. Malthus warned Young, apparently in vain, of the danger of too great an expansion of paper currency: "Too great stimulus by means of a paper circulation would probably cause a glut and defeat itself." *Cf.* Add. MSS. 35,133, ff. 318–319.

would send me your real opinion of them and of the calculations in particular. . . .[245]

Far more of a nuisance than Sinclair was the ever needy James Powell who wrote another begging letter early in December. This time his situation was really desperate:

In short so severely are we pressed, that we shall be obliged at Christmas, to turn into the Highway. . . . we shall be happy if you can get us an appointment to *sweep chimneys* or to *black shoes,* provided we can but pass our nights *without apprehension.*

There were extravagant praises for his eldest son, who could not get a job, but who wrote poetry samples of which were enclosed—mawkish, dull, and stilted to an extreme. Very interesting was one sentence: "Have the goodness to put your letter into the Post Office *anywhere* but at Bury."[246] Apparently there were people in Bury with whom he had no wish to renew acquaintance. The appeal to Young, however, was successful and apparently both Mary and Jane also contributed.

There were several incidents in 1816 which must have made the old man very happy. Shortly after his return to Bradfield in the summer there was a most cordial invitation to dinner from Orbell Ray Oakes:

I have just seen Peggy Metcalfe & she has promised to dine with me tomorrow & to make her, as well as ourselves truly happy, we hope you & Miss Young will join our *small* party. . . . I trust you will this time not disappoint us—If you know half the pleasure you will give us I think you would acquiesce in our humble petition.[247]

Since his health was good that summer, it is to be hoped that he did "acquiesce." There would have been good talk with Peggy Metcalfe and almost certainly reminiscences of happier times when Betsy was there to enliven the party and to add her gracious spirit.

When Sir John Sinclair returned from a short visit to France he brought a pretty story from Baron J. J. de Silvestre, secretary to the Royal Society of Agriculture:

He was in prison and brought to trial and told that his life should

[245] Add. MSS. 35,133, ff. 394, 407.
[246] *Ibid.,* ff. 411–412.
[247] *Ibid.,* ff. 354–355.

be saved if he could show that he had ever done anything really useful to the Republic. He replied that he had unquestionably done much good, for Arthur Young's "Travels through France" contained much highly important information, and in order to spread it through the Republic in a cheap form, "I published a useful abridgement," he said, "which has been much read, and has had important effects. I was pardoned and set at liberty," and then turning to Sir John, he said, "Tell your friend, Mr. Young, that he was thus the means of saving my life."[248]

When an Evangelical lady friend, Helen Johnston, and her father visited Napoleon on Elba, they heard that he desired to read Young's *Travels in France,* but had come to Elba without them. Fortunately Mr. Johnston had a copy with him "and presented them to the Emperor, who expressed much pleasure at receiving them, and Mr. Johnston afterwards heard that he had read them eagerly and with much approbation."[249] When Charlotte Barrett heard this story she wrote Young: "I wish poor Bonaparte would read Baxteriana as well as your Tour. . . ."[250] The thought of Bonaparte reading *Baxteriana* is indeed amusing!

Finally there was a letter from John Powell, Fellow of Trinity College, Cambridge, enclosing a printed syllabus for a proposed course of lectures on agriculture at Cambridge. The tone of the letter was that of a pupil to his master:

From the manner in which you have expressed yourself on this subject in the eleventh Vol. of your Annals, I doubt not but that such an attempt will meet your approbation and support; and I have taken the liberty of sending you herewith a prospectus of the plan I mean to pursue. . . . Should any thing occur to you on this subject, any hint or communication will be esteemed a particular favor. . . .[251]

[248] Young, 1898: pp. 464–465.
[249] *Ibid.,* pp. 467–468. In the *Autobiography* the name is spelled without a T, but a letter from the lady, Add. MSS. 25,133, f. 325, shows that the name was Johnston.
[250] Add. MSS. 35,133, f. 324.
[251] *Ibid.,* f. 165.

XII. Last Years: 1817-1820

Arthur Young's way of life during his last years did not differ sharply from the immediately preceding years. He remained in fairly good health, he divided his year pretty equally between Bradfield and London, he continued his intimacy with Marianne Francis and her family, and Evangelical religion occupied his chief attention. The material for these last years is, however, much less complete than for the earlier period. The *Autobiography* has two lines for 1817, two pages for 1818, and then stops completely. The British Museum manuscript letters addressed to Young are also very few in 1817 and 1818, although somewhat more numerous for 1819. Worst of all for his biographer the letters from Marianne Francis and her sister, Charlotte Barrett, have completely disappeared. Fortunately the Burney Papers in the Berg Collection in the New York Public Library contain many of Young's letters to Marianne for these years. The Board of Agriculture's records are also somewhat more complete than for the earlier period.

"The death of the Princess Charlotte this year created the greatest sensation ever known."[1] This inconsequential note is the *Autobiography's* only entry for 1817. Fortunately Young's fourteen letters to Marianne make it possible to follow his movements fairly clearly. He stayed at Bradfield until February 1. From that date to July 8 he was in London, then back at Bradfield for the balance of the year. During most of the year Young's health was excellent. Early in September he wrote Marianne that he was "in perfect health."[2]

Young's letters to Marianne contain only two references to the Reverend Arthur Young and his affairs. On June 3 he wrote:

I had lately a letter from Arthur, written in a more quiet, and tranquil, state of mind than common, with him: Mrs. Holderness, and her children, were with him, waiting the return of her vagabond husband; it was dated the first week of March.

[1] Young, 1898: p. 470.
[2] Berg Collection. Letter of September 5.

677

On July 14 Young wrote again:

Mr. Louis Way of Stanstead Park called on me in Town to make enquiries about Arthurs Karagos Estate, as he wants such a place for an establishment of converted Jews, it being sufficiently near to the Jewish districts in Poland.[3]

His letters showed that Jane Young was giving lessons in French and spent many evenings visiting Evangelical chapels.[4] In the spring she visited her father and sister, but accompanied her father-in-law to Bradfield for the summer.[5] During the summer and autumn Young was much worried over the bad health of his secretary, St. Croix. In July, Young's friends said he looked "like a ghost," and on September 29 Young wrote Marianne: "I am very sorry to tell you that Mr. St. Croix continues almost as bad as ever, so that we have all apprehensions of the final result."[6]

Although the Board of Agriculture met from February to June, 1817, its activities seem inconsequential. Its members were worried, like many Englishmen, by postwar unemployment and unrest. On February 25 Young proposed several premiums to encourage the employment of the agricultural poor. On March 4 the Board voted the Gold Medal, or £100, for the best essay "on the means of employing the industrious & unoccupied Poor," and also a first prize of the Gold Medal, or £50, and a second prize of £25 "To the person, who shall, during the present spring of 1817, cause to be dug by hand, for the production of any Crop of Corn or Pulse, Turnips or Cabbages, the greatest number of acres, not less than 10, never dug before."[7] It is hard to imagine Arthur Young supporting a premium for opening new lands by the spade. This was hardly the new agriculture! In June the treasurer of the Board was asked to invest some of its accumulated funds in exchequer bills.[8] In other words the Board had become so inactive that it could find no use for its annual appropriation.

---

[3] *Ibid.* Lewis Way actually visited the Crimea in 1818, but nothing came of his plans. For details *cf.* Bagshawe Muniments B/22/6/6; Gazley, 1956: pp. 399–400.

[4] Berg Collection. Letters of January 23, February 19.

[5] *Ibid.* Letters of May 14, June 19. For further details on Jane *cf. ibid.,* letters of March 15, August 13, September 5.

[6] *Ibid.*

[7] Minute Books, 7, ff. 13–14, 16.

[8] Letter Book, 2, f. 279.

Sir John Sinclair would never have permitted funds to accumulate unspent!

Young's correspondence in 1817 reveals a great concern with public problems. He criticized "the most licentious press that ever disgraced a Country" and bitterly attacked Cobbett who "sells fifty thousand two-penny numbers of rebellion and treason every week."[9] He was worried about low corn prices which meant reduced purchasing power of "the farmers and all dependent on them" and hence a lessened demand for cloth and hardware. He showed that Malthus had strongly influenced his thinking when he wrote that "no maxim in political economy can stand on a more secure basis than the certainty that population has always a decisive tendency to encrease too fast" and continued, "we have been long apt to consider building cottages and small houses a public benefit, but in general those who pull them down act far more advantageously for the general happiness of the community."[10] This was a striking reversal of his earlier beliefs, for Young had long advocated cheap cottages for the poor.

Young was still a strong advocate of a cheap paper currency. His letter of January 13 to Marianne deplored "the infatuation of Government not supplying the Kingdom with paper currency," and that of March 15 attacked "the neglect of Government who ought to force an extraordinary issue of paper."[11] The government's new silver and gold coinage was unnecessary, expensive, and deflationary. His interest in a cheap paper currency led Young to open a correspondence with Thomas Attwood of Birmingham, perhaps the leading exponent of such ideas at this time. In the following year Attwood used Young's letters as a point of departure in the title of his book, *Observations on Currency, Population, and Pauperism, in Two Letters to Arthur Young.*[12]

During 1817 Young was corresponding with the Archduke John of Austria who sent him Volume I of the transactions of the Imperial Royal Society at Vienna.[13] A note to Marianne reveals that he also received

[9] Berg Collection. Letter of February 19.
[10] Wakefield, 1885: pp. 64–65.
[11] Berg Collection.
[12] Thomas Attwood (1783–1856) became progressively more Radical, was prominent in Reform Bill agitation and even later as a Chartist.
[13] Minute Books, 7, ff. 13–14.

. . . a *Diploma* constituting me an Honorary Member: the whole in German, and *you must come and read it in English*. My Worldly Honours increase, while I am on the verge of another World. Last year the same from Brussels and Strelitz.[14]

On November 30 the archduke requested information from Young about a kind of potato which could be matured in the very short growing season in Alpine Austria.[15]

On July 28, 1817, Young wrote to the *Bury and Norwich Gazette* strongly protesting two statements by his great agricultural contemporary, T. W. Coke of Holkham in Norfolk. In the first place Coke had stated that Norfolk husbandry had been "execrable" when he had taken over his Holkham estate. To controvert this assertion, Young referred to the *Farmer's Tour through the East of England:*

Now, Sir, as to the first of these assertions, it contradicts so expressly the description which I gave of the Norfolk Husbandry in my Eastern Tour printed in the year 1771, in which I entered largely on the admirable system then common in the north-western district of that County, that it would be doing injustice to the memory of those excellent cultivators, who were then famous for their management, not to enter a negative to any reflections to their reputations. . . .

Then citing examples of excellent Norfolk husbandry from the Eastern Tour he concluded, "So much for one assertion." Coke's second statement to which Young objected was "that HE established the Wool Fair at Thetford." This assertion he disproved by an article in the *Annals*[16] and claimed the honor himself by pointing to "the minutes of Resolutions of a Meeting assembled by myself at the Angel, at Bury, for the express purpose of proposing this Wool Fair. They did me the honour of placing me in the chair, and came to the Resolutions proposed. . . ." At the end of his letter Young tried to remove any sting by complimenting Coke:

. . . this gentleman is of too estimable a character for me to desire to have any controversy with him; and I shall conclude this paper with an assurance, that I should not have written one line of it, but from a sense of justice due to many excellent characters long

[14] Berg Collection. Dated June, 1817.
[15] Add. MSS. 35,133, ff. 430–431.
[16] *Annals* 18: p. 613.

since departed. No man holds Mr. Coke in higher estimation than myself. . . .[17]

The documentation for Young's religious views and activities is found chiefly in his numerous letters to Marianne, whom he saw very little in 1817. When he came to London in February, Marianne was at Richmond, but she may have been in London at Kensington in late March and April. From late April to late May she was at Bath, helping to console her mother after the death of her younger brother, Dolph. The Barretts were at Bradfield from late June or early July until early December. Marianne spent the summer at Brighton with her mother, but she probably passed most of October and November at Bradfield.

Young's first letter to Marianne on January 23, 1817, contains two interesting paragraphs from a religious point of view:

. . . but I pay very little attention to politics, being convinced more and more every day I live that the next world ought to engross my attention, as I ought to consider my life as not worth at any time one months purchase. Prepare to meet thy God should be written in every room for those that have eyes and sounded for those who have none. . . .

There was a row at the Methodists meeting, a disturbance by 3 fellows who went merely to groan. . . . I must confess I do not like this groaning worship, as it carried some little appearance of hypocracy [*sic*]: . . . I am sorry that I must confess that I have some dislike to these seceders from our Church, and I had rather hear a dull Sermon from our dull Kedington than the gabble of the individuals composing a congregation.

In February, Young was upset by the rumor that Marianne might not be at Bradfield that coming summer:

Your brother Clem: called on me this morning . . . and tells me that your Mama has thoughts of going into Devonshire on account of a cough which is troublesome to Dolph: this News was near giving me a fit of the cholic. . . . I truly hope that this scheme will be in the Moon, for at my advanced age, I can expect to remain in this region of anxiety but a very short time: and to have a summer past, without seeing you will sit very hard indeed upon me.

Again his letter of March 15 was especially interesting religiously:

[17] Add. MSS. 35,133, f. 429. This is a clipping from the newspaper.

I have also to thank you for recommending Chalmer's Astronomical Sermons, which I bought, immediately; and am not a little pleased with them; having read the three first: I should have read the whole, but Jane will not suffer any one to read them to me, except herself: the first discourse is a magnificent one indeed. To convert the Sciences directly into religious instruction is the very best of all applications. . . . I . . . think there is an uncommon sweetness in Stuarts sermons, but he never gets into the heart, in order to cut and slash. . . . Stuart preaches as if he thought his whole congregation real Christians; but as that can . . . hardly be the case, I think he should introduce some flogging. . . .[18]

Young's letter of May 3 to Marianne was an attempt to console her and her mother after he heard of Dolph's death:

Most melancholy indeed is the affair . . . and I most cordially pity your dear Mama; but I hope, and trust, that she will take comfort in the circumstance of her deceased child having clearly manifested, by the desire of prayer, a state of mind fitting him for the World to which he is gone; and if this be impressed on your Mama's mind . . . she will not be ready to indulge in excessive grief, for it must be considered, as an aukward [sic] circumstance to lament that the person we most love is become an Angel in Heaven: in such a case, we are too selfish in preferring our own happiness to that of the friend we have lost.

His next letter on May 14, in response to Marianne's query about heaven, is a strange combination of orthodox beliefs with very modern views of the age of the universe, probably drawn from the work of James Hutton:

. . . that we shall know each other in the other world cannot be doubted from texts too numerous to quote. . . . If anything could be supposed to dash the felicity of heaven, it would be our missing those whom we loved on earth; but even this is provided against by . . . that text, *"in my Fathers house are many mansions:* and it does not follow that because we were acquainted on earth, we must be in the same mansion above. . . . when we consider the infinity of worlds that have existed so many thousands and millions of years and inhabited without doubt by myriads of accountable beings, to suppose that all are collected hereafter in one locality is not at all probable: the surface of this Globe, under a new system, may be

[18] Berg Collection. The reference in the last letter was to Dr. Thomas Chalmers, the great Scottish divine, who was just reaching the height of his influence.

a heaven, while the centre of it, may be a hell; if the presence of the Almighty be manifest, that must be a heaven; and this presence may thus be manifest in millions of heavens. . . . If your dear Mama is satisfied, as she ought to be, that her Son is happy, she should be quite content, and never for a single moment wish him back again.[19]

Young's letter of June 19 was a strange combination of his old vivid, amusing, descriptive style with mawkishly sentimental religious comments;

Yesterday, from all accounts the world was Bridge-mad; parading over the new Bridge to see the Duke of Wellington and the Prince Regent; and the Thames all alive with boats and barges: we hear of no accidents, but that of an elegantly dressed young Lady, with all her feathers slipping into the river, from the edge of a coal Barge; but it was only a cold-*Bath* in a hot day, as they got her out directly. The new-Bridge was a crowded promenade 'till two in the morning! I presume the Duke of Wellingtons self-love, vanity, and feelings, common to the Human breast, receive no slight gratification. . . . whether he has Christian feelings on these occasions, I know not:—if he is without them; he is much more to be pitied than envied! and a poor pious labourer whose pleasure is reading the word of God is a character, in my opinion, much more to be envied. . . . The precept of our Lord, to love not the world; nor the things of the World; should occur to Heroes, and Conquerors, in the moments of their trial. Many folks came to Town from one to two-hundred miles to be present at this navey [*sic*] show; and, *you* did *not* think it worth coming *ten* miles to Beds ready for you:—this, surely, is to the credit, of your Philosophical indifference! . . . . I am at present in Johnathan [*sic*] Edwards on justification; the doctrine of which is good and sound, but, like all his other works written in a very disagreeable style. But, as the Edinburgh Review is just come in, I suppose I shall find some amusement. . . .[20]

Young's letters in July, August, and September contain accounts of various goings-on at Bradfield that summer. On July 14 he wrote:

I have been incessantly employed unpacking and arranging 19 packages that came by the waggon, weighing above ½ a ton; I never move without resolving not to be plagued in future with so much baggage. . . . Last night we had our usual evening assembly, the Hall

[19] *Ibid.*
[20] *Ibid.*

and staircase as full as they could cram; . . . I described to them the wretched state of the poor in a large part of Europe and called on them to be greatful [*sic*] to the Almighty for the situation his goodness had placed them in. . . . The pond is clearing of weeds to make way for the boat in which Charlotte will exercise her rowing agility.

On August 13 he described a visit from William Wilberforce, who with his party "took their luncheon here last week, staying 3 hours, walking the round, and with Jane singing a Hymn in the Root House this is really a great event in the Annals of Bradfield." He also reported that Charlotte read the newspapers to him, but was "not so great a navigator as formerly, the weather however must not be forgotten. which has been rain, rain, every day. . . ." He was short of readers, probably partly because of St. Croix's illness, and had "only Bet Sturgeon till eight." He was reading some lady's account of Tripoli, part of which was interesting, "but I am quite tir.d with her continued description of dresses."[21]

On September 4 Charlotte Barrett apparently entertained a considerable number of the large Burney clan:

What do you think of a ball at Bradfield Cottage, they were last night with three Phillips, & Mr. Edwards, & Mary hopping, skipping, twisting, & reeling to Miss Burneys musick, as merrily as you could at Brighton itself. . . . Your sister expects Captn Burney every day, & I have ordered the flag to be hoisted on my ship the Nelson, to receive him. Mrs. Burney & Sally came on Saturday, and I have heard her perform once on the piano.[22]

Young gives no sign of disapproval of the party, but it is surprising to see such a strong Evangelical accepting such frivolity.

While at Bradfield, Marianne wrote to Mrs. Piozzi, describing a Bible Society meeting:

We had the anniversary of a Bible Socy held here, at Bradfield Hall, last Tuesday, 5 Clergymen attended, & spoke; & good Mr. Arth. Young, who is almost 80 years, & perfectly [word omitted], but his heart warm in the cause, took the chair. Farmers, poor people & rich poured

---

[21] *Ibid.* It was Agatha Sturgeon who later became the unmarried mother of the Rev. Arthur Young's two sons. The author on Tripoli was a "Miss Tully."

[22] *Ibid.* Letter of September 5. The Phillips children were those of Susan Burney, Mary was probably Young's daughter, now fifty-one years old, and Mr. Edwards his tenant, fifty-nine years old. The "Nelson" was, of course, his rowboat.

in, from all quarters. Nearly £ 50 were collected to supply the neighbouring poor with Bibles. . . .[23]

In 1817 *Oweniana* was published, a companion volume to *Baxteriana*, consisting of excerpts from the works of the eminent seventeenth-century Puritan divine John Owen (1616–1683). As in *Baxteriana*, the only interest of *Oweniana* to Young's biographer lies in the preface. The most interesting passage described his earlier religious apathy.

It is astonishing to reflect, in many former passages of my life, what system of error and deception it could be, that conducted me for many years together through a steady march of what I now consider as depravity, and a careless forgetfulness of God, of religion, and of a future life, yet without any feelings of remorse or apprehension. Such a condition cannot at present be clearly analyzed, and the less so, as, during my residence at Bradfield, I was tolerably regular in my attendance at Church, and never wholly absented from the Sacrament. I cannot but suppose that I acted thus, by way of not setting a bad example to the parish in what related to public worship. . . .

I have little doubt but that a species of infidelity was at the bottom of my conduct; I never formally rejected or disbelieved the authority of Scripture, but I did not value it sufficiently to examine the question of authenticity. . . .[24]

In 1818 Young was in London from early January until late June, and the rest of the year at Bradfield.[25] As in 1817, his letters to Marianne, and to Charlotte Barrett, constitute the chief sources for the year. He wrote seven to Marianne between February 24 and May 14 addressed to Richmond, and five late in the year addressed to Brighton. There is also some material in the manuscript records of the Board of Agriculture.

Young's health remained excellent. His letter of May 4 described a regimen which would have worn out most younger men:

I . . . am often in doubt whether I ought not to quit every thing like business and retire to solitary meditation; but by rising at 5

---

[23] Rylands MSS. 584 f. 151.

[24] Young, 1817: pp. iv-vi.

[25] Berg Collection. Letter of March 24 states that he reached London on January 2. Minute Books, **9**, f. 132, gives June 9 as date of last meeting of Board.

in the morning to read the word of God, and Mr. Wilsons Sermons,
I take care to secure no trifling portion of every day to something
better than business: and while through the Almighty's mercy to
me, I enjoy great health and strength, query whether I am not
bound to employ it in those duties which attach to the situation in
which his goodness has placed me? What say you to this? . . . but
I have plenty of reading, rarely less than 8 hours a day, not often
interrupted. . . .

Again he wrote on November 27:

I continue in good health, and free from all bodily pain, what a
mercy is that at 77, gratitude ought to vibrate from every fibre of
my heart. . . . You are very right in your conjecture that at 7 o'clock
I am fixed to the Sopha, for all my motions and quiescense are
pretty nearly as regular as the clock; I am soon asleep on the Sopha,
but it is not so in bed, and while I lay awake it demands some effort
to keep my mind pouring [*sic*] over the right subjects.[26]

The only news from the Reverend Arthur Young came in his
long letter of May 31 which described a recent interview with
Emperor Alexander I who in touring the Crimea had passed
very near Karagoss:

Nothing but the honour of conversing with so gracious and benevolent a Monarch induced me to venture on such a step: & indeed I
was most graciously received; and His Majesty had the goodness
to say, he was sorry he could not himself visit Karagoss. The interview passed within sight of Karagoss at the nearest post station
where the horses were changed . . . H. M. accosted me in English,
asked several questions about the Estate, how long I had lived there,
what English were with me, how I liked the country: whether you was
still alive; what improvements I had made there, whether I found
the climate suitable to English agriculture. . . . These questions I
readily answered, & apparently to the satisfaction of the Emperor. We
stood together, surrounded by a croud [*sic*] of people, for ten minutes,
in conversation till the horses being changed, H. M. drove off, & I
accompanyed H. M. by the side of his carriage till we came to the
bounds of the property still conversing all the time in the most affable
& condescending manner.[27]

Jane Young was with her father-in-law only during the spring.

[26] Berg Collection.
[27] Bagshawe Muniments B/22/6/6. Gazley, 1956: pp. 399–401 quotes nearly
all of this long letter.

On May 4 Young wrote that Jane would be either in France or
Yorkshire that summer and urged Marianne to spend her summer
at Bradfield, ". . . for the absence of Jane from Bradfield will
occasion a void that should be filled, and I know not by whom
if you do not perform that friendly office." In the late autumn
Jane was busy helping to settle her father's estate. Fearful that
the Rev. Arthur might somehow get the money, her father had
left Jane very substantial legacies "tied neck & heels into her
own hands."[28] Thus Jane was now well-to-do and certainly independent.

Arthur Young apparently spent a busy spring as secretary of
the Board of Agriculture. In February the Board voted (1) to
offer new premiums "and the Secretary was requested to attach to
them proper conditions," and (2) to publish a new volume of
Communications, "and that the Papers in the possession of the
Board be examined by the Secretary, and that he be requested to
report to the Board what Papers merit publication, discarding
such Papers, or parts thereof, as ought to be left out."[29] From
March until May he was reading the twenty-five prize essays
which had been submitted in 1817 "on the means of employing
the industrious & unoccupied Poor":

. . . it became my duty to read attentively, and give a short precis of
the Contents and merit—not that I was ordered to do so, but, to be
ready in case called upon: this has given me very steady employment,
notwithstanding I have at present two friends to read instead of one.[30]

On April 8 Young testified before the Select Committee of Parliament studying the salt duties:

I have been very busy making researches about the consumption of
Salt, and its various uses in agriculture, as I was yesterday examined
by the Salt Committee of the House of Commons and staid there
through the Examination also of Mr. Curwen M. P. and a member
of the Committee; this was a great event when compared with the
quiescent course of my life.[31]

---

[28] Berg Collection. Second letter dated December 23.
[29] Minute Books, **7,** ff. 70, 79.
[30] Berg Collection. Letter of March 5.
[31] *Ibid.* Letter of April 9 to Charlotte Barrett. This question is briefly discussed
in Smart, 1910: **1:** pp. 626–627, which shows clearly that the duty upon rock
salt as a cattle food was removed. Presumably the reference is to John Christian
Curwen, a Whig politician.

On May 15 the Committee on Papers for Printing resolved that "the Secretary's paper on the culture of Carrots be printed . . .," and at the last meeting on June 9 the Board voted:

The Board having taken into consideration the Recommendation of the Committee on Expenditure, in regard to presenting some mark of their respect to the Secretary for the Paper which he drew up on the culture of Carrots, Resolved unanimously that a Piece of Plate, to be chosen by himself, of the value of Thirty Guineas, be presented to Mr. Young for this Paper.[32]

Some of Young's letters commented on public affairs. On February 24 he drew Marianne's attention to the prevalence of juvenile delinquency in London:

Do you read a daily paper? The Morning Herald has been filled with such multiplied and horrible crimes that they make ones blood run cold; assuredly London grows worse and worse, and I attribute it very much to young offenders escaping punishment on account of their youth: you will see the history of the period by reading the Report of the Police Committee, which would entertain you much, vile as are the details. . . .

It is a little surprising to find the old Evangelical recommending to his young friend the "vile" details of the crime reports, which at this time left little to the imagination. Early in March he castigated Parliament for wasting time ". . . at a moment when every mind ought to be occupied in a new arrangement of the Poor Laws." On October 19 he reported the following incident to Marianne:

A man lately sold two pigs to Gosling of Cockfield who suspected from the cheapness that they were stolen, he stopt the man and took him to Mr. Colville when it came out that he had stolen them in Norfolk, he was committed to prison and will probably be sent to Botany-bay.

Here was the unreformed criminal law with a vengeance! That Young made no other comment shows how such a severe penalty for such a relatively trifling crime was accepted as a matter of course. The same letter commented on the Congress of Aix-la-Chapelle:

[32] Minute Books, 8, ff. 116, 132. That the Board was very flush is evident from a letter on November 14, asking the Treasurer, George Smith, to invest £2000 of their grant in exchequer bills. *Cf.* Letter Book, 2, f. 303.

I suppose the eyes of all your politicians are turned like those of the rest of Europe to Aix-la Chapelle, it rarely happens that Kings meet, except for mischief; but through the mercy of God I hope that this congress will establish peace on a solid foundation.[33]

Perhaps the most interesting feature of Young's letters in 1818 was his discussion of the books he was reading. Much of it was travel literature, especially about backward areas and savages. When Marianne apparently had recommended a collection of sermons by one of his favorite preachers, Young commented, "but while any thing about Savages and desert Islands is to be read I shall postpone the Sermons." He read several books about western North America, two on New Zealand, and one on a world tour. He borrowed the account of Captain Bligh's voyage from Marianne's mother "and I have just finished reading it; to my great entertainment." He was a little conscience stricken about spending so much time on travel books:

But you will . . . say, what have you to do, trembling on four score, with amusement in reading voyages unless it be the gt. voyage from time to eternity? But I think the case admits of some defence; for . . . what is a more proper subject for contemplation, than remarkable instances of Divine Providence, and the general wonders of creation so thickly sown around the globe?

As the author of very famous "tours," Young was critical of some of the travel books he read, although modern commentators might be fully as censorious of some of his works and on the same grounds:

I forget whether you have read Nicholas's Voyage to New Zealand"— I have just got to the end of the 1st Volume, and am disgusted with the manner in which every page is *spun out* in order to make two Volumes of a work that would hardly have made a good *one*. . . .[34]

.He also read a book on savings banks which "proved how much they ought to be encouraged," and the report of the British and Foreign Bible Society "the records of which are delightful, Bibles are spreading in all the languages of the Globe." He found Dean Isaac Milner's *History of the Church of* **Christ**

---

[33] Berg Collection.
[34] *Ibid*. Letters of March 5, 11, and 24.

"... though put together in a heavy manner, yet is very interesting by the importance of the subject, and the great care with which he has consulted a multitude of authorities." He also recommended to Marianne, Madame de Stael's book on the French Revolution: "you may depend on receiving much entertainment from it, and Buonaparte is described minutely in every point."[35]

While Charlotte Barrett, with her husband and older daughter, were touring France in the summer of 1818, Marianne and the younger daughter Hetty Barrett, and a maid, were at Bradfield. Naturally Young was keenly interested in Charlotte's trip and on July 23 wrote her, suggesting where she should go and what she should avoid, and especially urging her to keep a full journal for publication:

I must renew my former earnest recommendation that you keep a very minute journal of everything you see and hear, with lively and accurate transcripts of impressions around you . . . as I am confident that it will procure you an ample harvest of cash and fame. . . .

Twice in the autumn he urged Marianne to insist that Charlotte prepare the journal for publication: ". . . what a dressing I shall give her if she hesitates upon this point." Late in December he wrote to Charlotte herself:

It shall arrive in a very few days! . . . I believe I shall make a charge against you for a pair of shoes worn up [sic] in going to the Manger for your MSS, but the disappointments which my anxious expectation meets with is worse than wearing up a pair of Boots. . . . I am half crazy to read it, and if it does not come speedily I must have an effigy of you set up to the neck in the horsepond for you begin to want a good ducking, and the new pond now digging for your navigation, will I trust answer for that necessary operation. . . .[36]

There was relatively little worthy of note on religion in his letters. The less attractive side of his religious views was shown in his response to Marianne's request to write a letter of consolation to his old friend, Mme d'Arblay, on the death of her husband. Young's reply was certainly lacking in sympathy to say the least:

[35] *Ibid.* Letters of April 11, October 19, November 27.

[36] *Ibid.* Letter of December 23–24. There is no indication that Charlotte's journal was ever published. The "Manger" was the inn at Bradfield where Young collected his mail.

I thank you for the kindness of the hint of writing to Mrs d'Arblay, but I have not yet brought myself to do it, for what consolation can a Xtian offer in such a case, except it be to a real Xtian? what sort of figure should I make in offering philosophical comfort? I neither feel nor understand any thing of the matter. I should therefore first call upon her to become a Xtian, in order to be placed in a capability of receiving Xtian comfort: I must therefore turn it again in my mind, before I can set myself to work.[37]

Two of his letters contain rather amusing references to Sir John Sinclair. On February 24 he wrote:

Your lending Baxteriana to Sir John Sinclair is quite an anecdote; of course I should have presented it to him had I thought it possible for him to have read a single page in it; for, poor man, religion I conceive will never come into his mind, except he write a book upon the subject; and that indeed must be far distant.

In the autumn, there was just one sentence in a letter showing that Young's prediction about Sir John had come true much earlier than he had expected: "Sir J. Sinclair has sent me a printed prospectus of a new work of his entitled 'The Code of Religion!'"[38]

The materials for 1819 are still more scanty. The fairly large number of letters to Young are not very important. Nor are the relatively full records of the Board of Agriculture very significant. There is absolutely nothing in the *Autobiography* and there are only two letters from Young to Marianne and one to her sister. Even his movements during 1819 are not too clear. His letter to Marianne of January 4 shows him still at Bradfield but planning to go to London about January 15. He was at Sackville Street at least until the middle of June. Surely he spent the latter half of the year at Bradfield.

There are only two references to his health. On May 26 he wrote to Thomas Attwood: "My health has been of late but indifferent, so that I could not dictate the letters which I wish to send immediately. . . ."[39] On the other hand he wrote to Marianne on October 12:

[37] *Ibid.* Letter of May 14.
[38] *Ibid.* Sir John proposed a sort of compendium of all knowledge. His *Code of Agriculture* was completed and was quite a success, but the volume on religion was never finished.
[39] Wakefield, 1885: p. 69.

. . . if I say nothing about my health you may suppose that it is in the usual state. I believe I have often told you that I should be most ungrateful to my heavenly Father if I did not thank him with fervency for health, strength and freedom from bodily pain; What blessings are these at my advanced age. . . .[40]

It seems very doubtful whether Jane Young spent much of the year with her father-in-law. His January letter to Marianne predicted that Jane would not join him until May, while his October letter stated that Mary and St. Croix wished to be remembered to her, but Jane was not mentioned. Young's letter to Attwood provided the most definite information about the Rev. Arthur's Russian estate which we possess:

In answer to your enquiry relative to my son's property in the Crimea, I may inform you that it consists of about 9,600 acres, with a house for a residence and many attached conveniences, that it is within 9 miles of Caffa, . . . and that a small river passes through it turning several mills. I know not what price he fixes upon it at present, but the last that was communicated to me was 80,000 roubles. . . .[41]

On August 26 Young's old friend, the Earl of Egremont, replied very skeptically to Young's query about procuring an English living for the Rev. Arthur Young:

. . . but there is another question . . . whether his wishes are the same as yours, for I had some conversation upon the subject with him when I saw him last, & he told me that you thought that he ought to return to his Professional duties in England & that in deference to your opinion he should endeavour to sell his Russian Estate, but he seemed to me very much to prefer the thoughts of farming in the Crimea to preaching in England. . . . It is very probable that I could give him a Living immediately of about Two Hundred & Fifty pounds with a tolerable House, but then he would be obliged to reside there & not in Russia as he can now do with his Irish living.[42]

The Board of Agriculture met fairly regularly from January 26 to June 15. Rather surprising was the Board's motion on January 29 thanking Robert Owen for his pamphlet, *New View of Society*.[43] Even more astounding was the entry for June 8:

[40] Berg Collection.
[41] Wakefield, 1885: p. 70.
[42] Add. MSS. 35,133, ff. 440–441.
[43] Minute Books, **7**, f. 136.

Mr. Owen of New Lanark having attended the Board, and explained a model for the construction of Farms and the erection of manufacturies, for the Employment of the Agricultural and manufacturing Poor, the thanks of the Board were presented to him.[44]

It is a pity that Owen should not have commented on Young, or Young upon Owen. On February 9 the secretary was asked to "look over the Journal of Flanders-Husbandry, and to point out any parts of it that may merit particular attention,"[45] and on February 26 it was voted "that the Secretary be desired to draw up a paper explanatory of the Reasons which have induced the Board to print their volume in the 8to instead of the 4to form."[46]

All in all, the Board was in a sorry state of decline. Absenteeism among the members was all too general. The famous secretary was blind. On March 25 the president, Lord Macclesfield, wrote the Earl of Hardwicke that he could no longer give even irregular attention to Board matters and appealed to Hardwicke to take over again as President.[47] After the Board's annual meeting on April 6 had unanimously elected Hardwicke as president, although he had only attended three meetings during the previous year,[18] Young wrote him a graceful letter of congratulations:

It is with the greatest satisfaction to myself, and I am sure to every other person connected with the Board, that I can acquaint you, that your Lordship is unanimously elected to the chair: may you long fill it my Lord to your own satisfaction, as I am sure you will to that of all the members.[49]

Again the Board had more money than it knew how to use. The General Committee on March 26 proposed to ask for no grant at all, but it was finally decided to request half the annual grant.[50]

It is hardly surprising that Sir John Sinclair should have concluded that little could be expected of such a Board, and hence he suggested in the *Farmer's Journal* the establishment of a

---

[44] *Ibid.,* f. 199.
[45] *Ibid.,* f. 139.
[46] *Ibid.,* f. 148.
[47] Add. MSS. 35,652, f. 212.
[48] Minute Books, **7,** f. 161.
[49] Add. MSS. 35,652, f. 216.
[50] Minute Books, **7,** ff. 161, 165.

new, independent Board of Agriculture. Naturally Young was
upset and protested to Sinclair, who replied on October 20:

I am very much vexed that . . . the plan for establishing *An Independent Board of Agriculture* . . . should have given you a moment's
uneasiness. Since the unfortunate loss of your sight, several ideas have
been thrown out, respecting the appointment of a successor to you,
which I have always resisted; and would never have thought of, had
it not been for the necessity of placing the Board, *on a new footing,*
to enable it to do more good. But any attempt of that purpose,
would not succeed, without Mr. Webb-Hall's assistance, who, by his
zeal, industry & talents, has become a species of "Hunt" in agriculture. That was my reason for suggesting his being appointed secretary on the new plan, continuing the present income to you, on the
ground, of your eminent services to the cause of agriculture, which
is distinctly stated in the original proposal. I subjoin a copy of
another letter to the Editor of the "The Farmers Journal," containing
a proposal which I hope will be satisfactory to you on this subject. . . .

The enclosed revised proposal was as follows:

I am of opinion that it would be right to make Mr. Arthur Young
and Mr. Webb-Hall "Joint-secretaries." No man is more sensible than
myself, of Mr. Young's distinguished services in the cause of agriculture, and so long as he is desirous of holding any official situation connected with the Board, it would be a pity, that he should not be
continued in it.[51]

Thus the original plan had proposed Young's retirement with
full pay, while in the revised plan he would be associated as a
joint secretary with a much younger man, with whom there
could well be plenty of friction. It is doubtful whether Young
was mollified. Surely he must have been deeply hurt at the
original plan.

Two references have already been made to Young's letter of
May 26 to Thomas Attwood, thanking him for his publication,

which is of such importance that I hope and trust that it will have
due weight with those who have it in their power to give full efficiency
in practice to your highly commendable ideas. . . . The state of the
country at present I am afraid is very dubious, for the accounts which
we receive from various parts of it mention the bad circumstances
that many persons willing to work remain without employment.

[51] Add. MSS. 35,133, ff. 446–447. Webb-Hall was appointed secretary to the
old Board after Young's death. The reference is to "Orator" Hunt, the extreme
Radical of Peterloo fame.

When Attwood visited Young in London he wrote to his wife: "I have visited Arthur Young and find him an exceedingly sensible and generous old man. He urged me very much to write more, offering to undertake the printing free of expence, &c., &c., but I believe I must decline it."[52] Thus in the spring of 1819 Young had been listening to the socialist Robert Owen and was encouraging Thomas Attwood, the easy money Radical.

Young received two letters from Lord Sheffield and eight from William Wilberforce in 1819. Sheffield was advocating an extreme form of agricultural protection. He called Young "one of my dearest friends" and also "a kind of Father" of agriculture.[53] Wilberforce's letters were chiefly concerned with detailed plans for sending out plows and plowmen to improve agriculture among the former slaves of Hayti.

It is difficult to judge how much Young saw Marianne and her family in 1819. Almost certainly the Barretts were not at Bradfield during the summer. The paucity of letters would argue that Marianne was with him much of the year, but all the actual evidence supports the opposite conclusion. Young's letter to Charlotte Barrett of January 4 praised her journal which had finally reached him:

I have read your Journal with the greatest Satisfaction, the perusal has given me so much pleasure that I cannot for a moment doubt its complete success when published, and I say this after making great allowances for the great prejudice I feel, and must for ever feel, in favor of its authoress. I have put one or two Q for queries on the opposite page, which I will enquire about when we meet. . . .

The letter also thanked Charlotte for the cap for him and the scarf and "fril" for Mary, presumably presents from France. His letter to Marianne of the same date is not particularly revealing, but is worth a few quotations:

. . . I am extremely sorry to find that you are bilious, and much fear that you do not take proper advice for the management of your health. . . . Tomorrow the tenants dine here to pay their rents. . . . As I advance in reading Milner I am much entertained, he is very minute on the subject of Luther, which part of his work is very interesting.[54]

[52] Wakefield, 1885: pp. 69–70.
[53] Add. MSS. 35,133, ff. 439, 466–467.
[54] Berg Collection. Marianne referred to her bilious attacks in a letter to Mrs. Piozzi on March 23, Rylands MSS. 584, f. 158.

There are two very interesting references to Marianne in letters from Mme d'Arblay to her sister, Charlotte Broome. On March 8 she wrote from London:

I have had a visit from Marianne,—which, being arranged, went off serenely enough. I thought her, too, in manner & deportment, much improved, formed & fashioned with an air of being used to good company that becomes her much.

Her second letter of September 30, after a visit to Richmond, spoke of Marianne with much more cordiality:

My kind love, & many thanks to Marianne for her numerous attentions & daily augmenting hospitality of manner. Her kindness to Alex was so exactly of the right sort that I regret he has not such a cousin continually at his elbow. So does he, I assure you.[55]

Young's other letter to Marianne, on October 12, is probably the last of his letters to survive. At one point he expressed a very reactionary view towards the popular disturbances and the popular leader, "Orator" Hunt: "I have been much hurt at constantly reading in the Newspapers of a horrid spirit in so large a portion of our people, and I shall be out of all patience till I hear that Mr Hunt is hanged. . . ." On the other hand, his attitude towards Mme d'Arblay is much more friendly than in the letter of 1818 about her:

I am very sorry to hear that your good aunt d'Arblay is still in low spirits, I am sure that she must have sufficient Xtianity to produce a great measure of resignation to the will of the Almighty, and as I take for granted that her kind husband died a Xtian, it would be very wrong in her to be so selfish as to wish him back again: if he did not so die she has indeed reason to lament his loss: the longer I live the more clearly convinced I am that this great point of the state of heart in which our friends and relations quit the present life is that alone which should affect our minds on the death of all who are dear to us; there is nothing else worth the least attention. . . .

You hope I sometimes think of you in my prayers; I never pray in private without intending to think of and name all my friends, for in my opinion this is a very essential part of prayer. . . . I approve very much of Hannah More, and marked several passages for copy-

[55] Berg Collection.

ing to insert in my *Evangelical Dictionary for the use of Xtian families,*
a manuscript which I shall leave behind me. . . .[56]

Arthur Young died in London on April 12, 1820.[57] None of
his letters in that year have survived and the material, up to
his last illness and death, is very scanty. It is not certain even
when he came to London. Board meetings were called as early
as February 22, but no meeting before March 28 had a quorum,
and the minutes for that date read: "No member having attended, adjourned for the Easter recess. . . ."[58] Since all these
minutes were signed by William Cragg, the under-secretary, it
may be that Young had not yet reached London. Wilberforce's
letter to Mary, written after Young's death, apologized for not
having called, but gave as his excuse his belief that Young was
still at Bradfield.[59] He was in London, however, on March 30,
for Cragg's letter to Lord Hardwicke after Young's death, declared that he took a long walk that day.[60]

There are several sources for Young's last illness and death.
First there is Cragg's letter to Lord Hardwicke, dated April 13.
Secondly, there is Marianne's letter to Mrs. Piozzi, dated April
14.[61] Thirdly, there is Mary Young's letter to Charlotte Barrett,
dated April 19.[62] Fourthly, there is the obituary notice in the
Bury Gazette.[63] Finally, there is "A Biographical Memoir of
Arthur Young," by Dr. J. A. Paris, one of the physicians during
his last illness.[64] From these varied sources it is possible to
reconstruct the story fairly well.

[56] *Ibid.* This is the only reference to the *Evangelical Dictionary,* at least by that
name. Of course many of his letters for years had mentioned that he was having
excerpts copied from his reading.
[57] Miss M. Betham-Edwards has twice misdated his death as April 20. *Cf.*
Young, 1898: p. 472, and Young, 1924: p. 1.
[58] Minute Books, **8,** f. 2. Attempts were made to hold a meeting on February 22
and 29, and on March 7 and 14. There was no quorum even for the Annual
Meeting on March 25.
[59] Add. MSS. 35,133, ff. 472–473. The date was April 21, nine days after
Young's death. He stated that he had received a letter from Marianne dated
April 11, telling him that Young was in London, but it "had by mistake been
laid aside unread on its arrival & had not been perused till a few hours ago."
[60] Add. MSS. 35,562, f. 341.
[61] Rylands MSS. 584, f. 160.
[62] Berg Collection.
[63] *Bury Gazette,* May 17, 1820. His death had been noted in the April 19 issue.
[64] Paris, 1820: pp. 279–309.

According to Cragg, Young's long walk on March 30 had greatly fatigued him:

From that time he experienced a strong retention of urine accompanied with very great pain, so much so that it was deemed expedient to call in, on the Tuesday, W. Wilson the celebrated anatomist, who passed the Bougie & found two large conerations [*sic*]. From that time Mr. Young's sufferings were very great, & exhaustion was the consequence.

That he suffered very great pain during his last illness seems agreed upon by most of the witnesses. Marianne wrote to Mrs. Piozzi: "He was suffering, I grieve to say, under a most excruciating malady, the stone, which was almost immediately fatal." Mary reported to Charlotte Barrett:

His patience during the agonizing sufferings he endured was truly astonishing, indeed it was a lesson not easily I should think forgotten by those who were present—not one repining word escaped him. This is a very striking circumstance in so impatient a mind as his once was.

Dr. Paris, who was apparently the chief physician in the latter part of his illness, gave a more detailed account:

The disease, of which Mr. Young died, was not suspected until about a week of his death—a circumstance which received a very satisfactory explanation, from an examination' of the body, after death. He was attended by Mr. Wilson, Mr. Chilver, and myself, and although the incurable nature of his disease defied any hope of permanent relief, yet his sufferings were greatly palliated by the resources of art, and he died without entertaining the least suspicion of the malady under which he suffered. Pious resignation cheered him in his illness, and not a murmur of complaint was heard to escape his lips. On the 12th of April, in the year 1820, at his house in Sackville-street, after taking a glass of lemonade, and expressing himself calm and easy, he expired.[65]

Cragg added that Young died at "twenty minutes past one," that Dr. Paris was present, and that the doctor had attended his patient far beyond the call of duty:

His attentions to our revered friend were most praise-worthy, adding

[65] Paris also stated: "It is a very remarkable fact, that, during his whole life . . . he entertained the greatest horror of two diseases—blindness and the stone."

to the skill of an able Physician the kind assiduities of an affectionate friend, having when the Business of the day was over, spent hours with him watching the symptoms of the disorder, and devising means to alleviating the excrutiating [*sic*] pain.

Mary's letter shows that Charlotte Barrett spent a day or two with Young either during his illness or shortly before it. Marianne probably saw him the day he died, "little thinking it was for the last time," and it must have greatly comforted him to have her there frequently. On the other hand Mary wrote that Jane Young did not arrive until after the end.

Poor Mary of course had all the details of the funeral to arrange. She wrote:

. . . the severe and afflicting deprivation I have sustained in the death of my dear and highly venerated Father—true indeed it is, that *our* loss is his *gain*—but the bereavement must be long and deeply felt. . . . The last five & twenty years of his life was an example I wish I could follow.

On Wednesday morning, April 19, at six o'clock the body left London for the burial at Bradfield on Friday, April 21. Mr. St. Croix also left on Wednesday by the Bury coach to help with the arrangments. The obituary notice in the *Bury Gazette* briefly described the funeral:

His remains were interred in the church yard of Bradfield, near this town, where a small property has been in the possession of the family above 200 years. The funeral was attended by a large assemblage of poor from the surrounding country, all anxious to testify their sorrow for the loss of so benevolent a benefactor, his kindness must be long regretted both by—"the young who labour and the old who rest;" as few men, with so limited an income conferred greater benefits in their neighbourhood.

Marianne had also predicted that the poor of the Bradfield area would miss him greatly:

Poor dear Mr. Young!—it is a blessed transition for him to the place where there is none of the pain which racked him here; but his gain is the loss of his country, his friends, the poor, to whom he was a friend indeed, & all who knew him, except the fashionable world, from whom, in conformity to a high authority . . . he had long come out & separated himself before that heavy deprivation, loss of sight, overtook him.

Even more poignant was a sentence in Mary Young's letter to Charlotte Barrett of September 11, 1820: "My Poor Fathers loss is much felt here the Hall feels very strange on a Sunday, melancholy and deserted."[66]

His burial in Bradfield churchyard was in accord with his will of February 5, 1816,[67] which had directed that he be buried in the cheapest manner possible. Eventually his "affectionate and affected son" erected a very simple monument in the churchyard, in the form of a sarcophagus. It is in front of the little church facing the road and the avenue of fine trees leading to the Hall. On one of the side faces there is a summary of his career:

In Agriculture and Political Economy Pre-eminent
Distinguished for Public Value, Private Worth, and the serious
Performance of every moral duty, above all faithful but
Humble disciple of that Blessed Redeemer, on whose atoning
Blood alone, he relied for Salvation.
His natural genius, cultivated Talents, and benevolent exertions
Were disinterestedly and successfully devoted to the
Promotion of the Statistical, Commercial, Moral and
Religious Interests of his Country.[68]

The whole thing is so simple, so true, and in such good taste that it is hard to see how Arthur Young could have objected.

The will bequeathed Bradfield and all his other estates to Arthur and his heirs. To Mary went a life lease of a cottage, probably the one which the Barretts had rented, along with twenty acres of land, an annuity of £50 a year, and Old South Sea Annuities of nearly £2000 in value. She was also given considerable furniture and plate, some of it from her mother's family, the silver cup with an inscription, one third of the linen, crockery, and china, of books 4 folios, 12 octavos, and 20 duodecimos, and "my whiskey, harness & a horse if I have one."

A motion of the General Committee of the Board of Agriculture on April 28 reveals that Mary had asked to reside at 32 Sackville Street until June 3, a request "which the Committee most readily acceded to, as a mark of respect to the memory of her late father."[69] On April 19 Mary wrote Charlotte Barrett

[66] Berg Collection.
[67] This was examined at Somerset House.
[68] The monument also contains a bare statement of the birth and death of his wife.
[69] Minute Books, **8,** f. 6.

that she could not visit Richmond, for she had "so much packing before I can quit this house it will be out of my power to accept it."[70] In her letter to Charlotte of September 11 Mary told of the Rev. Arthur's plans for starting home, and continued:

In a previous letter he requested me to remain at the Hall, my cottage therefore continues untenanted tho I have removed to it the furniture left me by my dear Father. Mr. St. Croix will occupy it during the winter as I should be sorry to have depradations [*sic*] committed.[71]

On April 25, for the first time that year, the Board of Agriculture met with a quorum, and the following vote was passed:

The Under Secretary reported to the Board the melancholy event of the lamented death of their Secretary, Mr. Young. Ordered that the Under Secretary be instructed to make out & lay before the Committee on Friday next a Statement of such sums as are due from the Board to the Estate of the late Secretary; and that it referred to the Committee to consider the propriety of shewing some mark of respect in which the Board held his memory. Ordered that Mr. Garrard, who took a cast from Mr. Young, be desired to attend the Committee on Friday next.[72]

It transpired that somewhat over £50 was due to the estate of the deceased secretary.[73] On May 5 William Cragg's request for an increase in salary was refused, but he was voted a sum of £240, equal to a year's salary, "having taken into consideration the additional services performed by him during the infirmity and loss of sight of the late Secretary." At the same time Young's personal secretary, St. Croix, applied for a position with the Board, but was informed that there was no intention of adding to the establishment.[74] When the sculptor attended the meeting of the General Committee, he told them that he could make a marble bust of Young for £100.[75] Since this was more than the Board wished to spend, it was voted on June 13 to subscribe "to the extent of ten or fifteen pounds for a cast of the same . . . to

---

[70] Berg Collection. This letter is not entirely clear, but it may be that Mary did not go to Bradfield for the funeral.
[71] *Ibid.*
[72] Minute Books, **8**, f. 3.
[73] *Ibid.*, f. 6.
[74] *Ibid.*, ff. 9, 10.
[75] *Ibid.*, f. 12.

be placed in the Board Room."[76] When the Board was dissolved in 1822, George Webb-Hall, Young's successor as secretary, wrote to the Rev. Arthur Young:

This board being abt to be dissolved, I have ye pleasure to acquaint you "that ye Bust of ye late Secretary Arthur Young Esqr. was unanly [sic] ordered to be presented to his son." I therefore desire to know your pleasure where I am to deliver it. . . .[77]

On June 23, 1820, Sir John Sinclair brought to the Board's attention Young's work, the *Elements of Agriculture* "containing the result of all his enquiries." The secretary was therefore requested to enquire of Miss Young into the state of the manuscript and whether "this Board can in any manner facilitate such publication." After receiving Mary's reply, it was voted that the secretary "do enquire of Messrs. Longman & Co. . . . what will be the probable expense of printing the late Mr. Young's Mss. . . ."[78] Longman's reply apparently indicated that the cost of publishing this eight-volume manuscript would be prohibitive and Mary was so informed.[79]

To conclude, it remains but to quote some contemporary evaluations of Young and his work. Probably the summary which would have pleased him most was by Wilberforce in his letter of condolence to Mary: "May we be as fit as he was for ye awful call."[80] The obituary notice in the *Bury Gazette* was fairly long and very perceptive. It enumerated and evaluated his most important writings. Of his English tours it stated: ". . . it is believed that most of the modern improvements in agriculture originated from this work." It quoted Maria Edgeworth's comment on his Irish tour: ". . . it was the first faithful portrait of its inhabitants." It mentioned the *Annals* and George III's contributions to that periodical. It declared that his French travels ". . . gave the inhabitants a higher idea of the value of their own soil & climate than they before possessed." It listed

[76] *Ibid.*, f. 35. This was presumably a plaster cast. The sculptor was George Garrard, 1760–1826, famous for pictures of animals and for human busts. He was the painter of the famous Woburn sheepshearings. *Cf.* fig. 15 for reproduction.
[77] Letter Book, **2**, f. 398. I have been unable to find out what happened to Young's bust.
[78] Minute Books, **8**, ff. 41, 43.
[79] Letter Book, **2**, f. 370. This letter is not dated, but my guess is that it was written in March, 1821.
[80] Add. MSS. 35,133, f. 473.

the county reports which he had written and paid attention to his work as secretary of the Board of Agriculture. The summary of his personal characteristics was especially good:

The striking feature in this gentleman's personal qualities were an ardent industry, indefatigable perseverence, and a lively imagination. His manners and address were peculiarly pleasing, his conversation highly animated and instructive; his countenance strongly marked his decision of character, and the strength of his understanding.[81]

Best of all was a single sentence quoted by Dr. Paris from Richard Kirwan, the Irish economist: "To the labours of Mr. Arthur Young the world is more indebted for the diffusion of agricultural knowledge than to any writer that has yet appeared."[82] Kirwan here has hit upon his real achievement which justifies a full-length biography of the man. It should be noted that Kirwan said nothing about Young's agricultural innovations but rightly judged his greatness in the "diffusion" of the progress made by others. Young was the greatest agricultural publicist of all time, but his own writings happily coincided with a time of great technical advance. He was also, at his best, one of the greatest practitioners in the literary medium of the travel diary. Dr. Paris happily summarized his greatness in this field in commenting on the *Travels in France*:

His diary is written in a familiar and easy style, and his descriptions are so agreeably circumstantial and unreserved, and constantly enlivened with such smart and unaffected badinage, that the reader becomes one of the party, and cheerfully attends him through his route with all the familiarity of an old acquaintance, participates in all his embarrassments, laughs with him at the follies he witnesses, and partakes of all the amusements and the agreeable and instructive society, to which his celebrity introduced him.[83]

[81] *Bury Gazette*, May 17, 1820.
[82] Paris, 1820: p. 280. This is taken from the *Irish Transactions*.
[83] *Ibid.*, p. 299.

Bibliography

I. PRIMARY SOURCES

1. *Manuscripts*

For detailed critical evaluation of the following manuscript sources, see Gazley, 1955: pp. 410–428.

London

British Museum

Young Correspondence. Add. MSS. 35,126–35,133.

Young, *The Elements and Practice of Agriculture*. Add. MSS. 34,821–34,864.

Hardwicke Papers. Add. MSS. 35,643, 35,652, 35,697, 35,700.

Bentham Papers. Add. MSS. 33,541, 35,542.

Liverpool Papers. Add. MSS. 38,225.

Wilkes Papers. Add. MSS. 30,871.

Library of Royal Agricultural Society. Papers of the Board of Agriculture. Minute Books, Letter Books, Cash Book.

Library of the Society of Arts. Records of the Society. Minutes, committee reports, letters, unpublished transactions, etc.

Public Record Office. Chatham Papers. GD 8/193.

Library of the Royal Botanic Gardens at Kew. Banks Correspondence, vols. 1 and 2.

Manchester. John Rylands Library.

Bagshawe Muniments. B 3/10, B 3/16, B 22/6.

Thrale Correspondence. English MSS. 582–584. (Cited in footnotes as Rylands MSS.)

Spains Hall, Essex.

Ruggles Letters.

New York Public Library.

Burney MSS. in Henry W. and Albert A. Berg Collection.

Washington. Library of Congress.

Papers of George Washington. Vols. 234, 237, 241, 253, 258, 267.

New Haven, Yale University Library.

Banks Correspondence.

Boswell Papers.

San Francisco. Sutro Branch of the California State Library.

Banks Correspondence.

2. *Writings of Arthur Young*

(*Note*. Amery, G.D. 1924. "The Writings of Arthur Young." *Jour. Royal Agric. Soc.* 85: pp. 1–31. This is a very complete bibliography of Young's writings and includes his contributions to the *Annals of Agriculture*. For Young's writings, where there was more than one edition, I have given date of 1st edition and where necessary the date of the edition I have used. I have also listed some of the outstanding modern editions of his Irish and French tours.)

704

1758. *The Theatre of the Present War in North America:* . . . (London)
1759. *Reflections on the Present State of Affairs at Home and Abroad* (London)
1762. *The Universal Museum, or Gentlemen's and Ladies' Polite Magazine of History Politicks and Literature for 1762* (v. 1, London).
1764–6. *Museum Rusticum et Commerciale: or, Select Papers on Agriculture, Commerce, Arts and Manufactures* (6 v., London; 3 and 4 contain ten articles by Young, many re-printed as Sylvae in *The Farmer's Letters*).
1767. *A Letter to Lord Clive* . . . (London).
1767. *The Farmer's Letters to the People of England.* . . . *To which is added, Sylvae; or Occasional Tracts on Husbandry and Rural Oeconomics* (London).
1768. *A Six Weeks' Tour, through the Southern Counties of England and Wales* . . . (London; 2nd ed., London, 1769, was one used).
1769. *Letters concerning the Present State of the French Nation* . . . (London).
1769. *An Essay on the Management of Hogs* . . . (London; 2nd ed. 1770 includes *Culture of Coleseed*).
1769. *A Six Months' Tour through the North of England.* . . . (4 v., London; 2nd ed., London, 1771, was one used).
1770. *The Expediency of a Free Exportation of Corn at this time:* . . . (London).
1770. *The Farmer's Guide in Hiring and Stocking Farms* . . . (2 v., London).
1770. *Rural Economy; or, Essays on the Practical Parts of Husbandry. To which is added, the Rural Socrates* . . . (London).
1770. *A Course of Experimental Agriculture* . . . (paged as 4 v., bound as 2, London).
1771. *Proposals to the Legislature for Numbering the People* . . . (London).
1771. *The Farmer's Tour through the East of England* (4 v., London).
1771. *The Farmer's Kalendar* . . . (London).
1772. *Political Essays Concerning the Present State of the British Empire* (London).
1772. *Essays on the Spirit of Legislation* . . . *Translated from the Original French* (London).
1773. *Observations on the Present State of the Waste Lands of Great Britain* . . . (London).
1774. *Political Arithmetic. Containing Observations on the Present State of Great Britain and the Principles of her Policy in the Encouragement of Agriculture* (London).
1780. *A Tour in Ireland* . . . (London).
1925. *A Tour in Ireland,* ed. by Constantia Maxwell (Cambridge, England).
1783. *An Enquiry into the Legality and Expediency of Increasing the Royal Navy by Subscriptions for Building County Ships* (Bury St. Edmunds).
1784–1815. *Annals of Agriculture and other Useful Arts* (46 v. London).
1788. *The Question of Wool Truly Stated* . . . (London).
1788. *A Speech on the Wool Bill, that might have been spoken in the House of Commons* (London).
1792. *A Letter on Tithes to Arthur Young, with his remarks on it* . . . (London).
1792. *An Address Proposing a Loyal Association to the inhabitants of the Hundreds of Thedwastry and Thingoe* (no place, single sheet).
1793. *Travels during the Years 1787, 1788 and 1789, undertaken more particularly with a View of ascertaining the Cultivation, Wealth, Resources, and National Prosperity of the Kingdom of France* (2 v., Dublin).
1889. *Travels in France* . . . (with an introduction, biographical sketch, and notes by M. Betham-Edwards, London).
1915. *Travels in France and Italy* . . . (introduction signed T. Okey; Everyman's Library, London).

1929. *Travels in France* (ed. by Constantia Maxwell; Cambridge, England).
1931. *Voyages en France* . . . (première traduction compléte et critque par Henri Sée; 3 v., Paris).
1793. *The Example of France, a Warning to Britain* (London). 2nd. ed., Bury St. Edmunds, 1793, was one used.
1794. *General View of the Agriculture of the County of Suffolk* (London). Later and corrected editions in 1797, 1804, 1813.
1795. *An Idea of the Present State of France* . . . (London). I used this pamphlet as first published as an article in the *Annals* **23**: pp. 274–311.
1795. *The Constitution Safe without Reform: Containing some Remarks on a Book entitled The Commonwealth in Danger, by J. Cartwright* (Bury St. Edmunds). Since this pamphlet was not available I used the article with the same title in *Annals* **25**: pp. 246–293.
1797. *National Danger and the Means of Safety* (London). I have used this material as it first appeared in four articles in the *Annals* **26**: pp. 516–521; **27**: pp. 49–54, 528–538; **28**: pp. 177–187.
1798. *An Enquiry into the State of the Public Mind amongst the Lower Classes* . . . (London).
1799. *General View of the Agriculture of the County of Lincoln* (London).
1800. *The Question of Scarcity Plainly Stated, and Remedies considered* (London).
1801. *An Inquiry into the Propriety of Applying Wastes to the Better Maintenance and Support of the Poor* (Bury St. Edmunds).
1801. *Letters from His Excellency George Washington to Arthur Young* . . . (London).
1803–4. *Georgical Essays*, ed. by A. Hunter (6 v., York, contains five essays by Young.
1804. *General View of the Agriculture of Hertfordshire* (London).
1804. *General View of the Agriculture of the County of Norfolk* (London).
1805. *An Essay on Manures* (Bath and West of England Society, Letters and Papers, **10**: p. 97, Bath).
1807. *General View of the Agriculture of the County of Essex* (2 v., London).
1808. *General Report on Enclosures* (London).
1809. *On the Advantages that have Resulted from the Establishment of the Board of Agriculture* (London).
1809. *View of the Agriculture of Oxfordshire* (London).
1811. *On the Husbandry of Three Celebrated British Farmers, Messrs. Bakewell, Arbuthnot, and Ducket* (London).
1812. *An Enquiry into the Progressive Value of Money in England, as marked by the Price of Agricultural Products* . . . (bound with v. **46** of *Annals*, (London).
1815. *Baxteriana: Containing a Selection from the Writings of Baxter* (London).
1815. *An Enquiry into the Rise of Prices in Europe* . . . *compared with that which has taken place in England* . . . (bound with v. **46** of *Annals*, London).
1817 *Oweniana: or, Select Passages from the Works of Owen* (London).
1898. *The Autobiography of Arthur Young with Selections from his Correspondence* (ed. by M. Betham-Edwards, London).
1926. *Arthur Young on Industry and Economics; being excerpts from Arthur Young's observations of the state of Manufactures and his economic opinions on problems related to contemporary Industry* (arranged by Elizabeth P. Hunt, Philadelphia).

1932. *Tours in England and Wales. Selected from the Annals of Agriculture* (The London School of Economics and Political Science).

3. Contemporary Books, Pamphlets, and Articles

Anonymous. 1775. *American Husbandry* (2 v., London, by an American; re-edited, 1939, New York, by Harry J. Carman).

—————— 1795. "Biographical Anecdotes of Arthur Young, Esq." *The European Magazine and London Review* 28: pp. 363–365.

—————— 1795. *A Reply to a Pamphlet entitled, "An Idea of the Present State of France &c." By Arthur Young, Esq.*

—————— 1797–1813. *Communications to the Board of Agriculture, on subjects relative to the Husbandry and Internal Improvements of the Country* (7 v., London). (1819; n.s., 1 v.)

ACKERMANN, RUDOLPH. 1904. Microcosm of London (London). pp. 73–85. (original ed. 1809).

ARBUTHNOT, JOHN. 1773. *An Inquiry into the Connection between the present Price of Provisions, and the Size of Farms* . . . (London).

ATTWOOD, THOMAS. 1818. *Observations on Currency, Population and Pauperism, in two Letters to Arthur Young, Esq.* (Birmingham)

BAKER, JOHN W. 1771. *Considerations upon the Exportation of Corn* . . . (Dublin).

—————— 1771. *Experiments in Agriculture, made under the Direction of the* . . . *Dublin Society in the Year 1769* . . . (Dublin).

BERCHTOLD, LEOPOLD, 1789. *An Essay to Direct and Extend the Inquiries of Patriotic Travellers* . . . (2 v., London).

BENTHAM, JEREMY. 1962. *The Works of Jeremy Bentham* (ed. by John Bowring, 11 v., New York).

BURKE, EDMUND. 1800. *Thoughts and Details on Scarcity* (London).

CARTWRIGHT, JOHN. 1795. *The Commonwealth in Danger; with an Introduction containing Remarks on some late Writing of Arthur Young, Esq.* (London).

CHALMERS, GEORGE. 1782. *An Estimate of the Comparative Strength of Great Britain during the present and four preceding reigns* . . . (London).

—————— 1812. *An Historical View of the Domestic Economy of Great Britain and Ireland* . . . (Edinburgh, a new edition).

—————— 1817. *The Author of Junius Ascertained; From a Concatination of Circumstances; amounting to a Moral Demonstration* (London).

COMBER, THOMAS. 1770. *A Free and Candid Correspondence on the Farmer's Letters* . . . *with the Author, Arthur Young, Esq.* (London).

—————— 1772. *Real Improvements in Agriculture (On the Principles of A. Young, Esq.)* . . . (no place).

CULLUM, JOHN. 1813. *History of Hawstead* (2nd ed., London).

DAVY, HUMPHREY. 1839–1840. *The Collected Works of Sir Humphrey Davy* (ed. by John Davy; 7, 8, London).

DAY, THOMAS. 1788. *A Letter to Arthur Young, Esq. on the Bill* . . . *to Prevent the Exportation of Wool* (London).

HARTE, WALTER. 1770. *Essays on Husbandry* (2nd ed., London).

HOLDERNESS, MARY. 1827. *Journey from Riga to the Crimea* . . . (2nd ed., London).

HOWLETT, JOHN. 1781. *An Examination of Dr. Price's Essay on the Population of England and Wales* . . . (Maidstone).

—— 1786. *An Enquiry into the Influence which Enclosures have had upon the Population of England* (London).

KEIR, JAMES. 1791. *An Account of the Life and Writings of Thomas Day, Esq.* (London).

KENT, NATHANIEL. 1796. *General View of the Agriculture of the County of Norfolk* (London).

MALTHUS, THOMAS R. 1803. *An Essay on the Principle of Population* (London) 2: pp. 528–547.

MARSHALL, WILLIAM. 1808-1817. *A Review of the Reports of the Board of Agriculture* (5 v., York).

MITCHELL, JOHN. 1755. *A Map of the British and French Dominions in North America* (London).

—— 1757. *The Contest in America between Great Britain and France.* (London).

PARIS, J. A. 1820. "A Biographical Memoir of Arthur Young". *Quarterly Journal of Science, Literature and the Arts* 9: *pp.* 279–309.

PIGOTT, CHARLES. 1794. *The Female Jockey Club* (New York)

STEPHENSON, ROBERT. 1784. *Observations upon the Present State of the Linen Trade of Ireland . . . in which the Reports, Libel, and British Examination of Mr. J. Arbuthnot are . . . considered and refuted* (Dublin).

STONE, THOMAS. 1800. *A Letter to . . . Lord Somerville . . . late President of the Board of Agriculture, with a view to show the Inutility of the Plans and Researches of that Institution . . .* (London).

—— 1800. *A Review of the Corrected Survey of Lincolnshire by Arthur Young, Esq . . .* (2nd ed., London).

STUART, DANIEL. 1794. *Peace and Reform, against War and Corruption. In answer to a pamphlet, . . . The Example of France a Warning to Great Britain* (London).

SYMONDS, JOHN. 1789. *Observations upon the Expediency of Revising the Present English Version of the Four Gospels, and of the Acts of the Apostles* (Cambridge).

THOMSON, THOMAS. 1812. *History of the Royal Society, from its Institution to the End of the Eighteenth Century* (London).

WAKEFIELD, EDWARD. 1812. *Account of Ireland, Statistical and Political* (2 v., London).

WIMPEY, JOSEPH. 1770. *Thoughts upon Several Interesting Subjects* (London).

—— 1770. *A Defence of a Pamphlet Lately Published Entitled, Thoughts upon several Interesting Subjects . . .* (London).

—— 1775. *Rural Improvements . . .* (2nd ed., London).

4. Memoirs and Correspondence

MSS. of the Marquis of Ailesbury. 1898. (Historical MSS. Commission, 15th Report, Appendix, Part VII, London.)

Diary and Letters of Madame d'Arblay. 1842. (Ed. by her niece, 2 v., (Philadelphia.)

The Early Diary of Frances Burney, 1768–1778. 1889. (Ed. by her niece, 2 v., London).

The Correspondence of Edmund Burke. 1958–1970. (Ed. by Thomas W. Copeland, 9 v., Cambridge, England, and Chicago.)

Memoirs of Dr. Burney . . . 1832. (Ed. by his daughter, Madame d'Arblay, 3 v., London.)

Life and Correspondence of Major Cartwright. 1826. (Ed. by Frances D. Cartwright, 2 v.).

Memoirs of the Political and Private Life of James Caulfield, Earl of Charlemont.
1810. (Ed. by Francis Hardy, London.)

The MSS. and Correspondence of James, First Earl of Charlemont. 1891. (Historical MSS. Commission, 12th Report, Appendix, Part X, London.)

The Letters of Philip Dormer Stanhope, Earl of Chesterfield. 1845. (Ed. by Lord Mahon, 5 v., London.)

Lord Fife and his Factor; being the Correspondence of James, second Lord Fife, 1729–1809. 1925. (Ed. by Alistair and Henrietta Taylor, London.)

The Life and Correspondence of Philip Yorke, Earl of Hardwicke. 1913. (Ed. by Philip C. Yorke, 3 v., Cambridge, England.)

The Girlhood of Maria Josepha Holroyd [Lady Stanley of Alderley] Recorded in Letters of a hundred years ago: from 1776 to 1796. 1896. (Ed. by J. Adeane, London.)

Voyages en France de Francois de la Rochefoucauld (1781–1783). 1933–1938. (Ed. by Jean Marchand., 2 v., Paris.)

A Frenchman in England, 1784, Being the Mélanges sur l'Angleterre of François de la Rochefoucald. 1933. (Ed. by Jean Marchand, translated with notes by S. C. Roberts, Cambridge, England.)

The Correspondence of the Right Honourable Sir John Sinclair, Bart. . . . 1831. (2v, London.)

Memoirs of the Life and Works of the late Right Honourable Sir John Sinclair, Bart. 1837. (Ed. by his son, the Rev. John Sinclair, 2v., Edinburgh.)

The MSS. of the Marquis Townshend. 1887. (Historical MSS. Commission, 11th Report, Appendix, Part IV, London.)

The Russian Journals of Martha and Catherine Wilmot, 1803–1808. 1935. (Ed. by the Marchioness of Londonderry and H. M. Hyde, London.)

The Windham Papers; the Life and Correspondence of the Rt. Hon. William Windham. 1913. (Ed. by the Earl of Roseberry, Boston.)

5. *Newspapers and Periodicals*

Annual Register (London).

New Annual Register (London).

Bury and Norwich Post (*Bury Post* until 1787; Bury St. Edmunds, Suffolk).

Cobbett's Political Register (especially v. 13, London).

Gentleman's Magazine (London).

London Chronicle (London).

Monthly Magazine (London).

Monthly Review (London).

NANGLE, B. C. 1934. *The Monthly Review . . . Index of Contributors and Articles* (Oxford).

Quarterly Review (London).

The Times (London).

6. *Government Documents*

The Parliamentary Debates. 1812–1813. (**23, 25,** (London).

Parliamentary History of England. 1817. (**30,** London).

HOWELL, THOMAS B. 1819. *A Complete Calendar of State Trials . . .* (**26,** London).

II. SECONDARY SOURCES

ASHTON, THOMAS S. 1924. *Iron and Steel in the Industrial Revolution* (Manchester).

Barnes, Donald G. 1930. *A History of the English Corn Laws from 1660–1846* (New York).
———— 1939. *George III and William Pitt, 1783–1806* (Stanford University).
Baudrillart, Henri J. L. 1862. *Publicistes Modernes* (Paris), pp. 22–63.
Birrell, Augustine. 1905. *In the Name of the Bodleian* (New York), pp. 183–194.
Blunt, R. n. d. *Mrs. Montagu, "Queen of the Blues"* (Boston).
Bonar, James. 1931. *Theories of Population from Raleigh to Arthur Young* (London), pp. 219–240.
Bourde, André J. 1953. *The Influence of England on the French Agronomes, 1750–1789* (Cambridge, England).
Brown, Andrew C. 1936. *The Wilsons. A Banffshire Family of Factors* (Edinburgh).
Buttress, F. A., and R. W. G. Dennis. 1947. "The Early History of Seed Treatment in England." *Agric. Hist.* 21: pp. 93–103.
Cameron, Hector C. 1952. *Sir Joseph Banks . . . The Autocrat of the Philosophes* (London).
Childe-Pemberton, William S. 1924. *The Earl Bishop: the Life of Frederick Hervey, Bishop of Derry, Earl of Bristol* (2 v., New York).
Clarke, Ernest. 1891. "Agriculture and the House of Russell." *Jour. Roy. Agric. Soc. England,* 3rd ser., **2**: pp. 123–145.
———— 1897, "John, Fifteenth Lord Somerville." *Jour. Roy. Agric. Soc. England,* 3rd ser. **8**: pp. 1–20.
———— 1898. "The Board of Agriculture, 1793–1822." *Jour. Roy. Agric. Soc. England,* 3rd ser. **9**: pp. 1–41.
Cole, George D. H. 1924. *The Life of William Cobbett* (New York).
Cone, Carl B. 1945. "Edmund Burke, the Farmer." *Agric. Hist.* **19**: pp. 65–69.
Copinger, W. A. 1905. *The Manors of Suffolk. Notes on their History and Devolution* (7 v., London).
Coupland, Reginald. 1923. *Wilberforce. A Narrative* (Oxford).
Curtler, W. H. R. 1909. *A Short History of English Agriculture* (Oxford).
———— 1920. *The Enclosure and Redistribution of our Land* (Oxford).
Defries, Amelia D. 1938. *Sheep and Turnips: Being the Life and Times of Arthur Young . . .* (London).
Dreyfus, Ferdinand. 1903. *Un Philanthrope d'autrefois: La Rochefoucauld-Liancourt, 1747–1827* (Paris).
Farrer, Edmund. 1908. *Portraits in Suffolk Houses* (West) (London).
Fay, Charles R. 1928. *Great Britain from Adam Smith to the Present Day* (London).
———— 1932. *The Corn Laws and Social England* (Cambridge, England).
Ford, Franklin L. 1958. *Strasbourg in Transition, 1648–1789* (Cambridge, England).
Foster, James R. 1949. *History of the Pre-Romantic Novel in England* (London), pp. 154–158.
Fussell, George E. 1936. "English Agriculture from Arthur Young to William Cobbett." *Econ. Hist. Rev.* **6**: pp. 214–222.
———— 1937. "Animal Husbandry in Eighteenth Century England." *Agric. Hist.* **11**: pp. 96–116, 189–214.
———— 1943. "My Impressions of Arthur Young." *Agric. Hist.* **17**: pp. 135–144.
———— 1947. "The Farming Writers of Eighteenth Century England." *Agric. Hist* **21**: pp. 1–8.
———— 1951. "Impressions of Sir John Sinclair, First President of the Board of Agriculture." *Agric. Hist.* **25**: pp. 162–169.

———— 1952. *The Farmers' Tools, 1500–1900. The History of British Farm Implements, Tools and Machinery before the Tractor Came* (London).

FUSSELL, GEORGE E., and CONSTANCE GOODMAN. 1941–1942. "Crop Husbandry in Eighteenth Century England." *Agric. Hist.* 15: pp. 202–216; 16: pp. 41–63.

FUSSELL, GEORGE E., and K. R. 1955. *The English Countryman. His Life and Work, 1500–1900* (London).

GAGE, JOHN. 1838. *The History and Antiquities of Suffolk—Thingoe Hundred* (London).

GARNETT, RICHARD. 1898. *Edward Gibbon Wakefield* (London).

GARNIER, RUSSELL M., 1893. *History of the English Landed Interest: its Customs, Laws and Agriculture* (v. 2, London).

GAY, EDWIN F. 1927. "Arthur Young on English Roads." *Quart. Jour. Economics* 41: pp. 545–551.

GAZLEY, JOHN G. 1941. "Arthur Young and the Society of Arts." *Jour. Econ. Hist.* 1: pp. 129–152.

———— 1950. "Arthur Young, British Patriot." *Nationalism and Internationalism: Essays Inscribed to Carlton J. H. Hayes* (ed. by Edward M. Earle, New York), pp. 144–189.

———— 1955. "Arthur Young, Agriculturalist and Traveller, 1741–1820. . . ." *Bull. John Rylands Library* 37: pp. 393–428.

———— 1956. "The Reverend Arthur Young 1769–1827: Traveller in Russia and Farmer in the Crimea." *Bull. John Rylands Library* 38: pp. 360–405.

GIBBS, BRANDRETH T. 1857. *The Smithfield Club. A Condensed History of its Origins and Progress* . . .(London).

GILBOY, ELIZABETH W. 1934. *Wages in Eighteenth Century England* (Cambridge, Mass.).

GWYNN, STEPHEN L. 1939. *Henry Grattan and his Times* (London).

HAMMOND, JOHN L. and BARBARA. 1912. *The Village Labourer, 1760–1832* (London).

HASLAM, C. S. 1930. *The Biography of Arthur Young F. R. S. from his Birth until 1787* (Rugby).

HAWORTH, PAUL L. 1925. *George Washington, Country Gentleman; being an account of his Home Life and Agricultural Activities* (Indianapolis).

HEMLOW, JOYCE. 1958. *The History of Fanny Burney* (Oxford).

HIGGS, HENRY. 1900. "Arthur Young." *Dict. Natl. Biog.* 63: pp. 357–363.

———— 1935. *Bibliography of Economics, 1751–1775* (Cambridge, England).

HILL, CONSTANCE. 1904. *Juniper Hall, a Rendezvous of Certain Illustrious Personages during the French Revolution* . . . (London).

———— 1912. *Fanny Burney at the Court of Queen Charlotte* (London).

JOHNSON, R. BRIMLEY. 1926. *Fanny Burney and the Burneys* (New York).

KIRBY, J., 1829. *A Topographical and Historical Description of* . . . *Suffolk* (Woodbridge).

LACRETELLE, CHARLES DE. 1842. *Dix années d'éprouves pendant le révolution* (Paris).

LAVERGNE, LÉONCE. 1870. *Les Économistes français de dix-huitième siècle* (Paris). Appendix I. La Société d'agriculture de Paris.

LOEHR, RODNEY C. 1937. "The Influence of English Agriculture on American Agriculture, 1775–1825." *Agric. Hist.* 11: pp. 3–15.

———— 1940. "American Husbandry; a Commentary apropos of the Carman Edition." *Agric. Hist.* 14: pp. 104–109.

———— 1969. "Arthur Young and American Agriculture." *Agric. Hist.* 43: pp. 43–56.

MANTOUX, PAUL. 1927. *The Industrial Revolution in the Eighteenth Century* (London).

MARSHALL, DOROTHY. 1956. *English People in the Eighteenth Century* (New York).

MARTINENGO-CESARESCO, EVELYN. 1902. *Lombard Studies* (New York), pp. 233–252, "Arthur Young's North Italian Journey."

MAXWELL, CONSTANTIA E. 1932. *The English Traveller in France, 1698–1815* (London) pp. 136–151.

―――― 1937. *Dublin under the Georges, 1714–1830* (London).

―――― 1940. *Country and Town in Ireland under the Georges* (London).

MCDONALD, DONALD. 1908. *Agricultural Writers, from Sir Walter of Henley to Arthur Young, 1200–1800* (London).

MEAD, WILLIAM E. 1914. *The Grand Tour in the Eighteenth Century* (Boston).

MITCHISON, ROSALIND. 1959. "The Old Board of Agriculture (1793–1822)." *English Hist. Rev.* **74**: pp. 41–69.

―――― 1962. *Agricultural Sir John. The Life of Sir John Sinclair . . . 1754–1835* (London).

MORRISON, A. J. 1918. "The Historical Farmer in America." *South Atlantic Quart.* **17**: pp. 222–230.

PAGE, AUGUSTINE. 1843. *A Supplement to the Suffolk Traveller* (5 v., Ipswich).

PARISET, E. 1896. "Arthur Young et ses traducteurs." *La Revolution Française* **30**: pp. 65–75.

PELL, ALBERT. 1893. "Arthur Young." *Jour. Roy. Agric. Soc. England,* 3rd ser., **4**: pp. 1–23.

PROTHERO, ROWLAND E. [Lord Ernle]. 1927. *English Farming Past and Present* (4th ed., London).

QUINLAN, MAURICE J. 1941. *Victorian Prelude, a History of English Manners, 1700–1830* (New York).

RICHES, NAOMI. 1937. *The Agricultural Revolution in Norfolk* (Chapel Hill).

ROSE, J. HOLLAND. 1911. *William Pitt and the Great War* (London).

RUSSELL, PHILLIPS. 1936. *The Glittering Century* (New York).

RYE, WALTER. 1895. "Coke of Holkham." *Jour. Roy. Agric. Soc. England,* 3rd ser. **6**: pp. 1–14.

SEEBOHM, MARY E. [Mrs. Christie]. 1952. *The Evolution of the English Farm* (London).

SEINGUERLET, EUGÉNE 1881. *Strasbourg pendant la Révolution* (Paris).

SHELLABARGER, SAMUEL. 1935. *Lord Chesterfield* (London).

SINCLAIR, WILLIAM M. 1896. "Sir John Sinclair." *Jour. Roy. Agric. Soc. England,* 3rd ser. **7**: pp. 1–21.

SMART, WILLIAM. 1910. *Economic Annals of the Nineteenth Century, 1801–1820* (London) **1**.

SMITH, EDWARD. 1911. *The Life of Sir Joseph Banks, President of the Royal Society . . .* (London).

STEPHEN, LESLIE. 1898. *Studies of a Biographer* (London) **1**: pp. 188–226.

STEPHENSON, NATHANIEL W., and W. H. DUNN. 1940. *George Washington* (New York).

STEVENSON, FRANCIS S. 1893. "Arthur Young." *Westminster Rev.* **139**: pp. 109–120.

STIRLING, ANNA M. W. [PICKERING] 1912. *Coke of Norfolk and his Friends . . .* (London).

THORPE, THOMAS E. 1896. *Humphrey Davy, Poet and Philosopher* (New York).

TURBERVILLE, ARTHUR S., ed. 1933. *Johnson's England* 1: pp. 261–299, "Agriculture and Rural Life," by C. S. Orwin.

WAKEFIELD, C. M. 1885. *Life of Thomas Attwood* (London).

WALSH, ROBERT, JR., ed. 1819–1823. *Works of the British Poets with the Lives of the Authors* (Philadelphia). For Walter Harte, 29: pp. 323–330.

WEBB, SIDNEY J. [BARON PASSFIELD]. 1929. *English Poor Law History* (3 v., London).

WILBERFORCE, ROBERT J. and SAMUEL. 1838. *The Life of William Wilberforce* (5 v., London).

WILLIAMS, JUDITH B. 1926. *A Guide to the Printed Materials for English Social and Economic History, 1750–1850* (2v., New York).

WOOD, HENRY T. 1913. *A History of the Royal Society of Arts* (London).

WOODWARD, CARL R. 1969. "A Discussion of Arthur Young and American Agriculture." *Agric. Hist.* 43: pp. 57–67.

Index